HENRY SINCLAIR HAL
SAMUEL RATCLIFF KNIG

ELEMENTARY ALGEBRA
FOR SCHOOLS

(CONTAINING A CHAPTER ON GRAPHS)

WITH ANSWERS

Elibron Classics
www.elibron.com

Elibron Classics series.

© 2005 Adamant Media Corporation.

ISBN 1-4021-5906-4 (paperback)
ISBN 1-4021-2925-4 (hardcover)

This Elibron Classics Replica Edition is an unabridged facsimile of the edition published in 1906 by Macmillan and Co., Ltd., London.

Elibron and Elibron Classics are trademarks of Adamant Media Corporation. All rights reserved.

This book is an accurate reproduction of the original. Any marks, names, colophons, imprints, logos or other symbols or identifiers that appear on or in this book, except for those of Adamant Media Corporation and BookSurge, LLC, are used only for historical reference and accuracy and are not meant to designate origin or imply any sponsorship by or license from any third party.

ELEMENTARY ALGEBRA
FOR SCHOOLS

(CONTAINING A CHAPTER ON GRAPHS)

WITH ANSWERS

BY

H. S. HALL, M.A.
FORMERLY SCHOLAR OF CHRIST'S COLLEGE, CAMBRIDGE

AND

S. R. KNIGHT, B.A., M.B., Ch.B.
FORMERLY SCHOLAR OF TRINITY COLLEGE, CAMBRIDGE

EIGHTH EDITION
REVISED AND ENLARGED

London
MACMILLAN AND CO., LIMITED
NEW YORK: THE MACMILLAN COMPANY
1906

The right of Translation is reserved

First Edition, printed 1885; *Second Edition,* 1886; *Third Edition, February,* 1887; *Fourth Edition, October,* 1887; *Reprinted* 1889; *Fifth Edition, Revised and Enlarged, November,* 1889; *Sixth Edition, Revised and Enlarged,* 1890; *Reprinted* 1891, 1892, 1893, 1895, 1896 *(twice)*; *Seventh Edition, Revised and Enlarged,* 1897; *Reprinted,* 1898, 1899, 1900, 1901, 1902; *Eighth Edition, Revised and Enlarged,* 1903; *Reprinted* 1904, 1905, 1906.

GLASGOW: PRINTED AT THE UNIVERSITY PRESS
BY ROBERT MACLEHOSE AND CO. LTD.

PREFACE.

THE distinctive features of the Eighth Edition are :—

(1) A full treatment of Graphs, occupying more than 40 pages.

  In view of the fact that great diversity of opinion exists among teachers as to the place and importance of Graphs in an elementary course, the subject has been treated in one continuous chapter. References to this chapter will be found whenever a section of it may be suitably interpolated to illustrate other parts of the text.

(2) A new set of easy examples on Substitution in Chapter I.

(3) The greater part of Chapter VIII., on Simple Equations, has been re-written so as to bring the use of the fundamental axioms into greater prominence, and to urge the importance of verifying solutions.

(4) Chapter IX., on Symbolical Expression, has been enlarged. In particular the section on Formulæ has been illustrated by a new set of Examples.

(5) A section on Square Root by inspection has been inserted in Chapter XVI.

(6) In Chapter XVII., on Factors, a section on factorisation of trinomials, by completing the square, has been introduced. Also a large number of *easy* miscellaneous

PREFACE.

examples take the place of the Exercise XVII. 1 of earlier editions.

This is at the request of teachers who have frequently pointed out that at this stage a larger variety of easy examples would be preferred.

(7) Considerable additions to the chapters on Quadratic Equations. In particular a set of examples involving applications to Geometry will be found at the end of Chapter XXVII.

(8) The chapter on Logarithms has been re-written so as to introduce and explain the use of Four-Figure Tables. The Tables of Logarithms and Antilogarithms have been taken, with slight modifications, from those published by the Board of Education, South Kensington.

(9) An easy first course has been mapped out enabling teachers to postpone, if they wish, the harder cases of 'Long' Multiplication and Division, and the rules dependent on these processes.

H. S. HALL.

July, 1903.

SUGGESTIONS FOR A FIRST COURSE.

In the first thirty chapters an asterisk has been placed before all articles and examples which may conveniently be omitted in a first course. Notes are occasionally given suggesting the most suitable place for a section which may have to be postponed.

For those who wish to defer to a later stage all the rules dependent on 'Long' Multiplication and Division, so as to reach Quadratic Equations earlier, the following detailed course is recommended.

    Chap. I. Arts. 1-11, 13-15. [Omit Art. 12, Examples I. c.]
    Chap. II.-V. Arts. 16-40. [Omit all the rest of Chap. V., except Art. 44.]
    Chap. VI. Arts. 46-50. [Omit Arts. 51-55.]
    Chaps. VII.-XIII. Arts. 56-107. In connection with Chap. XIII. Arts. 417-424 on Elementary Graphs may be read.
    Chaps. XIV., XV. Arts. 108-113. [Omit Arts. 114, 115.]
    Chap. XVI. Arts. 116-118A. [Omit Arts. 119-124.]
    Chap. XVII. Arts. 125-136. [Omit Arts. 136A-137.]
    Chap. XVIII. Arts. 138, 139. [Omit Arts. 140-148.]
    Chap. XIX. [Omit Arts. 152, 153.]
    Chap. XX. [Omit Arts. 159-163.]
    Chap. XXI. [Omit Arts. 171, 172.]
    Chap. XXII. Arts. 173-179. [Omit Arts. 180-185.]
    [Chaps. XXIII., XXIV. may be taken later.]
    Chap. XXV. **Quadratic Equations.** In connection with this chapter Arts. 425-428, 437-440 may be read.

From this point the omitted sections must be taken at the discretion of the Teacher.

CONTENTS.

CHAP.		PAGE
I.	Definitions, Substitutions	1
II.	Negative Quantities. Addition of Like Terms	9
III.	Simple Brackets. Addition	13
IV.	Subtraction	20
	Miscellaneous Examples I.	23
V.	Multiplication	26
VI.	Division	38
VII.	Removal and Insertion of Brackets	46
VIII.	Simple Equations	52
IX.	Symbolical Expression	61
X.	Problems leading to Simple Equations	69
XI.	Highest Common Factor, Lowest Common Multiple of Simple Expressions	74
XII.	Elementary Fractions	76
	Miscellaneous Examples II.	80
XIII.	Simultaneous Equations	83
XIV.	Problems leading to Simultaneous Equations	93
XV.	Involution	98
XVI.	Evolution	102
XVII.	Resolution into Factors	112
	Miscellaneous Examples III.	126
XVIII.	Highest Common Factor	128
XIX.	Fractions	136
XX.	Lowest Common Multiple	144
XXI.	Addition and Subtraction of Fractions	148

CONTENTS.

CHAP.		PAGE
XXII.	Miscellaneous Fractions	160
	Miscellaneous Examples IV.	172
XXIII.	Harder Equations	178
XXIV.	Harder Problems	186
XXV.	Quadratic Equations	192
XXVI.	Simultaneous Quadratic Equations	202
XXVII.	Problems leading to Quadratic Equations	209
XXVIII.	Harder Fractions	215
XXIX.	Miscellaneous Theorems and Examples	222
XXX.	The Theory of Indices	240
XXXI.	Elementary Surds	253
XXXII.	Ratio, Proportion, and Variation	270
XXXIII.	Arithmetical Progression	287
XXXIV.	Geometrical Progression	294
XXXV.	Harmonical Progression	301
	Miscellaneous Examples V.	307
XXXVI.	Theory of Quadratic Equations	311
XXXVII.	Permutations and Combinations	319
XXXVIII.	Binomial Theorem	329
XXXIX.	Logarithms	340
XL.	Scales of Notation	349
XLI.	Exponential and Logarithmic Series	355
XLII.	Miscellaneous Equations	361
XLIII.	Interest and Annuities	367
XLIV.	Graphical Representation of Functions	371
	Miscellaneous Examples VI.	416
	Answers	437

ALGEBRA.

CHAPTER I.

DEFINITIONS. SUBSTITUTIONS.

1. ALGEBRA treats of quantities as in Arithmetic, but with greater generality; for while the quantities used in arithmetical processes are denoted by *figures* which have one single definite value, algebraical quantities are denoted by *symbols* which may have any value we choose to assign to them.

The symbols employed are letters, usually those of our own alphabet; and, though there is no restriction as to the numerical values a symbol may represent, it is understood that in the same piece of work it keeps the same value throughout. Thus, when we say "let $a=1$," we do not mean that a must have the value 1 always, but only in the particular example we are considering. Moreover, we may operate with symbols without assigning to them any particular numerical value at all; indeed it is with such operations that Algebra is chiefly concerned.

We begin with the definitions of Algebra, premising that the symbols $+$, $-$, $\times$, $\div$, $=$, will have the same meanings as in Arithmetic. Also, for the present it will be assumed that all the algebraical symbols employed denote integral numbers.

2. An **algebraical expression** is a collection of symbols; it may consist of one or more **terms**, which are separated from each other by the signs $+$ and $-$. Thus $7a+5b-3c-x+2y$ is an expression consisting of five terms.

Note. When no sign precedes a term the sign $+$ is understood.

3. Expressions are either **simple** or **compound**. A *simple expression* consists of *one* term, as $5a$. A *compound expression* consists of *two or more* terms. Compound expressions may be further distinguished. Thus an expression of *two* terms, as $3a-2b$, is called a **binomial** expression; one of *three* terms, as $2a-3b+c$, a **trinomial**; one of *more than three* terms a **multinomial**. Simple expressions are also spoken of as **monomials**.

4. When two or more quantities are multiplied together the result is called the **product**. One important difference between the notation of Arithmetic and Algebra should be here remarked. In Arithmetic the product of 2 and 3 is written 2×3, whereas in Algebra the product of a and b may be written in any of the forms $a \times b$, $a \cdot b$, or ab. The form ab is the most usual. Thus, if $a=2$, $b=3$, the product $ab = a \times b = 2 \times 3 = 6$; but in Arithmetic 23 means "twenty-three," or $2 \times 10 + 3$.

5. Each of the quantities multiplied together to form a product is called a **factor** of the product. Thus 5, a, b, are the factors of the product $5ab$.

6. When one of the factors of an expression is a numerical quantity, it is called the **coefficient** of the remaining factors. Thus, in the expression $5ab$, 5 is the coefficient. But the word coefficient is also used in a wider sense, and it is sometimes convenient to consider any factor, or factors, of a product as the coefficient of the remaining factors. Thus, in the product $6abc$, $6a$ may be appropriately called the coefficient of bc. A coefficient which is not merely numerical is sometimes called a **literal coefficient**.

Note. When the coefficient is unity it is usually omitted. Thus we do not write $1a$, but simply a.

7. If a quantity be multiplied by itself any number of times, the product is called a **power** of that quantity, and is expressed by writing the number of factors to the right of the quantity and above it. Thus

$a \times a$ is called the *second power* of a, and is written a^2;

$a \times a \times a$ .... *third power* of a, ... a^3;

and so on.

The number which expresses the power of any quantity is called its **index** or **exponent**. Thus 2, 5, 7 are respectively the indices of a^2, a^5, a^7.

Note. a^2 is usually read "a squared"; a^3 is read "a cubed"; a^4 is read "a to the fourth"; and so on.

When the index is unity it is omitted. Thus we do not write a^1, but simply a. Thus a, $1a$, a^1, $1a^1$ all have the same meaning.

8. The beginner must be careful to distinguish between *coefficient* and *index*.

Example 1. What is the difference in meaning between $3a$ and a^3?

By $3a$ we mean the product of the quantities 3 and a.

By a^3 we mean the third power of a; that is, the product of the quantities a, a, a.

Thus, if $a=4$,
$$3a = 3 \times a = 3 \times 4 = 12;$$
$$a^3 = a \times a \times a = 4 \times 4 \times 4 = 64.$$

Example 2. If $b=5$, distinguish between $4b^2$ and $2b^4$.

Here $\quad 4b^2 = 4 \times b \times b = 4 \times 5 \times 5 = 100;$

whereas $\quad 2b^4 = 2 \times b \times b \times b \times b = 2 \times 5 \times 5 \times 5 \times 5 = 1250.$

Example 3. If $a=4$, $x=1$, find the value of $5x^a$.

Here $\quad 5x^a = 5 \times x \times x \times x \times x = 5 \times 1 \times 1 \times 1 \times 1 = 5.$

Note. **The beginner should observe that every power of 1 is 1.**

9. In arithmetical multiplication the order in which the factors of a product are written is immaterial. For instance 3×4 means 4 sets of 3 units, and 4×3 means 3 sets of 4 units; in each case we have 12 units in all. Thus
$$3 \times 4 = 4 \times 3.$$

In a similar way,
$$3 \times 4 \times 5 = 4 \times 3 \times 5 = 4 \times 5 \times 3;$$

and it is easy to see that the same principle holds for the product of any number of arithmetical quantities.

In like manner in Algebra ab and ba each denote the product of the two quantities represented by the letters a and b, and have therefore the same value. Again, the expressions abc, acb, bac, bca, cab, cba have the same value, each denoting the product of the three quantities a, b, c. It is immaterial in what order the factors of a product are written; it is usual, however, to arrange them in alphabetical order.

Fractional coefficients which are greater than unity are usually kept in the form of improper fractions.

Example. If $a=6$, $x=7$, $z=5$, find the value of $\frac{13}{10}axz$.

Here $\quad \frac{13}{10}axz = \frac{13}{10} \times 6 \times 7 \times 5 = 273.$

EXAMPLES I. a.

If $a=7$, $b=2$, $c=1$, $x=5$, $y=3$, find the value of

1. $14x$. 2. x^3. 3. $3ax$. 4. a^3. 5. $5by$.
6. b^5. 7. $3b^2$. 8. $2xa$. 9. $6c^4$. 10. $4y^3$.
11. $7c^5$. 12. $9b^4$. 13. $8bcy$. 14. $7y^3$. 15. $8x^2$.

If $a=8$, $b=5$, $c=4$, $x=1$, $y=3$, find the value of

16. $9xy$. 17. $8b^3$. 18. $3x^5$. 19. x^8. 20. $7y^4$.
21. c^x. 22. b^y. 23. y^c. 24. x^b. 25. y^b.
26. a^y. 27. b^x. 28. a^c. 29. c^y. 30. $6bxy$.

If $a=5$, $b=1$, $c=6$, $x=4$, find the value of

31. $\frac{3}{8}x^3$. 32. $\frac{1}{10}ax$. 33. 3^x. 34. 2^c. 35. 8^b.
36. 7^x. 37. $\frac{7}{15}acx$. 38. $\frac{1}{8}bcx$. 39. $\frac{2}{9}c^3$. 40. $\frac{x^5}{64}$.

10. When several different quantities are multiplied together a notation similar to that of Art. 7 is adopted. Thus $aabbbbcddd$ is written $a^2b^4cd^3$. And conversely $7a^3cd^2$ has the same meaning as $7 \times a \times a \times a \times c \times d \times d$.

Example 1. If $x=5$, $y=3$, find the value of $4x^2y^3$.

$$4x^2y^3 = 4 \times 5^2 \times 3^3$$
$$= 4 \times 25 \times 27$$
$$= 2700.$$

Example 2. If $a=4$, $b=9$, $x=6$, find the value of $\frac{8bx^2}{27a^3}$.

$$\frac{8bx^2}{27a^3} = \frac{8 \times 9 \times 6^2}{27 \times 4^3} = \frac{8 \times 9 \times 36}{27 \times 64}$$
$$= \frac{3}{2} = 1\tfrac{1}{2}.$$

11. If one factor of a product is equal to 0, the whole product must be equal to 0, *whatever values the other factors may have.* A factor 0 is usually called a **zero factor**.

For instance, if $x=0$ then ab^3xy^2 contains a zero factor. Therefore $ab^3xy^2=0$ when $x=0$, whatever be the values of a, b, y.

Again, if $c=0$, then $c^3=0$; therefore $ab^2c^3=0$, whatever values a and b may have.

Note. Every power of 0 is 0.

EXAMPLES I. b.

If $a=7$, $b=2$, $c=0$, $x=5$, $y=3$, find the value of

1. $4ax^2$.
2. a^3b.
3. $8b^2y$.
4. $3xy^2$.
5. $\frac{3}{4}b^2x$.
6. $\frac{5}{6}b^3y^2$.
7. $\frac{2}{5}xy^4$.
8. a^3c.
9. a^2cy.
10. $8x^3y$.
11. $\frac{7}{20}ab^5x$.
12. $\frac{1}{9}x^2y^4$.

If $a=2$, $b=3$, $c=1$, $p=0$, $q=4$, $r=6$, find the value of

13. $\frac{3a^2r}{8b}$.
14. $\frac{8ab^2}{9q^2}$.
15. $\frac{6a^3c}{b^2}$.
16. $\frac{4cr^2}{9a^3}$.
17. $3a^2b^c$.
18. $\frac{5}{6}ba^r$.
19. $\frac{8b^q}{9a^r}$.
20. $5a^bc^r$.
21. $\frac{2a^2p}{7r}$.
22. 3^a2^b.
23. 2^ra^5.
24. c^bb^q.
25. $\frac{5a^rb^q}{64r^a}$.
26. $\frac{27a^q}{32}$.
27. $\frac{64}{q^r}$.
28. $\frac{b^r}{r^b}$.

[*The articles and examples marked with an asterisk may be postponed and taken in connection with* Chap. XVI.]

*12. Definition. The **square root** of any proposed expression is that quantity whose square, or second power, is equal to the given expression. Thus the square root of 81 is 9, because $9^2=81$.

The square root of a is denoted by $\sqrt[2]{a}$, or more simply $\sqrt{a}$.

Similarly the **cube, fourth, fifth,** etc., **root** of any expression is that quantity whose third, fourth, fifth, etc., power is equal to the given expression.

The roots are denoted by the symbols $\sqrt[3]{\ }$, $\sqrt[4]{\ }$, $\sqrt[5]{\ }$, etc.

Examples. $\sqrt[3]{27}=3$; because $3^3=27$. $\sqrt[5]{32}=2$; because $2^5=32$.

The symbol $\sqrt{\ }$ is sometimes called the **radical sign**.

Example 1. Find the value of $5\sqrt{(6a^3b^4c)}$, when $a=3$, $b=1$, $c=8$.
$$5\sqrt{(6a^3b^4c)} = 5 \times \sqrt{(6 \times 3^3 \times 1^4 \times 8)} = 5 \times \sqrt{(6 \times 27 \times 8)}$$
$$= 5 \times \sqrt{1296} = 5 \times 36 = 180.$$

Example 2. Find the value of $\sqrt[3]{\left(\dfrac{ab^4}{8x^3}\right)}$, when $a=9$, $b=3$, $x=5$.
$$\sqrt[3]{\left(\frac{ab^4}{8x^3}\right)} = \sqrt[3]{\left(\frac{9 \times 3^4}{8 \times 5^3}\right)} = \sqrt[3]{\left(\frac{9 \times 81}{8 \times 125}\right)}$$
$$= \sqrt[3]{\left(\frac{9 \times 9 \times 9}{1000}\right)} = \frac{9}{10}.$$

*EXAMPLES I. c.

If $a=8$, $c=0$, $k=9$, $x=4$, $y=1$, find the value of

1. $\sqrt{(2a)}$.
2. $\sqrt{(kx)}$.
3. $\sqrt{(2ax)}$.
4. $\sqrt{(2ak^2)}$.
5. $\sqrt[3]{(3k)}$.
6. $\sqrt[3]{(ax^3)}$.
7. $\sqrt[3]{(8x^3y^3)}$.
8. $\sqrt[3]{(cy^5)}$.
9. $2x\sqrt{(2ay)}$.
10. $5y\sqrt{(4kx)}$.
11. $3c\sqrt{(kx)}$.
12. $2xy\sqrt{(4y^5)}$.
13. $\sqrt{\left(\dfrac{8x^3}{ak}\right)}$.
14. $\sqrt{\left(\dfrac{25a}{2k}\right)}$.
15. $\sqrt{\left(\dfrac{16x}{49y^3}\right)}$.
16. $\sqrt{\left(\dfrac{ca^2}{16k}\right)}$.
17. $\sqrt[3]{\left(\dfrac{3a}{k^2}\right)}$.
18. $\sqrt[3]{\left(\dfrac{ax^3}{27y^3}\right)}$.
19. $\sqrt[3]{\left(\dfrac{ca}{3k}\right)}$.
20. $\sqrt[3]{\left(\dfrac{a^2k^2}{3x^3}\right)}$.
21. $\sqrt{\left(\dfrac{kax^2}{18y^3}\right)}$.

13. In the case of expressions which contain more than one term, each term can be dealt with singly by the rules already given, and by combining the terms the numerical value of the whole expression is obtained. When brackets ( ) are used, they will have the same meaning as in Arithmetic, indicating that the terms enclosed within them are to be considered as one quantity.

Example 1.   When $c=5$, find the value of $c^4 - 4c + 2c^3 - 3c^2$.

Here
$$c^4 = 5^4 = 5 \times 5 \times 5 \times 5 = 625;$$
$$4c = 4 \times 5 = 20;$$
$$2c^3 = 2 \times 5^3 = 2 \times 5 \times 5 \times 5 = 250;$$
$$3c^2 = 3 \times 5^2 = 3 \times 5 \times 5 = 75.$$

Hence the value of the expression
$$= 625 - 20 + 250 - 75 = 780.$$

Example 2.   If $a=7$, $b=3$, $c=2$, find the value of
$$a(b+c)^2 - c(a-b)^3.$$

The expression $= 7(3+2)^2 - 2(7-3)^3 = 7 \cdot 5^2 - 2 \cdot 4^3 = 175 - 128 = 47$.

Example 3.   When $a=5$, $b=3$, $c=1$, find the value of
$$a^2 \cdot \dfrac{a-b}{b+2c} - b^2 \cdot \dfrac{a-c}{(a+c)^2}.$$

The expression $= 5^2 \times \dfrac{5-3}{3+(2\times 1)} - 3^2 \times \dfrac{5-1}{(5+1)^2}$

$= 25 \times \dfrac{2}{5} - 9 \times \dfrac{4}{36}$

$= 10 - 1 = 9.$

DEFINITIONS. SUBSTITUTIONS.

14. By Art. 11 any term which contains a *zero factor* is itself zero, and may be called a *zero term*.

Example 1. If $a=2$, $b=0$, $x=5$, $y=3$, find the value of
$$5a^3 - ab^2 + 2x^2y + 3bxy.$$
The expression $= (5 \times 2^3) - 0 + (2 \times 5^2 \times 3) + 0$
$= 40 + 150 = 190$.

Note. The two zero terms do not affect the result.

Example 2. Find the value of $\frac{3}{5}x^2 - a^2y + 7abx - \frac{5}{2}y^3$, when $a=5$, $b=0$, $x=7$, $y=1$.

$\frac{3}{5}x^2 - a^2y + 7abx - \frac{5}{2}y^3 = \frac{3}{5} \cdot 7^2 - 5^2 \cdot 1 + 0 - \frac{5}{2} \cdot 1^3$
$= 29\frac{2}{5} - 25 - 2\frac{1}{2} = 1\frac{9}{10}$.

Example 3. Find the values of the expression $x^2 - 10x + 21$ when x has the values 0, 2, 3, 7, 8.

Here the following arrangement will be found convenient.

x	0	2	3	7	8
x^2	0	4	9	49	64
$10x$	0	20	30	70	80
$x^2 - 10x + 21$	21	5	0	0	5

Thus the required values are 21, 5, 0, 0, and 5.

15. In working examples the student should pay attention to the following hints.

1. Too much importance cannot be attached to neatness of style and arrangement. The beginner should remember that neatness is in itself conducive to accuracy.

2. The sign = should never be used except to connect quantities which are equal. Beginners should be particularly careful not to employ the sign of equality in any vague and inexact sense.

3. Unless the expressions are very short the signs of equality in the steps of the work should be placed one under the other.

4. It should be clearly brought out how each step follows from the one before it; for this purpose it will sometimes be advisable to add short verbal explanations; the importance of this will be seen later.

EXAMPLES I. d.

If $a=2$, $b=3$, $c=1$, $d=0$, find the numerical value of

1. $6a+5b-8c+9d$.
2. $3a-4b+6c+5d$.
3. $5a+3c-2b+8d$.
4. $ab+bc+ca-da$.
5. $6ab-3cd+2da-5cb+2db$.
6. $abc+bcd+cda+dab$.
7. $3abc-2bcd+2cda-4dab$.
8. $2bc+3cd-4da+5ab$.
9. $3bcd+5cda-7dab+abc$.
10. $a^2+b^2+c^2+d^2$.
11. $2a^2+3b^3-4c^4$.
12. $a^4+b^4-c^4$.

If $a=1$, $b=2$, $c=3$, $d=0$, find the numerical value of

13. $a^3+b^3+c^3+d^3$.
14. $\frac{1}{2}bc^3-a^3-b^3-\frac{3}{4}ab^3c$.
15. $3abc-b^2c-6a^3$.
16. $2a^2+2b^2+2c^2+2d^2-2bc-2cd-2da-2ab$.
17. $c^3+\frac{4}{5}ad^4-3a^3+b^2d$.
18. $a^2+2b^2+2c^2+d^2+2ab+2bc+\frac{2}{7}cd$.
19. $2c^2+2a^2+2b^2-4cb+6abcd$.
20. $13a^2+\frac{11}{9}c^4+20ab-16ac-16bc$.
21. $6ab-\frac{4}{3}ac^2-2a+\frac{1}{8}b^4-3d+\frac{4}{9}c^3$.
22. $a^2-c^2+b^2-d^2+2ab-2cd$.
23. $2ab-\frac{3}{4}b^3+3ac-2c-d+\frac{4}{15}ad$.
24. $125b^4c-9d^5+3abc^2d$.

If $a=2$, $b=1$, $c=3$, $x=4$, $y=6$, $z=0$, find the value of

25. $c^2(y-x)-b^2(c-a)$.
26. $(2a-c)(x+2y-z)$.
27. $\frac{2}{3}(c^2-z^2)+\frac{3}{5}(y^2-x^2)$.
28. $\frac{4}{9}(cy-2c^2)+\frac{3}{7}(xy-bc)$.
29. $\dfrac{a^2}{b^2}+\dfrac{b^2}{a^2}-\dfrac{2y}{x^2}$.
30. $\dfrac{a^2}{b^2}\cdot c^2+\dfrac{a^2}{b^2}+c^2$.
31. $\dfrac{(a+y)^2}{(x-z)^3}-\dfrac{6(c^2-a)}{7(a^2+x)}$.
32. $\dfrac{a^2-b^2}{a^2b^2}-\dfrac{(a+b+z)^2}{(b+c-z)^2}$.
33. $\dfrac{(a+b)^2}{(y-c)^2}-\dfrac{a(y-z)}{c(x+z)}$.
34. $\dfrac{(a+b+c)^2}{c(y-z)}-\dfrac{4(c-a)^3}{3(a+y)}$.

EXAMPLES I. e.

1. When x has the values 0, 3, 6, 8, 10 find the values of $x^2 - 9x + 20$.

2. Find the values of $3 + 2x + \dfrac{x^2}{4}$ when x has the values 0, 1, 2, 3, 4.

3. Shew that $y^2 - 15y + 56$ is 0 if $y = 7$, and also if $y = 8$. What is its value when $y = 10$?

4. Find the values of the expression $\dfrac{x^3}{100} + \dfrac{x^2}{10} + 2x$ when x has the values 2, 6, 8, 10.

5. Shew by substituting 10 for a and 7 for b that the two expressions
$$4(a-b) + 3(a+b), \quad 5(a+b) + 2(a-3b)$$
are equal.

Test the equality also when $a = 6$, $b = 0$.

6. Shew that $x^3 - 6x^2 + 11x - 6$ is 0 for each of the values $x = 1, 2, 3$. What is its value when $x = 10$?

7. Shew that the expression $x^3 - 13x^2 + 44x$ is equal to 32 when $x = 1, 4,$ or 8.

8. Shew that $x^3 + 10x$ is equal to $7x^2$ for each of the values $x = 0, 2, 5$. Which of the expressions is the greater, and by how much, when $x = 6$?

9. By substituting 3 for x and 2 for y shew that the expressions
$$6x^3 + 7x^2y - y^3 \quad \text{and} \quad (2x+y)(3x-y)(x+y)$$
are equal.

10. Find the value of $4x^2 + 4x - 3$ when $x = 2$, and when $x = \tfrac{1}{2}$.

11. When $x = 5$, shew that $4x^2 + 4x - 3$ is equal to $9(x + 8)$.

12. Shew that $6x^3 - 11x^2 + 3x$ is equal to 0 when $x = \tfrac{1}{3}$, and when $x = \tfrac{3}{2}$. Find its value in the form of a decimal when $x = \tfrac{1}{10}$.

Examples for Revision. (Oral.)

1. What do you understand by 63 and by 6.3?

2. What is meant by $45xy$ and $4.5xy$? If $x = 4$, $y = 5$, give the arithmetical value of each.

3. Which is the greater 245 or 2.4.5, and by how much?

4. Give the product of t and u in three ways.

5. If 5 boys have p marbles each, express algebraically how many they have in all. If $p=25$ what is the number?

6. If x cakes are to be shared equally among 6 boys, express algebraically how many each will have. If $x=42$ what is the number?

7. If 54 books are divided equally among c boys, express each boy's share algebraically. What is the arithmetical value if $c=6$?

8. What is the difference between "twice 3" and "3 squared"?

9. Give the expression for "thrice d," also that for the "cube of d." Give the arithmetical values if $d=2$.

10. Distinguish between "four times x" and "x to the fourth." Give the respective values when $x=3$.

11. The quantity c is to be multiplied by the quantity x. How is this expressed? Give the product if $c=7$ and $x=3$.

12. If x factors, each equal to c, are to be multiplied together, express this algebraically. What is the value if $x=3$ and the factor $c=7$?

13. The quantities a, b, c are to be added together. Express this algebraically. What is the answer if $a=5$, $b=7$, $c=11$?

14. The quantity r is to be taken from the quantity s. Give the algebraical expression that denotes this. What is the answer if $r=27$ and $s=41$?

15. A boy starts playing with x marbles and wins y. Express the number he then has. If $x=25$ and $y=9$, what number has he?

16. The same boy plays with his increased number and loses z. Express the number he then has. If $z=17$, how many has he left?

17. A farmer takes f sheep to market and sells g of them. How many has he left? What is the remainder if $f=64$ and $g=48$?

18. Another farmer takes k sheep to market and returns with l of them. How many has he sold? If $k=75$, and $l=32$, what is the number he has sold?

19. Give the sum and product of the three quantities a, b, c; and if $a=5$, $b=7$, $c=6$, give the arithmetical value of each.

20. If I walk y miles per hour for y hours, what is the algebraical expression for the length of my walk? If $y=4$, what is the answer?

CHAPTER II.

NEGATIVE QUANTITIES. ADDITION OF LIKE TERMS.

16. In the preceding examples the sum of the terms to be subtracted has never been greater than the sum of the terms to be added; that is to say, every operation has been capable of being worked by Arithmetic. But in an example that reduces to a result such as $4-9$ the subtraction cannot be arithmetically performed, yet as an algebraical result such an expression can be explained; and, moreover, a subtractive term may stand alone and its meaning be quite plain.

17. Algebraical quantities which are preceded by the sign $+$ are said to be **positive**; those to which the sign $-$ is prefixed are said to be **negative**. When no sign is prefixed the $+$ sign is to be understood. These signs are frequently used to denote a *quality* possessed by the quantities to which they are attached, as explained in the following illustrations:

(i.) Suppose a trader gains £100 and then loses £70, the result of his trading is a *gain* of £30, that is $+£100-£70=+£30$; and the $+£30$ denotes that he is £30 better off than when he began.

But if he had first gained £70 and then lost £70, the loss would exactly balance the gain, that is $+£70-£70=£0$. Thus he would be in the same position as when he began.

If, however, he had first gained £70 and then lost £100, the result of his trading would be a *loss* of £30, that is $+£70-£100=-£30$, and the $-£30$ denotes that he is £30 worse off than when he began, or that he now has a *debt* of £30.

Thus we see that the $-£30$ denotes a quantity *equal in magnitude, but opposite in character* to the $+£30$.

(ii.) Again, suppose a man to row 60 yards up a stream, and then to drift down with the current for 40 yards, his position relative to the starting point would be $+60$ yards -40 yards $=+20$ yards, the $+20$ yards denoting the distance he was *up* stream from his starting point.

10   ALGEBRA.   [CHAP.

If he had rowed 40 yards up stream and then drifted down 60 yards, his position relative to the starting point would be $+40$ yards -60 yards $= -20$ yards, the -20 yards denoting the distance he was *down* stream from his starting point.

Thus we see that -20 yards denotes a distance *equal in magnitude, but opposite in direction* to that denoted by $+20$ yards.

(iii.) On a Centigrade thermometer $15°$ C. means $15°$ *above* the freezing point, and $-15°$ C. denotes $15°$ *below* freezing point.

From the above examples it will be understood that $+5$, for example, will denote a quantity *greater* than 0 by 5 units, whereas -5 will denote a quantity that is *less* than 0 by 5 units, the two quantities being of the same *absolute value* but *of opposite character*.

EXAMPLES II. a.

1. A trader gains £20, loses £42, and then gains £10. Express algebraically the result of his three transactions.

2. Two cricket counties play 16 matches; one wins 10 and loses 6, and the other wins 7 and loses 9. Express the two results, allowing a gain of one point for a win and a loss of one point for a defeat.

3. In the night a Centigrade thermometer falls to $-8°$, and in the day-time it rises to $12°$. How many degrees are there between the readings?

4. A Centigrade thermometer rises to $9°$ in the day-time and falls $15°$ during the night; what is the night reading?

5. A snail climbs 6 feet vertically upwards from a given point on a wall, slips down 15 feet, and then climbs 6 feet upwards again. Express algebraically his final position from his starting point.

6. Two men each fire 20 shots at a mark and agree to register 4 points for every hit and to deduct 3 points for every miss. One hits the mark 12 times, the other 8 times. Express algebraically their separate scores.

7. Each of three football teams plays 20 matches during the season. The A team wins 9 and loses 5, the B team wins 6 and loses 8, and the C team wins 9 and loses 9, the other games being drawn. If one point be allowed for a win, and one point deducted for a loss, place the three teams in order of merit and give the expressions that denote the results of the season's play.

Addition of Like Terms.

18. Definition. When terms do not differ, or when they differ only in their numerical coefficients, they are called **like**, otherwise they are called **unlike**. Thus $3a$, $7a$; $5a^2b$, $2a^2b$; $3a^3b^2$, $-4a^3b^2$ are pairs of like terms; and $4a$, $3b$; $7a^2$, $9a^2b$ are pairs of unlike terms.

The rules for adding like terms are

Rule I. *The sum of a number of like terms is a like term.*

Rule II. *If all the terms are positive, add the coefficients.*

Example. Find the value of $8a + 5a$.

Here we have to increase 8 things by 5 like things, and the aggregate is 13 of such things;

for instance, $\qquad\qquad$ 8 lbs. + 5 lbs. = 13 lbs.

Hence also, $\qquad\qquad 8a + 5a = 13a$.

Similarly, $\qquad 8a + 5a + a + 2a + 6a = 22a$.

Rule III. *If all the terms are negative, add the coefficients numerically and prefix the minus sign to the sum.*

Example. To find the sum of $-3x$, $-5x$, $-7x$, $-x$.

Here we have to express, as one subtractive quantity, the *sum*, or total, of four subtractive quantities of like character. To subtract in succession 3, 5, 7, 1 like things would have the same effect as to take away $3 + 5 + 7 + 1$, or 16, such things in one operation.

Thus the sum of $-3x$, $-5x$, $-7x$, $-x$ is $-16x$.

Rule IV. *If the terms are not all of the same sign, add together separately the coefficients of all the positive terms and the coefficients of all the negative terms; the difference of these two results, preceded by the sign of the greater, will give the coefficient of the sum required.*

Example 1. Find the sum of $17x$ and $-8x$.

A gain of 17 followed by a loss of 8 would give as a result a gain of 9, for the difference of 17 and 8 is 9, and the gain, or positive term, is the greater.

Thus the sum of $17x$ and $-8x = 9x$.

Example 2. The sum of $-17x$ and $8x = -9x$.

Example 3. Find the sum of $8a$, $-9a$, $-a$, $3a$, $4a$, $-11a$, a.

The sum of the coefficients of the positive terms is 16.
 ,, ,, ,, negative ,, 21.

The difference of these is 5 and the sign of the greater is negative; hence the required sum is $-5a$.

When a number of quantities are connected together by the signs + and −, the value of the result is the same in whatever order the terms are taken.

For example, in a series of combined losses and gains, the result is the same in whatever order the gains and losses are taken.

We may, therefore, add or subtract the terms in the most convenient order, which is usually that stated in Rule IV. above. This process is called **collecting terms**.

19. When quantities are connected by the signs + and −, the resulting expression is called their **algebraical sum**.

Thus $11a - 27a + 13a = -3a$ states that the algebraical sum of $11a$, $-27a$, $13a$ is equal to $-3a$.

Note. The sum of two quantities numerically equal but with opposite signs is zero. Thus the sum of $5a$ and $-5a$ is 0.

EXAMPLES II. b.

Find the sum of

1. $5a, 7a, 11a, a, 23a$.
2. $4x, x, 3x, 7x, 9x$.
3. $7b, 10b, 11b, 9b, 2b$.
4. $6c, 8c, 2c, 15c, 19c, 100c, c$.
5. $-3x, -5x, -11x, -7x$.
6. $-5b, -6b, -11b, -18b$.
7. $-3y, -7y, -y, -2y, -4y$.
8. $-c, -2c, -50c, -13c$.
9. $-11b, -5b, -3b, -b$.
10. $5x, -x, -3x, 2x, -x$.
11. $26y, -11y, -15y, y, -3y, 2y$.
12. $5f, -9f, -3f, 21f, -30f$.
13. $2s, -3s, s, -s, -5s, 5s$.
14. $7y, -11y, 16y, -3y, -2y$.
15. $5x, -7x, -2x, 7x, 2x, -5x$.
16. $7ab, -3ab, -5ab, 2ab, ab$.

Find the value of

17. $-9x^2 + 11x^2 + 3x^2 - 4x^2$.
18. $3a^2x - 18a^2x + a^2x$.
19. $3a^3 - 7a^3 - 8a^3 + 2a^3 - 11a^3$.
20. $4x^3 - 5x^3 - 8x^3 - 7x^3$.
21. $4a^2b^2 - a^2b^2 - 7a^2b^2 + 5a^2b^2 - a^2b^2$.
22. $-9x^4 - 4x^4 - 12x^4 + 13x^4 - 7x^4$.
23. $7abcd - 11abcd - 41abcd + 2abcd$.
24. $\frac{1}{2}x - \frac{1}{3}x + x + \frac{2}{3}x$.
25. $\frac{3}{2}a + \frac{3}{5}a - \frac{1}{2}a$.
26. $-5b + \frac{1}{4}b - \frac{3}{2}b + 2b - \frac{1}{2}b + \frac{7}{4}b$.
27. $-\frac{5}{3}x^2 - 2x^2 - \frac{2}{3}x^2 + x^2 + \frac{1}{2}x^2 + \frac{11}{6}x^2$.
28. $-ab - \frac{1}{3}ab - \frac{1}{3}ab - \frac{1}{4}ab - \frac{1}{6}ab + ab + \frac{5}{12}ab$.
29. $\frac{2}{3}x - \frac{3}{4}x + \frac{5}{6}x - 2x + \frac{11}{6}x - \frac{1}{3}x + x$.
30. $-\frac{5}{3}x^2 - \frac{3}{4}x^2 - \frac{4}{3}x^2 - \frac{1}{4}x^2 - x^2$.

CHAPTER III.

SIMPLE BRACKETS. ADDITION.

20. WHEN a number of arithmetical quantities are connected together by the signs $+$ and $-$, the value of the result is the same in whatever order the terms are taken. This also holds in the case of algebraical quantities.

Thus $a-b+c$ is equivalent to $a+c-b$, for in the first of the two expressions b is taken from a, and c added to the result; in the second c is added to a, and b taken from the result. Similar reasoning applies to all algebraical expressions. Hence we may write the terms of an expression in any order we please.

Thus it appears that the expression $a-b$ may be written in the equivalent form $-b+a$.

To illustrate this we may suppose, as in Art. 17, that a represents a gain of a pounds, and $-b$ a loss of b pounds: it is clearly immaterial whether the gain precedes the loss or the loss precedes the gain.

21. Brackets ( ) are used to indicate that the terms enclosed within them are to be considered as one quantity. The full use of brackets will be considered in Chap. VII.; here we shall deal only with the simpler cases.

$8+(13+5)$ means that 13 and 5 are to be added and their sum added to 8. It is clear that 13 and 5 may be added separately or together without altering the result.

Thus $\qquad 8+(13+5)=8+13+5=26.$

Similarly $a+(b+c)$ means that the sum of b and c is to be added to a.

Thus $\qquad a+(b+c)=a+b+c.$

$8+(13-5)$ means that to 8 we are to add the excess of 13 over 5; now if we add 13 to 8 we have added 5 too much, and must therefore take 5 from the result.

Thus $\qquad 8+(13-5)=8+13-5=16.$

Similarly $a+(b-c)$ means that to a we are to add b, diminished by c.

Thus $\qquad a+(b-c)=a+b-c\ \dotfill(1).$

In like manner,
$$a+b-c+(d-e-f)=a+b-c+d-e-f\ \dotfill(2).$$
Conversely,
$$a+b-c+d-e-f=a+b-c+(d-e-f)\ \dotfill(3).$$
Again, $\quad a-b+c=a+c-b,\qquad\qquad$ [Art. 20.]
$\qquad\qquad\ =$ the sum of a and $c-b$,
$\qquad\qquad\ =$ the sum of a and $-b+c,\qquad$ [Art. 20.]
therefore $\quad a-b+c=a+(-b+c)\ \dotfill(4).$

By considering the results (1), (2), (3), (4) we are led to the following rule:

Rule. *When an expression within brackets is preceded by the sign $+$, the brackets can be removed without making any change in the expression.*

Conversely: *Any part of an expression may be enclosed within brackets and the sign $+$ prefixed, the sign of every term within the brackets remaining unaltered.*

Thus the expression $a-b+c-d+e$ may be written in any of the following ways,
$$a+(-b+c-d+e),$$
$$a-b+(c-d+e),$$
$$a-b+c+(-d+e).$$

22. The expression $a-(b+c)$ means that from a we are to take the sum of b and c. The result will be the same whether b and c are subtracted separately or in one sum. Thus
$$a-(b+c)=a-b-c.$$

Again, $a-(b-c)$ means that from a we are to subtract the excess of b over c. If from a we take b we get $a-b$; but by so doing we shall have taken away c too much, and must therefore add c to $a-b$. Thus
$$a-(b-c)=a-b+c.$$
In like manner,
$$a-b-(c-d-e)=a-b-c+d+e.$$

Accordingly the following rule may be enunciated:

Rule. *When an expression within brackets is preceded by the sign* $-$, *the brackets may be removed if the sign of every term within the brackets be changed.*

Conversely: *Any part of an expression may be enclosed within brackets and the sign* $-$ *prefixed, provided the sign of every term within the brackets be changed.*

Thus the expression $a-b+c+d-e$ may be written in any of the following ways,
$$a-(+b-c-d+e),$$
$$a-b-(-c-d+e),$$
$$a-b+c-(-d+e).$$

We have now established the following results:

I. *Additions and subtractions may be made in any order.*

Thus $a+b-c+d-e-f = a-c+b+d-f-e$
$$= a-c-f+d+b-e.$$

This is known as the **Commutative Law for Addition and Subtraction.**

II. *The terms of an expression may be grouped in any manner.*

Thus $a+b-c+d-e-f = (a+b)-c+(d-e)-f$
$$= a+(b-c)+(d-e)-f = a+b-(c-d)-(e+f).$$

This is known as the **Associative Law for Addition and Subtraction.**

Addition of Unlike Terms.

23. When two or more *like* terms are to be added together we have seen that they may be collected and the result expressed as a *single* like term. If, however, the terms are *unlike* they cannot be collected; thus in finding the sum of two unlike quantities a and b, all that can be done is to connect them by the sign of addition and leave the result in the form $a+b$.

Also by the rules for removing brackets, $a+(-b)=a-b$; that is, the algebraic sum of a and $-b$ is written in the form $a-b$.

It will be observed that in Algebra the word *sum* is used in a wider sense than in Arithmetic. Thus, in the language of Arithmetic, $a-b$ signifies that b is to be subtracted from a, and bears that meaning only; but in Algebra it is also taken to mean the sum of the two quantities a and $-b$ without any regard to the relative magnitudes of a and b.

Example 1. Find the sum of $3a-5b+2c$; $2a+3b-d$; $-4a+2b$.

$$\begin{aligned}\text{The sum} &= (3a-5b+2c)+(2a+3b-d)+(-4a+2b)\\ &= 3a-5b+2c+2a+3b-d-4a+2b)\\ &= 3a+2a-4a-5b+3b+2b+2c-d\\ &= a+2c-d,\end{aligned}$$

by collecting like terms.

The addition is, however, more conveniently effected by the following rule:

Rule. *Arrange the expressions in lines so that the like terms may be in the same vertical columns: then add each column beginning with that on the left.*

$$\begin{array}{l}3a-5b+2c\\ 2a+3b\quad\;\;-d\\ -4a+2b\\ \hline a\quad\;\;+2c-d\end{array}$$

The algebraical sum of the terms in the first column is a, that of the terms in the second column is zero. The single terms in the third and fourth columns are brought down without change.

Example 2. Add together $-5ab+6bc-7ac$; $8ab+3ac-2ad$; $-2ab+4ac+5ad$; $bc-3ab+4ad$.

$$\begin{array}{l}-5ab+6bc-7ac\\ 8ab\quad\quad\;\;+3ac-2ad\\ -2ab\quad\quad+4ac+5ad\\ -3ab+\;bc\quad\quad\;\;+4ad\\ \hline -2ab+7bc\quad\quad\;\;+7ad\end{array}$$

Here we first rearrange the expressions so that like terms are in the same vertical columns, and then add up each column separately.

EXAMPLES III. a.

Find the sum of

1. $a+2b-3c$; $-3a+b+2c$; $2a-3b+c$.
2. $3a+2b-c$; $-a+3b+2c$; $2a-b+3c$.
3. $-3x+2y+z$; $x-3y+2z$; $2x+y-3z$.
4. $-x+2y+3z$; $3x-y+2z$; $2x+3y-z$.
5. $4a+3b+5c$; $-2a+3b-8c$; $a-b+c$.
6. $-15a-19b-18c$; $14a+15b+8c$; $a+5b+9c$.
7. $25a-15b+c$; $13a-10b+4c$; $a+20b-c$.
8. $-16a-10b+5c$; $10a+5b+c$; $6a+5b-c$.
9. $5ax-7by+cz$; $ax+2by-cz$; $-3ax+2by+3cz$.
10. $20p+q-r$; $p-20q+r$; $p+q-20r$.

Add together the following expressions:

11. $-5ab+6bc-7ca$; $8ab-4bc+3ca$; $-2ab-2bc+4ca$.
12. $15ab-27bc-6ca$; $14ab-18bc+10ca$; $45bc-3ca-49ab$.
13. $5ab+bc-3ca$; $ab-bc+ca$; $-ab+2ca+bc$.
14. $pq+qr-rp$; $-pq+qr+rp$; $pq-qr+rp$.
15. $x+y+z$; $2x+3y-2z$; $3x-4y+z$.
16. $2a-3b+c$; $15a-21b-8c$; $24b+7c+3a$.
17. $4xy-9yz+2zx$; $-25xy+24yz-zx$; $23xy-15yz+zx$.
18. $17ab-13bc+8ca$; $-5ab+9bc-7ca$; $-7bc-ca+2ab$.
19. $47x-63y+z$; $-25x+15y-3z$; $-22x+15z+48y$.
20. $-17b-2c+23a$; $-9a+15b+7c$; $-13a+3b-4c$.

Dimension and Degree.
Ascending and Descending Powers.

24. Each of the letters composing a term is called a **dimension** of the term, and the number of letters involved is called the **degree** of the term. Thus the product abc is said to be *of three dimensions*, or *of the third degree*; and ax^4 is said to be *of five dimensions*, or *of the fifth degree*.

A numerical coefficient is not counted. Thus $8a^2b^5$ and a^2b^5 are each of *seven* dimensions.

The **degree of an expression** is the degree of the term of highest dimensions contained in it; thus a^4-8a^3+3a-5 is *an expression of the fourth degree*, and $a^2x-7b^2x^3$ is *an expression of the fifth degree*. But it is sometimes useful to speak of the dimensions of an expression with regard to some one of the letters it involves. For instance the expression ax^3-bx^2+cx-d is said to be of *three dimensions in x*.

A compound expression is said to be **homogeneous** when all its terms are of the same dimensions. Thus $8a^6-a^4b^2+9ab^5$ is a *homogeneous expression of six dimensions*.

25. Different powers of the same letter are unlike terms; thus the result of adding together $2x^3$ and $3x^2$ cannot be expressed by a single term, but must be left in the form $2x^3+3x^2$.

Similarly the algebraical sum of $5a^2b^2$, $-3ab^3$, and $-b^4$ is $5a^2b^2-3ab^3-b^4$. This expression is in its simplest form and cannot be abridged.

In adding together several algebraical expressions containing terms with different powers of the same letter, it will be found convenient to arrange all expressions in *descending* or *ascending* powers of that letter. This will be made clear by the following examples.

Example 1. Add together $3x^3 + 7 + 6x - 5x^2$; $2x^2 - 8 - 9x$; $4x - 2x^3 + 3x^2$; $3x^3 - 9x - x^2$; $x - x^2 - x^3 + 4$.

$$\begin{array}{r} 3x^3 - 5x^2 + 6x + 7 \\ 2x^2 - 9x - 8 \\ -2x^3 + 3x^2 + 4x \\ 3x^3 - x^2 - 9x \\ -x^3 - x^2 + x + 4 \\ \hline 3x^3 - 2x^2 - 7x + 3 \end{array}$$

In writing down the first expression we put in the first term the highest power of x, in the second term the next highest power, and so on till the last term, in which x does not appear. The other expressions are arranged in the same way, so that in each column we have *like powers of the same letter*.

Example 2. Add together $3ab^2 - 2b^3 + a^3$; $5a^2b - ab^2 - 3a^3$; $8a^3 + 5b^3$; $9a^2b - 2a^3 + ab^2$.

$$\begin{array}{r} -2b^3 + 3ab^2 \qquad\quad + a^3 \\ - ab^2 + 5a^2b - 3a^3 \\ 5b^3 \qquad\qquad\qquad + 8a^3 \\ ab^2 + 9a^2b - 2a^3 \\ \hline 3b^3 + 3ab^2 + 14a^2b + 4a^3 \end{array}$$

Here each expression contains powers of two letters, and is arranged according to *descending* powers of b, and *ascending* powers of a.

EXAMPLES III. b.

Find the sum of

1. $2ab + 3ca + 6abc$; $-5ab + 2bc - 5abc$; $3ab - 2bc - 3ca$.
2. $2x^2 - 2xy + 3y^2$; $4y^2 + 5xy - 2x^2$; $x^2 - 2xy - 6y^2$.
3. $3a^2 - 7ab - 4b^2$; $-6a^2 + 9ab - 3b^2$; $4a^2 + ab + 5b^2$.
4. $x^2 + xy - y^2$; $-z^2 + yz + y^2$; $-x^2 + xz + z^2$.
5. $-x^2 - 3xy + 3y^2$; $3x^2 + 4xy - 5y^2$; $x^2 + xy + y^2$.
6. $x^3 - x^2 + x - 1$; $2x^2 - 2x + 2$; $-3x^3 + 5x + 1$.
7. $2x^3 - x^2 - x$; $4x^3 + 8x^2 + 7x$; $-6x^3 - 6x^2 + x$.
8. $9x^2 - 7x + 5$; $-14x^2 + 15x - 6$; $20x^2 - 40x - 17$.
9. $10x^3 + 5x + 8$; $3x^3 - 4x^2 - 6$; $2x^3 - 2x - 3$.
10. $a^3 - ab + bc$; $ab + b^3 - ca$; $ca - bc + c^3$.
11. $5a^3 - 3c^3 + d^3$; $b^3 - 2a^3 + 3d^3$; $4c^3 - 2a^3 - 3d^3$.

ADDITION.

Find the sum of

12. $6x^3 - 2x + 1$; $2x^3 + x + 6$; $x^2 - 7x^3 + 2x - 4$.
13. $a^3 - a^2 + 3a$; $3a^3 + 4a^2 + 8a$; $5a^3 - 6a^2 - 11a$.
14. $x^2 + y^2 - 2xy$; $2z^2 - 3y^2 - 4yz$; $2x^2 - 2z^2 - 3xz$.
15. $x^3 - 2y^3 + x$; $y^3 - 2x^3 + y$; $x^2 + 2y^2 - x + y^3$.
16. $x^3 + 3x^2y + 3xy^2$; $-3x^2y - 6xy^2 - x^3$; $3x^2y + 4xy^2$.
17. $a^3 + 5ab^2 + b^3$; $b^3 - 10ab^2 - a^3$; $5ab^2 - 2b^3 + 2a^2b$.
18. $x^5 - 4x^4y - 5x^3y^3$; $3x^4y + 2x^3y^3 - 6xy^4$; $3x^3y^3 + 6xy^4 - y^5$.
19. $a^3 - 4a^2b + 6abc$; $a^2b - 10abc + c^3$; $b^3 + 3a^2b + abc$.
20. $x^3 - 4x^2y + 6xy^2$; $2x^2y - 3xy^2 + 2y^3$; $y^3 + 3x^2y + 4xy^2$.

Add together the following expressions:

21. $\frac{1}{2}a - \frac{1}{3}b$; $-a + \frac{2}{3}b$; $\frac{3}{4}a - b$.
22. $-\frac{1}{3}a - \frac{1}{4}b$; $-\frac{2}{3}a + \frac{3}{4}b$; $-2a - b$.
23. $-2a + \frac{5}{2}c$; $-\frac{1}{3}a - 2b$; $\frac{8}{3}b - 3c$.
24. $-1\frac{3}{8}a - \frac{11}{4}c$; $2a - 3b$; $\frac{11}{5}b - c$.
25. $\frac{2}{3}x^2 + \frac{1}{3}xy - \frac{1}{4}y^2$; $-x^2 - \frac{2}{3}xy + 2y^2$; $\frac{2}{3}x^2 - xy - \frac{5}{4}y^2$.
26. $3a^2 - \frac{2}{5}ab - \frac{1}{2}b^2$; $-\frac{3}{2}a^2 + 2ab - \frac{2}{3}b^2$; $-\frac{2}{3}a^2 - ab + b^2$.
27. $\frac{5}{8}x^2 - \frac{1}{3}xy + \frac{3}{10}y^2$; $-\frac{3}{4}x^2 + \frac{14}{15}xy - y^2$; $\frac{1}{2}x^2 - xy + \frac{1}{5}y^2$.
28. $-\frac{3}{4}x^3 + 5ax^2 - \frac{5}{8}a^2x$; $x^3 - \frac{37}{8}ax^2 + \frac{1}{2}a^2x$; $-\frac{1}{2}x^3 + \frac{3}{4}a^2x$.
29. $\frac{3}{8}x^2 - \frac{5}{3}xy - 7y^2$; $\frac{2}{3}xy + \frac{18}{5}y^2$; $-\frac{5}{8}x^2 + 4y^2$.
30. $\frac{1}{2}a^3 - 2a^2b - \frac{3}{5}b^3$; $\frac{3}{2}a^2b - \frac{3}{4}ab^2 + 2b^3$; $-\frac{3}{2}a^3 + ab^2 + \frac{1}{2}b^3$.

CHAPTER IV.

SUBTRACTION.

26. The simplest cases of Subtraction have already come under the head of addition of *like* terms, of which some are negative. [Art. 18.]

Thus
$$5a - 3a = 2a,$$
$$3a - 7a = -4a,$$
$$-3a - 6a = -9a.$$

Also, by the rule for removing brackets [Art. 22],
$$3a - (-8a) = 3a + 8a$$
$$= 11a,$$
and
$$-3a - (-8a) = -3a + 8a$$
$$= 5a.$$

Subtraction of Unlike Terms.

27. The method is shewn in the following example.

Example. Subtract $3a - 2b - c$ from $4a - 3b + 5c$.

The difference
$$= 4a - 3b + 5c - (3a - 2b - c)$$
$$= 4a - 3b + 5c - 3a + 2b + c$$
$$= 4a - 3a - 3b + 2b + 5c + c$$
$$= a - b + 6c.$$

The expression to be subtracted is first enclosed in brackets with a minus sign prefixed, then on removal of the brackets the like terms are combined by the rules already explained in Art. 18.

It is, however, more convenient to arrange the work as follows, the signs of all the terms in the lower line being changed.

$$4a - 3b + 5c$$
$$-3a + 2b + c$$
by *addition*, $\overline{a - b + 6c}$

The like terms are written in the same vertical column, and each column is treated separately.

Rule. *Change the sign of every term in the expression to be subtracted, and add to the other expression.*

Note. It is not necessary that in the expression to be subtracted the signs should be *actually* changed; the operation of changing signs ought to be performed mentally.

SUBTRACTION.

Example 1. From $5x^2 + xy$ take $2x^2 + 8xy - 7y^2$.

$$\begin{array}{l} 5x^2 + xy \\ 2x^2 + 8xy - 7y^2 \\ \hline 3x^2 - 7xy + 7y^2 \end{array}$$

In the first column we combine mentally $5x^2$ and $-2x^2$, the algebraic sum of which is $3x^2$. In the last column the sign of the term $-7y^2$ has to be changed before it is put down in the result.

Example 2. Subtract $3x^2 - 2x$ from $1 - x^3$.

Terms containing different powers of the same letter being *unlike* must stand in different columns.

$$\begin{array}{l} -x^3 +1 \\ 3x^2 - 2x \\ \hline -x^3 - 3x^2 + 2x + 1 \end{array}$$

In the first and last columns, as there is nothing to be subtracted, the terms are put down without change of sign. In the second and third columns each sign has to be changed.

The re-arrangement of terms in the first line is not *necessary*, but it is convenient, because it gives the result of subtraction in descending powers of x.

EXAMPLES IV. a.

Subtract

1. $4a - 3b + c$ from $2a - 3b - c$.
2. $a - 3b + 5c$ from $4a - 8b + c$.
3. $2x - 8y + z$ from $15x + 10y - 18z$.
4. $15a - 27b + 8c$ from $10a + 3b + 4c$.
5. $-10x - 14y + 15z$ from $x - y - z$.
6. $-11ab + 6cd$ from $-10bc + ab - 4cd$.
7. $4a - 3b + 15c$ from $25a - 16b - 18c$.
8. $-16x - 18y - 15z$ from $-5x + 8y + 7z$.
9. $ab + cd - ac - bd$ from $ab + cd + ac + bd$.
10. $-ab + cd - ac + bd$ from $ab - cd + ac - bd$.

From

11. $3ab + 5cd - 4ac - 6bd$ take $3ab + 6cd - 3ac - 5bd$.
12. $yz - zx + xy$ take $-xy + yz - zx$.
13. $-2x^3 - x^2 - 3x + 2$ take $x^3 - x + 1$.
14. $-8x^2y + 15xy^2 + 10xyz$ take $4x^2y - 6xy^2 - 5xyz$.
15. $\frac{1}{2}a - b + \frac{1}{3}c$ take $\frac{1}{3}a + \frac{1}{2}b - \frac{1}{2}c$.
16. $\frac{3}{4}x + y - z$ take $\frac{1}{2}x - \frac{1}{2}y - \frac{1}{3}z$.
17. $-a - 3b$ take $\frac{3}{2}a + \frac{1}{3}b - \frac{1}{2}c$.
18. $\frac{1}{2}x - \frac{3}{7}y + \frac{1}{10}z$ take $-\frac{1}{2}x + \frac{4}{7}y - \frac{1}{10}z$.
19. $-\frac{2}{3}x - \frac{3}{5}y - 5z$ take $\frac{2}{3}x - \frac{3}{5}y - \frac{11}{3}z$.
20. $-\frac{1}{2}x + \frac{2}{3}y - \frac{1}{6}$ take $\frac{1}{3}x - \frac{3}{2}y - \frac{1}{6}$.

EXAMPLES IV. b.

From

1. $3xy - 5yz + 8zx$ take $-4xy + 2yz - 10zx$.
2. $-8x^2y^2 + 15x^3y + 13xy^3$ take $4x^2y^2 + 7x^3y - 8xy^3$.
3. $-8 + 6ab + a^2b^2$ take $4 - 3ab - 5a^2b^2$.
4. $a^2bc + b^2ca + c^2ab$ take $3a^2bc - 5b^2ca - 4c^2ab$.
5. $-7a^2b + 8ab^2 + cd$ take $5a^2b - 7ab^2 + 6cd$.
6. $-8x^2y + 5xy^2 - x^2y^2$ take $8x^2y - 5xy^2 + x^2y^2$.
7. $10a^2b^2 + 15ab^2 + 8a^2b$ take $-10a^2b^2 + 15ab^2 - 8a^2b$.
8. $4x^2 - 3x + 2$ take $-5x^2 + 6x - 7$.
9. $x^3 + 11x^2 + 4$ take $8x^2 - 5x - 3$.
10. $-8a^2x^2 + 5x^2 + 15$ take $9a^2x^2 - 8x^2 - 5$.

Subtract

11. $x^3 - x^2 + x + 1$ from $x^3 + x^2 - x + 1$.
12. $3xy^2 - 3x^2y + x^3 - y^3$ from $x^3 + 3x^2y + 3xy^2 + y^3$.
13. $b^3 + c^3 - 2abc$ from $a^3 + b^3 - 3abc$.
14. $7xy^2 - y^3 - 3x^2y + 5x^3$ from $8x^3 + 7x^2y - 3xy^2 - y^3$.
15. $x^4 + 5 + x - 3x^3$ from $5x^4 - 8x^3 - 2x^2 + 7$.
16. $a^3 + b^3 + c^3 - 3abc$ from $7abc - 3a^3 + 5b^3 - c^3$.
17. $1 - x + x^5 - x^4 - x^3$ from $x^4 - 1 + x - x^2$.
18. $7a^4 - 8a^2 + 3a^5 + a$ from $a^2 - 5a^3 - 7 + 7a^5$.
19. $10a^2b + 8ab^2 - 8a^3b^3 - b^4$ from $5a^2b - 6ab^2 - 7a^3b^3$.
20. $a^3 - b^3 + 8ab^2 - 7a^2b$ from $-8ab^2 + 15a^2b + b^3$.

From

21. $\frac{1}{2}x^2 - \frac{1}{3}xy - \frac{3}{2}y^2$ take $-\frac{3}{2}x^2 + xy - y^2$.
22. $\frac{2}{3}a^2 - \frac{5}{2}a - 1$ take $-\frac{2}{3}a^2 + a - \frac{1}{2}$.
23. $\frac{1}{3}x^2 - \frac{1}{2}x + \frac{1}{6}$ take $\frac{1}{3}x - 1 + \frac{1}{2}x^2$.
24. $\frac{3}{8}x^2 - \frac{2}{3}ax$ take $\frac{1}{3} - \frac{1}{4}x^2 - \frac{5}{6}ax$.
25. $\frac{3}{4}x^3 - \frac{1}{3}xy^2 - y^2$ take $\frac{1}{2}x^2y - \frac{5}{6}y^2 - \frac{1}{3}xy^2$.
26. $\frac{1}{8}a^3 - 2ax^2 - \frac{1}{3}a^2x$ take $\frac{1}{3}a^2x + \frac{1}{4}a^3 - \frac{3}{2}ax^2$.

MISCELLANEOUS EXAMPLES I.

1. Simplify   (1) $4x - 2x^2 - (2x - 3x^2)$;
   (2) $3a - 4b - (3b + a) - (5a - 8b)$.

2. To the sum of $2a - 3b - 2c$ and $2b - a + 7c$ add the sum of $a - 4c + 7b$ and $c - 6b$.

3. When $x=3$, $y=2$, $z=0$, find the value of
   (1) $x^2 + \frac{3}{2}y^3 - xyz^3$;   (2) $\frac{1}{4}x^3y^4 + \frac{5z^2}{6}$.

4. Define *index*, *coefficient*. In the expressions $4x^2 + 3x$, $2x^3 + x^2$, $x^2 + 7x$, find (1) the sum of the indices, (2) the sum of the coefficients.

5. From $5x^3 + 3x - 1$ take the sum of
   $2x - 5 + 7x^2$ and $3x^2 + 4 - 2x^3 + x$.

6. Subtract $3a - 7a^3 + 5a^2$ from the sum of
   $2 + 8a^2 - a^3$ and $2a^3 - 3a^2 + a - 2$.

---

7. Distinguish between *like* and *unlike* terms. Pick out the like terms in the expression
$$a^3 - 3ab + b^2 - 2a^3 - a^2 + 3b^2 + 5ab + 7a^2.$$

8. Write down in as many ways as possible the result of adding together x, y, and z.

9. Subtract $5x^2 + 3x - 1$ from $2x^3$, and add the result to
$$3x^2 + 3x - 1.$$

10. If the number of pounds I possess is represented by $+a$, what will $-a$ denote?

11. Write down in algebraical symbols the result of diminishing $2a$ by the sum of $3b$ and $5c$.

12. When $x=1$, $y=2$, $z=3$, find the value of the sum of $5x^2$, $-2x^3z$, $3y^4$. Also find the value of $2z^y - 3y^x$.

---

13. Add the sum of $2y - 3y^2$ and $1 - 5y^3$ to the remainder left when $1 - 2y^2 + y$ is subtracted from $5y^3$.

14. Explain clearly why $x - (y - z) = x - y + z$.

15. If $x=4$, $y=3$, $z=2$, $a=0$, find the value of
$$3x^2 - 2yz - ax + 5ax^2y.$$

16. Simplify $2a - b - (3a - 2b) + (2a - 3b) - (a - 2b)$.

17. Find the algebraical sum of the like terms in the expression
$5a^3 - 4a^2b + b^3 + 6a^3b + 7ab^2 - 3a^2b + 4ab^3 + 8a^2b$.

18. A boy works $x+y$ sums, of which only $y-2z$ are right; how many are wrong?

---

19. In the expression $3a^3 - 7a^2b + b^4$, point out the highest power, the lowest power, the positive terms, and the coefficient of a^2.

20. Take $x^2 - y^2$ from $3xy - 4y^2$, and add the remainder to the sum of $4xy - x^2 - 3y^2$ and $2x^2 + 6y^2$.

21. If $x=1$, $y=3$, $z=5$, $w=0$, find the value of
$$\sqrt{(3xy)} + \sqrt{(5xz)} + \sqrt{(3yw)}.$$

22. What is the *degree* of a term in an algebraical expression? In the expression $4x^6 - 3x^5a^2 + a^8$, what is the degree of the negative term?

23. Find the sum of $5a - 7b + c$ and $3b - 9a$, and subtract the result from $c - 4b$.

24. If $x=3$, $y=4$, $p=8$, $q=10$, find the value of
$$xyp + \frac{2y}{p-y} + 2q.$$

---

25. If x represents the date 10 A.D. what will $-3x$ stand for?

26. Add together $3x^2 - 7x + 5$ and $2x^3 + 5x - 3$, and diminish the result by $3x^2 + 2$.

27. In the expression
$$4a^2b^3 - b^4 + 3a^3b^2 + 5b^5 - ab^3x + 2x^3ab + abx^4 - a^2b^3,$$
point out which terms are *like*, and which are homogeneous. What is the degree of the expression?

28. Express in algebraical symbols the excess of the sum of a and b over c diminished by d.

29. A man walks $2a - b$ miles due North from a fixed point O, and then walks a distance $3a + 2b$ miles due South; what is his final position with regard to O?

30. What expression must be added to $5x^2 - 7x + 2$ to produce $7x^2 - 1$?

CHAPTER V.

MULTIPLICATION.

[*Part of this chapter may be taken at a later stage. See remark on page* 33. *The easy graphical work in Arts.* 411–420 *may be studied after* Examples v. b.]

28. MULTIPLICATION in its primary sense signifies repeated addition.

Thus $\quad 3 \times 4 = 3$ taken 4 times
$$= 3+3+3+3.$$

Here the multiplier contains four units, and the number of times we take 3 is the same as the number of units in 4.

Again $\quad a \times b = a$ taken b times
$$= a+a+a+\ldots,$$
the number of terms being b.

Also $3 \times 4 = 4 \times 3$; and so long as a and b denote positive whole numbers, it is easy to show that $a \times b = b \times a$.

29. When the quantities to be multiplied together are not positive whole numbers, we may define multiplication as *an operation performed on one quantity which when performed on unity produces the other*. For example, to multiply $\frac{4}{5}$ by $\frac{3}{7}$, we perform on $\frac{4}{5}$ that operation which when performed on unity gives $\frac{3}{7}$; that is, we must divide $\frac{4}{5}$ into seven equal parts and take three of them. Now each part will be equal to $\frac{4}{5 \times 7}$, and the result of taking three of such parts is expressed by $\frac{4 \times 3}{5 \times 7}$.

Hence $\quad \dfrac{4}{5} \times \dfrac{3}{7} = \dfrac{4 \times 3}{5 \times 7}.$

Also, by the last article,
$$\frac{4 \times 3}{5 \times 7} = \frac{3 \times 4}{7 \times 5} = \frac{3}{7} \times \frac{4}{5}$$

$$\therefore \frac{4}{5} \times \frac{3}{7} = \frac{3}{7} \times \frac{4}{5}.$$

The reasoning is clearly general, and we may now say that $a \times b = b \times a$, where a and b are any positive quantities, integral or fractional.

In the same way it easily follows that
$$abc = a \times b \times c$$
$$= (a \times b) \times c = (b \times a) \times c = bac$$
$$= b \times (a \times c) = b \times c \times a = bca;$$
that is, *the factors of a product may be taken in any order.* This is the **Commutative Law for Multiplication.**

Example. $2a \times 3b \times c = 2 \times 3 \times a \times b \times c = 6abc.$

30. Again, *the factors of a product may be grouped in any way we please.*

Thus $abcd = a \times b \times c \times d$
$$= (ab) \times (cd) = a \times (bc) \times d = a \times (bcd).$$

This is the **Associative Law for Multiplication.**

31. Since, by definition, $a^3 = aaa$, and $a^5 = aaaaa$,
$$\therefore a^3 \times a^5 = aaa \times aaaaa = aaaaaaaa = a^8 = a^{3+5};$$
that is, *the index of a letter in a product is the sum of its indices in the factors of the product.* This is the **Index Law for Multiplication.**

Again, $5a^2 = 5aa$, and $7a^3 = 7aaa$.
$$\therefore 5a^2 \times 7a^3 = 5 \times 7 \times aaaaa = 35a^5.$$

When the expressions to be multiplied together contain powers of different letters, a similar method is used.

Example. $5a^3b^2 \times 8a^2bx^3 = 5aaabb \times 8aabxxx$
$$= 40a^5b^3x^3.$$

Note. The beginner must be careful to observe that in this process of multiplication *the indices of one letter cannot combine in any way with those of another.* Thus the expression $40a^5b^3x^3$ admits of no further simplification.

32. **Rule.** *To multiply two simple expressions together, multiply the coefficients together and prefix their product to the product of the different letters, giving to each letter an index equal to the sum of the indices that letter has in the separate factors.*

The rule may be extended to cases where more than two expressions are to be multiplied together.

Example 1. Find the product of x^2, x^3, and x^8.

The product $= x^2 \times x^3 \times x^8 = x^{2+3} \times x^8 = x^{2+3+8} = x^{13}$.

The product of three or more expressions is called **the continued product.**

Example 2. Find the continued product of $5x^2y^3$, $8y^2z^5$, and $3xz^4$.

The product $= 5x^2y^3 \times 8y^2z^5 \times 3xz^4 = 120x^3y^5z^9$.

Multiplication of a Compound Expression by a Simple Expression.

33. By definition,

$(a+b)m = m+m+m+\ldots$ taken $a+b$ times

$\qquad\qquad = (m+m+m+\ldots$ taken a times$)$,

together with $\quad (m+m+m+\ldots$ taken b times$)$

$\qquad\qquad = am + bm \ldots\ldots\ldots\ldots\ldots\ldots\ldots\ldots\ldots\ldots(1)$.

Also $\quad (a-b)m = m+m+m+\ldots$ taken $a-b$ times

$\qquad\qquad = (m+m+m+\ldots$ taken a times$)$,

diminished by $\quad (m+m+m+\ldots$ taken b times$)$

$\qquad\qquad = am - bm \ldots\ldots\ldots\ldots\ldots\ldots\ldots\ldots\ldots\ldots(2)$.

Similarly $\quad (a-b+c)m = am - bm + cm$.

Thus it appears that *the product of a compound expression by a single factor is the algebraic sum of the partial products of each term of the compound expression by that factor*. This is known as the **Distributive Law for Multiplication.**

Note. It should be observed that for the present a, b, c denote positive whole numbers, and that a is supposed greater than b.

Examples. $3(2a + 3b - 4c) = 6a + 9b - 12c$.

$(4x^2 - 7y - 8z^3) \times 3xy^2 = 12x^3y^2 - 21xy^3 - 24xy^2z^3$.

EXAMPLES V. a.

Find the value of

1. $5x^2 \times 7x^5$.
2. $4a^3 \times 5a^8$.
3. $7ab \times 8a^3b^2$.
4. $6xy^2 \times 5x^3$.
5. $8a^3b \times b^5$.
6. $2abc \times 3ac^3$.
7. $2a^3b^3 \times 2a^3b^3$.
8. $5a^2b \times 2a$.
9. $4a^2b^3 \times 7a^5$.
10. $5a^4b^3 \times x^2y^2$.
11. $x^3y^3 \times 6a^2x^4$.
12. $abc \times xyz$.
13. $3a^4b^7x^3 \times 5a^3bx$.
14. $4a^3bx \times 7b^2x^4$.
15. $5a^2x \times 8cx$.
16. $5x^3y^3 \times 6a^3x^3$.
17. $2x^2y \times x^5y^7$.
18. $3a^3x^4y^7 \times a^2x^5y^9$.

Multiply together:

19. $ab+bc$ and a^3b.
20. $5ab-7bx$ and $4a^2bx^3$.
21. $5x+3y$ and $2x^2$.
22. $a^2+b^2-c^2$ and a^3b.
23. $bc+ca-ab$ and abc.
24. $5a^2+3b^2-2c^2$ and $4a^2bc^3$.
25. $5x^2y+xy^2-7x^2y^2$ and $3x^3$.
26. $6x^3-5x^2y+7xy^2$ and $8x^2y^3$.
27. $6a^3bc-7ab^2c^2$ and a^2b^2.

Multiplication of Compound Expressions.

34. If in Art. 33 we write $c+d$ for m in (1), we have
$$(a+b)(c+d) = a(c+d)+b(c+d)$$
$$= (c+d)a+(c+d)b \qquad \text{[Art. 29.]}$$
$$= ac+ad+bc+bd \quad \ldots\ldots\ldots\ldots(3).$$

Again, from (2)
$$(a-b)(c+d) = a(c+d)-b(c+d)$$
$$= (c+d)a-(c+d)b$$
$$= ac+ad-(bc+bd)$$
$$= ac+ad-bc-bd \quad \ldots\ldots\ldots\ldots(4).$$

Similarly, by writing $c-d$ for m in (1),
$$(a+b)(c-d) = a(c-d)+b(c-d)$$
$$= (c-d)a+(c-d)b$$
$$= ac-ad+bc-bd \quad \ldots\ldots\ldots\ldots(5).$$

Also, from (2)
$$(a-b)(c-d) = a(c-d)-b(c-d)$$
$$= (c-d)a-(c-d)b$$
$$= ac-ad-(bc-bd)$$
$$= ac-ad-bc+bd \quad \ldots\ldots\ldots\ldots(6).$$

If we consider each term on the right-hand side of (6), and the way in which it arises, we find that
$$(+a) \times (+c) = +ac.$$
$$(-b) \times (-d) = +bd.$$
$$(-b) \times (+c) = -bc.$$
$$(+a) \times (-d) = -ad.$$

These results enable us to state what is known as the **Rule of Signs** in multiplication.

Rule of Signs. *The product of two terms with like signs is positive; the product of two terms with unlike signs is negative.*

35. The rule of signs, and especially the use of the negative multiplier, will probably present some difficulty to the beginner Perhaps the following numerical instances may be useful in illustrating the interpretation that may be given to multiplication by a negative quantity.

To multiply 3 by -4 we must do to 3 what is done to unity to obtain -4. Now -4 means that unity is taken 4 times and the result made negative; therefore $3 \times (-4)$ implies that 3 is to be taken 4 times and the product made negative.

But 3 taken 4 times gives $+12$;

$$\therefore 3 \times (-4) = -12.$$

Similarly -3×-4 indicates that -3 is to be taken 4 times, and the sign changed; the first operation gives -12, and the second $+12$.

Thus $(-3) \times (-4) = +12$.

Hence, *multiplication by a negative quantity indicates that we are to proceed just as if the multiplier were positive, and then change the sign of the product.*

Note on Arithmetical and Symbolical Algebra.

36. Arithmetical Algebra is that part of the science which deals solely with symbols and operations arithmetically intelligible. Starting from purely arithmetical definitions, we are enabled to prove certain fundamental laws.

Symbolical Algebra assumes these laws to be true in every case, and thence finds what meaning must be attached to symbols and operations which under unrestricted conditions no longer bear an arithmetical meaning. Thus the results of Arts. 33 and 34 were proved from arithmetical definitions which require the symbols to be positive whole numbers, such that $a > b$ and $c > d$. By the principles of symbolical Algebra we assume these results to be universally true when all restrictions are removed, and accept the interpretation to which we are led thereby.

Henceforth we are able to apply the Law of Distribution and the Rule of Signs without any restriction as to the symbols used. [See Art. 33, Note.]

37. To familiarize the beginner with the principles we have just explained we add a few examples in substitutions where some of the symbols denote negative quantities.

Example 1. If $a=-4$, find the value of a^3.

Here $\quad a^3=(-4)^3=(-4)\times(-4)\times(-4)=-64.$

By repeated applications of the rule of signs it may easily be shewn that any *odd* power of a negative quantity is *negative*, and any *even* power of a negative quantity is positive.

Example 2. If $a=-1$, $b=3$, $c=-2$, find the value of $-3a^4bc^3$.

Here $-3a^4bc^3 = -3\times(-1)^4\times 3\times(-2)^3$ | We write down at
$=-3\times(+1)\times 3\times(-8)$ | once $(-1)^4=+1$, and
$=72.$ | $(-2)^3=-8.$

EXAMPLES V. b.

If $a=-2$, $b=3$, $c=-1$, $x=-5$, $y=4$, find the value of

1. $3a^2b$.   2. $8abc^2$.   3. $-5c^3$.   4. $6a^2c^2$.
5. $4c^3y$.   6. $3a^2c$.   7. $-b^2c^2$.   8. $3a^3c^2$.
9. $-7a^3bc$.   10. $-2a^4bx$.   11. $-4a^2c^4$.   12. $3c^3x^3$.
13. $5a^2x^2$.   14. $-7c^4xy$.   15. $-8ax^3$.   16. $4c^5x^3$.
17. $-5a^2b^2c^2$.   18. $-7a^3c^3$.   19. $8c^4x^3$.   20. $7a^5c^4$.

If $a=-4$, $b=-3$, $c=-1$, $f=0$, $x=4$, $y=1$, find the value of

21. $3a^2+bx-4cy$.   22. $2ab^2-3bc^2+2fx$.
23. $fa^2-2b^3-cx^3$.   24. $3a^2y^3-5b^2x-2c^3$.
25. $2a^3-3b^3+7cy^4$.   26. $3b^2y^4-4b^2f-6c^4x$.
27. $2\sqrt{(ac)}-3\sqrt{(xy)}+\sqrt{(b^2c^4)}$.   28. $3\sqrt{(acx)}-2\sqrt{(b^2y)}-6\sqrt{(c^2y)}$.
29. $7\sqrt{(a^2x)}-3\sqrt{(b^4c^2)}+5\sqrt{(f^2x)}$.
30. $3c\sqrt{(3bc)}-5\sqrt{(4c^2y^3)}-2cy\sqrt{(3bc^5)}$.

38. The following examples further illustrate the rule of signs and the law of indices.

Example 1. Multiply $4a$ by $-3b$.

By the rule of signs the product is negative; also $4a\times 3b=12ab$;

$\therefore\quad 4a\times(-3b)=-12ab.$

Example 2. Multiply $-5ab^3x$ by $-ab^3x$.

Here the absolute value of the product is $5a^2b^6x^2$, and by the rule of signs the product is positive;

$\therefore\quad (-5ab^3x)\times(-ab^3x)=5a^2b^6x^2.$

Example 3. Find the continued product of $3a^2b$, $-2a^3b^2$, $-ab^4$.

$3a^2b\times(-2a^3b^2)=-6a^5b^3$;
$(-6a^5b^3)\times(-ab^4)=+6a^6b^7.$

Thus the complete product is $6a^6b^7$.

This result, however, may be written down at once: for

$3a^2b\times 2a^3b^2\times ab^4=6a^6b^7,$

and by the rule of signs the required product is positive.

Example 4. Multiply $6a^3 - \frac{5}{3}a^2b - \frac{4}{5}ab^2$ by $-\frac{3}{4}ab^2$.

The product is the algebraical sum of the partial products formed according to the rule enunciated in Art. 37; thus
$$(6a^3 - \tfrac{5}{3}a^2b - \tfrac{4}{5}ab^2) \times (-\tfrac{3}{4}ab^2) = -\tfrac{9}{2}a^4b^2 + \tfrac{5}{4}a^3b^3 + \tfrac{3}{5}a^2b^4.$$

EXAMPLES V. c.

Multiply together:

1. ax and $-3ax$.
2. $-2abx$ and $-7abx$.
3. a^2b and $-ab^2$.
4. $6x^2y$ and $-10xy$.
5. $-abcd$ and $-3a^2b^3c^4d^5$.
6. xyz and $-5x^2y^3z$.
7. $3xy + 4yz$ and $-12xyz$.
8. $ab - bc$ and a^2bc^3.
9. $-x - y - z$ and $-3x$.
10. $a^2 - b^2 + c^2$ and abc.
11. $-ab + bc - ca$ and $-abc$.
12. $-2a^2b - 4ab^2$ and $-7a^2b^2$.
13. $5x^2y - 6xy^2 + 8x^2y^2$ and $3xy$.
14. $-7x^3y - 5xy^3$ and $-8x^3y^3$.
15. $-5xy^2z + 3xyz^2 - 8x^2yz$ and xyz.
16. $4x^2y^2z^2 - 8xyz$ and $-12x^3yz^3$.
17. $-13xy^2 - 15x^2y$ and $-7x^3y^3$.
18. $8xyz - 10x^3yz^3$ and $-xyz$.
19. $abc - a^2bc - ab^2c$ and $-abc$.
20. $-a^2bc + b^2ca - c^2ab$ and $-ab$.

Find the product of

21. $2a - 3b + 4c$ and $-\frac{3}{2}a$.
22. $3x - 2y - 4$ and $-\frac{5}{8}x$.
23. $\frac{2}{3}a - \frac{1}{6}b - c$ and $\frac{3}{8}ax$.
24. $\frac{6}{7}a^2x^2 - \frac{3}{2}ax^3$ and $-\frac{7}{3}a^3x$.
25. $-\frac{5}{3}a^2x^2$ and $-\frac{3}{2}a^2 + ax - \frac{3}{5}x^2$.
26. $-\frac{7}{2}xy$ and $-3x^2 + \frac{2}{7}xy$.
27. $-\frac{3}{2}x^3y^2$ and $-\frac{1}{3}x^2 + 2y^2$.
28. $-\frac{4}{7}x^5y^3$ and $\frac{7}{4}x^3 - \frac{4}{7}y^3$.

39. The results of Art. 33 may be extended to the case where both of the expressions to be multiplied together contain two or more terms. For instance
$$(a - b + c)m = am - bm + cm;$$
replacing m by $x - y$, we have
$$(a - b + c)(x - y) = a(x - y) - b(x - y) + c(x - y)$$
$$= (ax - ay) - (bx - by) + (cx - cy)$$
$$= ax - ay - bx + by + cx - cy.$$

We may now state the general rule for multiplying together any two compound expressions.

Rule. *Multiply each term of the first expression by each term of the second. When the terms multiplied together have like signs, prefix to the product the sign +, when unlike prefix −; the algebraical sum of the partial products so formed gives the complete product.* This process is called **Distributing the Product.**

40. It should be noticed that the product of $a+b$ and $x-y$ is briefly expressed by $(a+b)(x-y)$, in which the brackets indicate that the expression $a+b$ taken as a whole is to be multiplied by the expression $x-y$ taken as a whole. By the above rule, the value of the product is the algebraical sum of the partial products $+ax$, $+bx$, $-ay$, $-by$; the sign of each product being determined by the rule of signs.

Example 1. Multiply $x+8$ by $x+7$.

$$\begin{aligned}\text{The product} &= (x+8)(x+7) \\ &= x^2 + 8x + 7x + 56 \\ &= x^2 + 15x + 56.\end{aligned}$$

The operation is more conveniently arranged as follows:

$$\begin{array}{r} x + 8 \\ x + 7 \\ \hline x^2 + 8x \\ + 7x + 56 \\ \hline \text{by addition,}\quad x^2 + 15x + 56 \end{array}$$

We begin on the left and work to the right, placing the second result one place to the right, so that like terms may stand in the same vertical column.

Example 2. Multiply $2x - 3y$ by $4x - 7y$.

$$\begin{array}{r} 2x - 3y \\ 4x - 7y \\ \hline 8x^2 - 12xy \\ - 14xy + 21y^2 \\ \hline \text{by addition,}\quad 8x^2 - 26xy + 21y^2. \end{array}$$

EXAMPLES V. d.

Find the product of

1. $x+5$ and $x+10$.
2. $x+5$ and $x-5$.
3. $x-7$ and $x-10$.
4. $x-7$ and $x+10$.
5. $x+7$ and $x-10$.
6. $x+7$ and $x+10$.
7. $x+6$ and $x-6$.
8. $x+8$ and $x-4$.
9. $x-12$ and $x-1$.
10. $x+12$ and $x-1$.
11. $x-15$ and $x+15$.
12. $x-15$ and $-x+3$.
13. $-x-2$ and $-x-3$.
14. $-x+7$ and $x-7$.
15. $-x+5$ and $-x-5$.
16. $x-13$ and $x+14$.
17. $x-17$ and $x+18$.
18. $x+19$ and $x-20$.
19. $-x-16$ and $-x+16$.
20. $-x+21$ and $x-21$.
21. $2x-3$ and $x+8$.
22. $2x+3$ and $x-8$.

MULTIPLICATION. 33

Find the product of

23. $x-5$ and $2x-1$.   24. $2x-5$ and $x-1$.
25. $3x-5$ and $2x+7$.   26. $3x+5$ and $2x-7$.
27. $5x-6$ and $2x+3$.   28. $5x+6$ and $2x-3$.
29. $3x-5y$ and $3x+5y$.   30. $3x-5y$ and $3x-5y$.
31. $a-2b$ and $a+3b$.   32. $a-7b$ and $a+8b$.
33. $3a-6b$ and $a-8b$.   34. $a-9b$ and $a+5b$.
35. $x+a$ and $x-b$.   36. $x-a$ and $x+b$.
37. $x-2a$ and $x+3b$.   38. $ax-by$ and $ax+by$.
39. $xy-ab$ and $xy+ab$.   40. $2pq-3r$ and $2pq+3r$.

[*With the exception of* Art. 44, *the rest of this chapter may be postponed and taken after* Chapter XIV.]

*41. We shall now give a few examples of greater difficulty.

Example 1. Find the product of $3x^2-2x-5$ and $2x-5$.

$$\begin{array}{l} 3x^2-2x-5 \\ 2x-5 \\ \hline 6x^3-4x^2-10x \\ -15x^2+10x+25 \\ \hline 6x^3-19x^2+25 \end{array}$$

Each term of the first expression is multiplied by $2x$, the first term of the second expression; then each term of the first expression is multiplied by -5; like terms are placed in the same columns and the results added.

Example 2. Multiply $a-b+3c$ by $a+2b$.

$$\begin{array}{l} a-b+3c \\ a+2b \\ \hline a^2-ab+3ac \\ 2ab-2b^2+6bc \\ \hline a^2+ab+3ac-2b^2+6bc \end{array}$$

*42. When the coefficients are fractional we use the ordinary process of Multiplication, combining the fractional coefficients by the rules of Arithmetic.

Example. Multiply $\tfrac{1}{3}a^2-\tfrac{1}{2}ab+\tfrac{2}{3}b^2$ by $\tfrac{1}{2}a+\tfrac{1}{3}b$.

$$\begin{array}{l} \tfrac{1}{3}a^2-\tfrac{1}{2}ab+\tfrac{2}{3}b^2 \\ \tfrac{1}{2}a+\tfrac{1}{3}b \\ \hline \tfrac{1}{6}a^3-\tfrac{1}{4}a^2b+\tfrac{1}{3}ab^2 \\ \phantom{\tfrac{1}{6}a^3}+\tfrac{1}{9}a^2b-\tfrac{1}{6}ab^2+\tfrac{2}{9}b^3 \\ \hline \tfrac{1}{6}a^3-\tfrac{5}{36}a^2b+\tfrac{1}{6}ab^2+\tfrac{2}{9}b^3 \end{array}$$

* 43. If the expressions are not arranged according to powers, ascending or descending, of some common letter, a rearrangement will be found convenient.

Example 1: Find the product of $2a^2+4b^2-3ab$ and $3ab-5a^2+4b^2$.

$$\begin{array}{r}2a^2-\ 3ab\ +4b^2\\ -\ 5a^2+\ 3ab\ +4b^2\\ \hline -10a^4+15a^3b-20a^2b^2\\ +\ 6a^3b-\ 9a^2b^2+12ab^3\\ 8a^2b^2-12ab^3+16b^4\\ \hline -10a^4+21a^3b-21a^2b^2\qquad\quad+16b^4\end{array}$$

The rearrangement is not *necessary*, but convenient, because it makes the collection of like terms more easy.

Example 2. Multiply $2xz-z^2+2x^2-3yz+xy$ by $x-y+2z$.

$$\begin{array}{l}2x^2+\ xy\ +2xz\ -3yz\ -z^2\\ \quad x\ -\ y\ +2z\\ \hline 2x^3+\ x^2y+2x^2z-3xyz-\ xz^2\\ \quad-2x^2y\qquad\quad-2xyz\qquad-xy^2+3y^2z+\ yz^2\\ \qquad\quad 4x^2z+2xyz+4xz^2\qquad\qquad-6yz^2-2z^3\\ \hline 2x^3-\ x^2y+6x^2z-3xyz+3xz^2-xy^2+3y^2z-5yz^2-2z^3\end{array}$$

* EXAMPLES V. e.

Multiply together

1. $a+b+c,\ a+b-c$.
2. $a-2b+c,\ a+2b-c$.
3. $a^2-ab+b^2,\ a^2+ab+b^2$.
4. $x^2+3y^2,\ x+4y$.
5. $x^3-2x^2+8,\ x+2$.
6. $x^4-x^2y^2+y^4,\ x^2+y^2$.
7. $x^2+xy+y^2,\ x-y$.
8. $a^2-2ax+4x^2,\ a^2+2ax+4x^2$.
9. $16a^2+12ab+9b^2,\ 4a-3b$.
10. $a^2x-ax^2+x^3-a^3,\ x+a$.
11. $x^2+x-2,\ x^2+x-6$.
12. $2x^3-3x^2+2x,\ 2x^2+3x+2$.
13. $-a^5+a^4b-a^3b^2,\ -a-b$.
14. $x^3-7x+5,\ x^2-2x+3$.
15. $a^3+2a^2b+2ab^2,\ a^2-2ab+2b^2$.
16. $4x^2+6xy+9y^2,\ 2x-3y$.
17. $x^2-3xy-y^2,\ -x^2+xy+y^2$.
18. $b^3-a^2b^2+a^3,\ a^3+a^2b^2+b^3$.
19. $x^2-2xy+y^2,\ x^2+2xy+y^2$.
20. $ab+cd+ac+bd,\ ab+cd-ac-bd$.
21. $-3a^2b^2+4ab^3+15a^3b,\ 5a^2b^2+ab^3-3b^4$.
22. $27x^3-36ax^2+48a^2x-64a^3,\ 3x+4a$.
23. $a^2-5ab-b^2,\ a^2+5ab+b^2$.
24. $x^2-xy+x+y^2+y+1,\ x+y-1$.

Multiply together

25. $a^2+b^2+c^2-bc-ca-ab$, $a+b+c$.
26. $-x^3y+y^4+x^2y^2+x^4-xy^3$, $x+y$.
27. $x^{12}-x^9y^2+x^6y^4-x^3y^6+y^8$, x^3+y^2.
28. $3a^2+2a+2a^3+1+a^4$, a^2-2a+1.
29. $-ax^2+3axy^2-9ay^4$, $-ax-3ay^2$.
30. $-2x^3y+y^4+3x^2y^2+x^4-2xy^3$, $x^2+2xy+y^2$.
31. $\frac{1}{2}a^2+\frac{1}{3}a+\frac{1}{4}$, $\frac{1}{2}a-\frac{1}{3}$. 32. $\frac{1}{2}x^2-2x+\frac{3}{2}$, $\frac{1}{2}x+\frac{1}{3}$.
33. $\frac{2}{3}x^2+xy+\frac{3}{2}y^2$, $\frac{1}{3}x-\frac{1}{2}y$. 34. $\frac{3}{2}x^2-ax-\frac{2}{3}a^2$, $\frac{3}{4}x^2-\frac{1}{2}ax+\frac{1}{3}a^2$.
35. $\frac{1}{2}x^2-\frac{2}{3}x-\frac{3}{4}$, $\frac{1}{2}x^2+\frac{2}{3}x-\frac{3}{4}$.
36. $\frac{2}{3}ax+\frac{2}{3}x^2+\frac{1}{3}a^2$, $\frac{3}{4}a^2+\frac{3}{2}x^2-\frac{3}{2}ax$.

44. Products written down by inspection. Although the result of multiplying together two binomial factors, such as $x+8$ and $x-7$, can always be obtained by the methods already explained, it is of the utmost importance that the student should soon learn to write down the product rapidly *by inspection*.

This is done by observing in what way the coefficients of the terms in the product arise, and noticing that they result from the combination of the numerical coefficients in the two binomials which are multiplied together; thus

$$(x+8)(x+7)=x^2+8x+7x+56$$
$$=x^2+15x+56.$$

$$(x-8)(x-7)=x^2-8x-7x+56$$
$$=x^2-15x+56.$$

$$(x+8)(x-7)=x^2+8x-7x-56$$
$$=x^2+x-56.$$

$$(x-8)(x+7)=x^2-8x+7x-56$$
$$=x^2-x-56.$$

In each of these results we notice that:

1. The product consists of three terms.

2. The first term is the product of the first terms of the two binomial expressions.

3. The third term is the product of the second terms of the two binomial expressions.

4. The middle term has for its coefficient the sum of the numerical quantities (taken with their proper signs) in the second terms of the two binomial expressions.

The intermediate step in the work may be omitted, and the products written down at once, as in the following examples:

$$(x+2)(x+3) = x^2 + 5x + 6.$$
$$(x-3)(x+4) = x^2 + x - 12.$$
$$(x+6)(x-9) = x^2 - 3x - 54.$$
$$(x-4y)(x-10y) = x^2 - 14xy + 40y^2.$$
$$(x-6y)(x+4y) = x^2 - 2xy - 24y^2.$$

By an easy extension of these principles we may write down the product of *any* two binomials.

Thus
$$(2x+3y)(x-y) = 2x^2 + 3xy - 2xy - 3y^2$$
$$= 2x^2 + xy - 3y^2.$$
$$(3x-4y)(2x+y) = 6x^2 - 8xy + 3xy - 4y^2$$
$$= 6x^2 - 5xy - 4y^2.$$
$$(x+4)(x-4) = x^2 + 4x - 4x - 16$$
$$= x^2 - 16.$$
$$(2x+5y)(2x-5y) = 4x^2 + 10xy - 10xy - 25y^2$$
$$= 4x^2 - 25y^2.$$

EXAMPLES V. f.

Write down the values of the following products:

1. $(x+8)(x-5)$.
2. $(x+6)(x-1)$.
3. $(x-3)(x+10)$.
4. $(x-1)(x+5)$.
5. $(x+7)(x-9)$.
6. $(x-10)(x-8)$.
7. $(x-4)(x+11)$.
8. $(x-2)(x+4)$.
9. $(x+2)(x-2)$.
10. $(a-1)(a+1)$.
11. $(a+9)(a-5)$.
12. $(a-3)(a+12)$.
13. $(a-8)(a+4)$.
14. $(a-8)(a+8)$.
15. $(a-6)(a+13)$.
16. $(a+3)(a+3)$.
17. $(a-11)(a+11)$.
18. $(a-8)(a-8)$.
19. $(x-3a)(x+2a)$.
20. $(x+6a)(x-5a)$.
21. $(x+3a)(x-3a)$.
22. $(x+4y)(x-2y)$.
23. $(x+7y)(x-7y)$.
24. $(x-3y)(x-3y)$.
25. $(a+3b)(a+3b)$.
26. $(a-5b)(a+10b)$.
27. $(a-9b)(a-8b)$.
28. $(2x-5)(x+2)$.
29. $(2x-5)(x-2)$.
30. $(2x+3)(x-3)$.
31. $(3x-1)(x+1)$.
32. $(2x+5)(2x-1)$.
33. $(3x+7)(2x-3)$.
34. $(4x-3)(2x+3)$.
35. $(3x+8)(3x-8)$.
36. $(2x-5)(2x-5)$.
37. $(3x-2y)(3x+y)$.
38. $(3x+2y)(3x+2y)$.
39. $(2x+7y)(2x-5y)$.
40. $(5x+3a)(5x-3a)$.
41. $(2x-5a)(x+5a)$.
42. $(2x+a)(2x+a)$.

The Method of Detached Coefficients.

***45.** When two compound expressions contain powers of one letter only, the labour of multiplication may be lessened by using **detached coefficients**, that is, by writing down the coefficients only, multiplying them together in the ordinary way, and then inserting the successive powers of the letter at the end of the operation. In using this method the expressions must be arranged according to ascending or descending powers of the common letter, and zero coefficients must be used to represent terms corresponding to missing powers of that letter.

Example. Multiply $2x^3 - 4x^2 - 5$ by $3x^2 + 4x - 2$.

$$\begin{array}{r}2 - 4 + 0 - 5\\ 3 + 4 - 2\\ \hline 6 - 12 + 0 - 15\\ 8 - 16 + 0 - 20\\ -4 + 8 - 0 + 10\\ \hline 6 - 4 - 20 - 7 - 20 + 10\end{array}$$

Here we insert a zero coefficient to represent the power of x which is absent in the multiplicand. In the product the highest power of x is clearly x^5, and the others follow in descending order.

Thus the product is

$$6x^5 - 4x^4 - 20x^3 - 7x^2 - 20x + 10.$$

The method of detached coefficients may also be used to multiply two compound expressions which are homogeneous and contain powers of two letters.

Example. Multiply $3a^4 + 2a^3b + 4ab^3 + 2b^4$ by $2a^2 - b^2$.

$$\begin{array}{r}3 + 2 + 0 + 4 + 2\\ 2 + 0 - 1\\ \hline 6 + 4 + 0 + 8 + 4\\ -3 - 2 - 0 - 4 - 2\\ \hline 6 + 4 - 3 + 6 + 4 - 4 - 2\end{array}$$

We write a zero coefficient to represent the term containing a^2b^2 which is absent in the first expression. Similarly, the term containing ab is represented by a zero coefficient in the second expression.

It is easily seen how the powers of a and b arise in the successive terms, and the complete product is

$$6a^6 + 4a^5b - 3a^4b^2 + 6a^3b^3 + 4a^2b^4 - 4ab^5 - 2b^6.$$

Note. Beginners should on no account attempt to use detached coefficients until they are well practised in the ordinary full process of multiplication.

CHAPTER VI.

Division.

[*If preferred, the articles in this chapter marked with an asterisk may be postponed and taken after* Chapter xv.]

46. When a quantity a is divided by the quantity b, the **quotient** is defined to be that which when multiplied by b produces a. This operation of division is denoted by $a \div b$, $\dfrac{a}{b}$ or a/b; in each of these modes of expression a is called the **dividend**, and b the **divisor**.

Division is thus the inverse of multiplication, and
$$(a \div b) \times b = a.$$
This statement may also be expressed verbally as follows:
$$\text{quotient} \times \text{divisor} = \text{dividend}.$$

Since Division is the inverse of Multiplication, it follows that the Laws of Commutation, Association, and Distribution, which have been established for Multiplication, hold for Division.

47. *The* **Rule of Signs** *holds for division.*

Thus
$$ab \div a = \frac{ab}{a} = \frac{a \times b}{a} = b.$$
$$-ab \div a = \frac{-ab}{a} = \frac{a \times (-b)}{a} = -b.$$
$$ab \div (-a) = \frac{ab}{-a} = \frac{(-a) \times (-b)}{-a} = -b.$$
$$-ab \div (-a) = \frac{-ab}{-a} = \frac{(-a) \times b}{-a} = b.$$

Hence in division as well as multiplication
like signs produce $+$,
unlike signs produce $-$.

Division of Simple Expressions.

48. The method is shewn in the following examples:

Example 1. Since the product of 4 and x is $4x$, it follows that when $4x$ is divided by x the quotient is 4,
or otherwise, $\qquad 4x \div x = 4$.

Example 2. Divide $27a^5$ by $9a^3$.

The quotient $= \dfrac{27a^5}{9a^3} = \dfrac{27aaaaa}{9aaa}$ $\quad\big|\quad$ We remove from the divisor and dividend the factors common to both, just as in arithmetic.
$\phantom{\text{The quotient}}= 3aa = 3a^2.$

Therefore $\qquad 27a^5 \div 9a^3 = 3a^2$.

Example 3. Divide $35a^3b^2c^3$ by $7ab^2c^2$.

The quotient $= \dfrac{35aaa \cdot bb \cdot ccc}{7a \cdot bb \cdot cc} = 5aa \cdot c = 5a^2c$.

We see, in each case, that *the index of any letter in the quotient is the difference of the indices of that letter in the dividend and divisor.* This is called the **Index Law for Division.**

The rule may now be stated:

Rule. *The index of each letter in the quotient is obtained by subtracting the index of that letter in the divisor from that in the dividend.*

To the result so obtained prefix with its proper sign the quotient of the coefficient of the dividend by that of the divisor.

Example 4. Divide $45a^6b^2x^4$ by $-9a^3bx^2$.

The quotient $= (-5) \times a^{6-3}b^{2-1}x^{4-2}$
$\phantom{\text{The quotient}} = -5a^3bx^2$.

Example 5. $\quad -21a^2b^3 \div (-7a^2b^2) = 3b$.

Note. If we apply the rule to divide any power of a letter by the same power of the letter we are led to a curious conclusion.

Thus, by the rule $\quad a^3 \div a^3 = a^{3-3} = a^0$;

but also $\qquad\qquad a^3 \div a^3 = \dfrac{a^3}{a^3} = 1$.

$\qquad\qquad\therefore\quad a^0 = 1$.

This result will appear somewhat strange to the beginner, but its full significance will be explained in the chapter on the Theory of Indices.

Division of a Compound Expression by a Simple Expression.

49. Rule. *To divide a compound expression by a single factor, divide each term separately by that factor, and take the algebraic sum of the partial quotients so obtained.*

This follows at once from Art. 33.

Examples. (1) $(9x - 12y + 3z) \div -3 = -3x + 4y - z$.

(2) $(36a^3b^2 - 24a^2b^5 - 20a^4b^2) \div 4a^2b = 9ab - 6b^4 - 5a^2b$.

(3) $(2x^2 - 5xy + \frac{3}{2}x^2y^3) \div -\frac{1}{2}x = -4x + 10y - 3xy^3$.

EXAMPLES VI. a.

Divide

1. $3x^3$ by x^2.
2. $27x^4$ by $-9x^3$.
3. $-35x^6$ by $7x^3$.
4. abx^2 by $-ax$.
5. x^3y^3 by x^2y.
6. a^4x^3 by $-a^2x^3$.
7. $4a^2b^2c^3$ by ab^2c^2.
8. $12a^6b^6c^6$ by $-3a^4b^3c$.
9. $-a^5c^9$ by $-ac^3$.
10. $15x^5y^7z^4$ by $5x^2y^2z^2$.
11. $-16x^3y^2$ by $-4xy^2$.
12. $-48a^9$ by $-8a^3$.
13. $35a^{11}$ by $7a^7$.
14. $63a^7b^8c^3$ by $9a^5b^5c^3$.
15. $7a^2bc$ by $-7a^2bc$.
16. $28a^4b^3$ by $-4a^3b$.
17. $16b^2yx^2$ by $-2xy$.
18. $-50y^3x^3$ by $-5x^3y$.
19. $x^2 - 2xy$ by x.
20. $x^3 - 3x^2 + x$ by x.
21. $x^6 - 7x^5 + 4x^4$ by x^2.
22. $10x^7 - 8x^6 + 3x^4$ by x^3.
23. $15x^5 - 25x^4$ by $-5x^3$.
24. $27x^6 - 36x^5$ by $9x^5$.
25. $-24x^6 - 32x^4$ by $-8x^3$.
26. $34x^3y^2 - 51x^2y^3$ by $17xy$.
27. $a^2 - ab - ac$ by $-a$.
28. $a^3 - a^2b - a^2b^2$ by a^2.
29. $3x^3 - 9x^2y - 12xy^2$ by $-3x$.
30. $4x^4y^4 - 8x^3y^2 + 6xy^3$ by $-2xy$.
31. $-3a^2 + \frac{9}{2}ab - 6ac$ by $-\frac{3}{2}a$.
32. $\frac{1}{2}x^5y^2 - 3x^3y^4$ by $-\frac{3}{2}x^3y^2$.
33. $-\frac{5}{2}x^2 + \frac{5}{3}xy + \frac{10}{3}x$ by $-\frac{5}{6}x$.
34. $-2a^5x^3 + \frac{7}{2}a^4x^4$ by $\frac{7}{3}a^3x$.
35. $\frac{1}{4}a^2x - \frac{1}{16}abx - \frac{3}{8}acx$ by $\frac{3}{8}ax$.

Division of Compound Expressions.

50. To divide one compound expression by another.

Rule. 1. *Arrange divisor and dividend in ascending or descending powers of some common letter.*

2. *Divide the term on the left of the dividend by the term on the left of the divisor, and put the result in the quotient.*

3. *Multiply the* WHOLE *divisor by this quotient, and put the product under the dividend.*

4. *Subtract and bring down from the dividend as many terms as may be necessary.*

Repeat these operations till all the terms from the dividend are brought down.

Example 1. Divide $x^2 + 11x + 30$ by $x + 6$.

Arrange the work thus:
$$x+6 \;)\; x^2 + 11x + 30 \;($$

divide x^2, the first term of the dividend, by x, the first term of the divisor; the quotient is x. Multiply the *whole* divisor by x, and put the product $x^2 + 6x$ under the dividend. We then have

$$x+6\;)\;x^2+11x+30\;(\;x$$
$$\underline{x^2+\;6x}$$

by subtraction $\quad\quad 5x + 30$.

On repeating the process above explained we find that the next term in the quotient is $+5$.

The entire operation is more compactly written as follows:

$$x+6\;)\;x^2+11x+30\;(\;x+5$$
$$\underline{x^2+\;6x}$$
$$5x+30$$
$$\underline{5x+30}$$

The reason for the rule is this: the dividend may be divided into as many parts as may be convenient, and the complete quotient is found by taking the sum of all the partial quotients. Thus $x^2 + 11x + 30$ is divided by the above process into two parts, namely $x^2 + 6x$, and $5x + 30$, and each of these is divided by $x + 6$; thus we obtain the complete quotient $x + 5$.

Example 2. Divide $24x^2 - 65xy + 21y^2$ by $8x - 3y$.

$$8x-3y\;)\;24x^2-65xy+21y^2\;(\;3x-7y$$
$$\underline{24x^2-\;9xy}$$
$$-56xy+21y^2$$
$$\underline{-56xy+21y^2}$$

EXAMPLES VI. b.

Divide

1. $x^2 + 3x + 2$ by $x + 1$.
2. $x^2 - 7x + 12$ by $x - 3$.
3. $a^2 - 11a + 30$ by $a - 5$.
4. $a^2 - 49a + 600$ by $a - 25$.
5. $3x^2 + 10x + 3$ by $x + 3$.
6. $2x^2 + 11x + 5$ by $2x + 1$.
7. $5x^2 + 11x + 2$ by $x + 2$.
8. $2x^2 + 17x + 21$ by $2x + 3$.
9. $5x^2 + 16x + 3$ by $x + 3$.
10. $3x^2 + 34x + 11$ by $3x + 1$.
11. $4x^2 + 23x + 15$ by $4x + 3$.
12. $6x^2 - 7x - 3$ by $2x - 3$.
13. $3x^2 + x - 14$ by $x - 2$.
14. $3x^2 - x - 14$ by $x + 2$.
15. $6x^2 - 31x + 35$ by $2x - 7$.
16. $4x^2 + x - 14$ by $x + 2$.
17. $12a^2 - 7ax - 12x^2$ by $3a - 4x$.
18. $15a^2 + 17ax - 4x^2$ by $3a + 4x$.
19. $12a^2 - 11ac - 36c^2$ by $4a - 9c$.
20. $9a^2 + 6ac - 35c^2$ by $3a + 7c$.
21. $60x^2 - 4xy - 45y^2$ by $10x - 9y$.
22. $96x^2 - 15y^2 - 4xy$ by $12x - 5y$.
23. $7x^3 + 96x^2 - 28x$ by $7x - 2$.
24. $100x^3 - 3x - 13x^2$ by $3 + 25x$.
25. $27x^3 + 9x^2 - 3x - 10$ by $3x - 2$.
26. $16a^3 - 46a^2 + 39a - 9$ by $8a - 3$.
27. $15 + 3a - 7a^2 - 4a^3$ by $5 - 4a$.
28. $16 - 96x + 216x^2 - 216x^3 + 81x^4$ by $2 - 3x$.

*51. The process of Art. 50 is applicable to cases in which the divisor consists of more than two terms.

Example 1. Divide $6x^5 - x^4 + 4x^3 - 5x^2 - x - 15$ by $2x^2 - x + 3$.

$$
\begin{array}{r}
2x^2 - x + 3 \,)\, 6x^5 - x^4 + 4x^3 - 5x^2 - x - 15 \,(\, 3x^3 + x^2 - 2x - 5 \\
\underline{6x^5 - 3x^4 + 9x^3} \\
2x^4 - 5x^3 - 5x^2 \\
\underline{2x^4 - x^3 + 3x^2} \\
-4x^3 - 8x^2 - x \\
\underline{-4x^3 + 2x^2 - 6x} \\
-10x^2 + 5x - 15 \\
\underline{-10x^2 + 5x - 15}
\end{array}
$$

Example 2. Divide $2a^3 + 10 - 16a - 39a^2 + 15a^4$ by $2 - 4a - 5a^2$.

Arrange the expressions in *ascending* powers of a and use detached coefficients as in Art. 45.

$$
\begin{array}{r}
2 - 4 - 5 \,)\, 10 - 16 - 39 + 2 + 15 \,(\, 5 + 2 - 3 \\
\underline{10 - 20 - 25} \\
4 - 14 + 2 \\
\underline{4 - 8 - 10} \\
-6 + 12 + 15 \\
\underline{-6 + 12 + 15}
\end{array}
$$

Thus the quotient is $5 + 2a - 3a^2$.

*52. We add a few harder cases worked out in full.

Example 1.   Divide x^4+4a^4 by $x^2+2xa+2a^2$.

$$x^2+2xa+2a^2 \,)\, x^4+4a^4 \quad (\, x^2-2xa+2a^2$$
$$\underline{x^4+2x^3a+2x^2a^2}$$
$$-2x^3a-2x^2a^2$$
$$\underline{-2x^3a-4x^2a^2-4xa^3}$$
$$2x^2a^2+4xa^3+4a^4$$
$$\underline{2x^2a^2+4xa^3+4a^4}$$

Example 2.   Divide $a^3+b^3+c^3-3abc$ by $a+b+c$.

$$a+b+c \,)\, a^3-3abc+b^3+c^3\,(\,a^2-ab-ac+b^2-bc+c^2$$
$$\underline{a^3+a^2b+a^2c}$$
$$-a^2b-a^2c-3abc$$
$$\underline{-a^2b-ab^2-abc}$$
$$-a^2c+ab^2-2abc$$
$$\underline{-a^2c-abc-ac^2}$$
$$ab^2-abc+ac^2+b^3$$
$$\underline{ab^2+b^3+b^2c}$$
$$-abc+ac^2-b^2c$$
$$\underline{-abc-b^2c-bc^2}$$
$$ac^2+bc^2+c^3$$
$$\underline{ac^2+bc^2+c^3}$$

Note. In the above example the dividend and successive remainders are arranged in *descending* powers of a.

The result of this important division will be referred to later.

*53. When the coefficients are fractional the ordinary process may still be employed.

Example.   Divide $\frac{1}{4}x^3+\frac{1}{72}xy^2+\frac{1}{12}y^3$ by $\frac{1}{2}x+\frac{1}{3}y$.

$$\tfrac{1}{2}x+\tfrac{1}{3}y\,)\,\tfrac{1}{4}x^3+\tfrac{1}{72}xy^2+\tfrac{1}{12}y^3\,(\,\tfrac{1}{2}x^2-\tfrac{1}{3}xy+\tfrac{1}{4}y^2$$
$$\underline{\tfrac{1}{4}x^3+\tfrac{1}{6}\,x^2y}$$
$$-\tfrac{1}{6}\,x^2y+\tfrac{1}{72}xy^2$$
$$\underline{-\tfrac{1}{6}\,x^2y-\tfrac{1}{9}\,xy^2}$$
$$\tfrac{1}{8}\,xy^2+\tfrac{1}{12}y^3$$
$$\underline{\tfrac{1}{8}\,xy^2+\tfrac{1}{12}y^3}$$

*54.   In the examples given hitherto the divisor has been exactly contained in the dividend. When the division is not exact the work should be carried on until the remainder is of lower dimensions [Art. 24] than the divisor.

*EXAMPLES VI. c.

[Examples 1–20 will furnish practice in the use of Detached Coefficients as explained in Art. 51.]

Divide

1. $x^3 - x^2 - 9x - 12$ by $x^2 + 3x + 3$.
2. $2y^3 - 3y^2 - 6y - 1$ by $2y^2 - 5y - 1$.
3. $6m^3 - m^2 - 14m + 3$ by $3m^2 + 4m - 1$.
4. $6a^5 - 13a^4 + 4a^3 + 3a^2$ by $3a^3 - 2a^2 - a$.
5. $x^4 + x^3 + 7x^2 - 6x + 8$ by $x^2 + 2x + 8$.
6. $a^4 - a^3 - 8a^2 + 12a - 9$ by $a^2 + 2a - 3$.
7. $a^4 + 6a^3 + 13a^2 + 12a + 4$ by $a^2 + 3a + 2$.
8. $2x^4 - x^3 + 4x^2 + 7x + 1$ by $x^2 - x + 3$.
9. $x^5 - 5x^4 + 9x^3 - 6x^2 - x + 2$ by $x^2 - 3x + 2$.
10. $x^5 - 4x^4 + 3x^3 + 3x^2 - 3x + 2$ by $x^2 - x - 2$.
11. $30x^4 + 11x^3 - 82x^2 - 5x + 3$ by $2x - 4 + 3x^2$.
12. $30y + 9 - 71y^3 + 28y^4 - 35y^2$ by $4y^2 - 13y + 6$.
13. $6k^5 - 15k^4 + 4k^3 + 7k^2 - 7k + 2$ by $3k^3 - k + 1$.
14. $15 + 2m^4 - 31m + 9m^2 + 4m^3 + m^5$ by $3 - 2m - m^2$.
15. $2x^3 - 8x + x^4 + 12 - 7x^2$ by $x^2 + 2 - 3x$.
16. $x^5 - 2x^4 - 4x^3 + 19x^2$ by $x^3 - 7x + 5$.
17. $192 - x^4 + 128x + 4x^2 - 8x^3$ by $16 - x^2$.
18. $14x^4 + 45x^3y + 78x^2y^2 + 45xy^3 + 14y^4$ by $2x^2 + 5xy + 7y^2$.
19. $x^5 - x^4y + x^3y^2 - x^3 + x^2 - y^3$ by $x^3 - x - y$.
20. $x^5 + x^4y - x^3y^2 + x^3 - 2xy^2 + y^3$ by $x^2 + xy - y^2$.
21. $a^9 - b^9$ by $a^3 - b^3$.   22. $x^9 - y^9$ by $x^2 + xy + y^2$.
23. $x^7 - 2y^{14} - 7x^5y^4 - 7xy^{12} + 14x^3y^8$ by $x - 2y^2$.
24. $a^3 + 3a^2b + b^3 - 1 + 3ab^2$ by $a + b - 1$.
25. $x^8 - y^8$ by $x^3 + x^2y + xy^2 + y^3$.   26. $a^{12} - b^{12}$ by $a^2 - b^2$.
27. $a^{12} + 2a^6b^6 + b^{12}$ by $a^4 + 2a^2b^2 + b^4$.
28. $1 - a^3 - 8x^3 - 6ax$ by $1 - a - 2x$.

Find the quotient of

29. $\frac{1}{8}a^3 - \frac{9}{4}a^2x + \frac{27}{2}ax^2 - 27x^3$ by $\frac{1}{2}a - 3x$.
30. $\frac{1}{27}a^3 - \frac{1}{12}a^2 + \frac{1}{16}a - \frac{1}{64}$ by $\frac{1}{3}a - \frac{1}{4}$.
31. $\frac{3}{4}a^2c^3 + \frac{6}{125}a^5$ by $\frac{1}{5}a^2 + \frac{1}{2}ac$.
32. $\frac{9}{16}a^4 - \frac{3}{4}a^3 - \frac{7}{4}a^2 + \frac{4}{3}a + \frac{16}{9}$ by $\frac{3}{2}a^2 - \frac{8}{3} - a$.
33. $36x^2 + \frac{1}{9}y^2 + \frac{1}{4} - 4xy - 6x + \frac{1}{3}y$ by $6x - \frac{1}{3}y - \frac{1}{2}$.
34. $\frac{8}{27}a^5 - \frac{243}{512}ax^4$ by $\frac{2}{3}a - \frac{3}{4}x$.

* 55. The following examples in division may be easily verified; they are of great importance, and should be carefully noticed.

$$\text{I.} \begin{cases} \dfrac{x^2-y^2}{x-y}=x+y, \\ \dfrac{x^3-y^3}{x-y}=x^2+xy+y^2, \\ \dfrac{x^4-y^4}{x-y}=x^3+x^2y+xy^2+y^3, \end{cases}$$

and so on; the divisor being $x-y$, the terms in the quotient *all positive,* and the index in the dividend *either odd or even.*

$$\text{II.} \begin{cases} \dfrac{x^3+y^3}{x+y}=x^2-xy+y^2, \\ \dfrac{x^5+y^5}{x+y}=x^4-x^3y+x^2y^2-xy^3+y^4, \\ \dfrac{x^7+y^7}{x+y}=x^6-x^5y+x^4y^2-x^3y^3+x^2y^4-xy^5+y^6, \end{cases}$$

and so on; the divisor being $x+y$, the terms in the quotient *alternately positive and negative,* and the index in the dividend *always odd.*

$$\text{III.} \begin{cases} \dfrac{x^2-y^2}{x+y}=x-y, \\ \dfrac{x^4-y^4}{x+y}=x^3-x^2y+xy^2-y^3, \\ \dfrac{x^6-y^6}{x+y}=x^5-x^4y+x^3y^2-x^2y^3+xy^4-y^5, \end{cases}$$

and so on; the divisor being $x+y$, the terms in the quotient *alternately positive and negative,* and the index in the dividend *always even.*

IV. The expressions x^2+y^2, x^4+y^4, x^6+y^6, ... (where the index is *even*, and the terms *both positive*) are *never* divisible by $x+y$ or $x-y$.

All these different cases may be more concisely stated as follows:

(1) x^n-y^n is divisible by $x-y$ if n be *any* whole number.
(2) x^n+y^n is divisible by $x+y$ if n be any *odd* whole number.
(3) x^n-y^n is divisible by $x+y$ if n be any *even* whole number.
(4) x^n+y^n is never divisible by $x+y$ or $x-y$, when n is an *even* whole number.

CHAPTER VII.

REMOVAL AND INSERTION OF BRACKETS.

56. QUANTITIES are enclosed within brackets to indicate that they must all be operated upon in the same way. Thus in the expression $2a - 3b - (4a - 2b)$ the brackets indicate that the expression $4a - 2b$ *treated as a whole* has to be subtracted from $2a - 3b$. When we wish to enclose within brackets part of an expression already enclosed within brackets it is usual to employ brackets of different forms. The brackets in common use are ( ), { }, [ ]. Sometimes a line called a **vinculum** is drawn over the symbols to be connected; thus $a - \overline{b+c}$ is used with the same meaning as $a - (b+c)$, and hence $a - \overline{b+c} = a - b - c$.

Removal of Brackets.

57. To remove brackets it is usually best to begin with the inside pair, and in dealing with each pair in succession we apply the rules already given in Arts. 21, 22.

Example 1. Simplify, by removing brackets, the expression
$$a - 2b - [4a - 6b - \{3a - c + (5a - 2b - \overline{3a - c + 2b})\}].$$

Removing the brackets one by one, we have
$$a - 2b - [4a - 6b - \{3a - c + (5a - 2b - 3a + c - 2b)\}]$$
$$= a - 2b - [4a - 6b - \{3a - c + 5a - 2b - 3a + c - 2b\}]$$
$$= a - 2b - [4a - 6b - 3a + c - 5a + 2b + 3a - c + 2b]$$
$$= a - 2b - 4a + 6b + 3a - c + 5a - 2b - 3a + c - 2b$$
$$= 2a, \text{ by collecting like terms.}$$

Example 2. Simplify the expression
$$-[-2x - \{3y - (2x - 3y) + (3x - 2y)\} + 2x].$$

The expression $= -[-2x - \{3y - 2x + 3y + 3x - 2y\} + 2x]$
$$= -[-2x - 3y + 2x - 3y - 3x + 2y + 2x]$$
$$= 2x + 3y - 2x + 3y + 3x - 2y - 2x$$
$$= x + 4y.$$

EXAMPLES VII. a.

Simplify by removing brackets

1. $a-(b-c)+a+(b-c)+b-(c+a)$.
2. $a-[b+\{a-(b+a)\}]$.
3. $a-[2a-\{3b-(4c-2a)\}]$.
4. $\{a-(b-c)\}+\{b-(c-a)\}-\{c-(a-b)\}$.
5. $2a-(5b+[3c-a])-(5a-[b+c])$.
6. $-\{-[-(a-\overline{b-c})]\}$.
7. $-[a-\{b-(c-a)\}]-[b-\{c-(a-b)\}]$.
8. $-(-(-(-x)))-(-(-y))$.
9. $-[-\{-(b+c-a)\}]+[-\{-(c+a-b)\}]$.
10. $-5x-[3y-\{2x-(2y-x)\}]$.
11. $-(-(-a))-(-(-(-x)))$.
12. $3a-[a+b-\{a+b+c-(a+b+c+d)\}]$.
13. $-2a-[3x+\{3c-(4y+3x+2a)\}]$.
14. $3x-[5y-\{6z-(4x-7y)\}]$.
15. $-[5x-(11y-3x)]-[5y-(3x-6y)]$.
16. $-[15x-\{14y-(15z+12y)-(10x-15z)\}]$.
17. $8x-\{16y-[3x-(12y-x)-8y]+x\}$.
18. $-[x-\{z+(x-z)-(z-x)-z\}-x]$.
19. $-[a+\{a-(a-x)-(a+x)-a\}-a]$.
20. $-[a-\{a+(x-a)-(x-a)-a\}-2a]$.

58. A coefficient placed before any bracket indicates that every term of the expression within the bracket is to be multiplied by that coefficient.

Note. The line between the numerator and denominator of a fraction is a kind of vinculum. Thus $\dfrac{x-5}{3}$ is equivalent to $\tfrac{1}{3}(x+5)$.

Again, an expression of the form $\sqrt{(x+y)}$ is often written $\sqrt{x+y}$, the line above being regarded as a vinculum indicating the square root of the compound expression $x+y$ *taken as a whole*.

Thus $\sqrt{25+144}=\sqrt{169}=13$,

whereas $\sqrt{25}+\sqrt{144}=5+12=17$.

59. Sometimes it is advisable to simplify in the course of the work.

Example. Find the value of
$$84-7[-11x-4\{-17x+3(8-\overline{9-5x})\}].$$

$$\begin{aligned}
\text{The expression} &= 84-7[-11x-4\{-17x+3(8-9+5x)\}] \\
&= 84-7[-11x-4\{-17x+3(5x-1)\}] \\
&= 84-7[-11x-4\{-17x+15x-3\}] \\
&= 84-7[-11x-4\{-2x-3\}] \\
&= 84-7[-11x+8x+12] \\
&= 84-7[-3x+12] \\
&= 84+21x-84 \\
&= 21x.
\end{aligned}$$

When the beginner has had a little practice the number of steps may be considerably diminished.

EXAMPLES VII. b.

Simplify by removing brackets

1. $a-[2b+\{3c-3a-(a+b)\}+2a-(b+3c)]$.
2. $a+b-(c+a-[b+c-(a+b-\{c+a-(b+c-a)\})])$.
3. $a-(b-c)-[a-b-c-2\{b+c-3(c-a)-d\}]$.
4. $2x-(3y-4z)-\{2x-(3y+4z)\}-\{3y-(4z+2x)\}$.
5. $b+c-(a+b-[c+a-(b+c-\{a+b-(c+a-b)\})])$.
6. $3b-\{5a-[6a+2(10a-b)]\}$.
7. $a-(b-c)-[a-b-c-2\{b+c\}]$.
8. $3a^2-[6a^2-\{8b^2-(9c^2-2a^2)\}]$.
9. $b-(c-a)-[b-a-c-2\{c+a-3(a-b)-d\}]$.
10. $-20(a-d)+3(b-c)-2[b+c+d-3\{c+d-4(d-a)\}]$.
11. $-4(a+d)+24(b-c)-2[c+d+a-3\{d+a-4(b+c)\}]$.
12. $-10(a+b)-[c+a+b-3\{a+2b-(c+a-b)\}]+4c$.
13. $a-2(b-c)-[-\{-(4a-b-c-2\{a+b+c\})\}]$.
14. $8(b-c)-[-\{a-b-3(c-b+a)\}]$.
15. $2(3b-5a)-7[a-6\{2-5(a-b)\}]$.
16. $6\{a-2[b-3(c+d)]\}-4\{a-3[b-4(c-d)]\}$.
17. $5\{a-2[a-2(a+x)]\}-4\{a-2[a-2(a+x)]\}$.
18. $-10\{a-6[a-(b-c)]\}+60\{b-(c+a)\}$.

VII.] REMOVAL AND INSERTION OF BRACKETS. 49

19. $-3\{-2[-4(-a)]\}+5\{-2[-2(-a)]\}.$

20. $-2\{-[-(x-y)]\}+\{-2[-(x-y)]\}.$

21. $\dfrac{1}{4}\{a-5(b-a)\}-\dfrac{3}{2}\left\{\dfrac{1}{3}\left(b-\dfrac{a}{3}\right)-\dfrac{2}{9}\left[a-\dfrac{3}{4}\left(b-\dfrac{4a}{5}\right)\right]\right\}.$

22. $35\left[\dfrac{3x-4y}{5}-\dfrac{1}{10}\left\{3x-\dfrac{5}{7}(7x-4y)\right\}\right]+8(y-2x).$

23. $\dfrac{3}{8}\left\{\dfrac{4}{3}(a-b)-8(b-c)\right\}-\left\{\dfrac{b-c}{2}-\dfrac{c-a}{3}\right\}-\dfrac{1}{2}\left\{c-a-\dfrac{2}{3}(a-b)\right\}.$

24. $\dfrac{1}{2}x-\dfrac{1}{2}\left(\dfrac{2}{3}y-\dfrac{1}{2}z\right)-\left[x-\left\{\dfrac{1}{2}x-\left(\dfrac{1}{3}y-\dfrac{1}{4}z\right)\right\}-\left(\dfrac{2}{3}y-\dfrac{1}{2}z\right)\right].$

Insertion of Brackets.

60. The converse operation of inserting brackets is important. The rules for doing this have been enunciated in Arts. 21, 22; for convenience we repeat them.

(1) *Any part of an expression may be enclosed within brackets and the sign + prefixed, the sign of every term within the brackets remaining unaltered.*

(2) *Any part of an expression may be enclosed within brackets and the sign − prefixed, provided the sign of every term within the brackets be changed.*

Examples.
$$a-b+c-d-e = a-b+(c-d-e).$$
$$a-b+c-d-e = a-(b-c)-(d+e).$$
$$x^2-ax+bx-ab = (x^2-ax)+(bx-ab).$$
$$xy-ax-by+ab = (xy-by)-(ax-ab).$$

61. The terms of an expression can be bracketed in various ways.

Example. The expression $ax-bx+cx-ay+by-cy$

may be written $\quad (ax-bx)+(cx-ay)+(by-cy),$

or $\quad\quad\quad\quad\quad\; (ax-bx+cx)-(ay-by+cy),$

or $\quad\quad\quad\quad\quad\; (ax-ay)-(bx-by)+(cx-cy).$

62. Whenever a factor is common to every term within a bracket, it may be removed and placed outside as a multiplier of the expression within the bracket.

Example 1. In the expression
$$ax^3 - cx + 7 - dx^2 + bx - c - dx^3 + bx^2 - 2x$$
bracket together the powers of x so as to have the sign $+$ before each bracket.

The expression $= (ax^3 - dx^3) + (bx^2 - dx^2) + (bx - cx - 2x) + (7 - c)$
$= x^3(a - d) + x^2(b - d) + x(b - c - 2) + (7 - c)$
$= (a - d)x^3 + (b - d)x^2 + (b - c - 2)x + 7 - c.$

In this last result the compound expressions $a - d$, $b - d$, $b - c - 2$ are regarded as the coefficients of x^3, x^2, and x respectively.

Example 2. In the expression $-a^2x - 7a + a^2y + 3 - 2x - ab$ bracket together the powers of a so as to have the sign $-$ before each bracket.

The expression $= -(a^2x - a^2y) - (7a + ab) - (2x - 3)$
$= -a^2(x - y) - a(7 + b) - (2x - 3)$
$= -(x - y)a^2 - (7 + b)a - (2x - 3).$

EXAMPLES VII. c.

In the following expressions bracket the powers of x so that the signs before all the brackets shall be positive:

1. $ax^4 + bx^2 + 5 + 2bx - 5x^2 + 2x^4 - 3x.$
2. $3bx^2 - 7 - 2x + ab + 5ax^3 + cx - 4x^2 - bx^3.$
3. $2 - 7x^3 + 5ax^2 - 2cx + 9ax^3 + 7x - 3x^2.$
4. $2cx^5 - 3abx + 4dx - 3bx^4 - a^2x^5 + x^4.$

In the following expressions bracket the powers of x so that the signs before all the brackets shall be negative:

5. $ax^2 + 5x^3 - a^2x^4 - 2bx^3 - 3x^2 - bx^4.$
6. $7x^3 - 3c^2x - abx^5 + 5ax + 7x^5 - abcx^3.$
7. $ax^2 + a^2x^3 - bx^2 - 5x^2 - cx^3.$
8. $3b^2x^4 - bx - ax^4 - cx^4 - 5c^2x - 7x^4.$

Simplify the following expressions, and in each result re-group the terms according to powers of x:

9. $ax^3 - 2cx - [bx^2 - \{cx - dx - (bx^3 + 3cx^2)\} - (cx^2 - bx)].$
10. $5ax^3 - 7(bx - cx^2) - \{6bx^2 - (3ax^2 + 2ax) - 4cx^3\}.$
11. $ax^2 - 3\{-ax^3 + 3bx - 4[\tfrac{1}{6}cx^3 - \tfrac{2}{3}(ax - bx^2)]\}.$
12. $x^5 - 4bx^4 - \tfrac{1}{6}\left[12ax - 4\left\{3bx^4 - 9\left(\dfrac{cx}{2} - bx^5\right) - \dfrac{3}{2}ax^4\right\}\right].$
13. $x\{x - b - x(a - bx)\} + ax - x\{x - x(ax - b)\}.$

63. In certain cases of addition, multiplication, etc., of expressions which involve literal coefficients, the results may be more conveniently written by grouping the terms according to powers of some common letter.

Example 1. Add together $ax^3 - 2bx^2 + 3$, $bx - cx^3 - x^2$ and
$$x^3 - ax^2 + cx.$$

The sum $= ax^3 - 2bx^2 + 3 + bx - cx^3 - x^2 + x^3 - ax^2 + cx$
$= ax^3 - cx^3 + x^3 - ax^2 - 2bx^2 - x^2 + bx + cx + 3$
$= (a - c + 1)x^3 - (a + 2b + 1)x^2 + (b + c)x + 3.$

Example 2. Multiply $ax^2 - 2bx + 3c$ by $px - q$.

The product $= (ax^2 - 2bx + 3c)(px - q)$
$= apx^3 - 2bpx^2 + 3cpx - aqx^2 + 2bqx - 3cq$
$= apx^3 - (2bp + aq)x^2 + (3cp + 2bq)x - 3cq.$

EXAMPLES VII. d.

Add together the following expressions, and in each case arrange the result according to powers of x:

1. $ax^3 - 2cx$, $bx^2 - cx^3$, $cx^2 - x$.
2. $x^2 - x - 1$, $ax^2 - bx^3$, $bx + x^3$.
3. $a^2x^3 - 5x$, $2ax^2 - 5ax^3$, $2x^3 - bx^2 - ax$.
4. $ax^2 + bx - c$, $qx - r - px^2$, $x^2 + 2x + 3$.
5. $px^3 - qx$, $qx^2 - px$, $q - x^3$, $px^2 + qx^3$.

Multiply together the following expressions, and in each case arrange the result according to powers of x:

6. $ax^2 + bx + 1$ and $cx + 2$.
7. $cx^2 - 2x + 3$ and $ax - b$.
8. $ax^2 - bx - c$ and $px + q$.
9. $2x^2 - 3x - 1$ and $bx + c$.
10. $ax^2 - 2bx + 3c$ and $x - 1$.
11. $px^2 - 2x - q$ and $ax - 3$.
12. $x^3 + ax^2 - bx - c$ and $x^3 - ax^2 - bx + c$.
13. $ax^3 - x^2 + 3x - b$ and $ax^3 + x^2 + 3x + b$.
14. $x^4 - ax^3 - bx^2 + cx + d$ and $x^4 + ax^3 - bx^2 - cx + d$.

CHAPTER VIII.

Simple Equations.

64. An **equation** is a statement that two algebraical expressions are equal.

Thus (i.) $x+3+x+4=2x+7$, (ii.) $4x+2=14$ are equations.

The parts of an equation separated by the sign of equality are called **members** or **sides** of the equation, and are distinguished as the *right side* and the *left side*.

65. If the two expressions are *always* equal, for *any* values we give to the symbols, the equation is called an **identical equation**, or briefly an **identity**. Thus equation (i.) above is an *identity*, as is easily seen by collecting the terms on the left side.

If two expressions are only equal for a particular value or values of the symbols, the equation is called an **equation of condition**, or more usually an *equation*, simply.

Thus the statement $4x+2=14$ will be found to be true only when $x=3$.

This, then, is an equation in the ordinary sense of the term, and the value 3 is said to **satisfy** the equation. The object of the present chapter is to shew how to find the values which satisfy equations of the simpler kinds.

66. The letter whose value it is required to find in any equation is called the **unknown quantity.** The process of finding its value is called **solving the equation.** The value so found is called the **root** or **solution** of the equation.

67. An equation which, when reduced to a simple form, involves no power of the unknown quantity higher than the first is called a **simple equation.** It is usual to denote the unknown quantity by x.

68. The process of solving a simple equation depends only on the following **axioms**:

1. If to equals we add equals the sums are equal.
2. If from equals we take equals the remainders are equal.
3. If equals are multiplied by equals the products are equal.
4. If equals are divided by equals the quotients are equal.

CHAP. VIII.] SIMPLE EQUATIONS. 53

Example 1. To solve the equation $7x = 14$.
Dividing both sides by 7, (Axiom 4) we get
$$x = 2.$$

Example 2. Solve the equation $\frac{x}{2} = -6$.
Multiplying both sides by 2, (Axiom 3) we get
$$x = -12.$$

Example 3. Solve the equation
$$7x - 2x - x = 10 - 23 - 15.$$
By collecting terms on each side, we get
$$4x = -28.$$
Dividing by 4, (Axiom 4) we get
$$x = -7.$$

EXAMPLES. (*Oral.*)

Find the values which satisfy the following equations:

1. $3x = 18$.   2. $4x = 12$.   3. $6x = 12$.   4. $7x = -7$.
5. $3x = 21$.   6. $11x = 55$.   7. $13x = 39$.   8. $14x = -42$.
9. $7x = -35$.   10. $-5x = 30$.   11. $-2x = -12$.   12. $-3x = 21$.
13. $3x = 0$.   14. $-4x = 0$.   15. $2x = 11$.   16. $9x = 15$.
17. $51x = 39$.   18. $3x = -7$.   19. $28x = 35$.   20. $34x = -51$.
21. $\frac{x}{3} = 7$.   22. $\frac{x}{7} = -3$.   23. $-\frac{x}{5} = 4$.   24. $\frac{x}{6} = 0$.
25. $8x + 5x - 3x = 17 - 9 + 33 - 11$.
26. $5x - 7x + 8x = 12 - 5 + 7 + 10$.
27. $-3x - 12x + 5x = 29 - 2 + 6 - 13$.
28. $4x - 15x - 9x + 27x = -28 + 8 - 60 + 17$.

69. In the preceding examples the terms have been so arranged that those involving the unknown quantity have been on one side of the equation and the numerical quantities on the other. We can always arrive at this arrangement by the aid of the axioms.

Example. Solve the equation $3x - 8 = x + 12$.
Subtracting x from both sides, we get
$$3x - x - 8 = 12. \qquad \text{[Axiom 2.]}$$
Adding 8 to both sides, we have
$$3x - x = 12 + 8; \qquad \text{[Axiom 1.]}$$
$$\therefore 2x = 20;$$
dividing by 2,
$$x = 10. \qquad \text{[Axiom 4.]}$$

70. Beginners should **verify**, that is, prove the correctness of their solutions by substituting, in both sides, the value obtained for the unknown quantity.

In the last equation $3x - 8 = x + 12$,
if $x = 10$,
the left side $= 3 \times 10 - 8 = 22$,
and the right side $= 10 + 12 = 22$.

Since these two results are equal the solution is correct.

71. In the following examples some preliminary reduction is necessary.

Example 1. Solve $5(x-3) - 7(6-x) = 24 - 3(8-x) - 3$.

Removing brackets, $5x - 15 - 42 + 7x = 24 - 24 + 3x - 3$;
collecting terms, $12x - 57 = 3x - 3$.

Subtracting $3x$ from each side, we get
$$9x - 57 = -3. \qquad \text{[Axiom 2.]}$$
Adding 57 to each side, we have
$$9x = 54. \qquad \text{[Axiom 1.]}$$
Dividing by 9, $x = 6.$ [Axiom 4.]

[*Verification.* When $x = 6$,
the left side $= 5(6-3) - 7(6-6)$
$= 5 \times 3 - 0 = 15.$
The right side $= 24 - 3(8-6) - 3$
$= 24 - 3 \times 2 - 3$
$= 24 - 9 = 15.$

Thus the solution is correct.]

Example 2. Solve $\dfrac{4x}{5} - \dfrac{3}{10} = \dfrac{x}{5} + \dfrac{x}{4}$.

Here it is convenient to begin by clearing the equation of fractional coefficients. This can be done by multiplying every term on each side of the equation by the least common multiple of the denominators. [Axiom 3.]

Hence, multiplying throughout by 20,
$$16x - 6 = 4x + 5x.$$
Subtracting $9x$ from each side,
$$7x - 6 = 0.$$
Adding 6 to each side, $7x = 6.$
Dividing by 7, $x = \dfrac{6}{7}.$

[*Verification.* When $x = \dfrac{6}{7}$,

the left side $= \dfrac{4}{5} \times \dfrac{6}{7} - \dfrac{3}{10} = \dfrac{48-21}{70} = \dfrac{27}{70}.$

The right side $= \dfrac{1}{5} \times \dfrac{6}{7} + \dfrac{1}{4} \times \dfrac{6}{7} = \dfrac{24+30}{140}$

$= \dfrac{54}{140} = \dfrac{27}{70}.$

Thus the solution is correct.]

72. The preceding examples have been worked out very fully in every detail for the purpose of impressing on beginners the importance of shewing clearly the meaning of every step of their work in solving simple equations. Each step should occupy a separate line, and each successive process should be referred to one of the fundamental axioms; the object in each case being to gradually reduce the equation until it consists of a single term containing x on one side, and a single known term on the other. The required root is then found by dividing each side by the coefficient of x.

Orderly arrangement should be studied throughout, and in particular, the signs of equality in the several lines should be written neatly in column.

In order to furnish the requisite practice in *method* and *arrangement*, we shall now give an exercise containing easy equations which are free from difficulty in the way of reduction, and which involve little actual work.

EXAMPLES VIII. (1).

Find the value of x which satisfies each of the following equations, and in each case verify the solution.

1. $7x - 4 = 17.$
2. $3x - 5 = 10.$
3. $2x + 15 = 23.$
4. $5x - 9 = 21.$
5. $7x = 18 - 2x.$
6. $3x = 25 - 2x.$
7. $4x - 3 = 2x + 1.$
8. $5x + 2 = 6x - 1.$
9. $3x + 2 = 4x - 3.$
10. $4x - 3 = 3x + 4.$
11. $8x - 9 = 33 - 4x.$
12. $5x + 3 = 15 - x.$
13. $2x + 15 = 27 - 4x.$
14. $7x + 11 = 3x + 27.$
15. $15 - 5x = 24 - 8x.$
16. $9x + 21 - 4x = 46.$
17. $5x + 7 + 4x + 11 + 3x = 24.$
18. $0 = 9 - 6x - 19 + 10x.$
19. $7 - 3x = 5 + 4x + 11 - 16x.$
20. $-3x - 5 = -7x + 1.$
21. $6x + 7 - 19 = 7x + 13 - 3x - 21.$
22. $3x + 4 + 10x - 17 = 14 - 23x + 16 - 7x.$

Solve and verify the following equations:

23. $\dfrac{x}{3} = \dfrac{5}{6}$.
24. $\dfrac{x}{5} = \dfrac{4}{3}$.
25. $\dfrac{2x}{3} = \dfrac{5}{12}$.
26. $\dfrac{4x}{5} = \dfrac{7}{15}$.
27. $\dfrac{7x}{6} = \dfrac{4}{9}$.
28. $\dfrac{3x}{8} = \dfrac{5}{9}$.
29. $\dfrac{1}{2}x - \dfrac{1}{4}x = x - 9$.
30. $\dfrac{x}{3} - \dfrac{1}{2} = \dfrac{x}{5} + 1\dfrac{1}{2}$.
31. $\dfrac{x}{3} - 2\dfrac{1}{2} = \dfrac{4x}{9} - \dfrac{2x}{3}$.
32. $\dfrac{1}{8}x + \dfrac{1}{6}x - x = \dfrac{5}{6} - \dfrac{1}{2}x$.

73. After enough practice to enforce the reasons for the several steps, the solutions may be presented in a shorter form.

When any term is brought over from one side of an equation to the other it is said to be **transposed**.

We shall now shew that any term may be transposed from one side of an equation to the other by simply writing it down on the opposite side *with its sign changed*.

Consider the equation $3x - 8 = x + 12$.

Subtracting x from each side, we get
$$3x - x - 8 = 12.$$
Adding 8 to each side, we have
$$3x - x = 12 + 8.$$
Thus we see that $+x$ has been removed from one side, and appears as $-x$ on the other; and -8 has been removed from one side and appears as $+8$ on the other.

Similar steps may be employed in all cases.

It appears from this that *we may change the sign of every term in an equation*; for this is equivalent to transposing all the terms, and then making the two sides change places.

Example. Take the equation $-3x - 12 = x - 24$.

Transposing, $\qquad -x + 24 = 3x + 12,$

or $\qquad\qquad\qquad 3x + 12 = -x + 24,$

which is the original equation with the sign of every term changed.

74. We can now give a general rule for solving any simple equation with one unknown quantity.

Rule. *First, if necessary, clear of fractions; then transpose all the terms containing the unknown quantity to one side of the equation, and the known quantities to the other. Collect the terms on each side; divide both sides by the coefficient of the unknown quantity and the value required is obtained.*

SIMPLE EQUATIONS.

Example 1. Solve $5x-(4x-7)(3x-5)=6-3(4x-9)(x-1)$.

Here the products $(4x-7)(3x-5)$ and $(4x-9)(x-1)$ must be multiplied out, or written down by inspection as in Art. 44, before any further reduction can be made.

Forming the products, we have
$$5x-(12x^2-41x+35)=6-3(4x^2-13x+9);$$
and by removing brackets,
$$5x-12x^2+41x-35=6-12x^2+39x-27.$$

The term $-12x^2$ may be removed from each side without altering the equality; thus
$$5x+41x-35=6+39x-27.$$
Transposing, $\quad 5x+41x-39x=6-27+35;$
collecting terms, $\quad\quad\quad 7x=14;$
$$\therefore\ x=2.$$

Note. Since the minus sign before a bracket affects every term within it, in the first line of work we do not remove the brackets until we have formed the products.

Example 2. Solve $7x-5[x-\{7-6(x-3)\}]=3x+1$.

Removing brackets, we have
$$7x-5[x-\{7-6x+18\}]=3x+1,$$
$$7x-5[x-25+6x]=3x+1,$$
$$7x-5x+125-30x=3x+1;$$
transposing, $\quad 7x-5x-30x-3x=1-125;$
collecting terms, $\quad\quad -31x=-124;$
$$\therefore\ x=4.$$

EXAMPLES VIII. a.

[*It is recommended that Nos. 1–16 of the following examples should be solved in full by reference to the axioms. In the rest of the exercise the solutions may be shortened by transposition of terms.*]

Solve the following equations and verify the solutions in Examples 1 to 20.

1. $3x+15=x+25.$
2. $2x-3=3x-7.$
3. $3x+4=5(x-2).$
4. $2x+3=16-(2x-3).$
5. $8(x-1)+17(x-3)=4(4x-9)+4.$
6. $15(x-1)+4(x+3)=2(7+x).$
7. $5x-6(x-5)=2(x+5)+5(x-4).$
8. $8(x-3)-(6-2x)=2(x+2)-5(5-x).$

Solve the following equations:

9. $7(25-x)-2x=2(3x-25)$.
10. $3(169-x)-(78+x)=29x$.
11. $5x-17+3x-5=6x-7-8x+115$.
12. $7x-39-10x+15=100-33x+26$.
13. $118-65x-123=15x+35-120x$.
14. $157-21(x+3)=163-15(2x-5)$.
15. $179-18(x-10)=158-3(x-17)$.
16. $97-5(x+20)=111-8(x+3)$.
17. $x-[3+\{x-(3+x)\}]=5$.
18. $5x-(3x-7)-\{4-2x-(6x-3)\}=10$.
19. $14x-(5x-9)-\{4-3x-(2x-3)\}=30$.
20. $25x-19-[3-\{4x-5\}]=3x-(6x-5)$.
21. $(x+1)(2x+1)=(x+3)(2x+3)-14$.
22. $(x+1)^2-(x^2-1)=x(2x+1)-2(x+2)(x+1)+20$.
23. $2(x+1)(x+3)+8=(2x+1)(x+5)$.
24. $6(x^2-3x+2)-2(x^2-1)=4(x+1)(x+2)-24$.
25. $2(x-4)-(x^2+x-20)=4x^2-(5x+3)(x-4)-64$.
26. $(x+15)(x-3)-(x^2-6x+9)=30-15(x-1)$.
27. $2x-5\{3x-7(4x-9)\}=66$.
28. $20(2-x)+3(x-7)-2[x+9-3\{9-4(2-x)\}]=22$.
29. $x+2-[x-8-2\{8-3(5-x)-x\}]=0$.
30. $3(5-6x)-5[x-5\{1-3(x-5)\}]=23$.
31. $(x+1)(2x+3)=2(x+1)^2+8$.
32. $3(x-1)^2-3(x^2-1)=x-15$.
33. $(3x+1)(2x-7)=6(x-3)^2+7$.
34. $x^2-8x+25=x(x-4)-25(x-5)-16$.
35. $x(x+1)+(x+1)(x+2)=(x+2)(x+3)+x(x+4)-9$.
36. $2(x+2)(x-4)=x(2x+1)-21$.
37. $(x+1)^2+2(x+3)^2=3x(x+2)+35$.
38. $4(x+5)^2-(2x+1)^2=3(x-5)+180$.
39. $84+(x+4)(x-3)(x+5)=(x+1)(x+2)(x+3)$.
40. $(x+1)(x+2)(x+6)=x^3+9x^2+4(7x-1)$.

75. The following examples illustrate the most useful methods of solving equations with fractional coefficients.

SIMPLE EQUATIONS.

Example 1. Solve $4 - \dfrac{x-9}{8} = \dfrac{x}{22} - \dfrac{1}{2}$.

Multiply by 88, the least common multiple of the denominators;
thus $\qquad\qquad 352 - 11(x-9) = 4x - 44$;
removing brackets, $\qquad 352 - 11x + 99 = 4x - 44$;
transposing, $\qquad\qquad -11x - 4x = -44 - 352 - 99$;
collecting terms and changing signs, $15x = 495$;
$$\therefore \ x = 33.$$

Note. Here $-\dfrac{x-9}{8}$ is equivalent to $-\dfrac{1}{8}(x-9)$, the *vinculum* or line between the numerator and denominator having the same effect as a bracket. [Art. 58.]

76. In certain cases it will be found more convenient not to multiply throughout by the L.C.M. of the denominator, but to clear of fractions in two or more steps.

Example 2. Solve $\dfrac{x-4}{3} + \dfrac{2x-3}{35} = \dfrac{5x-32}{9} - \dfrac{x+9}{28}$.

Multiplying throughout by 9, we have
$$3x - 12 + \dfrac{18x - 27}{35} = 5x - 32 - \dfrac{9x + 81}{28};$$
transposing, $\qquad \dfrac{18x - 27}{35} + \dfrac{9x + 81}{28} = 2x - 20.$

Now clear of fractions by multiplying by $5 \times 7 \times 4$ or 140;
thus $\qquad\qquad 72x - 108 + 45x + 405 = 280x - 2800$;
$\qquad\qquad \therefore \ 2800 - 108 + 405 = 280x - 72x - 45x$;
$\qquad\qquad \therefore \quad 3097 = 163x$;
$\qquad\qquad \therefore \qquad x = 19.$

77. To solve equations whose coefficients are decimals, we may express the decimals as vulgar fractions, and proceed as before; but it is often found more simple to work entirely in decimals.

Example 1. Solve $\cdot\dot{6}x + \cdot 25 - \dfrac{1}{9}x = 1\cdot\dot{8} - \cdot 75x - \dfrac{1}{3}$.

Expressing the decimals as vulgar fractions, we have
$$\tfrac{2}{3}x + \tfrac{1}{4} - \tfrac{1}{9}x = 1\tfrac{8}{9} - \tfrac{3}{4}x - \tfrac{1}{3};$$
clearing of fractions. $\qquad 24x + 9 - 4x = 68 - 27x - 12$;
transposing, $\qquad\qquad 24x - 4x + 27x = 68 - 12 - 9$,
$\qquad\qquad\qquad 47x = 47$;
$$\therefore \ x = 1.$$

Example 2. Solve $\cdot 375x - 1\cdot 875 = \cdot 12x + 1\cdot 185$.

Transposing, $\quad\cdot 375x - \cdot 12x = 1\cdot 185 + 1\cdot 875$;

collecting terms, $\quad(\cdot 375 - \cdot 12)x = 3\cdot 06$,

that is, $\quad\cdot 255x = 3\cdot 06$;

$$\therefore x = \frac{3\cdot 06}{\cdot 255}$$

$$= 12.$$

EXAMPLES VIII. b.

Solve the following equations, and verify Nos. 1–16.

1. $\dfrac{x}{4} + \dfrac{x-5}{3} = 10$.

2. $\dfrac{x-5}{10} + \dfrac{x+5}{5} = 5$.

3. $\dfrac{x-2}{2} + \dfrac{x+10}{9} = 5$.

4. $\dfrac{x+19}{5} = 3 + \dfrac{x}{4}$.

5. $\dfrac{x-4}{7} = \dfrac{x-10}{5}$.

6. $\dfrac{x-1}{8} = 1 + \dfrac{x+1}{18}$.

7. $\dfrac{4(x+2)}{5} = 7 + \dfrac{5x}{13}$.

8. $\dfrac{x+4}{14} + \dfrac{x-4}{6} = 2$.

9. $\dfrac{x+20}{9} + \dfrac{3x}{7} = 6$.

10. $\dfrac{x-8}{7} + \dfrac{x-3}{3} + \dfrac{5}{21} = 0$.

11. $\dfrac{x+5}{6} - \dfrac{x+1}{9} = \dfrac{x+3}{4}$.

12. $\dfrac{4-5x}{6} - \dfrac{1-2x}{3} = \dfrac{13}{42}$.

13. $\dfrac{5(x+5)}{8} - \dfrac{2(x-3)}{7} = 5\tfrac{19}{28}$.

14. $\dfrac{4(x+2)}{3} - \dfrac{6(x-7)}{7} = 12$.

15. $1 + \dfrac{x}{2} - \dfrac{2x}{3} = \dfrac{3x}{4} - 4\tfrac{1}{2}$.

16. $\dfrac{x}{2} + \dfrac{x}{3} - \dfrac{x}{4} + \dfrac{x}{5} = 7\tfrac{5}{6}$.

17. $\dfrac{3}{16}(x-1) - \dfrac{5}{12}(x-4) = \dfrac{2}{5}(x-6) + \dfrac{5}{48}$.

18. $x + \dfrac{5}{3}(x-7) - \dfrac{6}{7}(x-8) = 3x - 14\tfrac{1}{3}$.

19. $\dfrac{3x}{4} - \dfrac{6}{17}(x+10) - (x-3) = \dfrac{x-7}{51} - 4\tfrac{3}{4}$.

20. $\dfrac{7x}{5} - \dfrac{1}{14}(x-11) = \dfrac{3}{7}(x-25) + 34$.

21. $3 + \dfrac{x}{4} = \dfrac{1}{2}\left(4 - \dfrac{x}{3}\right) - \dfrac{5}{6} + \dfrac{1}{3}\left(11 - \dfrac{x}{2}\right)$.

22. $\dfrac{1}{5}(x-8) + \dfrac{4+x}{4} + \dfrac{x-1}{7} = 7 - \dfrac{23-x}{5}$.

VIII.] SIMPLE EQUATIONS. 59

23. $\dfrac{1}{3}\left(\dfrac{x}{4}-3\right)+\dfrac{5x}{6}-\dfrac{5x}{4}=\dfrac{x-12}{5}-\dfrac{x+3}{3}.$

24. $x-\left(3x-\dfrac{2x-5}{10}\right)=\dfrac{1}{6}(2x-57)-\dfrac{5}{3}.$

25. $\dfrac{x}{4}-\dfrac{x+10}{5}+4\dfrac{3}{4}=x-1-\dfrac{x-2}{3}.$

26. $\cdot 5x-\cdot \dot{3}x=\cdot 25x-1.$ 27. $3+\dfrac{x}{\cdot 5}=7-\dfrac{x}{\cdot 2}.$

28. $2\cdot 25x-\cdot 125=3x+3\cdot 75.$ 29. $\cdot 2x-\cdot 1\dot{6}x=\cdot 6-\cdot \dot{3}.$

30. $\cdot 6x-\cdot 7x+\cdot 75x-\cdot 875x+15=0.$

31. $12\{3x-\cdot 25(x-4)-\cdot \dot{3}(5x+14)\}=47.$

32. $\dfrac{\cdot 25(x-3)+\cdot \dot{3}(x-4)}{\cdot 125}=5x-19.$

33. $\dfrac{x+\cdot 75}{\cdot 125}-\dfrac{x-\cdot 25}{\cdot 25}=15.$ 34. $\dfrac{5}{6}x+\cdot 25x-\cdot \dot{3}x=x-3.$

35. $\cdot 5x-\cdot \dot{2}x=\cdot \dot{3}x-1\cdot 5.$ 36. $1\cdot 5=\dfrac{\cdot 36}{\cdot 2}-\dfrac{\cdot 09x-\cdot 18}{\cdot 9}.$

[*Some of the examples in Miscellaneous Examples II., p. 80, will furnish further practice in Simple Equations.*]

78. Before concluding this chapter it will be worth while to draw attention to the following cases which occur so frequently in solving equations that the beginner should learn to write down the solution at sight.

Case I. Suppose $\dfrac{7x}{5}=\dfrac{4}{3}.$

Multiplying both sides by 5, we have

$$\left.\begin{array}{l}7x=\dfrac{4\times 5}{3}\\[6pt]\therefore\ x=\dfrac{4\times 5}{3\times 7}\end{array}\right\}\ \dotfill(1).$$

Case II. Suppose $\dfrac{5}{3x}=\dfrac{9}{7}.$

Multiplying both sides by $3x$, we have

$$\left.\begin{array}{l}5=\dfrac{9\times 3x}{7}\\[6pt]5\times 7=9\times 3x\\[6pt]\dfrac{5\times 7}{9\times 3}=x\end{array}\right\}\ \dotfill(2).$$

By a careful examination of the results in (1) and (2), the truth of the following principles will be evident:

Any factor of the numerator of one side of an equation may be transferred to the denominator of the other side, and any factor of the denominator of one side may be transferred to the numerator of the other side.

The ready application of these principles will be found very useful.

Example 1. If $\dfrac{3x}{14} = \dfrac{9}{35}$,

then $x = \dfrac{9 \times 14}{35 \times 3} = 1\dfrac{1}{5}$.

Example 2. If $\dfrac{2}{x} = -5$,

then $\dfrac{2}{5} = -x$; $\therefore x = -\dfrac{2}{5}$.

After a little practice the arithmetic should be performed mentally, and the intermediate steps omitted.

EXAMPLES VIII. c.

Write down the values of x which satisfy the following equations:

1. $\dfrac{2}{x} = \dfrac{3}{4}$.
2. $\dfrac{3}{7} = \dfrac{x}{14}$.
3. $\dfrac{3}{5} = \dfrac{6}{x}$.
4. $-\dfrac{x}{3} = 2$.
5. $\dfrac{x}{17} = \dfrac{29}{51}$.
6. $\dfrac{2x}{15} = \dfrac{8}{45}$.
7. $\dfrac{3}{2x} = -\dfrac{1}{8}$.
8. $-\dfrac{4x}{3} = -\dfrac{1}{2}$.
9. $\dfrac{5}{3x} = \dfrac{25}{27}$.
10. $\dfrac{5}{2x} = -\dfrac{10}{3}$.
11. $\dfrac{13}{21} = \dfrac{65x}{84}$.
12. $-\dfrac{7}{2} = \dfrac{1}{3x}$.
13. $\dfrac{3}{8} = \dfrac{x}{4}$.
14. $\dfrac{8}{21x} = -\dfrac{4}{7}$.
15. $\dfrac{36}{35} = \dfrac{9}{5x}$.
16. $\dfrac{5}{8} = \dfrac{15}{2x}$.
17. $\dfrac{x}{18} = \dfrac{9}{42}$.
18. $\dfrac{4}{3x} = \dfrac{16}{27}$.
19. $\dfrac{49}{15} = \dfrac{7}{3x}$.
20. $\dfrac{56}{15} = \dfrac{8x}{3}$.
21. $\dfrac{19x}{7} = \dfrac{57}{49}$.

CHAPTER IX.

Symbolical Expression.

79. In solving algebraical problems the chief difficulty of the beginner is to express the conditions of the question by means of symbols. A question proposed in algebraical symbols will frequently be found puzzling, when a similar arithmetical question would present no difficulty. Thus, the answer to the question "find a number greater than x by a" may not be self-evident to the beginner, who would of course readily answer an analogous arithmetical question, "find a number greater than 50 by 6." The process of addition which gives the answer in the second case supplies the necessary hint; and, just as the number which is greater than 50 by 6 is $50+6$, so the number which is greater than x by a is $x+a$.

80. The following examples will perhaps be the best introduction to the subject of this chapter. After the first we leave to the student the choice of arithmetical instances, should he find them necessary.

Example 1. By how much does x exceed 17?

Take a numerical instance; "by how much does 27 exceed 17?" The answer obviously is 10, which is equal to $27-17$.

Hence the excess of x over 17 is $x-17$.

Similarly the defect of x from 17 is $17-x$.

Example 2. If x is one *part* of 45 the other part is $45-x$.

Example 3. If x is one *factor* of 45 the other factor is $\dfrac{45}{x}$.

Example 4. How far can a man walk in a hours at the rate of 4 miles an hour?

In 1 hour he walks 4 miles,

In a hours he walks a times as far, that is, $4a$ miles.

Example 5. If £20 is divided equally among y persons, the share of each is the total sum divided by the number of persons, or $£\dfrac{20}{y}$.

Example 6. If 17 be divided by 6 the quotient is 2, and the remainder 5,

that is, $$\frac{17}{6} = 2 + \frac{5}{6}.$$

So if N be divided by D, and the quotient be Q and the remainder R, we have
$$\frac{N}{D} = Q + \frac{R}{D},$$
or $$N = QD + R.$$

Thus, if the divisor is x, the quotient y, and the remainder z, the dividend is $xy + z$.

Example 7. A and B are playing for money; A begins with £p and B with q shillings: after B has won £x, how many shillings has each?

What B has won A has lost,

∴ A has $20(p - x)$ shillings,

B has $q + 20x$ shillings.

EXAMPLES IX. a.

1. What must be added to x to make y?
2. By what must 3 be multiplied to make a?
3. What dividend gives b as the quotient when 5 is the divisor?
4. What is the defect of $2c$ from $3d$?
5. By how much does $3k$ exceed k?
6. If 100 be divided into two parts and one part be x what is the other?
7. If a be one factor of b, what is the other?
8. What number is less than 20 by c?
9. What is the price in pence of a oranges at tenpence a dozen?
10. What is the price in pence of 100 oranges when x cost sixpence?
11. If the difference of two numbers be 11, and if the smaller be x, what is the greater?
12. If the sum of two numbers be c and one of them is 20, what is the other?
13. What is the excess of 90 over x?
14. By how much does x exceed 30?
15. If 100 contains x five times, what is the value of x?
16. What is the cost in pounds of 40 books at x shillings each?

SYMBOLICAL EXPRESSION.

17. In x years a man will be 36 years old, what is his present age?

18. How old will a man be in a years if his present age is x years?

19. If x men take 5 days to reap a field, how long will one man take?

20. What value of x will make $5x$ equal to 20?

21. What is the price in shillings of 120 apples, when the cost of a score is x pence?

22. How many hours will it take to walk x miles at 4 miles an hour?

23. How far can I walk in x hours at the rate of y miles an hour?

24. In x days a man walks y miles, what is his rate per day?

25. How many minutes will it take to walk x miles at a miles an hour?

26. A train goes x miles an hour, how long does it take to go from Bristol to London, a distance of 120 miles?

27. How many miles is it between two places, if a train travelling p miles an hour takes 5 hours to perform the journey?

28. What is the velocity in feet per second of a train which travels 30 miles in x hours?

29. A man has a crowns and b florins, how many shillings has he?

30. If I spend x shillings out of a sum of £20, how many shillings have I left?

31. Out of a purse containing £a and b shillings a man spends c pence; express in pence the sum left.

32. By how much does $2x-5$ exceed $x+1$?

33. What number must be taken from $a-2b$ to leave $a-3b$?

34. If a bill is shared equally amongst x persons and each pays 3s. 4d., how many pence does the bill amount to?

35. If I give away c shillings out of a purse containing a sovereigns and b florins, how many shillings have I left?

36. In how many weeks will x horses eat 100 bushels of oats if one horse eats y bushels a week?

37. If I spend x shillings a week, how many pounds do I save out of a yearly income of £y?

38. A bookshelf contains x Latin, y Greek, and z English books: if there are 100 books, how many are there in other languages?

39. I have x pounds in my purse, y shillings in one pocket, and z pence in another: if I give away half-a-crown, how many pence have I left?

40. In a class of x boys, y work at Classics, z at Mathematics, and the rest are idle: what is the excess of workers over idlers?

81. We subjoin a few harder examples worked out in full.

Example 1. What is the present age of a man who x years hence will be m times as old as his son now aged y years?

In x years the son's age will be $y+x$ years; hence the father's age will be $m(y+x)$ years; therefore *now* the father's age is $m(y+x)-x$ years.

Example 2. Find the simple interest on £k in n years at f per cent.

Interest on £100 for 1 year is £f,

$\therefore$ .... £1 .... £$\dfrac{f}{100}$,

$\therefore$ .... £k .... £$\dfrac{kf}{100}$,

$\therefore$ Interest on £k for n years is £$\dfrac{nkf}{100}$.

Example 3. A room is x yards long, y feet broad, and a feet high; find how many square yards of carpet will be required for the floor, and how many square yards of paper for the walls.

(1) The area of the floor is $3xy$ square feet;

$\therefore$ the number of square yards of carpet required is $\dfrac{3xy}{9}=\dfrac{xy}{3}$.

(2) The perimeter of the room is $2(3x+y)$ feet;

$\therefore$ the area of the walls is $2a(3x+y)$ square feet;

$\therefore$ number of square yards of paper required is $\dfrac{2a(3x+y)}{9}$.

Example 4. The digits of a number beginning from the left are a, b, c; what is the number?

Here c is the digit in the units' place; b standing in the tens' place represents b tens; similarly a represents a hundreds.
The number is therefore equal to a hundreds $+ b$ tens $+ c$ units
$$=100a+10b+c.$$
If the digits of the number are inverted, a new number is formed which is symbolically expressed by
$$100c+10b+a.$$

Example 5. What is (1) the sum, (2) the product of three consecutive numbers of which the least is n?

The numbers consecutive to n are $n+1$, $n+2$;

$\therefore$ the sum $= n+(n+1)+(n+2)$
$$=3n+3.$$
And the product $= n(n+1)(n+2)$.

We may remark here that any *even* number may be denoted by $2n$, where n is *any* positive whole number; for this expression is exactly divisible by 2.

Similarly, any odd number may be denoted by $2n+1$; for this expression when divided by 2 leaves remainder 1.

Example 6. How many days will a men take to mow b acres if c boys can mow a acres in b days, and each man's work equals that of n boys?

Since c boys can mow a acres in b days;

$\therefore$ 1 boy .... bc days,

$\therefore$ n boys, or 1 man, ... $\dfrac{bc}{n}$ days,

$\therefore$ a men .... $\dfrac{bc}{an}$ days,

$\therefore$ a men .... 1 acre ... $\dfrac{bc}{a^2n}$ days;

therefore a men can mow b acres in $\dfrac{b^2c}{a^2n}$ days.

EXAMPLES IX. b.

1. Write down four consecutive numbers of which x is the least.

2. Write down three consecutive numbers of which y is the greatest.

3. Write down five consecutive numbers of which x is the middle one.

4. What is the next even number after $2n$?

5. What is the odd number next before $2x+1$?

6. Find the sum of three consecutive odd numbers of which the middle one is $2n+1$.

7. A man makes a journey of x miles. He travels a miles by coach, b by train, and finishes the journey by boat. How far does the boat carry him?

8. A horse eats a bushels and a donkey b bushels of corn in a week: how many bushels will they together consume in n weeks?

9. If a man was x years old 5 years ago, how old will he be y years hence?

10. A boy is x years old, and 5 years hence his age will be half that of his father. How old is the father now?

11. What is the age of a man who y years ago was m times as old as a child then aged x years?

12. A's age is double B's, B's is three times C's, and C is x years old: find A's age.

13. What is the interest on £1000 in b years at c per cent.?

14. What is the interest on £x in a years at 5 per cent.?

15. What is the interest on £$50a$ in a years at a per cent.?

16. What is the interest on £$24xy$ in x months at y per cent. per annum?

17. A room is x yards in length, and y feet in breadth: how many square feet are there in the area of the floor?

18. A square room measures x feet each way: how many square yards of carpet will be required to cover it?

19. A room is p feet long and x yards in width: how many yards of carpet two feet wide will be required for the floor?

20. What is the cost in pounds of carpeting a room a yards long, b feet broad, with carpet costing c shillings a square yard?

21. How many yards of carpet x inches wide will be required to cover the floor of a room y feet long and z feet broad?

22. A room is a yards long and b yards broad; in the middle there is a carpet c feet square: how many square yards of oil-cloth will be required to cover the rest of the floor?

23. How many miles can a person walk in 45 minutes if he walks a miles in x hours?

24. How long will it take a person to walk b miles if he walks 20 miles in c hours?

25. If a train travels a miles in b hours, how many feet does it move through in one second?

26. A train is running with a velocity of x feet per second: how many miles will it travel in y hours?

27. How long will x men take to mow y acres of corn, if each man mow z acres a day?

28. How many men will be required to do in x hours what y men do in xz hours?

29. What is the rate per cent. which will produce £y interest from a principal of £1000 in r years?

30. Find in how many years a principal of £a will produce £p interest at r per cent. per annum.

[*The following examples will assist the student in stating the conditions of a problem in equational form.*]

31. If y is the product of three consecutive numbers of which the greatest is p, express this fact by an equation.

32. The sum of three consecutive even numbers is equal to x. If the middle number is $2n$ express this by an equation.

33. The product of p and q is equal to five times the excess of a over b; express this by an equation.

34. If x is divided by y, the quotient is equal to 10 more than the sum of m and n; express this in algebraical symbols.

35. A man is x years older than his son, whose present age is a years; five years hence the father's age will be twice that of the son; express this in algebraical symbols. If the son is now 15, what is the father's age? If the father is now 53, how old is the son?

36. A has £p and B has q shillings; A hands £x to B and finds that he then has three times as much as B; express this fact by an equation.

37. A man who is p years old has a son whose age is q years; five years ago the father's age was seven times that of his son. Express this in algebraical symbols.

Formulæ.

82. In Example 6, Art. 80, we proved
$$\frac{N}{D} = Q + \frac{R}{D},$$
a result which gives in a single statement a general relation expressing the connection between a number, its divisor, and resulting quotient and remainder.

This is an example of a very important class of algebraical statements known as *formulæ*, the use and application of which we shall now briefly explain.

DEFINITION. **A formula** is a relation established by reasoning among certain quantities, any one of which may in turn be regarded as the unknown.

Thus in the formula above mentioned, if Q, R, and D are given quantities, we have an equation to find the corresponding value of N. Or, a question may be proposed as follows: "By what must 96 be divided so as to give a quotient 5, and a remainder 11?" Here we have given $N=96$, $Q=5$, $R=11$, and therefore from the formula we obtain
$$\frac{96}{D} = 5 + \frac{11}{D},$$
whence $D=17$, the required divisor.

83. A formula, it must be observed, includes all particular cases in one general statement: and so by the use of a single algebraical formula we are enabled briefly to express a whole class of results in a form at once simple, easily remembered, and easily applied. Experience will convince the student how much of the power and utility of Algebra lies in the ready application of formulæ to many kinds of problems.

It would be out of place here to make more than a passing allusion to other branches of Mathematics, or to Physical Science; but on account of the interest and importance of the subject, it may be useful to draw the reader's attention to a few of the more elementary formulæ he is likely to meet with in his other studies.

(1) If a triangle on a base b, has a height h, its area (A) is given by the formula
$$A = \tfrac{1}{2}hb.$$

(2) If a pyramid of height h stands on a base whose area is a^2, its volume (V) is given by the formula
$$V = \tfrac{1}{3}a^2 h.$$

In these cases any linear unit, inch, foot... being chosen, the superficial and solid units will be respectively the square and cubic inch, foot, ...; and in each of these formulæ if two of the three quantities be given, the third is easily obtained by Arithmetic.

Example. The Great Pyramid of Egypt stands on a square base each side of which is 764 feet; and its height is 480 feet. Find the number of cubic feet of stone used in its construction.

From the formula, $V = \tfrac{1}{3} \times (764)^2 \times 480$
$$= 160 \times 764 \times 764$$
$$= 93391360 \text{ cubic feet.}$$

84. We have in this chapter given several examples involving space, velocity, and time; and all these can be solved without difficulty by common sense reasoning. At the same time we may remark that they are only particular cases of the general formula $s = vt$, in which s denotes the space described by a body which moves with uniform velocity v for a time t.

In this formula, if t denotes the number of seconds the body has been in motion, and v the number of feet passed over in one second, then s is the space (in feet) described in t seconds.

Example. If a train has a velocity of 75 feet a second, how long will it take to cross a viaduct which is 300 yards in length?

Substituting the values of s and v (expressed in feet) in the formula, we get
$$900 = 75t,$$
$$t = \frac{900}{75}$$
$$= 12.$$

Therefore the time is 12 seconds.

85. Another very interesting case is that of a body falling vertically under the action of gravity.

It is proved in works on Dynamics that if a body fall freely from rest, and if s denote the space (in feet) described in t seconds,
$$s = \tfrac{1}{2}gt^2.$$

In this formula g denotes the number of feet per second by which the velocity is increased in each successive second in consequence of the earth's attraction, and it is found by experiment that $g = 32\cdot 2$ nearly.

Example 1. A stone dropped from the Clifton suspension bridge takes 4 seconds before it reaches the water. Find the height of the bridge above the river.

From the above formula, $s = \tfrac{1}{2} \times 32\cdot 2 \times (4)^2$
$$= 257\cdot 6,$$
and the height is therefore $257\cdot 6$ feet.

Example 2. How long will it take a stone to reach the bottom of a well $144\cdot 9$ feet deep?

From the formula, $144\cdot 9 = \tfrac{1}{2} \times 32\cdot 2 \times t^2$;
$$\therefore\ t^2 = \frac{144\cdot 9}{16\cdot 1} = 9;$$
$$\therefore\ t = 3.$$

Therefore the time is 3 seconds.

EXAMPLES IX. c.

1. From the formula for the area of a triangle in Art. 83, find
    (i) The area, when the base is 32 ft., and the height 17 ft.
    (ii) The base, when the area is 56 sq. ft., and the height 7 ft.
    (iii) The height (in chains and links), when the area is $5\cdot 985$ acres, and the base 17 chains 50 links.

2. By means of formula (2) in Art. 83, find
    (i) The volume of a pyramid of height 10 ft., on a base whose area is 15 sq. ft.
    (ii) The volume of a pyramid of height 6 ft., standing on a square base each of whose sides is $1\tfrac{1}{2}$ ft.
    (iii) The height of a pyramid whose volume is 20 cu. ft. and whose base has an area of 12 sq. ft.

3. By means of the formula $s=vt$ (Art. 84), find
   (i) How many miles a train will run in 84 minutes at 35 miles per hour.
   (ii) How long a train will take to run 56 miles at 42 miles per hour.
   (iii) The velocity in miles per hour of a train which travels 5500 yards in 5 minutes.

4. By means of the formula $s=\tfrac{1}{2}gt^2$ (Art. 85), find
   (i) The height of a flagstaff if a stone dropped from the top takes 3 seconds to reach the ground.
   (ii) How long it will take a stone to drop from a balloon whose height above the ground is 402 ft. 6 in.

5. The circumference (C) of a circle is π times the diameter (d); and the area (A) of a circle is π times the square of the radius (r). Express these two results by formulæ.

If $\pi = \tfrac{22}{7}$, find the circumferences and areas of circles whose radii are $3\tfrac{1}{2}$ inches and 1 ft. 9 in. respectively.

6. The surface S of a sphere of radius r is given by the formula
$$S = 4 \times \tfrac{22}{7} r^2.$$
Find (i) the surface of a sphere whose radius is 1·4 in.;
   (ii) the radius of a sphere whose surface is $38\tfrac{1}{2}$ sq. ft.

7. If a room is x feet long, y feet broad, and z feet high, find formulæ for (i) the perimeter, (ii) the area of the floor, (iii) the area of the walls.

8. From the formulæ of the last example find the perimeter, area of floor, and area of the walls of a room 18 ft. 8 in. long, 11 ft. 3 in. wide, and 12 ft. high.

9. From formula (iii) of Example 7, find the height of a room when the length and breadth are 17 ft. 9 in., 12 ft. 3 in. respectively, and the area of the walls is 630 sq. ft.

10. If a parallelogram on a base b has a height h, its area (A) is given by the formula
$$A = bh.$$
Find the area of parallelograms in which
   (i) the base = 5·5 cm., and the height = 4 cm.;
   (ii) the base = 2·4 in., and the height = 1·5 in.

11. The area of a parallelogram is 4·2 sq. in., and the base is 2·8 in. Find the height.

12. The area of a trapezium is equal to

$\frac{1}{2}$(sum of parallel sides) × (distance between them).

Express this in algebraical symbols, and apply the formula to find the area of a trapezium when the parallel sides are 6 ft. 4 in. and 7 ft. 2 in. and the distance between them is 4 ft.

13. Use the formula of Art. 80, Ex. 6, to find a number which when divided by 19 gives a quotient 17 and remainder 5.

14. By what number must 566 be divided so as to give a quotient 37 and remainder 11?

15. What is the present age of a man who 5 years hence will be three times as old as his son who is now 15? Verify the answer by substituting in the formula of Art. 81, Ex. 1.

16. In a right-angled triangle if a and b denote the lengths of the sides containing the right angle and c denotes the length of the hypotenuse, it is known that $c^2 = a^2 + b^2$.

By substitution find which of the following sets of numbers can be taken to represent the sides of a right-angled triangle.

    (i)   7, 24, 25.   (ii)   12, 35, 36.   (iii)   1·6, 6·3, 6·5.

17. The rectangle contained by two straight lines, one of which is divided into any number of parts, is equal to the sum of the rectangles contained by the undivided line and the several parts of the divided line.

Prove this by taking algebraical symbols to represent the undivided line and the segments of the divided line.

18. AB is a straight line divided into any two parts at O. Prove algebraically, as in the last example:

    (i) $AB^2 = AB \cdot AO + AB \cdot OB$.

    (ii) $AB \cdot AO = AO^2 + AO \cdot OB$.

Express these two results in a verbal form as in Example 17.

19. Prove algebraically the following theorems:

    (i)   If a straight line is divided into any two parts, the square on the whole line is equal to the sum of the squares on the two parts together with twice the rectangle contained by the two parts.

   (ii)   If a straight line is divided into any two parts, the sum of the squares on the whole line and on one of the parts is equal to twice the rectangle contained by the whole and that part, together with the square on the other part.

Express the results of these theorems in a form corresponding to (i) and (ii) of Example 18.

20. With the notation of Example 16, find the value of

    (i) c when $a=15$, $b=8$;

    (ii) a when $c=25$, $b=7$;

    (iii) b when $c=41$, $a=9$;

    (iv) a when $c=6{\cdot}5$, $b=6{\cdot}3$.

21. If $\pi=3{\cdot}1416$, $l=2{\cdot}0125$, $s=144{\cdot}9$, $g=32{\cdot}2$, $m=18{\cdot}75$, $v=5{\cdot}6$, find the values of

    (i) $\pi\sqrt{\dfrac{l}{g}}$;   (ii) $\sqrt{2gs}$;   (iii) $\dfrac{1}{2}mv^2$.

22. In the formula $F=\dfrac{mv^2}{gr}$, given $m=12{\cdot}075$, $r=3$, $g=32{\cdot}2$, $F=200$, find v.

23. In the formula $v^2-u^2=2as$, find the value of a when $v=50$, $u=10$, and $s=100$.

24. From the formula $s=\dfrac{n}{2}(a+l)$, find

    (i) the value of s, when $n=20$, $a=14$, $l=964$;

    (ii) the value of a, when $s=25{\cdot}2$, $n=12$, $l=3{\cdot}2$;

    (iii) the value of n, when $s=46{\cdot}8$, $a={\cdot}6$, $l=7{\cdot}2$;

    (iv) the value of l, when $s=-175{\cdot}5$, $a=13{\cdot}5$, $n=13$.

25. If $y=4+\tfrac{3}{10}x$, find the value of y when x has the values 0, 4, 8, 12, 16, 20.

There is a wall 20 ft. long, whose height at any point x ft. from one end is $4+\tfrac{3}{10}x$ feet. Draw the wall on a scale of 1 inch to 4 feet, marking on it the height at each end and at intervals of 4 ft.

CHAPTER X.

Problems leading to Simple Equations.

86. The principles of the last chapter may now be employed to solve various problems.

The method of procedure is as follows:

Represent the unknown quantity by a symbol x, and express in symbolical language the conditions of the question; we thus obtain a simple equation which can be solved by the methods already given in Chapter VIII.

Example 1. Find two numbers whose sum is 28, and whose difference is 4.

Let x be the smaller number, then $x+4$ is the greater.

Their sum is $x+(x+4)$, which is to be equal to 28.

Hence $\quad\quad\quad x+x+4=28$;
$$2x=24;$$
$$\therefore\ x=12,$$
and $\quad\quad\quad x+4=16$;

so that the numbers are 12 and 16.

The beginner is advised to test his solution by finding whether it satisfies the data of the question or not.

Example 2. Divide 60 into two parts, so that three times the greater may exceed 100 by as much as 8 times the less falls short of 200.

Let x be the greater part, then $60-x$ is the less.

Three times the greater part is $3x$, and its excess over 100 is
$$3x-100.$$

Eight times the less is $8(60-x)$, and its defect from 200 is
$$200-8(60-x).$$

Whence the symbolical statement of the question is
$$3x-100=200-8(60-x);$$
$$3x-100=200-480+8x,$$
$$480-100-200=8x-3x,$$
$$5x=180,$$
$$\therefore\ x=36,\text{ the greater part,}$$
and $\quad\quad\quad 60-x=24$, the less.

Example 3. Divide £47 between A, B, C, so that A may have £10 more than B, and B £8 more than C.

Suppose that C has x pounds; then B has $x+8$ pounds, and A has $x+8+10$ pounds.

Hence
$$x+(x+8)+(x+8+10)=47;$$
$$x+x+8+x+8+10=47,$$
$$3x=21;$$
$$\therefore x=7;$$
so that C has £7, B £15, A £25.

Example 4. A person spent £28. 4s. in buying geese and ducks; if each goose cost 7s., and each duck 3s., and if the total number of birds bought was 108 : how many of each did he buy?

In questions of this kind it is of essential importance to have all quantities expressed in the same denomination; in the present instance it will be convenient to express the money in shillings.

Let x be the number of geese, then $108-x$ is the number of ducks.

Since each goose costs 7 shillings, x geese cost $7x$ shillings.

And since each duck costs 3 shillings, $108-x$ ducks cost $3(108-x)$ shillings.

Therefore the amount spent is
$$7x+3(108-x) \text{ shillings};$$
but the question states that the amount is also £28. 4s., that is 564 shillings.

Hence
$$7x+3(108-x)=564;$$
$$7x+324-3x=564,$$
$$4x=240,$$
$$\therefore x=60, \text{ the number of geese,}$$
and
$$108-x=48, \text{ the number of ducks.}$$

Example 5. A is twice as old as B, ten years ago he was four times as old: what are their present ages?

Let B's age be x years, then A's age is $2x$ years.

Ten years ago their ages were respectively, $x-10$ and $2x-10$ years; thus we have
$$2x-10=4(x-10);$$
$$2x-10=4x-40,$$
$$2x=30;$$
$$\therefore x=15,$$
so that B is 15 years old, A 30 years.

Note. In the above examples the unknown quantity x represents a *number* of pounds, ducks, years, etc.; and the student must be careful to avoid beginning a solution with a supposition of the kind, "let $x=A$'s share" or "let $x=$ the ducks", or any statement so vague and inexact.

EXAMPLES X. a.

1. One number exceeds another by 5, and their sum is 29; find them.

2. The difference between two numbers is 8; if 2 be added to the greater the result will be three times the smaller: find the numbers.

3. Find a number such that its excess over 50 may be greater by 11 than its defect from 89.

4. A man walks 10 miles, then travels a certain distance by train, and then twice as far by coach. If the whole journey is 70 miles, how far does he travel by train?

5. What two numbers are those whose sum is 58, and difference 28?

6. If 288 be added to a certain number, the result will be equal to three times the excess of the number over 12: find the number.

7. Twenty-three times a certain number is as much above 14 as 16 is above seven times the number: find it.

8. Divide 105 into two parts, one of which diminished by 20 shall be equal to the other diminished by 15.

9. Find three consecutive numbers whose sum shall equal 84.

10. The sum of two numbers is 8, and one of them with 22 added to it is five times the other: find the numbers.

11. Find two numbers differing by 10 whose sum is equal to twice their difference.

12. A and B begin to play each with £60. If they play till A's money is double B's, what does A win?

13. Find a number such that if 5, 15, and 35 are added to it, the product of the first and third results may be equal to the square of the second.

14. The difference between the squares of two consecutive numbers is 121: find the numbers.

15. The difference of two numbers is 3, and the difference of their squares is 27: find the numbers.

16. Divide £380 between A, B, and C, so that B may have £30 more than A, and C may have £20 more than B.

17. A sum of £8. 17s. is made up of 124 coins which are either florins or shillings: how many are there of each?

18. If silk costs six times as much as linen, and I spend £9. 8s. in buying 23 yards of silk and 50 yards of linen: find the cost of each per yard.

19. A father is four times as old as his son; in 24 years he will only be twice as old: find their ages.

20. A is 25 years older than B, and A's age is as much above 20 as B's is below 85: find their ages.

21. *A*'s age is six times *B*'s, and fifteen years hence *A* will be three times as old as *B*: find their ages.

22. A sum of £4. 5s. was paid in crowns, half-crowns, and shillings. The number of half-crowns used was four times the number of crowns and twice the number of shillings: how many were there of each?

23. The sum of the ages of *A* and *B* is 30 years, and five years hence *A* will be three times as old as *B*: find their present ages.

24. In a cricket match the byes were double of the wides, and the remainder of the score was greater by three than twelve times the number of byes. If the whole score was 138, how were the runs obtained?

25. The length of a room exceeds its breadth by 3 feet; if the length had been increased by 3 feet, and the breadth diminished by 2 feet, the area would not have been altered: find the dimensions.

26. The length of a room exceeds its breadth by 8 feet; if each had been increased by 2 feet, the area would have been increased by 60 square feet: find the original dimensions of the room.

87. We add some problems which lead to equations with fractional coefficients.

Example 1. Find two numbers which differ by 4, and such that one-half of the greater exceeds one-sixth of the less by 8.

Let x be the smaller number, then $x+4$ is the greater.

One-half of the greater is represented by $\frac{1}{2}(x+4)$, and one sixth of the less by $\frac{1}{6}x$.

Hence $\frac{1}{2}(x+4) - \frac{1}{6}x = 8$;

multiplying by 6, $3x + 12 - x = 48$;

$\therefore 2x = 36$;

$\therefore x = 18$, the less number,

and $x+4 = 22$, the greater.

Example 2. *A* has £9, and *B* has 4 guineas; after *B* has won from *A* a certain sum, *A* has then five-sixths of what *B* has: how much did *B* win?

Suppose that *B* wins x *shillings*, *A* has then $180-x$ shillings, and *B* has $84+x$ shillings.

Hence $180 - x = \frac{5}{6}(84+x)$;

$1080 - 6x = 420 + 5x$,

$11x = 660$;

$\therefore x = 60$.

Therefore *B* wins 60 shillings, or £3.

EXAMPLES X. b.

1. Find a number such that the sum of its sixth and ninth parts may be equal to 15.

2. What is the number whose eighth, sixth, and fourth parts together make up 13?

3. There is a number whose fifth part is less than its fourth part by 3: find it.

4. Find a number such that six-sevenths of it shall exceed four-fifths of it by 2.

5. The fifth, fifteenth, and twenty-fifth parts of a number together make up 23: find the number.

6. Two consecutive numbers are such that one-fourth of the less exceeds one fifth of the greater by 1: find the numbers.

7. Two numbers differ by 28, and one is eight-ninths of the other: find them.

8. There are two consecutive numbers such that one-fifth of the greater exceeds one-seventh of the less by 3: find them.

9. Find three consecutive numbers such that if they be divided by 10, 17, and 26 respectively, the sum of the quotients will be 10.

10. A and B begin to play with equal sums, and when B has lost five-elevenths of what he had to begin with, A has gained £6 more than half of what B has left: what had they at first?

11. From a certain number 3 is taken, and the remainder is divided by 4; the quotient is then increased by 4 and divided by 5 and the result is 2: find the number.

12. In a cellar one-fifth of the wine is port and one-third claret; besides this it contains 15 dozen of sherry and 30 bottles of hock: how much port and claret does it contain?

13. Two-fifths of A's money is equal to B's, and seven-ninths of B's is equal to C's; in all they have £770: what have they each?

14. A, B, and C have £1285 between them: A's share is greater than five-sixths of B's by £25, and C's is four-fifteenths of B's: find the share of each.

15. A man sold a horse for £35 and half as much as he gave for it, and gained thereby ten guineas: what did he pay for the horse?

16. The width of a room is two thirds of its length. If the width had been 3 feet more, and the length 3 feet less, the room would have been square: find its dimensions.

17. What is the property of a person whose income is £430, when he has two-thirds of it invested at 4 per cent., one-fourth at 3 per cent., and the remainder at 2 per cent.?

18. I bought a certain number of apples at three a penny, and five-sixths of that number at four a penny; by selling them at sixteen for sixpence I gained $3\frac{1}{2}d.$: how many apples did I buy?

E.A. F

CHAPTER XI.

Highest Common Factor, Lowest Common Multiple of Simple Expressions.

Highest Common Factor.

88. DEFINITION. The **highest common factor** of two or more algebraical expressions is the expression of highest dimensions [Art. 24] which divides each of them without remainder.

The abbreviation H.C.F. is sometimes used instead of the words *highest common factor*.

89. In the case of *simple expressions* the highest common factor can be written down by inspection.

Example 1. The highest common factor of a^4, a^3, a^2, a^6 is a^2.

Example 2. The highest common factor of a^3b^4, ab^5c^2, a^2b^7c is ab^4; for a is the highest power of a that will divide a^3, a, a^2; b^4 is the highest power of b that will divide b^4, b^5, b^7; and c is not a *common* factor.

90. If the expressions have numerical coefficients, find by Arithmetic their greatest common measure, and prefix it as a coefficient to the algebraical highest common factor.

Example. The highest common factor of $21a^4x^3y$, $35a^2x^4y$, $28a^3xy^4$ is $7a^2xy$; for it consists of the product of

(1) the greatest common measure of the numerical coefficients;

(2) the highest power of each letter which divides every one of the given expressions.

EXAMPLES XI. a.

Find the highest common factor of

1. $4ab^2$, $2a^2b$.
2. $3x^2y^2$, x^3y^2.
3. $6xy^2z$, $8x^2y^3z^2$.
4. abc, $2ab^2c$.
5. $5a^3b^3$, $15abc^2$.
6. $9x^2y^2z^2$, $12xy^3z$.
7. $4a^2b^3c^2$, $6a^3b^2c^3$.
8. $7a^2b^4c^5$, $14ab^2c^3$.
9. $15x^4y^3z^2$, $12x^2yz^2$.
10. $8a^2x$, $6abxy$, $10abx^3y^2$.
11. $49ax^2$, $63ay^2$, $56az^2$.
12. $17ab^2c$, $34a^2bc$, $51abc^2$.
13. a^3x^2y, b^3xy^2, c^3x^2y.
14. $24a^2b^3c^3$, $64a^3b^3c^2$, $48a^3b^2c^3$.
15. $25xy^2z$, $100x^2yz$, $125xy$.
16. a^2bpxy, b^2qxy, a^3bxr^2.
17. $15a^5b^3c^7$, $60a^3b^7c^6$, $25a^4b^5c^2$.
18. $35a^2c^3b$, $42a^3cb^2$, $30ac^2b^3$.

Lowest Common Multiple.

91. DEFINITION. The **lowest common multiple** of two or more algebraical expressions is the expression of lowest dimensions which is divisible by each of them without remainder.

The abbreviation L.C.M. is sometimes used instead of the words *lowest common multiple*.

92. In the case of *simple expressions* the lowest common multiple can be written down by inspection.

Example 1. The lowest common multiple of a^4, a^3, a^2, a^6 is a^6.

Example 2. The lowest common multiple of a^3b^4, ab^5, a^2b^7 is a^3b^7; for a^3 is the lowest power of a that is divisible by each of the quantities a^3, a, a^2; and b^7 is the lowest power of b that is divisible by each of the quantities b^4, b^5, b^7.

93. If the expressions have numerical coefficients, find by Arithmetic their least common multiple, and prefix it as a coefficient to the algebraical lowest common multiple.

Example. The lowest common multiple of $21a^4x^3y$, $35a^2x^4y$, $28a^3xy^4$ is $420a^4x^4y^4$; for it consists of the product of

(1) the least common multiple of the numerical coefficients;
(2) the lowest power of each letter which is divisible by every power of that letter occurring in the given expressions.

EXAMPLES XI. b.

Find the lowest common multiple of

1. abc, $2a^2$.
2. x^3y^2, xyz.
3. $3x^2yz$, $4x^3y^3$.
4. $5a^2bc^3$, $4ab^2c$.
5. $3a^4b^2c^3$, $5a^2b^3c^5$.
6. $12ab$, $8xy$.
7. ac, bc, ab.
8. a^2c, bc^2, cb^2.
9. $2ab$, $3bc$, $4ca$.
10. $2x$, $3y$, $4z$.
11. $3x^2$, $4y^2$, $3z^2$.
12. $7a^2$, $2ab$, $3b^3$.
13. a^2bc, b^2ca, c^2ab.
14. $5a^2c$, $6cb^2$, $3bc^2$.
15. $2x^2y^3$, $3xy$, $4x^3y^4$.
16. $7x^4y$, $8xy^5$, $2x^3y^3$.
17. $35a^2c^3b$, $42a^3cb^2$, $30ac^2b^3$.
18. $66a^4b^2c^3$, $44a^3b^4c^2$, $24a^2b^3c^4$.

Find both the highest common factor and the lowest common multiple of

19. $2abc$, $3ca$, $4bca$.
20. $2xy$, $4yz$, $6zxy$.
21. $9abc$, $3b^2c$, cab.
22. $13a^2bc$, $39a^3bc^2$.
23. $17xyz^2$, $51x^2y$.
24. $15x^3y^3z$, $25xy^3z^2$.
25. $3ab$, $2bc$, $5cab$.
26. $17m^2n^4p^2$, $51m^4p^4$.
27. x^3y^2, y^2z^4, $z^4x^3y^5$.
28. $15p^3q^4$, $20m^2p^2q^3$, $30mp^3$.
29. $72k^2m^3n^4$, $108k^3m^2n^5$.

CHAPTER XII.

Elementary Fractions.

94. Definition. If a quantity x be divided into b equal parts, and a of these parts be taken, the result is called *the fraction $\frac{a}{b}$ of x*. If x be the unit, the fraction $\frac{a}{b}$ of x is called simply "the fraction $\frac{a}{b}$"; so that *the fraction $\frac{a}{b}$ represents a equal parts, b of which make up the unit*.

95. In this chapter we propose to deal only with the easier kinds of fractions, where the numerator and denominator are simple expressions. Their reduction and simplification will be performed by the usual arithmetical rules. The proofs of these rules will be given in Chapters XIX. and XXI.

Rule. To reduce a fraction to its lowest terms: *divide numerator and denominator by every factor which is common to them both, that is by their highest common factor.*

Dividing numerator and denominator of a fraction by a common factor is called *cancelling* that factor.

Examples. (1) $\dfrac{6a^2c}{9ac^2} = \dfrac{2a}{3c}$.

(2) $\dfrac{7x^2yz}{28x^3yz^2} = \dfrac{1}{4xz}$.

(3) $\dfrac{35a^5b^3c}{7ab^2c} = \dfrac{5a^4b}{1} = 5a^4b$.

EXAMPLES XII. a.

Reduce to lowest terms:

1. $\dfrac{3a}{6ab}$.
2. $\dfrac{4a^2}{16ab}$.
3. $\dfrac{2xy^2}{5x^2y}$.
4. $\dfrac{3abc}{15a^2b^2c}$.
5. $\dfrac{x^2yz^3}{x^3y^2z}$.
6. $\dfrac{15ab}{25bc}$.
7. $\dfrac{21x^2y^2}{28y^2z^2}$.
8. $\dfrac{8a^2b}{12b^2c}$.
9. $\dfrac{12mn^2p}{15m^2np^2}$.
10. $\dfrac{15m^2p^3}{18n^4p}$.
11. $\dfrac{abc^2}{a^3b^2c}$.
12. $\dfrac{3x^2yz^3}{5xy^4z^2}$.
13. $\dfrac{2xy^3z^4}{4x^2y^2z}$.
14. $\dfrac{5a^3b^2c^4}{15ab^4c}$.
15. $\dfrac{mn^4pq}{m^2n^3p^4}$.
16. $\dfrac{4m^3n^2p^5}{6m^4np^2}$.
17. $\dfrac{15ax^3y^2}{25a^2xy^6}$.
18. $\dfrac{39a^2b^4c^3}{52a^3b^5c^4}$.
19. $\dfrac{38k^2p^3m^4}{57k^3pm^2}$.
20. $\dfrac{46x^3y^4z^5}{69x^2y^3z^4}$.

Multiplication and Division of Fractions.

96. Rule. To multiply algebraical fractions: *as in Arithmetic, multiply together all the numerators to form a new numerator, and all the denominators to form a new denominator.*

Example 1. $\dfrac{2a}{3b} \times \dfrac{5x^2}{2a^2b} \times \dfrac{3b^2}{2x} = \dfrac{2a \times 5x^2 \times 3b^2}{3b \times 2a^2b \times 2x} = \dfrac{5x}{2a}$,

by cancelling like factors in numerator and denominator.

Example 2. $\dfrac{3a^2b}{5c^2} \times \dfrac{7bc}{3a^3} \times \dfrac{5ca}{7b^2} = 1$,

all the factors cancelling each other.

97. Rule. To divide one fraction by another: *invert the divisor and proceed as in multiplication.*

Example. $\dfrac{7a^3}{4x^3y^2} \times \dfrac{6c^3x}{5ab^2} \div \dfrac{28a^2c^2}{15b^2xy^2} = \dfrac{7a^3}{4x^3y^2} \times \dfrac{6c^3x}{5ab^2} \times \dfrac{15b^2xy^2}{28a^2c^2} = \dfrac{9c}{8x}$,

all the other factors cancelling each other.

EXAMPLES XII. b.

Simplify the following expressions:

1. $\dfrac{2ab}{3cd} \times \dfrac{c^2d^3}{ab^2}$.

2. $\dfrac{12a^2bc}{8ab^3} \times \dfrac{24ab^2}{36bc^2}$.

3. $\dfrac{15xyz^3}{a^2bc} \times \dfrac{3a^3x}{5yz}$.

4. $\dfrac{7a^2b^3}{9ax^2y} \times \dfrac{18x^2c}{15ac^4}$.

5. $\dfrac{8m^2n^3}{5x^2yz} \times \dfrac{15xyz^3}{16mn^2}$.

6. $\dfrac{21k^2p^3}{13mn^2} \times \dfrac{39n^3m^2}{28p^2k^3}$.

7. $\dfrac{3a^2b}{4b^3c} \times \dfrac{2c^2}{8a^3} \div \dfrac{6ac}{16b^2x}$.

8. $\dfrac{2x^2y}{3yz} \times \dfrac{5z^2x}{7xy^2} \div \dfrac{21x^2y^3z^2}{40xy^2z}$.

9. $\dfrac{7m^2p}{17x^2y} \times \dfrac{51y^3z}{21p^2n} \div \dfrac{m^2x^2}{pyz}$.

10. $\dfrac{26xk^2p^3}{58mp^4} \times \dfrac{2xk^3}{13pkm} \div \dfrac{2x^2k^4}{87m^2p^2}$.

11. $\dfrac{15b^2}{40c} \times \dfrac{27c^2}{81d^3} \div \dfrac{abc}{14d^3}$.

12. $\dfrac{b^2}{3c} \times \dfrac{4c^2}{5d^3} \div \dfrac{16a^2b^2c^2}{15d^5}$.

13. $\dfrac{8ax^2}{7by} \times \dfrac{49cy^2}{64dx^3}$.

14. $\dfrac{15abc}{16xyz} \times \dfrac{128x^3y^2z^2}{160a^2bc}$.

15. $\dfrac{45a^2b^3c^4}{27x^4y^3z} \times \dfrac{243xy^2z^3}{180a^2bc^3}$.

16. $\dfrac{104xyzk^2p}{28xy^2kp^2} \times \dfrac{56y^3z^5p}{26y^2z^6k}$.

17. $\dfrac{m^2}{8n} \times \dfrac{36p^3q^2}{81mn} \div \dfrac{15mpx^5}{27n^2x^3y}$.

18. $\dfrac{a^3}{b^3} \times \dfrac{xy^2}{ab} \times \dfrac{pb^2}{ax} \div \dfrac{ap}{b^2}$.

Reduction to a Common Denominator.

98. In order to find the sum or difference of any fractions, we must, as in Arithmetic, first reduce them to a common denominator; and it is most convenient to take the lowest common multiple of the denominators of the given fractions.

Example. Express with lowest common denominator the fractions
$$\frac{a}{3xy}, \frac{b}{6xyz}, \frac{c}{2yz}.$$

The lowest common multiple of the denominators is $6xyz$. Multiplying the numerator of each fraction by the factor which is required to **make** its denominator $6xyz$, we have the equivalent fractions
$$\frac{2az}{6xyz}, \frac{b}{6xyz}, \frac{3cx}{6xyz}.$$

Note. The same result would clearly be obtained by dividing the lowest common denominator by each of the denominators in turn, and multiplying the corresponding numerators by the respective quotients.

EXAMPLES XII. c.

Express as equivalent fractions with common denominator:

1. $\dfrac{2x}{a}, \dfrac{y}{2a}.$
2. $\dfrac{4x}{3y}, \dfrac{y}{x^2}.$
3. $\dfrac{a}{2b}, \dfrac{b}{c}.$
4. $\dfrac{a}{b}, \dfrac{c}{d}, 2.$
5. $\dfrac{2a}{b}, \dfrac{b}{3c}.$
6. $\dfrac{m}{4n}, \dfrac{p}{5n}.$
7. $\dfrac{k}{2x}, \dfrac{p}{3x}.$
8. $\dfrac{m}{3x}, \dfrac{n}{6x}.$
9. $\dfrac{a}{bc}, \dfrac{b}{ca}.$
10. $\dfrac{a}{x}, \dfrac{b}{x^2}.$
11. $\dfrac{2}{x}, \dfrac{3}{y}.$
12. $\dfrac{x}{y}, \dfrac{y}{x}, 3x.$
13. $\dfrac{2x}{3y}, \dfrac{3y}{2x}.$
14. $\dfrac{4a}{5b}, \dfrac{3a}{10c}.$
15. $\dfrac{3a}{7b}, \dfrac{5b}{21c}.$
16. $\dfrac{2}{a}, \dfrac{b}{3}, \dfrac{a}{9}.$

Addition and Subtraction of Fractions.

99. Rule. To add or subtract fractions: *express all the fractions with their lowest common denominator; form the algebraical sum of the numerators, and retain the common denominator.*

Example 1. Simplify $\dfrac{5x}{3} + \dfrac{3}{4}x - \dfrac{7x}{6}.$

The least common denominator is 12.

The expression $= \dfrac{20x + 9x - 14x}{12} = \dfrac{15x}{12} = \dfrac{5x}{4}.$

XII.]   ELEMENTARY FRACTIONS.   79

Example 2. Simplify $\dfrac{3ab}{5x} - \dfrac{ab}{2x} - \dfrac{1}{10}\cdot\dfrac{ab}{x}$.

The expression $= \dfrac{6ab - 5ab - ab}{10x} = \dfrac{0}{10x} = 0$.

Example 3. Simplify $\dfrac{2x}{a^2c^2} - \dfrac{y}{3ca^3}$.

The expression $= \dfrac{6ax - cy}{3a^3c^2}$, and admits of no further simplification.

Note. The beginner must be careful to distinguish between **erasing equal terms with different signs**, as in Example 2, and **cancelling equal factors** in the course of multiplication, or in reducing fractions to lowest terms. Moreover, in simplifying fractions he must remember that a factor can only be removed from numerator and denominator when it divides each *taken as a whole*.

Thus in $\dfrac{6ax - cy}{3a^3c^2}$, c cannot be cancelled because it only divides cy and not the *whole* numerator. Similarly a cannot be cancelled because it only divides $6ax$ and not the whole numerator. The fraction is therefore in its simplest form.

When no denominator is expressed the denominator 1 may be understood.

Example 4. $3x - \dfrac{a^2}{4y} = \dfrac{3x}{1} - \dfrac{a^2}{4y} = \dfrac{12xy - a^2}{4y}$.

If a fraction is not in its lowest terms it should be simplified before combining it with other fractions.

Example 5. $\dfrac{ax}{2} - \dfrac{x^2y}{3xy} = \dfrac{ax}{2} - \dfrac{x}{3} = \dfrac{3ax - 2x}{6}$.

EXAMPLES XII. d.

Simplify the following expressions:

1. $\dfrac{x}{2} + \dfrac{x}{3}$.
2. $\dfrac{y}{4} - \dfrac{y}{5}$.
3. $\dfrac{a}{3} - \dfrac{a}{4}$.
4. $\dfrac{2x}{3} - \dfrac{5}{x}$.
5. $\dfrac{x}{2} + \dfrac{y}{5}$.
6. $\dfrac{a}{4} - \dfrac{b}{6}$.
7. $\dfrac{m}{8} - \dfrac{n}{12}$.
8. $\dfrac{2m}{15} - \dfrac{n}{5}$.
9. $\dfrac{x}{7} - \dfrac{y}{21}$.
10. $\dfrac{a}{13} + \dfrac{b}{39}$.
11. $\dfrac{p}{16} - \dfrac{q}{48}$.
12. $\dfrac{5m}{12} - \dfrac{n}{36}$.
13. $\dfrac{2x}{3} + \dfrac{4x}{5}$.
14. $\dfrac{5x}{4} - \dfrac{4x}{5}$.
15. $\dfrac{5x}{6} - \dfrac{7x}{12}$.
16. $\dfrac{2a}{5} - \dfrac{4b}{15}$.

Simplify the following expressions:

17. $\dfrac{a}{2} - \dfrac{a}{3} + \dfrac{a}{5}$.
18. $\dfrac{x}{4} - \dfrac{x}{8} + \dfrac{x}{12}$.
19. $\dfrac{x}{3} + \dfrac{x}{6} - \dfrac{x}{9}$.
20. $\dfrac{2x}{3} - \dfrac{x}{6} + \dfrac{3x}{4}$.
21. $\dfrac{5x}{6} - \dfrac{x}{12} + \dfrac{x}{9}$.
22. $\dfrac{7x}{8} + \dfrac{x}{12} - \dfrac{x}{4}$.
23. $\dfrac{x}{a} - \dfrac{y}{b}$.
24. $\dfrac{3x^3}{ax^2} + \dfrac{2y}{3b}$.
25. $a + \dfrac{b}{c}$.
26. $x - \dfrac{y^2}{yz}$.
27. $\dfrac{a^3}{3a^2} - \dfrac{b^2}{a}$.
28. $a^2 + \dfrac{b^3}{a}$.
29. $\dfrac{3x^2}{6x} - \dfrac{y^2}{x^2}$.
30. $p^3 - \dfrac{k^5}{p^2}$.

MISCELLANEOUS EXAMPLES II.

(*Chiefly* on Chapters I.–VIII.)

[*The examples marked with an asterisk must be postponed by those who adopt the suggestions printed in italics on pages* 33 *and* 38.]

1. What expression must be added to $4x^3 - 3x^2 + 2$ to produce $4x^3 + 7x - 6$?

2. If $A = 6x - 3y + 2z$, $B = x + y + z$, and $C = 10x + y - 7z$, find the value of $A + 4B - C$.

3. If $x = 3$, $y = 4$, $z = 1$, find the value of
$$\sqrt{2xy + 4xz} + \sqrt{9y} + \dfrac{2xyz}{3}.$$

4. Simplify by removing brackets
$$a^2 + 2d^2 - (2e^2 - b^2) - \{(d^2 - c^2 - e^2) + (d^2 - e^2)\}.$$

*5. Multiply $x^3 + x^2 + 3x + 5$ by $x^2 - x - 2$.

6. Solve the equations:
  (1) $3 - 4x = 36x - 17$;   (2) $5x\ 15 = 17x + 21$.

*7. Divide $x^4 - 10x^2 + 9$ by $x^2 - 2x - 3$.

8. Simplify $7a - 4b - \{5a - 3[b - 2(a - b)]\}$.

9. In an examination A has $x + y$ marks, B has $2x - 3y$, and C has twice as many as A; how many marks have A, B, and C together?

10. Find the sum of $1 - 2x + x^2$, $3x - 2x^2$, $5x^2 - 7x - 2$, arranging the result in descending powers of x.

MISCELLANEOUS EXAMPLES. II.

11. Write down the following products:
(1) $(x+17)(x-3)$; (2) $(3x-8)(8x+3)$.

12. Solve the equations:
(1) $7x-3-(7-5x)=3-3x-(5x+8)$;
(2) $(5x+1)(x-2)-(4x-3)(3x-1)=10-(7x+2)(x+1)$.

---

13. From the sum of $3ab$, $-5ab$, $2ab$, $7ab$, $-9ab$, subtract the sum of $-8ab$, $6ab$, $-9ab$, $10ab$.

14. When $a=4$, $b=3$, $c=2$, find the numerical value of
$$\frac{2a+b(2c-a)}{3b-\sqrt{2c^3}}.$$

15. From what expression must $11a^2-5ab-7bc$ be subtracted so as to give for remainder $7b(a+c)+5a^2$?

*16. Multiply x^3+6x^2+8x-8 by x^2-2x+4.

17. Simplify
$$12a-[6a-2\{3a-4(b-a)\}-(9a+8b)].$$

18. Solve the equations:
(1) $3(2x-1)+2(3x-2)+3=4(x-5)$;
(2) $\tfrac{1}{3}(x+1)+\tfrac{1}{4}(x+3)=\tfrac{1}{5}(x+4)+16$.

Verify the solution in each case.

---

*19. Divide $3p^5+16p^4-33p^3+14p^2$ by p^2+7p.

20. Add together
$a+2b-(2c+d)$, $3a-(b-2c)+2d$, and $2a-[b-(2c-3d)]$.

21. To what expression must $7x^3-6x^2-5x$ be added so as to make $9x^3-6x-7x^2$?

22. What value of x will make the product of $x+1$ and $2x+1$ less than the product of $x+3$ and $2x+3$ by 14?

23. When $a=2$, $b=3$, $c=1$, $q=4$, $r=6$, find the value of
$$5a^b c^r - 3^a 2^b + 2^r a^5 - c^b b^q.$$

24. Solve the equation:
$$x - \frac{x-13}{9} = \frac{6x+1}{5} + \frac{2}{3}\left(6 - \frac{3x}{2}\right).$$

Shew also that $x=3$ does *not* satisfy the equation.

25. A horse can eat $3m+2n$ bushels of corn in a week; how many weeks will he be in eating $12m^2 - 7mn - 10n^2$ bushels?

26. Subtract the sum of
$$2x^3 - 3x + 4 \text{ and } -3x^2 + 2x - 7$$
from
$$4x^3 - 3x^2 + x - 6 - [2x^3 - (x-6)].$$

27. Find the value of $a^3 + b^3 + c^3 - 3abc$, when $a=1$, $b=4$, $c=-5$.

28. Solve the equations:

(1) $\dfrac{2x}{15} + \dfrac{x-6}{12} = \dfrac{3}{10}\left(\dfrac{x}{2} - 5\right)$;

(2) $\dfrac{2(x-1)}{5} + \dfrac{15}{2}\left(1 - \dfrac{x}{3}\right) + \dfrac{19}{10} = \dfrac{9}{5}\left(\dfrac{x}{6} - \dfrac{1}{3}\right)$.

*29. Divide $3y^6 - 37y^4 + 35y^3 + 7y^2 + 2$ by $y(y-1)(y+4) - 2$.

30. Divide £1120 between A and B so that for every half-crown that A receives B may receive a shilling.

---

31. Find the value of
$$(a+b)^2 + (b+c)^2 + (c+a)^2$$
when $a=-1$, $b=-2$, $c=3$.

32. Multiply $(2m^2 + 8)(m+2)$ by $3m - 6$.

*33. Divide the product of
$$x-2, \ x+3, \text{ and } 2x-7$$
by the sum of $\quad 3(x^2 - 2x - 2)$ and $5x - x^2 - 15$.

34. A man walks at the rate of a miles an hour for p hours; he then rides for q hours at the rate of b miles an hour. How far has he travelled, and how long would it have taken to ride the same distance at c miles an hour?

Also work out the result supposing $p=7$, $q=3$, $a=4$, $b=9$, $c=11$.

35. Solve the equations:

(1) $\dfrac{3x}{2} - \dfrac{5}{7} = 21x - \dfrac{1}{3}\left(2x + 10\dfrac{3}{14}\right)$;

(2) $3x - 4 - \dfrac{4(7x-9)}{15} = \dfrac{4}{5}\left(6 + \dfrac{x-1}{3}\right)$.

36. An egg-dealer bought a certain number of eggs at 1s. 4d. per score, and five times the number at 6s. 3d. per hundred. He sold the whole at 10d. per dozen, gaining £1. 7s. by the transaction. How many eggs did he buy?

CHAPTER XIII.

SIMULTANEOUS EQUATIONS.

[*In connection with this chapter the student may read* Chap. XLIV. Arts. 417–424.]

100. CONSIDER the equation $2x+5y=23$, which contains *two* unknown quantities.

From this we get $5y=23-2x$,

that is, $$y=\frac{23-2x}{5} \quad \ldots\ldots\ldots\ldots\ldots\ldots\ldots\ldots (1).$$

From this it appears that for every value we choose to give to x there will be one corresponding value of y. Thus we shall be able to find as many pairs of values as we please which satisfy the given equation.

For instance, if $x=1$, then from (1) $y=\frac{21}{5}$.

Again, if $x=-2$, then $y=\frac{27}{5}$; and so on.

But if also we have a second equation of the same kind, such as $3x+4y=24$,

we have from this $$y=\frac{24-3x}{4} \quad \ldots\ldots\ldots\ldots\ldots\ldots (2).$$

If now we seek values of x and y which satisfy *both* equations, the values of y in (1) and (2) must be identical.

Therefore $\quad \dfrac{23-2x}{5}=\dfrac{24-3x}{4}$.

Multiplying up, $\quad 92-8x=120-15x$;

$\therefore \quad 7x=28$;

$\therefore \quad x=4$.

Substituting this value in the first equation, we have

$8+5y=23$;

$\therefore \quad 5y=15$;

$\therefore \quad y=3,$

and $\quad x=4.$

Thus, if both equations are to be satisfied by the *same* values of x and y, there is only one solution possible.

101. Definition. When two or more equations are satisfied by the same values of the unknown quantities they are called **simultaneous equations.**

We proceed to explain the different methods for solving simultaneous equations. In the present chapter we shall confine our attention to the simpler cases in which the unknown quantities are involved in the first degree.

102. In the example already worked we have used the method of solution which best illustrates the meaning of the term *simultaneous equation*; but in practice it will be found that this is rarely the readiest mode of solution. It must be borne in mind that since the two equations are simultaneously true, *any* equation formed by combining them will be satisfied by the values of x and y which satisfy the original equations. Our object will always be to obtain an equation which involves *one only* of the unknown quantities.

103. The process by which we get rid of either of the unknown quantities is called **elimination**, and it must be effected in different ways according to the nature of the equations proposed.

Example 1. Solve
$$3x + 7y = 27 \quad \ldots\ldots (1),$$
$$5x + 2y = 16 \quad \ldots\ldots (2).$$

To eliminate x we multiply (1) by 5 and (2) by 3, so as to make the coefficients of x in both equations equal. This gives
$$15x + 35y = 135,$$
$$15x + 6y = 48;$$
subtracting, $\quad 29y = 87;$
$$\therefore y = 3.$$

To find x, substitute this value of y in *either* of the given equations.

Thus from (1) $\quad 3x + 21 = 27;$
$$\therefore x = 2,$$
and $\quad y = 3.$

Note. When one of the unknowns has been found, it is immaterial which of the equations we use to complete the solution. Thus, in the present example, if we substitute 3 for y in (2), we have
$$5x + 6 = 16;$$
$$\therefore x = 2, \text{ as before.}$$

Example 2. Solve
$$7x + 2y = 47 \quad \ldots\ldots(1),$$
$$5x - 4y = 1 \quad \ldots\ldots(2).$$

Here it will be more convenient to eliminate y.

Multiplying (1) by 2, $14x + 4y = 94$,
and from (2) $5x - 4y = 1$;
adding, $19x = 95$;
$$\therefore x = 5.$$

Substitute this value in (1),
$$\therefore 35 + 2y = 47;$$
$$\therefore y = 6,$$
and $x = 5.$

Note. *Add* when the coefficients of one unknown are equal and *unlike* in sign; *subtract* when the coefficients are equal and *like* in sign.

Example 3. Solve
$$2x = 5y + 1 \quad \ldots\ldots(1),$$
$$24 - 7x = 3y \quad \ldots\ldots(2).$$

Here we can eliminate x by substituting in (2) its value obtained from (1). Thus
$$24 - \frac{7}{2}(5y + 1) = 3y;$$
$$\therefore 48 - 35y - 7 = 6y;$$
$$\therefore 41 = 41y;$$
$$\therefore y = 1,$$
and from (1) $x = 3.$

104. Any one of the methods given above will be found sufficient; but there are certain arithmetical artifices which will frequently shorten the work.

Example 1. Solve
$$171x - 213y = 642 \quad \ldots\ldots(1),$$
$$114x - 326y = 244 \quad \ldots\ldots(2).$$

Noticing that 171 and 114 contain a common factor 57, we shall make the coefficients of x in the two equations equal to the *least common multiple* of 171 and 114 if we multiply (1) by 2 and (2) by 3.

Thus
$$342x - 426y = 1284,$$
$$342x - 978y = 732;$$
subtracting, $552y = 552;$
that is, $y = 1,$
and therefore from (1) $x = 5.$

Example 2. Solve $\quad 127x + 59y = 1928$ ....(1),
$\qquad\qquad\qquad\qquad 59x + 127y = 1792$............................(2).
By addition, $\qquad 186x + 186y = 3720$;
$\qquad\qquad\qquad\therefore\ x + y = 20$ ....(3).
Subtracting (2) from (1), $\ 68x - 68y = 136$;
$\qquad\qquad\qquad\therefore\ x - y = 2$ ....(4).

Thus, by an easy combination of (1) and (2), the problem is reduced to the solution of the equations (3) and (4). From these we obtain by addition $2x = 22$, and by subtraction $2y = 18$.

Therefore $\qquad\qquad x = 11,\ \text{and}\ y = 9$.

EXAMPLES XIII. a.

[Art. 421 *may be read in connection with these Examples.*]

Solve the equations:

1. $3x + 4y = 10,$
   $4x + y = 9.$
2. $x + 2y = 13,$
   $3x + y = 14,$
3. $4x + 7y = 29,$
   $x + 3y = 11.$
4. $2x - y = 9,$
   $3x - 7y = 19,$
5. $5x + 6y = 17,$
   $6x + 5y = 16.$
6. $2x + y = 10,$
   $7x + 8y = 53.$
7. $8x - y = 34,$
   $x + 8y = 53.$
8. $15x + 7y = 29,$
   $9x + 15y = 39.$
9. $14x - 3y = 39,$
   $6x + 17y = 35.$
10. $28x - 23y = 33,$
    $63x - 25y = 101.$
11. $35x + 17y = 86,$
    $56x - 13y = 17.$
12. $15x + 77y = 92,$
    $55x - 33y = 22.$
13. $5x - 7y = 0,$
    $7x + 5y = 74.$
14. $21x - 50y = 60,$
    $28x - 27y = 199.$
15. $39x - 8y = 99,$
    $52x - 15y = 80.$
16. $5x = 7y - 21,$
    $21x - 9y = 75.$
17. $6y - 5x = 18,$
    $12x - 9y = 0.$
18. $8x = 5y,$
    $13x = 8y + 1.$
19. $3x = 7y,$
    $12y = 5x - 1.$
20. $19x + 17y = 0,$
    $2x - y = 53.$
21. $93x + 15y = 123,$
    $15x + 93y = 201.$

105. We add a few cases in which, before proceeding to solve, it will be necessary to simplify the equations.

Example 1. Solve $5(x + 2y) - (3x + 11y) = 14$ ....(1),
$\qquad\qquad\qquad 7x - 9y - 3(x - 4y) = 38$ ....(2).
From (1) $\qquad 5x + 10y - 3x - 11y = 14$;
$\qquad\qquad\qquad\therefore\ 2x - y = 14$ ....(3).
From (2) $\qquad 7x - 9y - 3x + 12y = 38$;
$\qquad\qquad\qquad\therefore\ 4x + 3y = 38$ ....(4).
From (3) $\qquad 6x - 3y = 42.$

By addition $10x = 80$; whence $x = 8$. From (3) we obtain $y = 2$.

SIMULTANEOUS EQUATIONS.

Example 2. Solve $\quad 3x - \dfrac{y-5}{7} = \dfrac{4x-3}{2}$ ....(1).

$\qquad\qquad\qquad \dfrac{3y+4}{5} - \dfrac{1}{3}(2x-5) = y$ ....(2).

Clear of fractions. Thus
from (1) $\qquad 42x - 2y + 10 = 28x - 21;$
$\qquad\qquad \therefore\ 14x - 2y = -31$ ....(3).
From (2) $\qquad 9y + 12 - 10x + 25 = 15y;$
$\qquad\qquad \therefore\ 10x + 6y = 37$ ....(4).

Eliminating y from (3) and (4), we find that
$$x = -\dfrac{14}{13}.$$
Eliminating x from (3) and (4), we find that
$$y = \dfrac{207}{26}.$$

Note. Sometimes, as in the present instance, the value of the second unknown is more easily found by elimination than by substituting the value of the unknown already found.

EXAMPLES XIII. b.

Solve the equations:

1. $\dfrac{2x}{3} + y = 16,$
   $x + \dfrac{y}{4} = 14.$

2. $\dfrac{x}{5} + \dfrac{y}{2} = 5,$
   $x - y = 4.$

3. $\dfrac{5x}{6} - y = 3,$
   $x - \dfrac{5y}{6} = 8.$

4. $x - y = 5,$
   $\dfrac{x}{4} - \dfrac{y}{5} = 2.$

5. $\dfrac{x}{9} + \dfrac{y}{7} = 10,$
   $\dfrac{x}{3} + y = 50.$

6. $x = 3y,$
   $\dfrac{x}{3} + y = 34.$

7. $\dfrac{2}{5}x - \dfrac{1}{12}y = 3,$
   $4x - y = 20.$

8. $\dfrac{1}{2}x - \dfrac{1}{5}y = 4,$
   $\dfrac{1}{7}x + \dfrac{1}{15}y = 3.$

9. $2x + y = 0,$
   $\dfrac{1}{2}y - 3x = 8.$

10. $\dfrac{x}{7} + \dfrac{y}{5} = 1\dfrac{3}{7},$
    $x + \dfrac{y}{3} = 4\dfrac{2}{3}.$

11. $3x - 7y = 0,$
    $\dfrac{2}{7}x + \dfrac{5}{3}y = 7.$

12. $\dfrac{x}{5} - \dfrac{y}{4} = 0,$
    $3x + \dfrac{1}{2}y = 17.$

Solve the equations:

13. $\dfrac{x}{3}+\dfrac{y}{4}=3x-7y-37=0.$

14. $\dfrac{x+1}{10}=\dfrac{3y-5}{2}=\dfrac{x-y}{8}.$

15. $\dfrac{x+3}{5}=\dfrac{8-y}{4}=\dfrac{3(x+y)}{8}.$

16. $\dfrac{x}{13}-\dfrac{y}{7}=6x-10y-8=0.$

106. In order to solve simultaneous equations which contain two unknown quantities we have seen that we must have two equations. Similarly we find that in order to solve simultaneous equations which contain three unknown quantities we must have three equations.

Rule. *Eliminate one of the unknowns from any pair of the equations, and then eliminate the same unknown from another pair. Two equations involving two unknowns are thus obtained, which may be solved by the rules already given. The remaining unknown is then found by substituting in any one of the given equations.*

Example 1. Solve
$$6x+2y-5z=13 \quad \ldots\ldots\ldots\ldots\ldots\ldots(1),$$
$$3x+3y-2z=13 \quad \ldots\ldots\ldots\ldots\ldots\ldots(2),$$
$$7x+5y-3z=26 \quad \ldots\ldots\ldots\ldots\ldots\ldots(3).$$

Choose y as the unknown to be eliminated.

Multiply (1) by 3 and (2) by 2,
$$18x+6y-15z=39,$$
$$6x+6y-4z=26;$$
subtracting, $\quad 12x-11z=13 \quad \ldots\ldots\ldots\ldots\ldots\ldots(4).$

Again, multiply (1) by 5 and (3) by 2,
$$30x+10y-25z=65,$$
$$14x+10y-6z=52;$$
subtracting, $\quad 16x-19z=13 \quad \ldots\ldots\ldots\ldots\ldots\ldots(5).$

Multiply (4) by 4 and (5) by 3,
$$48x-44z=52,$$
$$48x-57z=39;$$
subtracting, $\quad 13z=13,$
$$z=1,$$
and from (4) $\quad x=2,$
from (1) $\quad y=3.$

Note. After a little practice the student will find that the solution may often be considerably shortened by a suitable combination of the proposed equations. Thus, in the present instance, by adding (1) and (2) and subtracting (3) we obtain $2x-4z=0$, or $x=2z$. Substituting in (1) and (2) we have two easy equations in y and z.

SIMULTANEOUS EQUATIONS.

Some modification of the foregoing rule may often be used with advantage.

Example 2. Solve
$$\frac{x}{2}-1=\frac{y}{6}+1=\frac{z}{7}+2,$$
$$\frac{y}{3}+\frac{z}{2}=13.$$

From the equation $\frac{x}{2}-1=\frac{y}{6}+1,$

we have $\quad 3x-y=12$ ....(1).

Also, from the equation $\frac{x}{2}-1=\frac{z}{7}+2,$

we have $\quad 7x-2z=42$ ....(2).

And, from the equation $\frac{y}{3}+\frac{z}{2}=13,$

we have $\quad 2y+3z=78$ ....(3).

Eliminating z from (2) and (3), we have
$$21x+4y=282;$$
and from (1) $\quad 12x-4y=48;$
whence $\quad x=10,\ y=18.$

Also by substitution in (2) we obtain $z=14$.

Example 3. Consider the equations
$$5x-3y-\ z=6 \ \text{...........................(1)},$$
$$13x-7y+3z=14\text{...........................(2)},$$
$$7x-4y=8 \ \text{...........................(3)}.$$

Multiplying (1) by 3 and adding to (2), we have
$$28x-16y=32,$$
or $\quad 7x-\ 4y=8.$

Thus the combination of equations (1) and (2) leads us to an equation which is identical with (3), and so to find x and y we have but a single equation $7x-4y=8$, the solution of which is indeterminate. [Art. 100.]

In this and similar cases the anomaly arises from the fact that the equations are not *independent*; in other words, one equation is deducible from the others, and therefore contains no new relation between the unknown quantities which is not already implied in the other equations.

E.A.

EXAMPLES XIII. c.

Solve the equations:

1. $x+2y+2z=11,$
   $2x+y+z=7,$
   $3x+4y+z=14.$

2. $x+3y+4z=14,$
   $x+2y+z=7,$
   $2x+y+2z=2.$

3. $x+4y+3z=17,$
   $3x+3y+z=16,$
   $2x+2y+z=11.$

4. $3x-2y+z=2,$
   $2x+3y-z=5,$
   $x+y+z=6.$

5. $2x+y+z=16,$
   $x+2y+z=9,$
   $x+y+2z=3.$

6. $x-2y+3z=2,$
   $2x-3y+z=1,$
   $3x-y+2z=9.$

7. $3x+2y-z=20,$
   $2x+3y+6z=70,$
   $x-y+6z=41.$

8. $2x+3y+4z=20,$
   $3x+4y+5z=26,$
   $3x+5y+6z=31.$

9. $3x-4y=6z-10,\ 4x\ y\ x=5,\ x=3y+2(z-1).$

10. $5x+2y=14,\ y-6z=-15,\ x+2y+z=0.$

11. $x-\dfrac{y}{5}=6,\ y-\dfrac{z}{7}=8,\ z-\dfrac{x}{2}=10.$

12. $\dfrac{y+z}{4}=\dfrac{z+x}{3}=\dfrac{x+y}{2},\ x+y+z=27.$

13. $\dfrac{y-z}{3}=\dfrac{y-x}{2}=5z-4x,\ y+z=2x+1.$

14. $2x+3y=5,\ 2z-y=1,\ 7x-9z=3.$

15. $\dfrac{1}{2}(x+z-5)=y-z$
    $=2x-11$
    $=9-(x+2z).$

16. $x+20=\dfrac{3y}{2}+10$
    $=2z+5$
    $=110-(y+z).$

*107. DEFINITION. If the product of two quantities be equal to unity, each is said to be the **reciprocal** of the other. Thus if $ab=1$, a and b are **reciprocals**. They are so called because $a=\dfrac{1}{b}$, and $b=\dfrac{1}{a}$; and consequently a is related to b exactly as b is related to a.

The reciprocals of x and y are $\dfrac{1}{x}$ and $\dfrac{1}{y}$ respectively, and in solving the following equations we consider $\dfrac{1}{x}$ and $\dfrac{1}{y}$ as the unknown quantities.

SIMULTANEOUS EQUATIONS.

Example 1. Solve
$$\frac{8}{x} - \frac{9}{y} = 1 \quad \dots (1),$$
$$\frac{10}{x} + \frac{6}{y} = 7 \quad \dots (2).$$

Multiply (1) by 2 and (2) by 3; thus
$$\frac{16}{x} - \frac{18}{y} = 2,$$
$$\frac{30}{x} + \frac{18}{y} = 21;$$

adding,
$$\frac{46}{x} = 23;$$

multiplying up,
$$46 = 23x,$$
$$x = 2;$$

and by substituting in (1), $y = 3$.

Example 2. Solve
$$\frac{1}{2x} + \frac{1}{4y} - \frac{1}{3z} = \frac{1}{4} \quad \dots (1),$$
$$\frac{1}{x} = \frac{1}{3y} \quad \dots (2),$$
$$\frac{1}{x} - \frac{1}{5y} + \frac{4}{z} = 2\frac{2}{15} \quad \dots (3),$$

clearing of fractional coefficients, we obtain

from (1)
$$\frac{6}{x} + \frac{3}{y} - \frac{4}{z} = 3 \quad \dots (4),$$

from (2)
$$\frac{3}{x} - \frac{1}{y} = 0 \quad \dots (5),$$

from (3)
$$\frac{15}{x} - \frac{3}{y} + \frac{60}{z} = 32 \quad \dots (6).$$

Multiply (4) by 15 and add the result to (6); we have
$$\frac{105}{x} + \frac{42}{y} = 77;$$

dividing by 7,
$$\frac{15}{x} + \frac{6}{y} = 11 \quad \dots (7);$$

from (5)
$$\frac{18}{x} - \frac{6}{y} = 0;$$

$$\therefore \frac{33}{x} = 11;$$

$$\therefore x = 3,$$
from (5) $\quad y = 1,$
from (4) $\quad z = 2.$

*EXAMPLES. XIII. d.

Solve the equations:

1. $\dfrac{5}{x}+\dfrac{6}{y}=3,$
   $\dfrac{15}{x}+\dfrac{3}{y}=4.$

2. $\dfrac{6}{x}-\dfrac{7}{y}=2,$
   $\dfrac{2}{x}+\dfrac{14}{y}=3.$

3. $\dfrac{12}{x}-\dfrac{4}{y}=2,$
   $\dfrac{3}{x}-\dfrac{2}{y}=0.$

4. $\dfrac{5}{x}+\dfrac{16}{y}=79,$
   $\dfrac{16}{x}-\dfrac{1}{y}=44.$

5. $\dfrac{21}{x}+\dfrac{12}{y}=5,$
   $\dfrac{1}{y}-\dfrac{1}{x}=\dfrac{1}{42}.$

6. $\dfrac{5}{x}+\dfrac{3}{y}=30,$
   $\dfrac{9}{x}-\dfrac{5}{y}=2.$

7. $\dfrac{8}{x}-\dfrac{9}{y}=7,$
   $6\left(\dfrac{1}{x}+\dfrac{1}{y}\right)=1.$

8. $\dfrac{25}{x}+\dfrac{24}{y}=1,$
   $20\left(\dfrac{2}{x}+\dfrac{3}{y}\right)=7.$

9. $\dfrac{4}{x}+\dfrac{27}{y}=42,$
   $\dfrac{14}{x}-\dfrac{15}{y}=1.$

10. $\dfrac{3}{x}+\dfrac{5}{y}=\dfrac{8}{15},$
    $9y-22x=\dfrac{3xy}{25}.$

11. $\dfrac{1}{4x}+\dfrac{1}{3y}=2,$
    $\dfrac{1}{y}-\dfrac{1}{2x}=1.$

12. $2y-x=4xy,$
    $\dfrac{4}{y}-\dfrac{3}{x}=9.$

13. $\dfrac{1}{x}-\dfrac{2}{y}+4=0,$
    $\dfrac{1}{y}-\dfrac{1}{z}+1=0,$
    $\dfrac{2}{z}+\dfrac{3}{x}=14.$

14. $\dfrac{1}{x}+\dfrac{1}{y}+\dfrac{1}{z}=36,$
    $\dfrac{1}{x}+\dfrac{3}{y}-\dfrac{1}{z}=28,$
    $\dfrac{1}{x}+\dfrac{1}{3y}+\dfrac{1}{2z}=20.$

15. $\dfrac{9}{x}-\dfrac{2}{y}=\dfrac{5}{z}-\dfrac{3}{x}=\dfrac{7}{y}+\dfrac{15}{2z}=4.$

CHAPTER XIV.

PROBLEMS LEADING TO SIMULTANEOUS EQUATIONS.

108. IN the Examples discussed in the last chapter we have seen that it is essential to have as many equations as there are unknown quantities to determine. Consequently in the solution of problems which give rise to simultaneous equations, it will always be necessary that the statement of the question should contain as many independent conditions as there are quantities to be determined.

Example 1. Find two numbers whose difference is 11, and one-fifth of whose sum is 9.

Let x be the greater number, y the less;

Then $\qquad x-y=11 \qquad\qquad\qquad$ (1),

Also $\qquad \dfrac{x+y}{5}=9,$

or $\qquad x+y=45 \qquad\qquad\qquad$ (2).

By addition $2x=56$; and by subtraction $2y=34$.
The numbers are therefore 28 and 17.

Example 2. If 15 lbs. of tea and 17 lbs. of coffee together cost £3. 5s. 6d., and 25 lbs. of tea and 13 lbs. of coffee together cost £4. 6s. 2d.; find the price of each per pound.

Suppose a pound of tea to cost x shillings,
and .... coffee ...... y ....

Then from the question, we have

$\qquad 15x+17y=65\tfrac{1}{2} \qquad\qquad\qquad$ (1),
$\qquad 25x+13y=86\tfrac{1}{6} \qquad\qquad\qquad$ (2).

Multiplying (1) by 5 and (2) by 3, we have

$\qquad 75x+85y=327\tfrac{1}{2},$
$\qquad 75x+39y=258\tfrac{1}{2}.$

Subtracting, $\qquad 46y=69,$
$\qquad\qquad y=1\tfrac{1}{2}.$

And from (1) $\qquad 15x+25\tfrac{1}{2}=65\tfrac{1}{2},$
whence $\qquad\qquad 15x=40;$
$\qquad\qquad \therefore \ x=2\tfrac{2}{3}.$

∴ the cost of a pound of tea is $2\tfrac{2}{3}$ shillings, or 2s. 8d.,
and the cost of a pound of coffee is $1\tfrac{1}{2}$ shillings, or 1s. 6d.

Example 3. A person spent 15s. 2d. in buying oranges at the rate of 3 for twopence and apples at fivepence a dozen; if he had bought five times as many oranges and a quarter of the number of apples he would have spent £2. 4s. 2d. How many of each did he buy?

Let x be the number of oranges, and y the number of apples.

$$x \text{ oranges cost } \frac{2x}{3} \text{ pence,}$$

$$y \text{ apples cost } \frac{5y}{12} \text{ pence,}$$

$$\therefore \quad \frac{2x}{3} + \frac{5y}{12} = 182 \quad\quad\quad\quad\quad\quad (1).$$

Again, $5x$ oranges cost $5x \times \frac{2}{3}$, or $\frac{10x}{3}$ pence,

and $\frac{y}{4}$ apples cost $\frac{y}{4} \times \frac{5}{12}$, or $\frac{5y}{48}$ pence,

$$\therefore \quad \frac{10x}{3} + \frac{5y}{48} = 530 \quad\quad\quad\quad\quad\quad (2).$$

Multiply (1) by 5 and subtract (2) from the result;

$$\therefore \quad \left(\frac{25}{12} - \frac{5}{48}\right) y = 380\,;$$

$$\therefore \quad \frac{95y}{48} = 380\,;$$

$$\therefore \quad y = 192,$$

and from (1) $\quad\quad\quad\quad\quad x = 153.$

Thus there were 153 oranges and 192 apples.

Example 4. If the numerator of a fraction is increased by 2 and the denominator by 1, it becomes equal to $\frac{5}{8}$; and, if the numerator and denominator are each diminished by 1, it becomes equal to $\frac{1}{2}$: find the fraction.

Let x be the numerator of the fraction, y the denominator; then the fraction is $\frac{x}{y}$.

From the first supposition,

$$\frac{x+2}{y+1} = \frac{5}{8} \quad\quad\quad\quad\quad\quad (1),$$

from the second,

$$\frac{x-1}{y-1} = \frac{1}{2} \quad\quad\quad\quad\quad\quad (2).$$

These equations give $x = 8$, $y = 15$.

Thus the fraction is $\frac{8}{15}$.

XIV.] PROBLEMS LEADING TO SIMULTANEOUS EQUATIONS. 95

Example 5. The middle digit of a number between 100 and 1000 is zero, and the sum of the other digits is 11. If the digits be reversed, the number so formed exceeds the original number by 495; find it.

Let x be the digit in the units' place,

y .... hundreds' place;

then, since the digit in the tens' place is 0, the number will be represented by $100y + x$. [Art. 81, Ex. 4.]

And if the digits are reversed the number so formed will be represented by $100x + y$.

$$\therefore 100x + y - (100y + x) = 495,$$

or $$100x + y - 100y - x = 495;$$

$$\therefore 99x - 99y = 495,$$

that is, $$x - y = 5 \quad \text{......................(1)},$$

Again, since the sum of the digits is 11, and the middle one is 0, we have $$x + y = 11 \quad \text{......................(2)}.$$

From (1) and (2) we find $x = 8$, $y = 3$.

Hence the number is 308.

EXAMPLES XIV.

1. Find two numbers whose sum is 34, and whose difference is 10.

2. The sum of two numbers is 73, and their difference is 37; find the numbers.

3. One third of the sum of two numbers is 14, and one half of their difference is 4; find the numbers.

4. One nineteenth of the sum of two numbers is 4, and their difference is 30; find the numbers.

5. Half the sum of two numbers is 20, and three times their difference is 18; find the numbers.

6. Six pounds of tea and eleven pounds of sugar cost £1. 3s. 8d., and eleven pounds of tea and six pounds of sugar cost £1. 18s. 8d. Find the cost of tea and sugar per pound.

7. Six horses and seven cows can be bought for £250, and thirteen cows and eleven horses can be bought for £461. What is the value of each animal?

8. A, B, C, D have £290 between them; A has twice as much as C, and B has three times as much as D; also C and D together have £50 less than A. Find how much each has.

9. A, B, C, D have £270 between them; A has three times as much as C, and B five times as much as D; also A and B together have £50 less than eight times what C has. Find how much each has.

10. Four times B's age exceeds A's age by twenty years, and one third of A's age is less than B's age by two years: find their ages.

11. One eleventh of A's age is greater by two years than one seventh of B's, and twice B's age is equal to what A's age was thirteen years ago: find their ages.

12. In eight hours A walks twelve miles more than B does in seven hours; and in thirteen hours B walks seven miles more than A does in nine hours. How many miles does each walk per hour?

13. In eleven hours C walks $12\frac{1}{2}$ miles less than D does in twelve hours; and in five hours D walks $3\frac{1}{4}$ miles less than C does in seven hours. How many miles does each walk per hour?

14. Find a fraction such that if 1 be added to its denominator it reduces to $\frac{1}{2}$, and reduces to $\frac{3}{5}$ on adding 2 to its numerator.

15. Find a fraction which becomes $\frac{1}{2}$ on subtracting 1 from the numerator and adding 2 to the denominator, and reduces to $\frac{1}{3}$ on subtracting 7 from the numerator and 2 from the denominator.

16. If 1 be added to the numerator of a fraction it reduces to $\frac{1}{5}$; if 1 be taken from the denominator it reduces to $\frac{1}{7}$: required the fraction.

17. If $\frac{2}{3}$ be added to the numerator of a certain fraction the fraction will be increased by $\frac{1}{21}$, and if $\frac{1}{2}$ be taken from its denominator the fraction becomes $\frac{2}{9}$: find it.

18. The sum of a number of two digits and of the number formed by reversing the digits is 110, and the difference of the digits is 6: find the numbers.

19. The sum of the digits of a number is 13, and the difference between the number and that formed by reversing the digits is 27: find the numbers.

20. A certain number of two digits is three times the sum of its digits, and if 45 be added to it the digits will be reversed: find the number.

21. A certain number between 10 and 100 is eight times the sum of its digits, and if 45 be subtracted from it the digits will be reversed: find the number.

22. A man has a number of pounds and shillings, and he observes that if the pounds were turned into shillings and the shillings into pounds he would gain £5. 14s.; but if the pounds were turned into half-sovereigns and the shillings into half-crowns he would lose £1. 13s. 6d. What sum has he?

23. In a bag containing black and white balls, half the number of white is equal to a third of the number of black; and twice the whole number of balls exceeds three times the number of black balls by four. How many balls did the bag contain?

XIV.] PROBLEMS LEADING TO SIMULTANEOUS EQUATIONS. 97

24. A number consists of three digits, the right-hand one being zero. If the left-hand and middle digits be interchanged the number is diminished by 180; if the left-hand digit be halved and the middle and right-hand digits be interchanged the number is diminished by 454. Find the number.

25. The wages of 10 men and 8 boys amount to £1. 17s.; if 4 men together receive 1s. more than 6 boys, what are the wages of each man and boy?

26. A grocer wishes to mix spice at 8s. a pound with another sort at 5s. a pound to make 60 pounds to be sold at 6s. a pound: what quantity of each must he take?

27. A traveller walks a certain distance; had he gone half a mile an hour faster, he would have walked it in four-fifths of the time: had he gone half a mile an hour slower, he would have been $2\frac{1}{2}$ hours longer on the road. Find the distance.

28. A man walks 35 miles partly at the rate of 4 miles an hour, and partly at 5; if he had walked at 5 miles an hour when he walked at 4, and vice versâ, he would have covered two miles more in the same time. Find the time he was walking.

29. Two persons, 27 miles apart, setting out at the same time are together in 9 hours if they walk in the same direction, but in 3 hours if they walk in opposite directions: find their rates of walking.

30. A family, consisting of three adults and five children, spends in food £1. 17s. 6d. a week. Distress, however, comes when they can afford only £1 per week, and the food of each adult is diminished by one-half, and of each child by one-third. Find the cost per week of an adult and of a child.

31. If I lend a sum of money at 6 per cent., the interest for a certain time exceeds the loan by £100; but if I lend it at 3 per cent., for a fourth of the time, the loan exceeds its interest by £425. How much do I lend?

32. A takes 3 hours longer than B to walk 30 miles; but if he doubles his pace he takes 2 hours less time than B: find their rates of walking.

CHAPTER XV.

INVOLUTION.

[*Arts. 41–45 should be studied here by those who have adopted the postponement suggested on page* 33.]

109. DEFINITION. Involution is the general name for multiplying an expression by itself so as to find its second, third, fourth, or any other power.

Involution may always be effected by actual multiplication. Here, however, we shall give some rules for writing down at once

(1) any power of a simple expression;

(2) the square and cube of any binomial;

(3) the square of any multinomial.

110. It is evident from the Rule of Signs that

(1) no *even* power of *any* quantity can be *negative*;

(2) any *odd* power of a quantity will have *the same sign* as the quantity itself.

Note. It is especially worthy of notice that the *square* of every expression, whether positive or negative, is *positive*.

111. From definition we have, by the rules of multiplication,

$$(a^2)^3 = a^2 \cdot a^2 \cdot a^2 = a^{2+2+2} = a^6.$$
$$(-x^3)^2 = (-x^3)(-x^3) = x^{3+3} = x^6.$$
$$(-a^5)^3 = (-a^5)(-a^5)(-a^5) = -a^{5+5+5} = -a^{15}.$$
$$(-3a^3)^4 = (-3)^4(a^3)^4 = 81a^{12}.$$

Hence we obtain a rule for raising a simple expression to any proposed power.

Rule. (1) *Raise the coefficient to the required power by Arithmetic, and prefix the proper sign found by the Rule of Signs.*

(2) *Multiply the index of every factor of the expression by the exponent of the power required.*

CHAP. XV.] INVOLUTION. 99

Examples.
$$(-2x^2)^5 = -32x^{10}.$$
$$(-3ab^3)^6 = 729a^6b^{18}.$$
$$\left(\frac{2ab^3}{3x^2y}\right)^4 = \frac{16a^4b^{12}}{81x^8y^4}.$$

It will be seen that in the last case the numerator and the denominator are operated upon separately.

EXAMPLES XV. a.

Write down the square of each of the following expressions:

1. $3ab^3$.
2. a^3c.
3. $7ab^2$.
4. $11b^2c^3$.
5. $4a^4b^5x^2$.
6. $5x^2y^5$.
7. $-2abc^2$.
8. $-3cx^3$.
9. $4xyz^3$.
10. $-\frac{2}{3}a^2b^3$.
11. $\frac{2x^2}{3y^3}$.
12. $-\frac{4}{3x^2y}$.
13. $-\frac{7ab}{3}$.
14. $\frac{3a^2b^3}{4c^5x^4}$.
15. $-\frac{1}{2xy}$.
16. $-2xy^2$.
17. $\frac{5ab^3}{2xy}$.
18. $13c^5x^3$.
19. $-\frac{1}{4a^4}$.
20. $-\frac{3a^5}{5x^3}$.

Write down the cube of each of the following expressions:

21. $2ab^2$.
22. $3x^3$.
23. $4x^4$.
24. $-3a^3b$.
25. $-5ab^2$.
26. $-b^3c^2x$.
27. $-6a^6$.
28. $-2a^7c^2$.
29. $\frac{1}{3y^2}$.
30. $-\frac{3x^5}{5a^3}$.
31. $7x^3y^4$.
32. $-\frac{2}{3}a^5$.

Write down the value of each of the following expressions:

33. $(3a^2b^3)^4$.
34. $(-a^2x)^6$.
35. $(-2x^3y)^5$.
36. $\left(\frac{1}{2a^2}\right)^7$.
37. $\left(\frac{3x^4}{2y^3}\right)^5$.
38. $\left(\frac{2x^3}{3y}\right)^8$.
39. $\left(-\frac{x^3}{3}\right)^7$.
40. $\left(-\frac{2x^5}{3a^4}\right)^6$.

112. By multiplication we have
$$(a+b)^2 = (a+b)(a+b)$$
$$= a^2 + 2ab + b^2 \dots\dots\dots\dots\dots(1),$$
$$(a-b)^2 = (a-b)(a-b)$$
$$= a^2 - 2ab + b^2 \dots\dots\dots\dots\dots(2).$$

These results are embodied in the following rules:

Rule 1. *The square of the sum of two quantities is equal to the sum of their squares increased by twice their product.*

Rule 2. *The square of the difference of two quantities is equal to the sum of their squares diminished by twice their product.*

Example 1. $(x+2y)^2 = x^2 + 2 \cdot x \cdot 2y + (2y)^2$
$= x^2 + 4xy + 4y^2.$

Example 2. $(2a^3 - 3b^2)^2 = (2a^3)^2 - 2 \cdot 2a^3 \cdot 3b^2 + (3b^2)^2$
$= 4a^6 - 12a^3b^2 + 9b^4.$

113. These rules may sometimes be conveniently applied to find the squares of numerical quantities.

Example 1. The square of $1012 = (1000 + 12)^2$
$= (1000)^2 + 2 \cdot 1000 \cdot 12 + (12)^2$
$= 1000000 + 24000 + 144$
$= 1024144.$

Example 2. The square of $98 = (100 - 2)^2$
$= (100)^2 - 2 \cdot 100 \cdot 2 + (2)^2$
$= 10000 - 400 + 4$
$= 9604.$

The work is considerably shortened by the omission of the first two steps.

114. We may now extend the rules of Art 112 thus:

$(a+b+c)^2 = \{(a+b)+c\}^2$
$= (a+b)^2 + 2(a+b)c + c^2$ [Art. 112. Rule 1.]
$= a^2 + b^2 + c^2 + 2ab + 2ac + 2bc.$

In the same way we may prove

$(a-b+c)^2 = a^2 + b^2 + c^2 - 2ab + 2ac - 2bc$
$(a+b+c+d)^2 = a^2 + b^2 + c^2 + d^2 + 2ab + 2ac + 2ad + 2bc + 2bd + 2cd.$

In each of these instances we observe that the square consists of

(1) the sum of the squares of the several terms of the given expression;

(2) twice the sum of the products two and two of the several terms, taken with their proper signs; that is, in each product the sign is + or − according as the quantities composing it have like or unlike signs.

Note. The *square terms* are always positive.

The same laws hold whatever be the number of terms in the expression to be squared.

Rule. *To find the square of any multinominal: to the sum of the squares of the several terms add twice the product (with the proper sign) of each term into each of the terms that follow it.*

INVOLUTION.

Ex. 1. $(x-2y-3z)^2 = x^2+4y^2+9z^2-2.x.2y-2.x.3z+2.2y.3z$
$= x^2+4y^2+9z^2-4xy-6xz+12yz$.

Ex. 2. $(1+2x-3x^2)^2 = 1+4x^2+9x^4+2.1.2x-2.1.3x^2-2.2x.3x^2$
$= 1+4x^2+9x^4+4x-6x^2-12x^3$
$= 1+4x-2x^2-12x^3+9x^4$,

by collecting like terms and rearranging.

EXAMPLES XV. b.

Write down the square of each of the following expressions:

1. $a+3b$.
2. $a-3b$.
3. $x-5y$.
4. $2x+3y$.
5. $3x-y$.
6. $3x+5y$.
7. $9x-2y$.
8. $5ab-c$.
9. $pq-r$.
10. $x-abc$.
11. $ax+2by$.
12. x^2-1.
13. $a-b-c$.
14. $a+b-c$.
15. $a+2b+c$.
16. $2a-3b+4c$.
17. $x^2-y^2-z^2$.
18. $xy+yz+zx$.
19. $3p-2q+4r$.
20. x^2-x+1.
21. $2x^2+3x-1$.
22. $x-y+a-b$.
23. $2x+3y+a-2b$.
24. $m-n-p-q$.
25. $\frac{1}{2}a-2b+\frac{c}{4}$.
26. $\frac{a}{3}-3b-\frac{3}{2}$.
27. $\frac{2}{3}x^2-x+\frac{3}{2}$.

115. By actual multiplication we have
$$(a+b)^3 = (a+b)(a+b)(a+b)$$
$$= a^3+3a^2b+3ab^2+b^3.$$
Also $(a-b)^3 = a^3-3a^2b+3ab^2-b^3$.

By observing the law of formation of the terms in these results we can write down the cube of any binomial.

Example 1. $(2x+y)^3 = (2x)^3+3(2x)^2y+3(2x)y^2+y^3$
$= 8x^3+12x^2y+6xy^2+y^3$.

Example 2. $(3x-2a^2)^3 = (3x)^3-3(3x)^2(2a^2)+3(3x)(2a^2)^2-(2a^2)^3$
$= 27x^3-54x^2a^2+36xa^4-8a^6$.

EXAMPLES XV c.

Write down the cube of each of the following expressions:

1. $x+a$.
2. $x-a$.
3. $x-2y$.
4. $2x+y$.
5. $3x-5y$.
6. $ab+c$.
7. $2ab-3c$.
8. $5a-bc$.
9. x^2+4y^2.
10. $4x^2-5y^2$.
11. $2a^3-3b^2$.
12. $5x^5-4y^4$.
13. $a-\frac{2b}{3}$.
14. $\frac{a}{3}+2$.
15. $\frac{x^2}{3}-3x$.
16. $\frac{a}{6}+2x$.

CHAPTER XVI.

EVOLUTION.

[Arts. 51–54 *should be studied here by those who have adopted the postponement suggested on page* 38.]

116. Definition. The **root** of any proposed expression is that quantity which being multiplied by itself the requisite number of times produces the given expression.

The operation of finding the root is called **Evolution**: it is the reverse of Involution.

117. By the Rule of Signs we see that

(1) any *even* root of a *positive* quantity may be either *positive* or *negative*;

(2) *no negative* quantity can have an *even* root;

(3) every *odd* root of a quantity has the same sign as the quantity itself.

Note. It is especially worthy of remark that every positive quantity has two square roots equal in magnitude, but opposite in sign.

Example. $\sqrt{9a^2x^6} = \pm 3ax^3$.

In the present chapter, however, we shall confine our attention to the positive root.

Examples. $\sqrt{a^6b^4} = a^3b^2$, because $(a^3b^2)^2 = a^6b^4$.

$\sqrt[3]{-x^9} = -x^3$, because $(-x^3)^3 = -x^9$.

$\sqrt[5]{c^{20}} = c^4$, because $(c^4)^5 = c^{20}$.

$\sqrt[4]{81x^{12}} = 3x^3$, because $(3x^3)^4 = 81x^{12}$.

118. From the foregoing examples we may deduce a general rule for extracting any proposed root of a simple expression:

Rule. (1) *Find the root of the coefficient by Arithmetic, and prefix the proper sign.*

(2) *Divide the exponent of every factor of the expression by the index of the proposed root.*

Examples. $\sqrt[3]{-64x^6} = -4x^2$.

$\sqrt[4]{16a^8} = 2a^2$.

$\sqrt{\dfrac{81x^{10}}{25c^4}} = \dfrac{9x^5}{5c^2}$.

CHAP. XVI.] EVOLUTION. 103

EXAMPLES XVI. a.

Write down the square root of the following expressions:

1. $4a^2b^4$. 2. $9x^6y^2$. 3. $25x^4y^6$. 4. $16a^4b^2c^6$.
5. $81a^6b^8$. 6. $100x^8$. 7. $a^{20}b^{16}c^4$. 8. $a^8b^2c^{12}$.
9. $64x^6y^{18}$. 10. $\dfrac{36}{a^{36}}$. 11. $\dfrac{a^{16}b^8}{16}$. 12. $\dfrac{289y^4}{25}$.
13. $\dfrac{324x^{12}}{169y^6}$. 14. $\dfrac{81a^{18}}{36b^{12}}$. 15. $\dfrac{256x^2y^4}{289p^{14}}$. 16. $\dfrac{400a^{40}b^{20}}{81x^{10}y^{18}}$.

Write down the cube root of the following expressions:

17. $27a^6b^3c^3$. 18. $-8a^{12}b^9$. 19. $64x^6y^3z^{12}$. 20. $-343a^{12}b^{18}$.
21. $-\dfrac{x^{12}y^9}{125}$. 22. $\dfrac{8x^9}{729y^{15}}$. 23. $\dfrac{125a^3b^6}{216x^6y^9}$. 24. $-\dfrac{27x^{27}}{64y^{63}}$.

Write down the value of each of the following expressions:

25. $\sqrt[4]{(a^8x^{12})}$. 26. $\sqrt[7]{(x^{14}y^{21})}$. 27. $\sqrt[5]{(32x^5y^{10})}$.
28. $\sqrt[6]{(729a^{18}b^6)}$. 29. $\sqrt[8]{(256a^8x^{64})}$. 30. $\sqrt[5]{(-x^{10}y^{15})}$.
31. $\sqrt[7]{\dfrac{128}{a^{63}b^{56}}}$. 32. $\sqrt[10]{\dfrac{a^{30}x^{50}}{b^{100}}}$. 33. $\sqrt[9]{\dfrac{a^{18}}{b^{27}c^{36}}}$.

118$_A$. By Art. 112, we can write down the square of any binomial.

Thus $(2x+3y)^2 = (2x)^2 + 2 \cdot 2x \cdot 3y + (3y)^2$.

Conversely, by observing the form of the terms of an expression, its square root may often be written down at once.

Example 1. Find the square root of $25x^2 - 40xy + 16y^2$.

The expression $= (5x)^2 - 2 \cdot 20xy + (4y)^2$
$= (5x)^2 - 2(5x)(4y) + (4y)^2$
$= (5x - 4y)^2$.

Thus the required square root is $5x - 4y$.

Example 2. Find the square root of $\dfrac{64a^2}{9b^2} + 4 + \dfrac{32a}{3b}$.

The expression $= \left(\dfrac{8a}{3b}\right)^2 + (2)^2 + 2\left(\dfrac{16a}{3b}\right)$
$= \left(\dfrac{8a}{3b}\right)^2 + 2\left(\dfrac{8a}{3b}\right)(2) + (2)^2$
$= \left(\dfrac{8a}{3b} + 2\right)^2$.

Thus the required square root is $\dfrac{8a}{3b} + 2$.

Example 3.   Find the square root of $4a^2+b^2+c^2+4ab-4ac-2bc$.

Arrange the terms in descending powers of a, and the other letters alphabetically; then

$$\text{the expression} = 4a^2 + 4ab - 4ac + b^2 - 2bc + c^2$$
$$= 4a^2 + 4a(b-c) + (b-c)^2$$
$$= (2a)^2 + 2 \cdot 2a(b-c) + (b-c)^2$$
$$= \{2a + (b-c)\}^2;$$

whence the required square root is $2a+b-c$,

Or we might proceed as follows:

$$\text{the expression} = (2a)^2 + b^2 + c^2 + 2 \cdot (2a)b - 2 \cdot (2a)c - 2 \cdot b \cdot c,$$

which is evidently the square root of $2a+b-c$.   [Art. 114.]

119. When the square root cannot be easily determined by inspection we must have recourse to the rule we are about to explain, which is quite general, and applicable to all cases. *But the student is advised to use methods of inspection wherever possible, in preference to rules.*

Since the square of $a+b$ is $a^2+2ab+b^2$, we have to discover a process by which a and b, the terms of the root, can be found when $a^2+2ab+b^2$ is given.

Now $\qquad a^2 + 2ab + b^2 = a^2 + b(2a+b),$

so that the expression is made up of

(1) the *square* of the *first* term of the root, together with

(2) the *product* of the *second* term of the root into an expression consisting of *this second term added to twice the first term of the root.*

By reversing the process we arrive at the following method of working:

$$\begin{array}{r|l}
 & a^2 + 2ab + b^2\,(\,a+b \\
 & a^2 \\ \hline
2a+b & 2ab + b^2 \\
 & 2ab + b^2 \\ \hline
\end{array}$$

Explanation. (1) The terms are first arranged according to the powers of one letter a.

(2) The square root of a^2 is written down as *the first term of the root*, and its square subtracted from the given expression.

(3) The first term of the remainder is *divided by twice the first term of the root* to obtain the second term of the root, that is, b.

(4) *The second term of the root is added to twice the term already found* to form the complete divisor $2a+b$.

Example 1. Find the square root of $9x^2 - 42xy + 49y^2$.

$$9x^2 - 42xy + 49y^2\ (\ 3x - 7y$$
$$9x^2$$
$$6x - 7y\ \overline{\big|\ -42xy + 49y^2}$$
$$-42xy + 49y^2$$

Explanation. The square root of $9x^2$ is $3x$, and this is the first term of the root.

By doubling this we obtain $6x$, which is the first term of the divisor. Dividing $-42xy$, the first term of the remainder, by $6x$ we get $-7y$, the new term in the root, which has to be annexed both to the root and divisor. We next multiply the complete divisor by $-7y$ and subtract the result from the first remainder. There is now no remainder, and the root has been found.

The rule can be extended so as to find the square root of any multinomial. The first two terms of the root will be obtained as before. When we have brought down *the second remainder*, the first part of the new divisor is obtained by doubling the terms of the root already found. We then divide the first term of the remainder by the first term of the new divisor, and set down the result as the next term in the root and in the divisor. We next multiply the complete divisor by the last term of the root and subtract the product from the last remainder. If there is now no remainder the root has been found; if there is a remainder we continue the process.

Example 2. Find the square root of
$$25x^2a^2 - 12xa^3 + 16x^4 + 4a^4 - 24x^3a.$$

Rearrange in descending powers of x.

$$16x^4 - 24x^3a + 25x^2a^2 - 12xa^3 + 4a^4\ (\ 4x^2 - 3xa + 2a^2$$
$$16x^4$$

$8x^2 - 3xa$ $\quad\overline{\big|\ -24x^3a + 25x^2a^2}$
$-24x^3a + 9x^2a^2$

$8x^2 - 6xa + 2a^2\qquad \overline{\big|\ 16x^2a^2 - 12xa^3 + 4a^4}$
$16x^2a^2 - 12xa^3 + 4a^4$

Explanation. When we have obtained two terms in the root, $4x^2 - 3xa$, we have a remainder
$$16x^2a^2 - 12xa^3 + 4a^4.$$

Doubling the terms of the root already found, we place the result, $8x^2 - 6xa$, as the first part of the divisor. Dividing $16x^2a^2$, the first term of the remainder, by $8x^2$, the first term of the divisor, we get $+2a^2$, which we annex both to the root and divisor. We now multiply the complete divisor by $2a^2$ and subtract. There is no remainder, and the root is found.

E.A. H

EXAMPLES XVI. b.

Find the square root of each of the following expressions:

1. $x^2 + 4xy + 4y^2$.
2. $9a^2 + 12ab + 4b^2$.
3. $x^2 - 10xy + 25y^2$.
4. $4x^2 - 12xy + 9y^2$.
5. $81x^2 + 18xy + y^2$.
6. $25x^2 - 30xy + 9y^2$.
7. $x^4 - 2x^2y^2 + y^4$.
8. $1 - 2a^3 + a^6$.
9. $a^4 - 2a^3 + 3a^2 - 2a + 1$.
10. $4x^4 - 12x^3 + 29x^2 - 30x + 25$.
11. $9x^4 - 12x^3 - 2x^2 + 4x + 1$.
12. $x^4 - 4x^3 + 6x^2 - 4x + 1$.
13. $4a^4 + 4a^3 - 7a^2 - 4a + 4$.
14. $1 - 10x + 27x^2 - 10x^3 + x^4$.
15. $4x^2 + 9y^2 + 25z^2 + 12xy - 30yz - 20xz$.
16. $16x^6 + 16x^7 - 4x^8 - 4x^9 + x^{10}$.
17. $x^6 - 22x^4 + 34x^3 + 121x^2 - 374x + 289$.
18. $25x^4 - 30ax^3 + 49a^2x^2 - 24a^3x + 16a^4$.
19. $4x^4 + 4x^2y^2 - 12x^2z^2 + y^4 - 6y^2z^2 + 9z^4$.
20. $6ab^2c - 4a^2bc + a^2b^2 + 4a^2c^2 + 9b^2c^2 - 12abc^2$.
21. $-6b^2c^2 + 9c^4 + b^4 - 12c^2a^2 + 4a^4 + 4a^2b^2$.
22. $4x^4 + 9y^4 + 13x^2y^2 - 6xy^3 - 4x^3y$.
23. $67x^2 + 49 + 9x^4 - 70x - 30x^3$.
24. $1 - 4x + 10x^2 - 20x^3 + 25x^4 - 24x^5 + 16x^6$.
25. $6acx^5 + 4b^2x^4 + a^2x^{10} + 9c^2 - 12bcx^2 - 4abx^7$.

[*If preferred, the remainder of this chapter may be postponed and taken after* Chap. XXIV.]

*120. When the expression whose root is required contains fractional terms, we may proceed as before, the fractional part of the work being performed by the rules explained in Chap. XII.

*121. There is one important point to be observed when an expression contains powers of a certain letter and also powers of its reciprocal. Thus in the expression

$$2x + \frac{1}{x^2} + 4 + x^3 + \frac{5}{x} + 7x^2 + \frac{8}{x^3},$$

the order of *descending* powers is

$$x^3 + 7x^2 + 2x + 4 + \frac{5}{x} + \frac{1}{x^2} + \frac{8}{x^3};$$

and the numerical quantity 4 stands between x and $\frac{1}{x}$.

The reason for this arrangement will appear in Chap. XXX.

Example. Find the square root of $24 + \dfrac{16y^2}{x^2} - \dfrac{8x}{y} + \dfrac{x^2}{y^2} - \dfrac{32y}{x}$.

Arrange the expression in descending powers of y.

$$\dfrac{16y^2}{x^2} - \dfrac{32y}{x} + 24 - \dfrac{8x}{y} + \dfrac{x^2}{y^2} \left(\dfrac{4y}{x} - 4 + \dfrac{x}{y} \right.$$

$$\dfrac{16y^2}{x^2}$$

$\dfrac{8y}{x} - 4 \quad \Big| -\dfrac{32y}{x} + 24$

$\qquad\qquad\Big| -\dfrac{32y}{x} + 16$

$\dfrac{8y}{x} - 8 + \dfrac{x}{y} \quad \Big| \ 8 - \dfrac{8x}{y} + \dfrac{x^2}{y^2}$

$\qquad\qquad\qquad\Big| \ 8 - \dfrac{8x}{y} + \dfrac{x^2}{y^2}$

Here the second term in the root, -4, arises from division of $-\dfrac{32y}{x}$ by $\dfrac{8y}{x}$, and the third term, $\dfrac{x}{y}$, arises from division of 8 by $\dfrac{8y}{x}$; thus $8 \div \dfrac{8y}{x} = 8 \times \dfrac{x}{8y} = \dfrac{x}{y}$.

*EXAMPLES XVI. c.

Find the square root of each of the following expressions:

1. $\dfrac{x^2}{4} - 3x + 9$.

2. $4 - \dfrac{4x}{y} + \dfrac{x^2}{y^2}$.

3. $\dfrac{x^2}{25} + \dfrac{2xy}{5} + y^2$.

4. $\dfrac{x^2}{y^2} + \dfrac{10x}{y} + 25$.

5. $\dfrac{x^2}{4y^2} - \dfrac{2x}{y} + 4$.

6. $\dfrac{x^2}{y^2} - \dfrac{2ax}{by} + \dfrac{a^2}{b^2}$.

7. $\dfrac{64x^2}{9y^2} + \dfrac{32x}{3y} + 4$.

8. $\dfrac{9x^2}{25} - 2 + \dfrac{25}{9x^2}$.

9. $\dfrac{a^4}{64} + \dfrac{a^3}{8} - a + 1$.

10. $x^4 + 2x^3 - x + \dfrac{1}{4}$.

11. $-3a^3 + \dfrac{25}{9} + a^4 - 5a + \dfrac{67}{12}a^2$.

12. $x^4 - 2x + \dfrac{1}{9} + \dfrac{29}{3}x^2 - 6x^3$.

13. $\dfrac{a^4}{4} + \dfrac{a^3}{x} + \dfrac{a^2}{x^2} - ax - 2 + \dfrac{x^2}{a^2}$.

14. $x^4 - 2x^3 + \dfrac{3x^2}{2} - \dfrac{x}{2} + \dfrac{1}{16}$.

15. $\dfrac{x^4}{4} + 4x^2 + \dfrac{ax^2}{3} + \dfrac{a^2}{9} - 2x^3 - \dfrac{4ax}{3}$.

Find the square root of each of the following expressions:

16. $\dfrac{9a^2}{x^2} - \dfrac{6a}{5x} + \dfrac{101}{25} - \dfrac{4x}{15a} + \dfrac{4x^2}{9a^2}$.

17. $16m^4 + \dfrac{16}{3}m^2n + 8m^2 + \dfrac{4}{9}n^2 + \dfrac{4}{3}n + 1$.

18. $4x^4 + 32x^2 + 96 + \dfrac{64}{x^2} + \dfrac{128}{x^2}$.

*122. *To find the cube root of a compound expression.*

Since the cube of $a+b$ is $a^3+3a^2b+3ab^2+b^3$, we have to discover a process by which a and b, the terms of the root, can be found when $a^3+3a^2b+3ab^2+b^3$ is given.

The first term a is the cube root of a^3.

Arrange the terms according to powers of one letter a; then the first term is a^3, and its cube root a. Set this down as the first term of the required root. Subtract a^3 from the given expression and the remainder is

$$3a^2b+3ab^2+b^3 \text{ or } (3a^2+3ab+b^2) \times b.$$

Thus b, the second term of the root, will be the quotient when the remainder is divided by $3a^2+3ab+b^2$.

This divisor consists of three terms:

(1) Three times the square of a, the term of the root already found.

(2) Three times the product of the first term a, and the new term b.

(3) The square of b.

The work may be arranged as follows:

$$
\begin{array}{rl}
& a^3+3a^2b+3ab^2+b^3\ (\,a+b \\
& a^3 \\ \hline
3(a)^2 = 3a^2 & \quad 3a^2b+3ab^2+b^3 \\
3 \times a \times b = +3ab & \\
(b)^2 = +b^2 & \\ \hline
3a^2+3ab+b^2 & \quad 3a^2b+3ab^2+b^3
\end{array}
$$

Example 1. Find the cube root of $8x^3 - 36x^2y + 54xy^2 - 27y^3$.

$$
\begin{array}{rl}
& 8x^3 - 36x^2y + 54xy^2 - 27y^3\ (\,2x-3y \\
& 8x^3 \\ \hline
3(2x)^2 = 12x^2 & \quad -36x^2y + 54xy^2 - 27y^3 \\
3 \times 2x \times (-3y) = -18xy & \\
(-3y)^2 = +9y^2 & \\ \hline
12x^2 - 18xy + 9y^2 & \quad -36x^2y + 54xy^2 - 27y^3
\end{array}
$$

XVI.] EVOLUTION. 109

Example 2. Find the cube root of $27 + 108x + 90x^2 - 80x^3 - 60x^4 + 48x^5 - 8x^6$.

$$27 + 108x + 90x^2 - 80x^3 - 60x^4 + 48x^5 - 8x^6 \,(\, 3 + 4x - 2x^2$$
$$27$$

$$\begin{array}{rl} 3 \times (3)^2 & = 27 \\ 3 \times 3 \times 4x & = +36x \\ (4x)^2 & = +16x^2 \end{array}$$

$$\begin{array}{|l} 108x + 90x^2 - 80x^3 \\ \hline 108x + 144x^2 + 64x^3 \end{array}$$

$$\begin{array}{rl} 3 \times (3 + 4x)^2 & = 27 + 72x + 48x^2 \\ 3 \times (3 + 4x) \times (-2x^2) & = -18x^2 - 24x^3 \\ (-2x^2)^2 & = + 4x^4 \end{array}$$

$$\overline{27 + 72x + 30x^2 - 24x^3 + 4x^4}$$

$$\begin{array}{|l} -54x^2 - 144x^3 - 60x^4 + 48x^5 - 8x^6 \\ \hline -54x^2 - 144x^3 - 60x^4 + 48x^5 - 8x^6 \end{array}$$

Explanation. When we have obtained two terms in the root, $3 + 4x$, we have a remainder

$$-54x^2 - 144x^3 - 60x^4 + 48x^5 - 8x^6.$$

Take 3 times the square of the root already found and place the result, $27 + 72x + 48x^2$, as the first part of the new divisor. Divide $-54x^2$, the first term of the remainder, by 27, the first term of the divisor; this gives a new term of the root, $-2x^2$. To complete the divisor we take 3 times the product of $(3 + 4x)$ and $-2x^2$, and also the square of $-2x^2$. Now multiply the complete divisor by $-2x^2$ and subtract; there is no remainder and the root is found.

*EXAMPLES XVI. d.

Find the cube root of each of the following expressions:

1. $a^3 + 3a^2 + 3a + 1$.
2. $x^3 + 6x^2 + 12x + 8$.
3. $a^3x^3 - 3a^2x^2y^2 + 3axy^4 - y^6$.
4. $8m^3 - 12m^2 + 6m - 1$.
5. $64a^3 - 144a^2b + 108ab^2 - 27b^3$.
6. $1 + 3x + 6x^2 + 7x^3 + 6x^4 + 3x^5 + x^6$.
7. $1 - 6x + 21x^2 - 44x^3 + 63x^4 - 54x^5 + 27x^6$.
8. $a^3 + 6a^2b - 3a^2c + 12ab^2 - 12abc + 3ac^2 + 8b^3 - 12b^2c + 6bc^2 - c^3$.
9. $8a^6 - 36a^5 + 66a^4 - 63a^3 + 33a^2 - 9a + 1$.
10. $y^6 - 3y^5 + 6y^4 - 7y^3 + 6y^2 - 3y + 1$.
11. $8x^6 + 12x^5 - 30x^4 - 35x^3 + 45x^2 + 27x - 27$.
12. $27x^6 - 54x^5a + 117x^4a^2 - 116x^3a^3 + 117x^2a^4 - 54xa^5 + 27a^6$.
13. $27x^6 - 27x^5 - 18x^4 + 17x^3 + 6x^2 - 3x - 1$.
14. $24x^4y^2 + 96x^2y^4 - 6x^5y + x^6 - 96xy^5 + 64y^6 - 56x^3y^3$.
15. $216 + 342x^2 + 171x^4 + 27x^6 - 27x^5 - 109x^3 - 108x$.

*123. We add some examples of cube root where fractional terms occur in the given expressions.

Example. Find the cube root of $54 - 27x^3 + \dfrac{8}{x^6} - \dfrac{36}{x^3}$.

Arrange the expression in *ascending* powers of x.

$$\dfrac{8}{x^6} - \dfrac{36}{x^3} + 54 - 27x^3 \left(\dfrac{2}{x^2} - 3x \right.$$

$$\dfrac{8}{x^6}$$

$$\begin{array}{rl} 3 \times \left(\dfrac{2}{x^2}\right)^2 & = \dfrac{12}{x^4} \\ 3 \times \dfrac{2}{x^2} \times (-3x) & = -\dfrac{18}{x} \\ (-3x)^2 & = +9x^2 \\ \hline & \dfrac{12}{x^4} - \dfrac{18}{x} + 9x^2 \end{array} \quad \begin{array}{l} -\dfrac{36}{x^3} + 54 - 27x^3 \\ \\ \\ \\ -\dfrac{36}{x^3} + 54 - 27x^3 \end{array}$$

XVI.] EVOLUTION. 111

*EXAMPLES XVI. e.

Find the cube root of each of the following expressions:

1. $\dfrac{x^3}{8} - \dfrac{3x^2}{4} + \dfrac{3x}{2} - 1.$

2. $\dfrac{x^3}{27} + \dfrac{2x^2}{3} + 4x + 8.$

3. $8x^3 - 4x^2y^2 + \dfrac{2}{3}xy^4 - \dfrac{y^6}{27}.$

4. $\dfrac{27x^3}{64y^3} - \dfrac{27x^2}{8y^2} + \dfrac{9x}{y} - 8.$

5. $x^3 - 9x + \dfrac{27}{x} - \dfrac{27}{x^3}.$

6. $\dfrac{x^6}{y^3} - 6x^4 + 12x^2y^3 - 8y^6.$

7. $\dfrac{x^3}{y^3} + \dfrac{6x^2}{y^2} + \dfrac{9x}{y} - 4 - \dfrac{9y}{x} + \dfrac{6y^2}{x^2} - \dfrac{y^3}{x^3}.$

8. $\dfrac{x^3}{27} - \dfrac{x^2}{3} + 2x - 7 + \dfrac{18}{x} - \dfrac{27}{x^2} + \dfrac{27}{x^3}.$

9. $\dfrac{x^3}{a^3} - \dfrac{12x^2}{a^2} + \dfrac{54x}{a} - 112 + \dfrac{108a}{x} - \dfrac{48a^2}{x^2} + \dfrac{8a^3}{x^3}.$

10. $\dfrac{64a^3}{x^3} - \dfrac{192a^2}{x^2} + \dfrac{240a}{x} - 160 + \dfrac{60x}{a} - \dfrac{12x^2}{a^2} + \dfrac{x^3}{a^3}.$

11. $\dfrac{6b}{a} + \dfrac{6a}{b} - 7 + \dfrac{a^3}{b^3} - \dfrac{3a^2}{b^2} - \dfrac{3b^2}{a^2} + \dfrac{b^3}{a^3}.$

12. $\dfrac{60x^4}{y^4} - \dfrac{80x^3}{y^3} - \dfrac{90x^2}{y^2} + \dfrac{8x^6}{y^6} + \dfrac{108x}{y} - 27 + \dfrac{48x^5}{y^5}.$

*124. The ordinary rules for extracting square and cube roots in Arithmetic are based upon the algebraical methods explained in the present chapter. The following example is given to illustrate the arithmetical process.

Example. Find the cube root of 614125.

Since 614125 lies between 512000 and 729000, that is, between $(80)^3$ and $(90)^3$, its cube root lies between 80 and 90 and therefore consists of two figures.

$$
\begin{array}{r}
a+b \\
614125\,(\,80+5=85 \\
512000 \\
\end{array}
$$

$$
\begin{array}{rr|r}
 & & 102125 \\
3a^2 = 3 \times (80)^2 = 19200 & & \\
3 \times a \times b = 3 \times 80 \times 5 = 1200 & & \\
b^2 = 5 \times 5 = 25 & & \\
\hline
20425 & & 102125 \\
\end{array}
$$

In Arithmetic the ciphers are usually omitted, and there are other modifications of the algebraical rules.

CHAPTER XVII.

Resolution into Factors.

125. Definition. When an algebraical expression is the product of two or more expressions each of these latter quantities is called a **factor** of it, and the determination of these quantities is called the **resolution** of the expression into its factors.

In this chapter we shall explain the principal rules by which the resolution of expressions into their component factors may be effected.

126. *When each of the terms which compose an expression is divisible by a common factor*, the expression may be simplified by dividing each term separately by this factor, and enclosing the quotient within brackets; the common factor being placed outside as a coefficient.

Example 1. The terms of the expression $3a^2 - 6ab$ have a common factor $3a$;

$$\therefore \ 3a^2 - 6ab = 3a(a - 2b).$$

Example 2. $\quad 5a^2bx^3 - 15abx^2 - 20b^3x^2 = 5bx^2(a^2x - 3a - 4b^2).$

EXAMPLES XVII. a.

Resolve into factors:

1. $a^3 - ax$.
2. $x^3 - x^2$.
3. $2a - 2a^2$.
4. $a^2 - ab^2$.
5. $7p^2 + p$.
6. $8x - 2x^2$.
7. $5ax - 5a^3x^2$.
8. $3x^2 + x^5$.
9. $x^2 + xy$.
10. $x^3 - x^2y$.
11. $5x - 25x^2y$.
12. $15 + 25x^2$.
13. $16x + 64x^2y$.
14. $15a^2 - 225a^4$.
15. $54 - 81x$.
16. $10x^3 - 25x^4y$.
17. $3x^3 - x^2 + x$.
18. $6x^3 + 2x^4 + 4x^5$.
19. $x^3 - x^2y + xy^2$.
20. $3a^4 - 3a^3b + 6a^2b^2$.
21. $2x^3y^3 - 6x^2y^2 + 2xy^3$.
22. $6x^3 - 9x^2y + 12xy^2$.
23. $5x^5 - 10a^2x^3 - 15a^3x^3$.
24. $7a - 7a^3 + 14a^4$.
25. $38a^3x^5 + 57a^4x^2$.

127. An expression may be resolved into factors *if the terms can be arranged in groups which have a compound factor common.*

Example 1. Resolve into factors $x^2 - ax + bx - ab$.

Noticing that the first two terms contain a common factor x, and the last two terms a common factor b, we enclose the first two terms in one bracket, and the last two in another. Thus

$$\begin{aligned}
x^2 - ax + bx - ab &= (x^2 - ax) + (bx - ab) \\
&= x(x - a) + b(x - a) \\
&= (x - a) \text{ taken } x \text{ times } plus \ (x - a) \text{ taken } b \text{ times} \\
&= (x - a) \text{ taken } (x + b) \text{ times} \\
&= (x - a)(x + b).
\end{aligned}$$

Example 2. Resolve into factors $6x^2 - 9ax + 4bx - 6ab$.

$$\begin{aligned}
6x^2 - 9ax + 4bx - 6ab &= (6x^2 - 9ax) + (4bx - 6ab) \\
&= 3x(2x - 3a) + 2b(2x - 3a) \\
&= (2x - 3a)(3x + 2b).
\end{aligned}$$

Example 3. Resolve into factors $12a^2 - 4ab - 3ax^2 + bx^2$.

$$\begin{aligned}
12a^2 - 4ab - 3ax^2 + bx^2 &= (12a^2 - 4ab) - (3ax^2 - bx^2) \\
&= 4a(3a - b) - x^2(3a - b) \\
&= (3a - b)(4a - x^2).
\end{aligned}$$

Note. In the first line of work it is sufficient to see that each pair contains some common factor. Thus, in the last example, by a different arrangement, we have

$$\begin{aligned}
12a^2 - 4ab - 3ax^2 + bx^2 &= (12a^2 - 3ax^2) - (4ab - bx^2) \\
&= 3a(4a - x^2) - b(4a - x^2) \\
&= (4a - x^2)(3a - b),
\end{aligned}$$

the same result as before, since it is immaterial in what order the factors of a product are written.

EXAMPLES XVII. b.

Resolve into factors:

1. $a^2 + ab + ac + bc$.
2. $a^2 - ac + ab - bc$.
3. $a^2c^2 + acd + abc + bd$.
4. $a^2 + 3a + ac + 3c$.
5. $2x + cx + 2c + c^2$.
6. $x^2 - ax + 5x - 5a$.
7. $5a + ab + 5b + b^2$.
8. $ab - by - ay + y^2$.
9. $ax - bx - az + bz$.
10. $pr + qr - ps - qs$.
11. $mx - my - nx + ny$.
12. $mx - ma + nx - na$.
13. $2ax + ay + 2bx + by$.
14. $3ax - bx - 3ay + by$.

Resolve into factors:

15. $6x^2 + 3xy - 2ax - ay$.
16. $mx - 2my - nx + 2ny$.
17. $ax^2 - 3bxy - axy + 3by^2$.
18. $x^2 + mxy - 4xy - 4my^2$.
19. $ax^2 + bx^2 + 2a + 2b$.
20. $x^2 - 3x - xy + 3y$.
21. $2x^4 - x^3 + 4x - 2$.
22. $3x^3 + 5x^2 + 3x + 5$.
23. $x^4 + x^3 + 2x + 2$.
24. $y^3 - y^2 + y - 1$.
25. $axy + bcxy - az - bcz$.
26. $f^2x^2 + g^2x^2 - ag^2 - af^2$.
27. $2ax^2 + 3axy - 2bxy - 3by^2$.
28. $amx^2 + bmxy - anxy - bny^2$.
29. $ax - bx + by + cy - cx - ay$.
30. $a^2x + abx + ac + aby + b^2y + bc$.

Trinomial Expressions.

128. Before proceeding to the next case of resolution into factors the student is advised to refer to Chap. v. Art. 44. Attention has there been drawn to the way in which, in forming the product of two binomials, the coefficients of the different terms combine so as to give a trinomial result. Thus, by Art. 44,

$$(x+5)(x+3) = x^2 + 8x + 15 \dots \dots (1),$$
$$(x-5)(x-3) = x^2 - 8x + 15 \dots \dots (2),$$
$$(x+5)(x-3) = x^2 + 2x - 15 \dots \dots (3),$$
$$(x-5)(x+3) = x^2 - 2x - 15 \dots \dots (4).$$

We now propose to consider the converse problem: namely, the resolution of a trinomial expression, similar to those which occur on the right-hand side of the above identities, into its component binomial factors.

By examining the above results, we notice that:

1. The first term of both the factors is x.

2. The *product* of the second terms of the two factors is equal to the *third term* of the trinomial; e.g. in (2) above we see that 15 is the product of -5 and -3; while in (3) -15 is the product of $+5$ and -3.

3. The *algebraic sum* of the second terms of the two factors is equal to the *coefficient* of x in the trinomial; e.g. in (4) the sum of -5 and $+3$ gives -2, the coefficient of x in the trinomial.

In applying these laws we will first consider a case where the *third term of the trinomial is positive*.

Example 1. Resolve into factors $x^2 + 11x + 24$.

The second terms of the factors must be such that their product is $+24$, and their sum $+11$. It is clear that they must be $+8$ and $+3$.

$$\therefore x^2 + 11x + 24 = (x+8)(x+3).$$

Example 2. Resolve into factors $x^2 - 10x + 24$.

The second terms of the factors must be such that their product is $+24$, and their sum -10. Hence they must *both* be *negative*, and it is easy to see that they must be -6 and -4.

$$\therefore\ x^2 - 10x + 24 = (x-6)(x-4).$$

Example 3. $\quad x^2 - 18x + 81 = (x-9)(x-9)$
$\qquad\qquad\qquad\qquad = (x-9)^2.$

Example 4. $\quad x^4 + 10x^2 + 25 = (x^2+5)(x^2+5)$
$\qquad\qquad\qquad\qquad = (x^2+5)^2.$

Example 5. Resolve into factors $x^2 - 11ax + 10a^2$.

The second terms of the factors must be such that their product is $+10a^2$, and their sum $-11a$. Hence they must be $-10a$ and $-a$.

$$\therefore\ x^2 - 11ax + 10a^2 = (x-10a)(x-a).$$

Note. In examples of this kind the student should always verify his results, by forming the product (*mentally*, as explained in Chap. v.) of the factors he has chosen.

EXAMPLES XVII. c.

Resolve into factors:

1. $a^2 + 3a + 2$.
2. $a^2 + 2a + 1$.
3. $a^2 + 7a + 12$.
4. $a^2 - 7a + 12$.
5. $x^2 - 11x + 30$.
6. $x^2 - 15x + 56$.
7. $x^2 - 19x + 90$.
8. $x^2 + 13x + 42$.
9. $x^2 - 21x + 110$.
10. $x^2 - 21x + 108$.
11. $x^2 - 21x + 80$.
12. $x^2 + 21x + 90$.
13. $x^2 - 19x + 84$.
14. $x^2 - 19x + 78$.
15. $x^2 - 18x + 45$.
16. $x^2 + 20x + 96$.
17. $x^2 - 26x + 165$.
18. $x^2 - 21x + 104$.
19. $x^2 + 23x + 102$.
20. $a^2 - 24a + 95$.
21. $a^2 - 32a + 256$.
22. $a^2 + 30a + 225$.
23. $a^2 + 54a + 729$.
24. $a^2 - 38a + 361$.
25. $a^2 - 14ab + 49b^2$.
26. $a^2 + 5ab + 6b^2$.
27. $m^2 - 13mn + 40n^2$.
28. $m^2 - 22mn + 105n^2$.
29. $x^2 - 23xy + 132y^2$.
30. $x^2 - 26xy + 169y^2$.
31. $x^4 + 8x^2 + 7$.
32. $x^4 + 9x^2y^2 + 14y^4$.
33. $x^2y^2 - 16xy + 39$.
34. $x^2 + 49xy + 600y^2$.
35. $x^2y^2 + 34xy + 289$.
36. $a^4b^4 + 37a^2b^2 + 300$.
37. $a^2 - 20abx + 75b^2x^2$.
38. $x^2 + 43xy + 390y^2$.
39. $a^2 - 29ab + 54b^2$.
40. $x^4 + 162x^2 + 6561$.
41. $12 - 7x + x^2$.
42. $20 + 9x + x^2$.
43. $132 - 23x + x^2$.
44. $88 + 19x + x^2$.
45. $130 + 31xy + x^2y^2$.
46. $143 - 24xa + x^2a^2$.
47. $204 - 29x^2 + x^4$.
48. $216 + 35x + x^2$.

129. Next consider a case where *the third term of the trinomial is negative.*

Example 1. Resolve into factors $x^2+2x-35$.

The second terms of the factors must be such that their product is -35, and their *algebraical sum* $+2$. Hence they must have *opposite* signs, and the greater of them must be *positive* in order to give its sign to their sum.

The required terms are therefore $+7$ and -5.
$$\therefore\; x^2+2x-35=(x+7)(x-5).$$

Example 2. Resolve into factors $x^2-3x-54$.

The second terms of the factors must be such that their product is -54, and their *algebraical sum* -3. Hence they must have *opposite* signs, and the greater of them must be *negative* in order to give its sign to their sum.

The required terms are therefore -9 and $+6$.
$$\therefore\; x^2-3x-54=(x-9)(x+6).$$

Remembering that in these cases the numerical quantities *must have opposite signs*, if preferred, the following method may be adopted.

Example 3. Resolve into factors $x^2y^2+23xy-420$.

Find two numbers whose product is 420, and whose *difference* is 23. These are 35 and 12; hence inserting the signs so that the positive may predominate, we have
$$x^2y^2+23xy-420=(xy+35)(xy-12).$$

EXAMPLES XVII. d.

Resolve into factors:

1. x^2-x-2.
2. x^2+x-2.
3. x^2-x-6.
4. x^2+x-6.
5. x^2-2x-3.
6. x^2+2x-3.
7. x^2+x-56.
8. $x^2+3x-40$.
9. $x^2-4x-12$.
10. a^2-a-20.
11. $a^2-4a-21$.
12. a^2+a-20.
13. $a^2-4a-117$.
14. $x^2+9x-36$.
15. $x^2+x-156$.
16. $x^2+x-110$.
17. $x^2-9x-90$.
18. $x^2-x-240$.
19. $a^2-12a-85$.
20. $a^2-11a-152$.
21. $x^2y^2-5xy-24$.
22. $x^2+7xy-60y^2$.
23. $x^2+ax-42a^2$.
24. $x^2-32xy-105y^2$.
25. $a^2-ay-210y^2$.
26. $x^2+18x-115$.
27. $x^2-20xy-96y^2$.
28. $x^2+16x-260$.
29. $a^2-11a-26$.
30. $a^2y^2+14ay-240$.

XVII.]   RESOLUTION INTO FACTORS.   117

31. $a^4 - a^2b^2 - 56b^4$.   32. $x^4 - 14x^2 - 51$.   33. $y^4 + 6x^2y^2 - 27x^4$.
34. $a^2b^2 - 3abc - 10c^2$.   35. $a^2 + 12abx - 28b^2x^2$.
36. $a^2 - 18axy - 243x^2y^2$.   37. $x^4 + 13a^2x^2 - 300a^4$.
38. $x^4 - a^2x^2 - 132a^4$.   39. $x^4 - a^2x^2 - 462a^4$.
40. $x^6 + x^3 - 870$.   41. $2 + x - x^2$.   42. $6 + x - x^2$.
43. $110 - x - x^2$.   44. $380 - x - x^2$.   45. $120 - 7ax - a^2x^2$.
46. $65 + 8xy - x^2y^2$.   47. $98 - 7x - x^2$.   48. $204 - 5x - x^2$.

[*For easy Miscellaneous Examples see page* 124$_A$.]

130. We proceed now to the resolution into factors of trinomial expressions when *the coefficient of the highest power is not unity*.

Again, referring to Chap. v. Art. 44, we may write down the following results:

$$(3x+2)(x+4) = 3x^2 + 14x + 8 \dots\dots\dots\dots(1),$$
$$(3x-2)(x-4) = 3x^2 - 14x + 8 \dots\dots\dots\dots(2),$$
$$(3x+2)(x-4) = 3x^2 - 10x - 8 \dots\dots\dots\dots(3),$$
$$(3x-2)(x+4) = 3x^2 + 10x - 8 \dots\dots\dots\dots(4).$$

The converse problem presents more difficulty than the cases we have yet considered.

Before endeavouring to give a general method of procedure, it will be worth while to examine in detail two of the identities given above.

Consider the result $3x^2 - 14x + 8 = (3x-2)(x-4)$.

The first term $3x^2$ is the product of $3x$ and x.
The third term $+8$ .... -2 and -4.

The middle term $-14x$ is the result of adding together the two products $3x \times -4$ and $x \times -2$.

Again, consider the result $3x^2 - 10x - 8 = (3x+2)(x-4)$.

The first term $3x^2$ is the product of $3x$ and x.
The third term -8 .... $+2$ and -4.

The middle term $-10x$ is the result of adding together the two products $3x \times -4$ and $x \times 2$; and its sign is negative because the greater of these two products is negative.

131. The beginner will frequently find that it is not easy to select the proper factors at the first trial. Practice alone will enable him to detect at a glance whether any pair he has chosen will combine so as to give the correct coefficients of the expression to be resolved.

Example. Resolve into factors $7x^2 - 19x - 6$.

Write down $(7x\ \ 3)(x\ \ 2)$ for a first trial, noticing that 3 and 2 must have opposite signs. These factors give $7x^2$ and -6 for the first and third terms. But since $7 \times 2 - 3 \times 1 = 11$, the combination fails to give the correct coefficient of the middle term.

Next try $(7x\ \ 2)(x\ \ 3)$.

Since $7 \times 3 - 2 \times 1 = 19$, these factors will be correct if we insert the signs so that the negative shall predominate.

Thus $\qquad 7x^2 - 19x - 6 = (7x + 2)(x - 3)$.

[Verify by mental multiplication.]

132. In actual work it will not be necessary to put down all these steps at length. The student will soon find that the different cases may be rapidly reviewed, and the unsuitable combinations rejected at once.

It is especially important to pay attention to the two following hints:

1. If the third term of the trinomial is positive, then the second terms of its factors have both the same sign, and this sign is the same as that of the middle term of the trinomial.

2. If the third term of the trinomial is negative, then the second terms of its factors have opposite signs.

Example 1. Resolve into factors $14x^2 + 29x - 15$ ....(1),
$\qquad\qquad\qquad\qquad\qquad\qquad 14x^2 - 29x - 15$ ....(2).

In each case we may write down $(7x\ \ 3)(2x\ \ 5)$ as a first trial, noticing that 3 and 5 must have opposite signs.

And since $7 \times 5 - 3 \times 2 = 29$, we have only now to insert the proper signs in each factor.

In (1) the positive sign must predominate,
in 2 the negative ....
Therefore $\qquad 14x^2 + 29x - 15 = (7x - 3)(2x + 5)$.
$\qquad\qquad\qquad 14x^2 - 29x - 15 = (7x + 3)(2x - 5)$.

Example 2. Resolve into factors $5x^2 + 17x + 6$ ....(1),
$\qquad\qquad\qquad\qquad\qquad\qquad 5x^2 - 17x + 6$ ....(2).

In (1) we notice that the factors which give 6 are both positive.
In (2) .... negative.
And therefore for (1) we may write $(5x+\ )(x+\ )$.
$\qquad\qquad\qquad\qquad$ (2) .... $(5x-\ )(x-\ )$.
And, since $5 \times 3 + 1 \times 2 = 17$, we see that
$\qquad\qquad 5x^2 + 17x + 6 = (5x + 2)(x + 3)$.
$\qquad\qquad 5x^2 - 17x + 6 = (5x - 2)(x - 3)$.

RESOLUTION INTO FACTORS.

Note. In each expression the third term 6 also admits of factors 6 and 1; but this is one of the cases referred to above which the student would reject at once as unsuitable.

Example 3. $9x^2 - 48xy + 64y^2 = (3x - 8y)(3x - 8y)$
$= (3x - 8y)^2.$

Example 4. $6 + 7x - 5x^2 = (3 + 5x)(2 - x).$

EXAMPLES XVII. e.

Resolve into factors:

1. $2x^2 + 3x + 1.$
2. $3x^2 + 5x + 2.$
3. $2x^2 + 5x + 2.$
4. $3x^2 + 10x + 3.$
5. $2x^2 + 9x + 4.$
6. $3x^2 + 8x + 4.$
7. $2x^2 + 7x + 6.$
8. $2x^2 + 11x + 5.$
9. $3x^2 + 11x + 6.$
10. $5x^2 + 11x + 2.$
11. $2x^2 + 3x - 2.$
12. $3x^2 + x - 2.$
13. $4x^2 + 11x - 3.$
14. $3x^2 + 14x - 5.$
15. $2x^2 + 15x - 8.$
16. $2x^2 - x - 1.$
17. $3x^2 + 7x - 6.$
18. $2x^2 + x - 28.$
19. $3x^2 + 13x - 30.$
20. $6x^2 + 7x - 3.$
21. $6x^2 - 7x - 3.$
22. $3x^2 + 7x + 4.$
23. $3x^2 + 23x + 14.$
24. $2x^2 - x - 15.$
25. $3x^2 + 19x - 14.$
26. $3x^2 - 19x - 14.$
27. $6x^2 - 31x + 35.$
28. $4x^2 + x - 14.$
29. $3x^2 - 13x + 14.$
30. $3x^2 + 41x + 26.$
31. $4x^2 + 23x + 15.$
32. $2x^2 - 5xy - 3y^2.$
33. $8x^2 - 38x + 35.$
34. $12x^2 - 23xy + 10y^2.$
35. $15x^2 + 224x - 15.$
36. $15x^2 - 77x + 10.$
37. $12x^2 - 31x - 15.$
38. $24x^2 + 22x - 21.$
39. $72x^2 - 145x + 72.$
40. $24x^2 - 29xy - 4y^2.$
41. $2 - 3x - 2x^2.$
42. $3 + 11x - 4x^2.$
43. $6 + 5x - 6x^2.$
44. $4 - 5x - 6x^2.$
45. $5 + 32x - 21x^2.$
46. $7 + 10x + 3x^2.$
47. $18 - 33x + 5x^2.$
48. $8 + 6x - 5x^2.$
49. $20 - 9x - 20x^2.$
50. $24 + 37x - 72x^2.$

The Difference of Two Squares.

133. By multiplying $a + b$ by $a - b$ we obtain the identity
$$(a+b)(a-b) = a^2 - b^2,$$
a result which may be verbally expressed as follows:

The product of the sum and the difference of any two quantities is equal to the difference of their squares.

Conversely, *the difference of the squares of any two quantities is equal to the product of the sum and the difference of the two quantities.*

Thus any expression which is the difference of two squares may at once be resolved into factors.

Example. Resolve into factors $25x^2 - 16y^2$.
$$25x^2 - 16y^2 = (5x)^2 - (4y)^2.$$
Therefore the first factor is the sum of $5x$ and $4y$,
and the second factor is the difference of $5x$ and $4y$.
$$\therefore 25x^2 - 16y^2 = (5x+4y)(5x-4y).$$

The intermediate steps may usually be omitted.

Example. $\qquad 1 - 49c^6 = (1+7c^3)(1-7c^3).$

The difference of the squares of two numerical quantities may be found by the formula $a^2 - b^2 = (a+b)(a-b)$.

Example. $\qquad (329)^2 - (171)^2 = (329+171)(329-171)$
$$= 500 \times 158$$
$$= 79000.$$

EXAMPLES XVII. f.

Resolve into factors:

1. $x^2 - 4$.
2. $a^2 - 81$.
3. $y^2 - 100$.
4. $c^2 - 144$.
5. $9 - a^2$.
6. $49 - c^2$.
7. $121 - x^2$.
8. $400 - a^2$.
9. $x^2 - 9a^2$.
10. $y^2 - 25x^2$.
11. $36x^2 - 25b^2$.
12. $9x^2 - 1$.
13. $36p^2 - 49q^2$.
14. $4k^2 - 1$.
15. $49 - 100k^2$.
16. $1 - 25x^2$.
17. $a^2 - 4b^2$.
18. $9x^2 - y^2$.
19. $p^2q^2 - 36$.
20. $a^2b^2 - 4c^2d^2$.
21. $x^4 - 9$.
22. $9a^4 - 121$.
23. $25x^2 - 64$.
24. $81a^4 - 49x^4$.
25. $x^6 - 25$.
26. $1 - 36a^6$.
27. $9x^4 - a^2$.
28. $81x^6 - 25a^2$.
29. $x^4a^2 - 49$.
30. $a^2 - 64x^6$.
31. $a^2b^2 - 9x^6$.
32. $x^6y^6 - 4$.
33. $1 - a^2b^2$.
34. $4 - x^2$.
35. $9 - 4a^2$.
36. $9a^4 - 25b^4$.
37. $x^4 - 16b^2$.
38. $x^2 - 25y^2$.
39. $1 - 100b^2$.
40. $25 - 64x^2$.
41. $121a^2 - 81x^2$.
42. $p^2q^2 - 64a^4$.
43. $64x^2 - 25z^6$.
44. $49x^4 - 16y^4$.
45. $81p^4z^6 - 25b^2$.
46. $16x^{16} - 9y^6$.
47. $36x^{36} - 49a^{14}$.
48. $1 - 100a^6b^4c^2$.
49. $25x^{10} - 16a^8$.
50. $a^2b^4c^6 - x^{16}$.

Find by resolving into factors the value of

51. $(575)^2 - (425)^2$.
52. $(121)^2 - (120)^2$.
53. $(750)^2 - (250)^2$.
54. $(339)^2 - (319)^2$.
55. $(753)^2 - (253)^2$.
56. $(101)^2 - (99)^2$.
57. $(1723)^2 - (277)^2$.
58. $(1639)^2 - (739)^2$.
59. $(1811)^2 - (689)^2$.
60. $(2731)^2 - (269)^2$.
61. $(8133)^2 - (8131)^2$.
62. $(10001)^2 - 1$.

134. When one or both of the squares is a compound quantity the same method is employed.

Example 1. Resolve into factors $(a+2b)^2 - 16x^2$.

The sum of $a+2b$ and $4x$ is $a+2b+4x$,
and their difference is $a+2b-4x$.
$$\therefore (a+2b)^2 - 16x^2 = (a+2b+4x)(a+2b-4x).$$

Example 2. Resolve into factors $x^2 - (2b-3c)^2$.

The sum of x and $2b-3c$ is $x+2b-3c$,
and their difference is $x-(2b-3c) = x-2b+3c$.
$$\therefore x^2 - (2b-3c)^2 = (x+2b-3c)(x-2b+3c).$$

If the factors contain like terms they should be collected so as to give the result in its simplest form.

Example 3. $(3x+7y)^2 - (2x-3y)^2$
$$= \{(3x+7y)+(2x-3y)\}\{(3x+7y)-(2x-3y)\}$$
$$= (3x+7y+2x-3y)(3x+7y-2x+3y)$$
$$= (5x+4y)(x+10y).$$

EXAMPLES XVII. g.

Resolve into factors:

1. $(a+b)^2 - c^2$.
2. $(a-b)^2 - c^2$.
3. $(x+y)^2 - 4z^2$.
4. $(x+2y)^2 - a^2$.
5. $(a+3b)^2 - 16x^2$.
6. $(x+5a)^2 - 9y^2$.
7. $(x+5c)^2 - 1$.
8. $(a-2x)^2 - b^2$.
9. $(2x-3a)^2 - 9c^2$.
10. $a^2 - (b-c)^2$.
11. $x^2 - (y+z)^2$.
12. $4a^2 - (y-z)^2$.
13. $9x^2 - (2a-3b)^2$.
14. $1 - (a-b)^2$.
15. $c^2 - (5a-3b)^2$.
16. $(a+b)^2 - (c+d)^2$.
17. $(a-b)^2 - (x+y)^2$.
18. $(7x+y)^2 - 1$.
19. $(a+b)^2 - (m-n)^2$.
20. $(a-n)^2 - (b+m)^2$.
21. $(b-c)^2 - (a-x)^2$.
22. $(4a+x)^2 - (b+y)^2$.
23. $(a+2b)^2 - (3x+4y)^2$.
24. $1 - (7a-3b)^2$.
25. $(a-b)^2 - (x-y)^2$.
26. $(a-3x)^2 - 16y^2$.
27. $(2a-5x)^2 - 1$.
28. $(a+b-c)^2 - (x-y+z)^2$.
29. $(3a+2b)^2 - (c+x-2y)^2$.

Resolve into factors and simplify:

30. $(x+y)^2 - x^2$.
31. $x^2 - (y-x)^2$.
32. $(x+3y)^2 - 4y^2$.
33. $(24x+y)^2 - (23x-y)^2$.
34. $(5x+2y)^2 - (3x-y)^2$.
35. $9x^2 - (3x-5y)^2$.
36. $(7x+3)^2 - (5x-4)^2$.
37. $(3a+1)^2 - (2a-1)^2$.
38. $16a^2 - (3a+1)^2$.
39. $(2a+b-c)^2 - (a-b+c)^2$.
40. $(x-7y+z)^2 - (7y-z)^2$.
41. $(x+y-8)^2 - (x-8)^2$.
42. $(2x+a-3)^2 - (3-2x)^2$.

135. By suitably grouping together the terms, compound expressions can often be expressed as the difference of two squares, and so be resolved into factors.

Example 1. Resolve into factors $a^2 - 2ax + x^2 - 4b^2$.
$$\begin{aligned}a^2 - 2ax + x^2 - 4b^2 &= (a^2 - 2ax + x^2) - 4b^2 \\ &= (a-x)^2 - (2b)^2 \\ &= (a-x+2b)(a-x-2b).\end{aligned}$$

Example 2. Resolve into factors $9a^2 - c^2 + 4cx - 4x^2$.
$$\begin{aligned}9a^2 - c^2 + 4cx - 4x^2 &= 9a^2 - (c^2 - 4cx + 4x^2) \\ &= (3a)^2 - (c-2x)^2 \\ &= (3a+c-2x)(3a-c+2x).\end{aligned}$$

Example 3. Resolve into factors $2bd - a^2 - c^2 + b^2 + d^2 + 2ac$.

Here the terms $2bd$ and $2ac$ suggest the proper preliminary arrangement of the expression. Thus
$$\begin{aligned}2bd - a^2 - c^2 + b^2 + d^2 + 2ac &= b^2 + 2bd + d^2 - a^2 + 2ac - c^2 \\ &= b^2 + 2bd + d^2 - (a^2 - 2ac + c^2) \\ &= (b+d)^2 - (a-c)^2 \\ &= (b+d+a-c)(b+d-a+c).\end{aligned}$$

Example 4. Resolve into factors $x^4 + x^2y^2 + y^4$.
$$\begin{aligned}x^4 + x^2y^2 + y^4 &= (x^4 + 2x^2y^2 + y^4) - x^2y^2 \\ &= (x^2 + y^2)^2 - (xy)^2 \\ &= (x^2 + y^2 + xy)(x^2 + y^2 - xy) \\ &= (x^2 + xy + y^2)(x^2 - xy + y^2).\end{aligned}$$

This result is very important and will be referred to again in Chapter XXVIII.

EXAMPLES XVII. h.

Resolve into factors:

1. $x^2 + 2xy + y^2 - a^2$.
2. $a^2 - 2ab + b^2 - x^2$.
3. $x^2 - 6ax + 9a^2 - 16b^2$.
4. $4a^2 + 4ab + b^2 - 9c^2$.
5. $x^2 + a^2 + 2ax - y^2$.
6. $2ay + a^2 + y^2 - x^2$.
7. $x^2 - a^2 - 2ab - b^2$.
8. $y^2 - c^2 + 2cx - x^2$.
9. $1 - x^2 - 2xy - y^2$.
10. $c^2 - x^2 - y^2 + 2xy$.
11. $x^2 + y^2 + 2xy - 4x^2y^2$.
12. $a^2 - 4ab + 4b^2 - 9a^2c^2$.
13. $x^2 + 2xy + y^2 - a^2 - 2ab - b^2$.
14. $a^2 - 2ab + b^2 - c^2 - 2cd - d^2$.
15. $x^2 - 4ax + 4a^2 - b^2 + 2by - y^2$.
16. $y^2 + 2by + b^2 - a^2 - 6ax - 9x^2$.

17. $x^2-2x+1-a^2-4ab-4b^2$.   18. $9a^2-6a+1-x^2-8dx-16d^2$.
19. $x^2-a^2+y^2-b^2-2xy+2ab$.   20. $a^2+b^2-2ab-c^2-d^2-2cd$.
21. $4x^2-12ax-c^2-k^2-2ck+9a^2$.
22. $a^2+6bx-9b^2x^2-10ab-1+25b^2$.
23. $a^4-25x^6+8a^2x^2-9+30x^3+16x^4$.
24. $x^4-x^2-9-2a^2x^2+a^4+6x$.
25. $a^4+a^2b^2+b^4$.   26. $x^4+4x^2y^2+16y^4$.   27. $p^4+9p^2q^2+81q^4$.
28. $c^4+3c^2d^2+4d^4$.   29. $x^4+y^4-11x^2y^2$.   30. $4m^4-5m^2n^2+n^4$.

The Sum or Difference of Two Cubes.

136. If we divide a^3+b^3 by $a+b$ the quotient is a^2-ab+b^2; and if we divide a^3-b^3 by $a-b$ the quotient is a^2+ab+b^2.

We have therefore the following identities:
$$a^3+b^3=(a+b)(a^2-ab+b^2);$$
$$a^3-b^3=(a-b)(a^2+ab+b^2).$$

These results enable us to resolve into factors any expression which can be written as the sum or the difference of two cubes.

Example 1. $\quad 8x^3-27y^3=(2x)^3-(3y)^3$
$$=(2x-3y)(4x^2+6xy+9y^2).$$

Note. The middle term $6xy$ is the *product* of $2x$ and $3y$.

Example 2. $\quad 64a^3+1=(4a)^3+(1)^3$
$$=(4a+1)(16a^2-4a+1).$$

We may usually omit the intermediate step and write down the factors at once.

Examples. $343a^6-27x^3=(7a^2-3x)(49a^4+21a^2x+9x^2)$.
$\qquad\qquad 8x^9+729=(2x^3+9)(4x^6-18x^3+81)$.

EXAMPLES XVII. k.

Resolve into factors:

1. x^3-y^3.   2. x^3+y^3.   3. x^3-1.   4. $1+a^3$
5. $8x^3-y^3$.   6. x^3+8y^3.   7. $27x^3+1$.   8. $1-8y^3$.
9. $a^3b^3-c^3$. 10. $8x^3+27y^3$. 11. $1-343x^3$. 12. $64+y^3$.
13. $125+a^3$.   14. $216-a^3$.   15. a^3b^3+512. 16. $1000y^3-1$.

Resolve into factors:

17. $x^3 + 64y^3$.
18. $27 - 1000x^3$.
19. $a^3b^3 + 216c^3$.
20. $343 - 8x^3$.
21. $a^3 + 27b^3$.
22. $27x^3 - 64y^3$.
23. $125x^3 - 1$.
24. $216p^3 - 343$.
25. $x^3y^3 + z^3$.
26. $a^3b^3c^3 - 1$.
27. $343x^3 + 1000y^3$.
28. $729a^3 - 64b^3$.
29. $8a^3b^3 + 125x^3$.
30. $x^3y^3 - 216z^3$.
31. $x^6 - 27y^3$.
32. $64x^6 + 125y^3$.
33. $8x^3 - z^6$.
34. $216x^6 - b^3$.
35. $a^3 + 343b^3$.
36. $a^6 + 729b^3$.
37. $8x^3 - 729y^6$.
38. $p^3q^3 - 27x^3$.
39. $z^3 - 64y^6$.
40. $x^3y^3 - 512$.

136$_A$. In Arts. 128 to 132 we have discussed the factorisation of trinomials by trial. And in Arts. 133 to 135 we have shewn how any expression which is the difference of two squares can be written down as the product of two factors. We shall now explain a general method by which any expression of the form $x^2 + px + q$ or $ax^2 + bx + c$ can be expressed as the difference of two squares.

By Art. 112 we have the following identities:

$$x^2 + 2ax + a^2 = (x+a)^2, \quad x^2 - 2ax + a^2 = (x-a)^2.$$

So that if a trinomial is a perfect square, and *its highest power x^2 has unity for its coefficient*, we must always have the term without x equal to *the square of half the coefficient of x*. If therefore the first two terms (containing x^2 and x) of such a trinomial are given, the square may be completed by adding the square of half the coefficient of x.

Thus $x^2 + 6x$ is made a perfect square if we add to it $\left(\dfrac{6}{2}\right)^2$, or 9; and it then becomes $x^2 + 6x + 9$, or $(x+3)^2$.

Similarly to make $x^2 - 7x$ a perfect square we must add $\left(-\dfrac{7}{2}\right)^2$, or $\dfrac{49}{4}$, and we then have $x^2 - 7x + \dfrac{49}{4}$, or $\left(x - \dfrac{7}{2}\right)^2$.

Note. The added term is always positive.

Example 1. Find the factors of $x^2 + 6x + 5$.

The expression may be written $(x^2 + 6x + 9) + 5 - 9$;
that is,
$$x^2 + 6x + 5 = (x+3)^2 - 4$$
$$= (x+3+2)(x+3-2)$$
$$= (x+5)(x+1).$$

Example 2. Find the factors of $x^2 - 7x - 228$.

$$x^2 - 7x - 228 = \left(x^2 - 7x + \frac{49}{4}\right) - 228 - \frac{49}{4}$$
$$= \left(x - \frac{7}{2}\right)^2 - \frac{961}{4}$$
$$= \left(x - \frac{7}{2} + \frac{31}{2}\right)\left(x - \frac{7}{2} - \frac{31}{2}\right)$$
$$= (x+12)(x-19).$$

Example 3. Find the factors of $3x^2 - 13x + 14$.

$$3x^2 - 13x + 14 = 3\left(x^2 - \frac{13}{3}x + \frac{14}{3}\right)$$
$$= 3\left\{x^2 - \frac{13}{3}x + \left(\frac{13}{6}\right)^2 + \frac{14}{3} - \frac{169}{36}\right\}$$
$$= 3\left\{\left(x - \frac{13}{6}\right)^2 - \frac{1}{36}\right\}$$
$$= 3\left(x - \frac{13}{6} + \frac{1}{6}\right)\left(x - \frac{13}{6} - \frac{1}{6}\right)$$
$$= 3\left(x - \frac{7}{3}\right)(x - 2)$$
$$= (3x - 7)(x - 2).$$

As the process of completing the square is quite general and applicable to all cases, it may conveniently be used when factorisation by trial would prove uncertain and tedious. For example, if the factors of $24x^2 + 118x - 247$ were required, it would probably be best to apply the general method at once.

136$_B$. The following exercise contains easy miscellaneous examples of the different cases explained in this chapter.

EXAMPLES XVII 1. (Miscellaneous.)

(*On Arts.* 128, 129.)

Resolve into factors:

1. $x^2 - 3x + 2$.
2. $a^2 + 7a + 10$.
3. $b^2 + b - 12$.
4. $y^2 - 4y - 21$.
5. $c^2 + 12c + 11$.
6. $x^2 - 4x - 5$.
7. $n^2 + 12n + 20$.
8. $y^2 + 9y - 10$.
9. $p^2 - 2pq - 24q^2$.
10. $y^2 + y - 110$.
11. $z^2 - 9z - 90$.
12. $k^2 - 14k + 48$.

Resolve into factors:

13. $a^2 + 18a + 81$.
14. $b^2 - 24b - 81$.
15. $c^2 + 30c + 81$.
16. $x^2 - 14x + 49$.
17. $y^2 + 10yz + 21z^2$.
18. $z^2 + 2z - 63$.
19. $n^2 + 11n + 24$.
20. $p^2 - 5p - 24$.
21. $l^2 + 9l - 36$.
22. $a^2b^2 - 4ab + 4$.
23. $a^2b^2 + 10ab + 16$.
24. $b^2 - 4b - 45$.
25. $m^2 + 3m - 88$.
26. $n^2 - 12n - 45$.
27. $p^2 + 10p - 39$.
28. $x^2y^2 - xy - 72$.
29. $z^2 - z - 20$.
30. $x^2 + xy - 56y^2$.
31. $a^2 - 11ab - 26b^2$.
32. $a^2b^2 - ab - 56$.
33. $y^4 + y^2 - 156$.
34. $z^4 - 7z^2 - 78$.
35. $y^4 - 2y^2 - 35$.
36. $x^2 + 6xy - 91y^2$.

(On Arts. 125–132.)

Resolve into two or more factors:

37. $m^3n^2 - 3m^2n^3$.
38. $10x^3 + 25x^4y$.
39. $y^2 - 2y - 15$.
40. $(a+b)x + (a+b)y$.
41. $x^2 - xz + xy - yz$.
42. $3c^2 + c - 2$.
43. $2b^2 + 11b + 5$.
44. $x^2 - 6xy + 9y^2$.
45. $3x^2 - 10x + 3$.
46. $c^2d^2 - cd - 2$.
47. $6x^2 + 7x - 3$.
48. $4(a-b) - c(a-b)$.
49. $a^4 + a^3 + 2a + 2$.
50. $2c^3d - 6c^2d^2 + 2c^2d^3$.
51. $x^3y + 2x^2y - 63xy$.
52. $6y^2 - 7y - 3$.
53. $4x^2 - 12x + 9$.
54. $3 - 5p - 12p^2$.
55. $16 + 8pq + p^2q^2$.
56. $4z^3 + 5z^2 - 6z$.
57. $a^3 + a^2 - 42a$.
58. $2m^4 - m^3 + 4m - 2$.
59. $a^4 - 3a^3 - a^3b + 3a^2b$.
60. $14 - 5x - x^2$.
61. $17 - 18z + z^2$.
62. $2m^4 - 11m^2 - 21$.
63. $5x^2 + 7xy - 6y^2$.
64. $6m^6 + 17m^3 - 45$.
65. $9m^2 - 24m + 16$.

(On Arts. 125–136$_A$.)

66. $25 - 81a^2$.
67. $a^4b^4 - 9$.
68. $27 + l^3$.
69. $1 - 64m^3$.
70. $k^4 - 25l^2$.
71. $p^3q^3 - 1$.
72. $8z^3 + 1$.
73. $1 - 64x^2$.
74. $250p^3 + 2$.
75. $100a^2b^4 - 4$.
76. $729 + c^3d^3$.
77. $(a+x)^2 - 1$.
78. $16 - (b-c)^2$.
79. $9x^3 - 4xy^2$.
80. $p^2 - pq - 20q^2$.
81. $l^3 - l^2 - 42l$.
82. $a^2b^2c^2 - 81d^2$.
83. $64x^6 - 27y^3$.
84. $x^2 + 2x - 323$.
85. $x^4 - 289$.
86. $l^2 + l - 272$.
87. $1000z^3 - 27$.
88. $a^2 + 10a - 299$.
89. $a^2 - b^2 - 2bc - c^2$.
90. $1 - x^2 + 6xy - 9y^2$.
91. $x^4 + y^4 - 7x^2y^2$.
92. $a^4 + 3a^2 + 4$.
93. $b^2 - 2b - 783$.

137. Miscellaneous cases of resolution into factors.

Example 1. Resolve into factors $16a^4 - 81b^4$.
$$16a^4 - 81b^4 = (4a^2 + 9b^2)(4a^2 - 9b^2)$$
$$= (4a^2 + 9b^2)(2a + 3b)(2a - 3b).$$

Example 2. Resolve into factors $x^6 - y^6$.
$$x^6 - y^6 = (x^3 + y^3)(x^3 - y^3)$$
$$= (x + y)(x^2 - xy + y^2)(x - y)(x^2 + xy + y^2).$$

Note. When an expression can be arranged either as the difference of two squares, or as the difference of two cubes, it will be found simplest to first use the rule for the difference of two squares.

Example 3. Resolve into factors $28x^4y + 64x^3y - 60x^2y$.
$$28x^4y + 64x^3y - 60x^2y = 4x^2y(7x^2 + 16x - 15)$$
$$= 4x^2y(7x - 5)(x + 3).$$

Example 4. Resolve into factors $x^3p^2 - 8y^3p^2 - 4x^3q^2 + 32y^3q^2$.
The expression $= p^2(x^3 - 8y^3) - 4q^2(x^3 - 8y^3)$
$$= (x^3 - 8y^3)(p^2 - 4q^2)$$
$$= (x - 2y)(x^2 + 2xy + 4y^2)(p + 2q)(p - 2q).$$

Example 5. Resolve into factors $4x^2 - 25y^2 + 2x + 5y$.
$$4x^2 - 25y^2 + 2x + 5y = (2x + 5y)(2x - 5y) + 2x + 5y$$
$$= (2x + 5y)(2x - 5y + 1).$$

EXAMPLES XVII. 1. (*Continued.*)

Resolve into two or more factors:

94. $x^6 - 64$.   95. $729y^6 - 64x^6$.   96. $x^8 - 1$.
97. $729a^7b - ab^7$.   98. $a^8x^6 - 64a^2y^6$.   99. $a^{12} - b^{12}$.
100. $x^4 + 4x^2y^2z^2 + 4y^4z^4$.   101. $a^3b^3 + 512$.   102. $2x^2 + 17x + 35$.
103. $500x^2y - 20y^3$.   104. $(a + b)^4 - 1$.   105. $(c + d)^3 - 1$.
106. $1 - (x - y)^3$.   107. $x^2 - 6x - 247$.   108. $a^2 - 22a + 279$.
109. $250(a - b)^3 + 2$.   110. $(c + d)^3 + (c - d)^3$.
111. $8(x + y)^3 - (2x - y)^3$.   112. $x^2 - 4y^2 + x - 2y$.
113. $a^2 - b^2 + a - b$.   114. $(a + b)^2 + a + b$.
115. $a^3 + b^3 + a + b$.   116. $a^2 - 9b^2 + a + 3b$.
117. $4(x - y)^3 - (x - y)$.   118. $x^4y - x^2y^3 - x^3y^2 + xy^4$.

[*Miscellaneous Examples* IV., p. 174, *and* Chapter XXVIII. *will furnish further practice in Resolution into Factors.*]

MISCELLANEOUS EXAMPLES III.

1. Subtract $3x^3 - 7x + 1$ from $2x^2 - 5x - 3$, then subtract the difference from zero, and add this last result to $2x^2 - 2x^3 - 4$.

2. Simplify
$$2\{3a - (4b - 5c)\} + 4\{4a - (5b - 2c)\} + 4\{5a - 3(b - c)\}.$$

3. Find the product of
$$a^3 - 2a^2c + 2ac^2 - c^3$$
and
$$a^3 + 2a^2c + 2ac^2 + c^3.$$

4. Solve the equations:

(1) $\dfrac{x}{2} + \dfrac{x}{3} = \dfrac{x}{4} + 7$;

(2) $9x + 5y = 75$,
$7x - 4y = 11$.

5. Find the square root of $8x^4 + 16x^2 + 1 - 8x - 2x^3 + x^6$.

6. Find a number whose third, fourth, sixth, and eighth parts together make up 63.

---

7. If $a = 4$, $b = 3$, $c = 2$, find the value of $\dfrac{a^2 - b^2}{b + c} + \dfrac{b^2 - c^2}{c + a} + \dfrac{c^2 - a^2}{a + b}$.

8. Divide $x^4 + \dfrac{9}{4}x^3 + \dfrac{21}{8}x^2 + \dfrac{33}{16}x + \dfrac{5}{16}$ by $x^2 + \dfrac{3x}{2} + \dfrac{1}{4}$.

9. Add $5x^2 - 6x$ to the excess of 1 over $3x^2 - 5x + 1$.

10. Find the factors of (1) $a^2x^2 - 2ax - 15$; (2) $4m^4 - 81p^2q^2$.

11. Solve the equations:

(1) $13x + 11y = 18$,
$11x + 13y = 30$.

(2) $57x + 52y = 181$,
$76x - 39y = 458$.

12. A train which travels a miles in b hours is p times as fast as a coach. If the coach takes m hours to cover the distance between two places, how many miles are they apart?

---

13. Find the continued product of $3x^2 - 2x + 3$, $4x + 5$, $7x - 2$.

14. Solve the equations:

(1) $\dfrac{5x}{7} - \dfrac{4}{5}\left(x - \dfrac{3}{4}\right) - \dfrac{2}{21}\left(x + \dfrac{7}{2}\right) + 1 = 0$;

(2) $2\left(\dfrac{5x}{3} - 1\right) + \dfrac{11}{5}\left(1 + \dfrac{14x}{33}\right) = \dfrac{2x + 7}{5} - 7$.

15. Write down the square of $x^2 + 7x - 11$.

MISCELLANEOUS EXAMPLES. III.

16. Resolve into factors:
  (1) $x^2 + 2ax - bx - 2ab$;   (2) $x^4 + 10x^2y - 56y^2$.

17. Find the H.C.F. and L.C.M. of $49bc^3$, $21a^2b^2$, $56ca^3$, $63abc^2$.

18. A has £50, and B has £6; after B has won from A a certain sum he then has five-ninths of what A has: how much did B win?

---

19. Simplify $\dfrac{15a^2p^3}{56mk^2} \times \dfrac{49ak^5}{40p^2m^3} \div \dfrac{7a^3k^3}{64m^4}$.

20. Shew that $a(a-1)(a-2)(a-3) = (a^2 - 3a + 1)^2 - 1$.

21. Express by means of symbols:
  (1) The excess of m over n is greater than a by c;
  (2) Three times the square of ab together with the cube of c is equal to p times the sum of m and n.

22. Solve $\dfrac{x}{4}\left(3 - \dfrac{8}{x}\right) - \dfrac{7}{8}\left(7 - \dfrac{3x}{4}\right) = 15\left(\dfrac{1}{3} - \dfrac{x}{64}\right)$,

and shew that $x = 2$ does *not* satisfy the equation.

23. Divide the product of $3x^2 - 2xy - y^2$ and $2x - y$ by $x - y$.

24. What is the price of apples per dozen, and of eggs per score, when 60 apples and 100 eggs together cost 8s. 4d., and 72 apples cost as much as 30 eggs?

---

25. Express the product $(2x^2 - 13x + 15)(x^2 - 4x - 5)(2x^2 - x - 3)$ in simple factors, and thence write down its square root as the product of three binomial factors.

26. If $x = 6$, $y = 7$, $z = 8$, find the value of
$$x - (y - z) - 2[x + z - 3\{-2(y - 1)\}] + 4\left[\dfrac{x}{2} - \left(3 - \dfrac{9}{2}y\right)\right].$$

27. Divide $6x^5 + 57x^4y + 128x^3y^2 - 60x^2y^3 - 130xy^4 + 63y^5$
by $3x^3 + 15x^2y + 7xy^2 - 9y^3$.

28. Solve the equations:
$$4x + 2y + z = 14, \quad 3x - y + 2z = 3, \quad x + 7y - z = 23.$$

29. Resolve into two or more factors:
  (1) $x^3y - 4xy^3$;   (2) $2m^4 + m^2n^2 - 3n^4$.

30. In how many days will a men do $\dfrac{1}{m}$th of a piece of work, the whole of which can be done by b men in c days?

If $m = 4$, $a = 24$, $b = 14$, $c = 18$, what is the numerical value of the answer?

CHAPTER XVIII.

Highest Common Factor.

138. Definition. The **highest common factor** of two or more algebraical expressions is the *expression of highest dimensions* which divides each of them without remainder.

Note. The term *greatest common measure* is sometimes used instead of *highest common factor*; but, strictly speaking, the term *greatest common measure* ought to be confined to arithmetical quantities; for the highest common factor is not necessarily the greatest common measure in all cases, as will appear later. [Art. 145.]

In Chap. XI. we have explained how to write down by inspection the highest common factor of two or more *simple* expressions. An analogous method will enable us readily to find the highest common factor of *compound* expressions which are given as the product of factors, or which can be easily resolved into factors.

Example 1. Find the highest common factor of
$$4cx^3 \text{ and } 2cx^3 + 4c^2x^2.$$

It will be easy to pick out the common factors if the expressions are arranged as follows:
$$4cx^3 = 4cx^3,$$
$$2cx^3 + 4c^2x^2 = 2cx^2(x + 2c);$$
therefore the H.C.F. is $2cx^2$.

Example 2. Find the highest common factor of
$$3a^2 + 9ab, \quad a^3 - 9ab^2, \quad a^3 + 6a^2b + 9ab^2.$$

Resolving each expression into its factors, we have
$$3a^2 + 9ab = 3a(a + 3b),$$
$$a^3 - 9ab^2 = a(a + 3b)(a - 3b),$$
$$a^3 + 6a^2b + 9ab^2 = a(a + 3b)(a + 3b);$$
therefore the H.C.F. is $a(a + 3b)$.

CHAP. XVIII.]   HIGHEST COMMON FACTOR.   129

139. When there are two or more expressions containing different powers of the same *compound* factor, the student should be careful to notice that the highest common factor must contain the highest power of the compound factor which is common to all the given expressions.

Example 1. The highest common factor of
$$x(a-x)^2, \quad a(a-x)^3, \quad \text{and} \quad 2ax(a-x)^5 \quad \text{is} \quad (a-x)^2.$$

Example 2. Find the highest common factor of
$$ax^2 + 2a^2x + a^3, \quad 2ax^2 - 4a^2x - 6a^3, \quad 3(ax+a^2)^2.$$

Resolving the expressions into factors, we have
$$ax^2 + 2a^2x + a^3 = a(x^2 + 2ax + a^2)$$
$$= a(x+a)^2 \quad \ldots\ldots\ldots\ldots\ldots (1),$$
$$2ax^2 - 4a^2x - 6a^3 = 2a(x^2 - 2ax - 3a^2)$$
$$= 2a(x+a)(x-3a) \quad \ldots\ldots (2),$$
$$3(ax+a^2)^2 = 3a^2(x+a)^2 \quad \ldots\ldots\ldots\ldots\ldots (3).$$

Therefore from (1), (2), (3), by inspection, the highest common factor is $a(x+a)$.

EXAMPLES XVIII. a.

Find the highest common factor of

1. $a^2 + ab, \quad a^2 - b^2.$
2. $(x+y)^2, \quad x^2 - y^2.$
3. $2x^2 - 2xy, \quad x^3 - x^2y.$
4. $6x^2 - 9xy, \quad 4x^2 - 9y^2.$
5. $x^3 + x^2y, \quad x^3 + y^3.$
6. $a^3b - ab^3, \quad a^5b^2 - a^2b^5.$
7. $a^3 - a^2x, \quad a^3 - ax^2, \quad a^4 - ax^3.$
8. $a^2 - 4x^2, \quad a^2 + 2ax.$
9. $a^2bx + ab^2x, \quad a^2b - b^3.$
10. $2x^2y - 6xy^2, \quad x^2 - 9y^2.$
11. $a^2 - x^2, \quad a^2 - ax, \quad a^2x - ax^2.$
12. $4x^2 + 2xy, \quad 12x^2y - 3y^3.$
13. $20x - 4, \quad 50x^2 - 2.$
14. $6bx + 4by, \quad 9cx + 6cy.$
15. $x^2 + x, \quad (x+1)^2, \quad x^3 + 1.$
16. $xy - y, \quad x^4y - xy.$
17. $x^2 - 2xy + y^2, \quad (x-y)^3.$
18. $x^3 + a^2x, \quad x^4 - a^4.$
19. $x^3 + 8y^3, \quad x^2 + xy - 2y^2.$
20. $x^4 - 27a^3x, \quad (x-3a)^2.$
21. $x^2 + 3x + 2, \quad x^2 - 4.$
22. $x^2 - x - 20, \quad x^2 - 9x + 20.$
23. $x^2 - 18x + 45, \quad x^2 - 9.$
24. $2x^2 - 7x + 3, \quad 3x^2 - 7x - 6.$
25. $12x^2 + x - 1, \quad 15x^2 + 8x + 1.$
26. $2x^2 - x - 1, \quad 3x^2 - x - 2.$
27. $c^2x^2 - d^2, \quad acx^2 - bcx + adx - bd.$
28. $x^5 - xy^2, \quad x^3 + x^2y + xy + y^2.$
29. $a^3x - a^2bx - 6ab^2x, \quad a^2bx^2 - 4ab^2x^2 + 3b^3x^2.$
30. $2x^2 + 9x + 4, \quad 2x^2 + 11x + 5, \quad 2x^2 - 3x - 2.$
31. $3x^4 + 8x^3 + 4x^2, \quad 3x^5 + 11x^4 + 6x^3, \quad 3x^4 - 16x^3 - 12x^2.$

[*If preferred, the remainder of this chapter may be taken after* Chap. XXV.]

*140. The highest common factor should always be found by inspection if possible, but it may happen that the expressions cannot be readily resolved into factors. In such cases we adopt a method analogous to that used in Arithmetic, for finding the greatest common measure of two or more numbers.

*141. We shall now illustrate the algebraical process of finding the highest common factor by examples, postponing for the present the complete proof of the rules we use. But we shall *enunciate* two principles, which the student should bear in mind in reading the examples which follow.

I. *If an expression contains a certain factor, any multiple of the expression is divisible by that factor.*

II. *If two expressions have a common factor, it will divide their sum and their difference; and also the sum and the difference of any multiples of them.*

Example. Find the highest common factor of
$$4x^3 - 3x^2 - 24x - 9 \text{ and } 8x^3 - 2x^2 - 53x - 39.$$

$$
\begin{array}{r|lr|l}
x & 4x^3 - 3x^2 - 24x - 9 & 8x^3 - 2x^2 - 53x - 39 & 2 \\
 & 4x^3 - 5x^2 - 21x & 8x^3 - 6x^2 - 48x - 18 & \\ \hline
2x & 2x^2 - 3x - 9 & 4x^2 - 5x - 21 & 2 \\
 & 2x^2 - 6x & 4x^2 - 6x - 18 & \\ \hline
3 & 3x - 9 & x - 3 & \\
 & 3x - 9 & & \\
\end{array}
$$

Therefore the H.C.F. is $x - 3$.

Explanation. First arrange the given expressions according to descending or ascending powers of x. The expressions so arranged having their first terms of the same order, we take for divisor that whose highest power has the smaller coefficient. Arrange the work in parallel columns as above. When the first remainder $4x^2 - 5x - 21$ is made the divisor we put the quotient x to the *left* of the dividend. Again, when the second remainder $2x^2 - 3x - 9$ is in turn made the divisor, the quotient 2 is placed to the *right*; and so on. As in Arithmetic, the last divisor $x - 3$ is the highest common factor required.

*142. This method is only useful to determine the *compound* factor of the highest common factor. Simple factors of the given expressions must be first removed from them, and the highest common factor of these, if any, must be observed and multiplied into the *compound* factor given by the rule.

XVIII.] HIGHEST COMMON FACTOR. 131

Example. Find the highest common factor of
$$24x^4 - 2x^3 - 60x^2 - 32x \text{ and } 18x^4 - 6x^3 - 39x^2 - 18x.$$

We have $24x^4 - 2x^3 - 60x^2 - 32x = 2x(12x^3 - x^2 - 30x - 16)$,
and $18x^4 - 6x^3 - 39x^2 - 18x = 3x(6x^3 - 2x^2 - 13x - 6)$.

Also $2x$ and $3x$ have the common factor x. Removing the simple factors $2x$ and $3x$, and *reserving* their common factor x, we continue as in Art. 141.

$$
\begin{array}{r|lr|lr}
2x & 6x^3 - 2x^2 - 13x - 6 & & 12x^3 - x^2 - 30x - 16 & 2 \\
 & 6x^3 - 8x^2 - 8x & & 12x^3 - 4x^2 - 26x - 12 & \\
\cline{2-2}\cline{4-4}
2 & 6x^2 - 5x - 6 & & 3x^2 - 4x - 4 & x \\
 & 6x^2 - 8x - 8 & & 3x^2 + 2x & \\
\cline{2-2}\cline{4-4}
 & 3x + 2 & & -6x - 4 & -2 \\
 & & & -6x - 4 &
\end{array}
$$

Therefore the H.C.F. is $x(3x+2)$.

*143. So far the process of Arithmetic has been found exactly applicable to the algebraical expressions we have considered. But in many cases certain modifications of the arithmetical method will be found necessary. These will be more clearly understood if it is remembered that, at every stage of the work, the remainder must contain as a factor of itself the highest common factor we are seeking. [See Art. 141, I. & II.].

Example 1. Find the highest common factor of
$$3x^3 - 13x^2 + 23x - 21 \text{ and } 6x^3 + x^2 - 44x + 21.$$

$$
\begin{array}{r|lr}
3x^3 - 13x^2 + 23x - 21 & 6x^3 + x^2 - 44x + 21 & 2 \\
 & 6x^3 - 26x^2 + 46x - 42 & \\
\cline{2-2}
 & 27x^2 - 90x + 63 &
\end{array}
$$

Here on making $27x^2 - 90x + 63$ a divisor, we find that it is not contained in $3x^3 - 13x^2 + 23x - 21$ with an *integral* quotient. But noticing that $27x^2 - 90x + 63$ may be written in the form $9(3x^2 - 10x + 7)$, and also bearing in mind that every remainder in the course of the work contains the H.C.F., we conclude that the H.C.F. we are seeking is contained in $9(3x^2 - 10x + 7)$. But the two original expressions have no *simple* factors, therefore their H.C.F. can have none. We may therefore *reject* the factor 9 and go on with divisor $3x^2 - 10x + 7$.

132   ALGEBRA.   [CHAP.

Resuming the work, we have

$$\begin{array}{r|rr|r}
x & 3x^3-13x^2+23x-21 & 3x^2-10x+7 & x \\
 & 3x^3-10x^2+7x & 3x^2-7x & \\
\cline{2-3}
-1 & -3x^2+16x-21 & -3x+7 & -1 \\
 & -3x^2+10x-7 & -3x+7 & \\
\cline{2-3}
 & 2)\,6x-14 & & \\
\cline{2-2}
 & 3x-7 & &
\end{array}$$

Therefore the H.C.F. is $3x-7$.

The factor 2 has been removed on the same grounds as the factor 9 above.

Example 2. Find the highest common factor of

$$2x^3+x^2-x-2 \dots\dots\dots\dots\dots\dots\dots(1),$$

and

$$3x^3-2x^2+x-2 \dots\dots\dots\dots\dots\dots\dots(2),$$

As the expressions stand we cannot begin to divide one by the other without using a fractional quotient. The difficulty may be obviated by *introducing* a suitable factor, just as in the last case we found it useful to remove a factor when we could no longer proceed with the division in the ordinary way. The given expressions have no common *simple* factor, hence their H.C.F. cannot be affected if we multiply either of them by any simple factor.

Multiply (2) by 2, and use (1) as a divisor:

$$\begin{array}{r|rr|r}
 & 2x^3+x^2-x-2 & 6x^3-4x^2+2x-4 & 3 \\
 & 7 & 6x^3+3x^2-3x-6 & \\
\cline{2-3}
-2x & 14x^3+7x^2-7x-14 & -7x^2+5x+2 & \\
 & 14x^3-10x^2-4x & 17 & \\
\cline{2-3}
17x & 17x^2-3x-14 & -119x^2+85x+34 & -7 \\
 & 17x^2-17x & -119x^2+21x+98 & \\
\cline{2-3}
14 & 14x-14 & 64)\,64x-64 & \\
 & 14x-14 & x-1 &
\end{array}$$

Therefore the H.C.F. is $x-1$.

After the first division the factor 7 is introduced because the first remainder $-7x^2+5x+2$ will not divide $2x^3+x^2-x-2$. At the next stage the factor 17 is introduced for a similar reason, and finally the factor 64 is removed as explained in Example 1.

Note. Here the highest common factor might have been more easily obtained by arranging the expressions in *ascending* powers of x. In this case it will be found that there is no need to introduce a numerical factor in the course of the work. Detached coefficients, as explained in Art. 45, may also be used with advantage here, and will often effect a considerable saving of labour.

*144. From the last two examples it appears that we may multiply or divide either of the given expressions, or any of the remainders which occur in the course of the work, by any factor which does not divide both of the given expressions.

*145. Let the two expressions in Example 2, Art. 143, be written in the form
$$2x^3+x^2-x-2=(x-1)(2x^2+3x+2),$$
$$3x^3-2x^2+x-2=(x-1)(3x^2+x+2).$$
Then their highest common factor is $x-1$, and therefore $2x^2+3x+2$ and $3x^2+x+2$ *have no algebraical common divisor*. If, however, we put $x=6$, then
$$2x^3+x^2-x-2=460,$$
and $$3x^3-2x^2+x-2=580;$$
and the greatest common measure of 460 and 580 is 20; whereas 5 is the numerical value of $x-1$, the algebraical highest common factor. Thus the numerical values of the algebraical highest common factor and of the arithmetical greatest common measure do not in this case agree.

The reason may be explained as follows: when $x=6$, the expressions $2x^2+3x+2$ and $3x^2+x+2$ become equal to 92 and 116 respectively, and have a common arithmetical factor 4; whereas the expressions have no algebraical common factor.

It will thus often happen that the highest common factor of two expressions, and their numerical greatest common measure, when the letters have particular values, are not the same; for this reason the term *greatest common measure* is inappropriate when applied to algebraical quantities.

*EXAMPLES XVIII. b.

Find the highest common factor of the following expressions:

1. $x^3+2x^2-13x+10, \quad x^3+x^2-10x+8$.
2. $x^3-5x^2-99x+40, \quad x^3-6x^2-86x+35$.
3. $x^3+2x^2-8x-16, \quad x^3+3x^2-8x-24$.
4. $x^3+4x^2-5x-20, \quad x^3+6x^2-5x-30$.
5. $x^3-x^2-5x-3, \quad x^3-4x^2-11x-6$.
6. $x^3+3x^2-8x-24, \quad x^3+3x^2-3x-9$.
7. $a^3-5a^2x+7ax^2-3x^3, \quad a^3-3ax^2+2x^3$.
8. $x^4-2x^3-4x-7, \quad x^4+x^3-3x^2-x+2$.
9. $2x^3-5x^2+11x+7, \quad 4x^3-11x^2+25x+7$.

Find the highest common factor of the following expressions:

10. $2x^3 + 4x^2 - 7x - 14,\ \ 6x^3 - 10x^2 - 21x + 35.$
11. $3x^4 - 3x^3 - 2x^2 - x - 1,\ \ 9x^4 - 3x^3 - x - 1.$
12. $2x^4 - 2x^3 + x^2 + 3x - 6,\ \ 4x^4 - 2x^3 + 3x - 9.$
13. $3x^3 - 3ax^2 + 2a^2x - 2a^3,\ \ 3x^3 + 12ax^2 + 2a^2x + 8a^3.$
14. $2x^3 - 9ax^2 + 9a^2x - 7a^3,\ \ 4x^3 - 20ax^2 + 20a^2x - 16a^3.$
15. $10x^3 + 25ax^2 - 5a^3,\ \ 4x^3 + 9ax^2 - 2a^2x - a^3.$
16. $6a^3 + 13a^2x - 9ax^2 - 10x^3,\ \ 9a^3 + 12a^2x - 11ax^2 - 10x^3.$
17. $24x^4y + 72x^3y^2 - 6x^2y^3 - 90xy^4,\ \ 6x^4y^2 + 13x^3y^3 - 4x^2y^4 - 15xy^5.$
18. $4x^5a^2 + 10x^4a^3 - 60x^3a^4 + 54x^2a^5,\ \ 24x^5a^3 + 30x^3a^5 - 126x^2a^6.$
19. $4x^5 + 14x^4 + 20x^3 + 70x^2,\ \ 8x^7 + 28x^6 - 8x^5 - 12x^4 + 56x^3.$
20. $72x^3 - 12ax^2 + 72a^2x - 420a^3,\ \ 18x^3 + 42ax^2 - 282a^2x + 270a^3.$
21. $9x^4 + 2x^2y^2 + y^4,\ \ 3x^4 - 8x^3y + 5x^2y^2 - 2xy^3.$
22. $x^5 - x^3 - x + 1,\ \ x^7 + x^6 + x^4 - 1.$
23. $1 + x + x^3 - x^5,\ \ 1 - x^4 - x^6 + x^7.$
24. $6 - 8a - 32a^2 - 18a^3,\ \ 20 - 35a - 95a^2 - 40a^3.$
25. $9x^2 - 15x^3 - 45x^4 - 12x^5,\ \ 42x - 49x^2 - 203x^3 - 84x^4.$
26. $3x^5 - 5x^3 + 2,\ \ 2x^5 - 5x^2 + 3.$
27. $4x^5 - 6x^3 - 28x,\ \ 6x^4 + 10x^3 - 17x^2 - 35x - 14.$

*146. The statements of Art. 141 may be proved as follows.

I. If F divides A it will also divide mA.

For suppose $A = aF$, then $mA = maF$.

Thus F is a factor of mA.

II. If F divides A and B, then it will divide $mA \pm nB$.

For suppose $A = aF,\ B = bF$,

then
$$mA \pm nB = maF \pm nbF$$
$$= F(ma \pm nb).$$

Thus F divides $mA \pm nB$.

*147. We may now enunciate and prove the rule for finding the highest common factor of any two compound algebraical expressions.

We suppose that any simple factors are first removed. [See Example, Art. 142.]

Let A and B be the two expressions after the simple factors have been removed. Let them be arranged in descending or ascending powers of some common letter; also let the highest power of that letter in B be not less than the highest power in A.

Divide B by A; let p be the quotient, and C the remainder. Suppose C to have a *simple* factor m. Remove this factor, and so obtain a new divisor D. Further, suppose that in order to make A divisible by D it is necessary to multiply A by a *simple* factor n. Let q be the next quotient and E the remainder. Finally, divide D by E; let r be the quotient, and suppose that there is no remainder. Then E will be the H.C.F. required.

The work will stand thus:

$$\begin{array}{r} A)B(p \\ pA \\ \hline m)C \\ D)nA(q \\ qD \\ \hline E)D(r \\ rE \\ \hline \end{array}$$

First, to shew that E is a common factor of A and B.

By examining the steps of the work, it is clear that E divides D, therefore also qD; therefore $qD + E$, therefore nA; therefore A, since n is a *simple* factor.

Again, E divides D, therefore mD, that is, C. And since E divides A and C, it also divides $pA + C$, that is, B. Hence E divides both A and B.

Secondly, to show that E is the *highest* common factor.

If not, let there be a factor X of higher dimensions than E.

Then X divides A and B, therefore $B - pA$, that is, C; therefore D (since m is a *simple* factor); therefore $nA - qD$, that is, E.

Thus X divides E; which is impossible since by hypothesis, X is of higher dimensions than E.

Therefore E is the highest common factor.

*148. The highest common factor of three expressions A, B, C may be obtained as follows.

First determine F the highest common factor of A and B; next find G the highest common factor of F and C; then G will be the required highest common factor of A, B, C.

For F contains *every* factor which is common to A and B, and G is the highest common factor of F and C. Therefore G is the highest common factor of A, B, C.

CHAPTER XIX.

FRACTIONS.

[*On first reading the subject, the student may omit the general proofs of the rules given in this chapter.*

The articles and examples marked with an asterisk must be omitted by those who adopt the suggestion printed at the top of page 130.]

149. In Chapter XII. we discussed the simpler kinds of fractions, using the ordinary arithmetical rules. We here propose to give proofs of those rules, and shew that they are applicable to algebraical fractions.

DEFINITION. If a quantity x be divided into b equal parts, and a of these parts be taken, the result is called *the fraction* $\frac{a}{b}$ *of* x. If x be the unit, the fraction $\frac{a}{b}$ of x is called simply "the fraction $\frac{a}{b}$"; so that *the fraction* $\frac{a}{b}$ *represents* a *equal parts,* b *of which make up the unit.*

Note. This definition requires that a and b should be positive whole numbers. In Art. 155 we shall adopt a definition which will enable us to remove this restriction.

150. *To prove that* $\frac{a}{b} = \frac{ma}{mb}$, *where* a, b, m *are positive integers.*

By $\frac{a}{b}$ we mean a equal parts, b of which make up the unit ... (1);

by $\frac{ma}{mb}$ .... ma .... mb .... (2).

But $\qquad\qquad b$ parts in (1) $= mb$ parts in (2);

$\qquad\qquad\therefore$ 1 part .... $= m$ ....

$\qquad\qquad\therefore$ a parts .... $= ma$ ....

that is, $\qquad\qquad\qquad\qquad \frac{a}{b} = \frac{ma}{mb}.$

Conversely, $\qquad\qquad\qquad \frac{ma}{mb} = \frac{a}{b}.$

Hence, *the value of a fraction is not altered if we multiply or divide the numerator and denominator by the same quantity.*

Reduction to Lowest Terms.

151. An algebraical fraction may be changed into an equivalent fraction by dividing numerator and denominator by any common factor; if this factor be the highest common factor the resulting fraction is said to be **reduced to its lowest terms.**

Example 1. Reduce to lowest terms $\dfrac{24a^3c^2x^2}{18a^3x^2 - 12a^2x^3}$.

The expression $= \dfrac{24a^3c^2x^2}{6a^2x^2(3a - 2x)} = \dfrac{4ac^2}{3a - 2x}$.

Example 2. Reduce to lowest terms $\dfrac{6x^2 - 8xy}{9xy - 12y^2}$.

The expression $= \dfrac{2x(3x - 4y)}{3y(3x - 4y)} = \dfrac{2x}{3y}$.

Note. The beginner should be careful not to begin cancelling until he has expressed both numerator and denominator in the most convenient form, by resolution into factors where necessary.

EXAMPLES XIX. a.

Reduce to lowest terms:

1. $\dfrac{3a^2 - 6ab}{2a^2b - 4ab^2}$.
2. $\dfrac{abx + bx^2}{acx + cx^2}$.
3. $\dfrac{ax}{a^2x^2 - ax}$.
4. $\dfrac{15a^2b^2c}{100(a^3 - a^2b)}$.
5. $\dfrac{4x^2 - 9y^2}{4x^2 + 6xy}$.
6. $\dfrac{20(x^3 - y^3)}{5x^2 + 5xy + 5y^2}$.
7. $\dfrac{x(2a^2 - 3ax)}{a(4a^2x - 9x^3)}$.
8. $\dfrac{x^3 - 2xy^2}{x^4 - 4x^2y^2 + 4y^4}$.
9. $\dfrac{(xy - 3y^2)^2}{x^3y^2 - 27y^5}$.
10. $\dfrac{x^2 - 5x}{x^2 - 4x - 5}$.
11. $\dfrac{3x^2 + 6x}{x^2 + 4x + 4}$.
12. $\dfrac{5a^3b + 10a^2b^2}{3a^2b^2 + 6ab^3}$.
13. $\dfrac{x^3y + 2x^2y + 4xy}{x^3 - 8}$.
14. $\dfrac{3a^4 + 9a^3b + 6a^2b^2}{a^4 + a^3b - 2a^2b^2}$.
15. $\dfrac{x^4 - 14x^2 - 51}{x^4 - 2x^2 - 15}$.
16. $\dfrac{x^2 + xy - 2y^2}{x^3 - y^3}$.
17. $\dfrac{2x^2 + 17x + 21}{3x^2 + 26x + 35}$.
18. $\dfrac{a^2x^2 - 16a^2}{ax^2 + 9ax + 20a}$.
19. $\dfrac{3x^2 + 23x + 14}{3x^2 + 41x + 26}$.
20. $\dfrac{27a + a^4}{18a - 6a^2 + 2a^3}$.

*152. When the factors of the numerator and denominator cannot be determined by inspection, the fraction may be reduced to its lowest terms by dividing both numerator and denominator by the highest common factor, which may be found by the rules given in Chap. XVIII.

Example. Reduce to lowest terms $\dfrac{3x^3 - 13x^2 + 23x - 21}{15x^3 - 38x^2 - 2x + 21}$.

First Method. The H.C.F. of numerator and denominator is $3x - 7$.

Dividing numerator and denominator by $3x - 7$, we obtain as respective quotients $x^2 - 2x + 3$ and $5x^2 - x - 3$.

Thus $\dfrac{3x^3 - 13x^2 + 23x - 21}{15x^3 - 38x^2 - 2x + 21} = \dfrac{(3x - 7)(x^2 - 2x + 3)}{(3x - 7)(5x^2 - x - 3)} = \dfrac{x^2 - 2x + 3}{5x^2 - x - 3}$.

This is the simplest solution for the beginner; but in this and similar cases we may often effect the reduction without actually going through the process of finding the highest common factor.

Second Method. By Art. 141, the H.C.F. of numerator and denominator must be a factor of their sum $18x^3 - 51x^2 + 21x$, that is, of $3x(3x - 7)(2x - 1)$. If there be a common divisor it must clearly be $3x - 7$; hence arranging numerator and denominator so as to shew $3x - 7$ as a factor,

$$\text{the fraction } = \frac{x^2(3x - 7) - 2x(3x - 7) + 3(3x - 7)}{5x^2(3x - 7) - x(3x - 7) - 3(3x - 7)}$$
$$= \frac{(3x - 7)(x^2 - 2x + 3)}{(3x - 7)(5x^2 - x - 3)}$$
$$= \frac{x^2 - 2x + 3}{5x^2 - x - 3}.$$

*153. If either numerator or denominator can readily be resolved into factors we may use the following method.

Example. Reduce to lowest terms $\dfrac{x^3 + 3x^2 - 4x}{7x^3 - 18x^2 + 6x + 5}$.

The numerator $= x(x^2 + 3x - 4) = x(x + 4)(x - 1)$.

Of these factors the only one which can be a common divisor is $x - 1$. Hence, arranging the denominator,

$$\text{the fraction} = \frac{x(x + 4)(x - 1)}{7x^2(x - 1) - 11x(x - 1) - 5(x - 1)}$$
$$= \frac{x(x + 4)(x - 1)}{(x - 1)(7x^2 - 11x - 5)} = \frac{x(x + 4)}{7x^2 - 11x - 5}.$$

*EXAMPLES XIX. b.

Reduce to lowest terms:

1. $\dfrac{a^3 - a^2b - ab^2 - 2b^3}{a^3 + 3a^2b + 3ab^2 + 2b^3}.$

2. $\dfrac{x^3 - 5x^2 + 7x - 3}{x^3 - 3x + 2}.$

3. $\dfrac{a^3 + 2a^2 - 13a + 10}{a^3 + a^2 - 10a + 8}.$

4. $\dfrac{2x^3 + 5x^2y - 30xy^2 + 27y^3}{4x^3 + 5xy^2 - 21y^3}.$

5. $\dfrac{4a^3 + 12a^2b - ab^2 - 15b^3}{6a^3 + 13a^2b - 4ab^2 - 15b^3}.$

6. $\dfrac{1 + 2x^2 + x^3 + 2x^4}{1 + 3x^2 + 2x^3 + 3x^4}.$

7. $\dfrac{x^2 - 2x + 1}{3x^3 + 7x - 10}.$

8. $\dfrac{3a^3 - 3a^2b + ab^2 - b^3}{4a^2 - 5ab + b^2}.$

9. $\dfrac{4x^3 + 3ax^2 + a^3}{x^4 + ax^3 + a^3x + a^4}.$

10. $\dfrac{4x^3 - 10x^2 + 4x + 2}{3x^4 - 2x^3 - 3x + 2}.$

11. $\dfrac{16x^4 - 72x^2a^2 + 81a^4}{4x^2 + 12ax + 9a^2}.$

12. $\dfrac{6x^3 + x^2 - 5x - 2}{6x^3 + 5x^2 - 3x - 2}.$

13. $\dfrac{5x^3 + 2x^2 - 15x - 6}{7x^3 - 4x^2 - 21x + 12}.$

14. $\dfrac{4x^4 + 11x^2 + 25}{4x^4 - 9x^2 + 30x - 25}.$

15. $\dfrac{3x^3 - 27ax^2 + 78a^2x - 72a^3}{2x^3 + 10ax^2 - 4a^2x - 48a^3}.$

16. $\dfrac{ax^3 - 5a^2x^2 - 99a^3x + 40a^4}{x^4 - 6ax^3 - 86a^2x^2 + 35a^3x}.$

Multiplication and Division of Fractions.

154. Rule I. To multiply a fraction by an integer: *multiply the numerator by that integer; or, if the denominator be divisible by the integer, divide the denominator by it.*

The rule may be proved as follows:

(1) $\dfrac{a}{b}$ represents a equal parts, b of which make up the unit;

$\dfrac{ac}{b}$ represents ac equal parts, b of which make up the unit;

and the number of parts taken in the second fraction is c times the number taken in the first;

that is $\qquad \dfrac{a}{b} \times c = \dfrac{ac}{b}.$

(2) $\qquad \dfrac{a}{bd} \times d = \dfrac{ad}{bd}$, by the preceding case,

$$= \dfrac{a}{b}.$$

[Art. 151.]

155. By the preceding article

$$\frac{a}{b} \times b = \frac{ab}{b} = a,$$

that is, the fraction $\frac{a}{b}$ is that which must be multiplied by b in order to obtain a. But, by Art. 46, the quantity which must be multiplied by b in order to obtain a is the quotient resulting from the division of a by b; we may therefore define a fraction thus:

the fraction $\frac{a}{b}$ *is the quotient of* a *divided by* b.

156. Rule II. To divide a fraction by an integer: *divide the numerator, if it be divisible, by the integer; or if the numerator be not divisible, multiply the denominator by that integer.*

The rule may be proved as follows:

(1) $\frac{ac}{b}$ represents ac equal parts, b of which make up the unit;

$\frac{a}{b}$ represents a equal parts, b of which make up the unit.

The number of parts taken in the first fraction is c times the number taken in the second. Therefore the second fraction is the quotient of the first fraction divided by c;

that is
$$\frac{ac}{b} \div c = \frac{a}{b}.$$

(2) But if the numerator be not divisible by c, we have

$$\frac{a}{b} = \frac{ac}{bc};$$

$$\therefore \frac{a}{b} \div c = \frac{ac}{bc} \div c$$

$$= \frac{a}{bc}, \text{ by the preceding case.}$$

157. Rule III. To multiply together two or more fractions: *multiply together all the numerators to form a new numerator, and all the denominators to form a new denominator.*

To find the value of $\quad \frac{a}{b} \times \frac{c}{d}.$

Let $\quad x = \frac{a}{b} \times \frac{c}{d}$

Multiplying each side by $b \times d$, we have

$$x \times b \times d = \frac{a}{b} \times \frac{c}{d} \times b \times d$$

$$= \frac{a}{b} \times b \times \frac{c}{d} \times d \qquad \text{[Art. 29.]}$$

$$= a \times c \qquad \text{[Art. 154.]}$$

$$\therefore x \times bd = ac.$$

Dividing each side by bd, we have

$$x = \frac{ac}{bd};$$

$$\therefore \frac{a}{b} \times \frac{c}{d} = \frac{ac}{bd}.$$

Similarly $\quad \dfrac{a}{b} \times \dfrac{c}{d} \times \dfrac{e}{f} = \dfrac{ace}{bdf};$

and so for any number of fractions.

158. Rule IV. To divide one fraction by another: *invert the divisor, and proceed as in multiplication.*

Since division is the inverse of multiplication, we may define the quotient x, when $\dfrac{a}{b}$ is divided by $\dfrac{c}{d}$, to be such that

$$x \times \frac{c}{d} = \frac{a}{b}.$$

Multiplying by $\dfrac{d}{c}$ we have $x \times \dfrac{c}{d} \times \dfrac{d}{c} = \dfrac{a}{b} \times \dfrac{d}{c};$

$$\therefore x = \frac{ad}{bc}.$$

Hence $\quad \dfrac{a}{b} \div \dfrac{c}{d} = \dfrac{ad}{bc} = \dfrac{a}{b} \times \dfrac{d}{c},\qquad$ [Art. 157.]

which proves the rule.

Example 1. Simplify $\dfrac{2a^2 + 3a}{4a^3} \times \dfrac{4a^2 - 6a}{12a + 18}.$

$$\frac{2a^2 + 3a}{4a^3} \times \frac{4a^2 - 6a}{12a + 18} = \frac{a(2a+3)}{4a^3} \times \frac{2a(2a-3)}{6(2a+3)}$$

$$= \frac{2a-3}{12a},$$

by cancelling those factors which are common to both numerator and denominator.

Example 2. Simplify $\dfrac{6x^2 - ax - 2a^2}{ax - a^2} \times \dfrac{x - a}{9x^2 - 4a^2} \div \dfrac{2x + a}{3ax + 2a^2}$.

The expression $= \dfrac{6x^2 - ax - 2a^2}{ax - a^2} \times \dfrac{x - a}{9x^2 - 4a^2} \times \dfrac{3ax + 2a^2}{2x + a}$

$= \dfrac{(3x - 2a)(2x + a)}{a(x - a)} \times \dfrac{x - a}{(3x + 2a)(3x - 2a)} \times \dfrac{a(3x + 2a)}{2x + a}$

$= 1$,

since all the factors cancel each other.

EXAMPLES XIX. c.

Simplify

1. $\dfrac{14x^2 - 7x}{12x^3 + 24x^2} \div \dfrac{2x - 1}{x^2 + 2x}$.

2. $\dfrac{a^2b^2 + 3ab}{4a^2 - 1} \div \dfrac{ab + 3}{2a + 1}$.

3. $\dfrac{x^2 - 4a^2}{ax + 2a^2} \times \dfrac{2a}{x - 2a}$.

4. $\dfrac{a^2 - 121}{a^2 - 4} \div \dfrac{a + 11}{a + 2}$.

5. $\dfrac{16x^2 - 9a^2}{x^2 - 4} \times \dfrac{x - 2}{4x - 3a}$.

6. $\dfrac{25a^2 - b^2}{9a^2x^2 - 4x^2} \times \dfrac{x(3a + 2)}{5a + b}$.

7. $\dfrac{x^2 + 5x + 6}{x^2 - 1} \times \dfrac{x^2 - 2x - 3}{x^2 - 9}$.

8. $\dfrac{x^2 + 3x + 2}{x^2 + 9x + 20} \times \dfrac{x^2 + 7x + 12}{x^2 + 5x + 6}$.

9. $\dfrac{2x^2 + 5x + 2}{x^2 - 4} \times \dfrac{x^2 + 4x}{2x^2 + 9x + 4}$.

10. $\dfrac{2x^2 + 13x + 15}{4x^2 - 9} \div \dfrac{2x^2 + 11x + 5}{4x^2 - 1}$.

11. $\dfrac{x^2 - 14x - 15}{x^2 - 4x - 45} \div \dfrac{x^2 - 12x - 45}{x^2 - 6x - 27}$.

12. $\dfrac{2x^2 - x - 1}{2x^2 + 5x + 2} \times \dfrac{4x^2 + x - 14}{16x^2 - 49}$.

13. $\dfrac{b^4 - 27b}{2b^2 + 5b} \times \dfrac{4b^2 - 25}{2b^2 - 11b + 15}$.

14. $\dfrac{x^3 - 6x^2 + 36x}{x^2 - 49} \div \dfrac{x^4 + 216x}{x^2 - x - 42}$.

15. $\dfrac{64p^2q^2 - z^4}{x^2 - 4} \times \dfrac{(x - 2)^2}{8pq + z^2} \div \dfrac{x^2 - 4}{(x + 2)^2}$.

16. $\dfrac{x^2 - x - 20}{x^2 - 25} \times \dfrac{x^2 - x - 2}{x^2 + 2x - 8} \div \dfrac{x + 1}{x^2 + 5x}$.

17. $\dfrac{x^2 - 18x + 80}{x^2 - 5x - 50} \times \dfrac{x^2 - 6x - 7}{x^2 - 15x + 56} \times \dfrac{x + 5}{x - 1}$.

18. $\dfrac{x^2 - 8x - 9}{x^2 - 17x + 72} \times \dfrac{x^2 - 25}{x^2 - 1} \div \dfrac{x^2 + 4x - 5}{x^2 - 9x + 8}$.

19. $\dfrac{4x^2+x-14}{6xy-14y} \times \dfrac{4x^2}{x^2-4} \times \dfrac{x-2}{4x-7} \div \dfrac{2x^2+4x}{3x^2-x-14}.$

20. $\dfrac{x^2+x-2}{x^2-x-20} \times \dfrac{x^2+5x+4}{x^2-x} \div \left(\dfrac{x^2+3x+2}{x^2-2x-15} \times \dfrac{x+3}{x^2}\right).$

21. $\dfrac{4x^2-16x+15}{2x^2+3x+1} \times \dfrac{x^2-6x-7}{2x^2-17x+21} \times \dfrac{4x^2-1}{4x^2-20x+25}.$

22. $\dfrac{x^4-8x}{x^2-4x-5} \times \dfrac{x^2+2x+1}{x^3-x^2-2x} \div \dfrac{x^2+2x+4}{x-5}.$

23. $\dfrac{(a+b)^2-c^2}{a^2+ab-ac} \times \dfrac{a}{(a+c)^2-b^2} \times \dfrac{(a-b)^2-c^2}{ab-b^2-bc}.$

24. $\dfrac{a^2+2ab+b^2-c^2}{a^2-b^2-c^2-2bc} \times \dfrac{a^2-2ac+c^2-b^2}{b^2-2bc+c^2-a^2}.$

25. $\dfrac{x^2-64}{x^2+24x+128} \times \dfrac{x^2+12x-64}{x^3-64} \div \dfrac{x^2-16x+64}{x^2+4x+16}.$

26. $\dfrac{(a^2+ax)^3}{a^2-x^2} \times \dfrac{(a-x)^2}{a^5+a^2x^3} \times \dfrac{a^2-ax+x^2}{a^3+2a^2x+ax^2}.$

27. $\dfrac{m^3+4m^2n+4mn^2}{3m^2n-5mn^2-2n^3} \times \dfrac{m^2-4n^2}{9m^2-3mn+n^2} \div \dfrac{(m+2n)^3}{27m^3+n^3}.$

28. $\dfrac{1+8x^3}{(2-x)^2} \times \dfrac{4x-x^3}{1-4x^2} \div \dfrac{(1-2x)^2+2x}{2-5x+2x^2}.$

29. $\dfrac{x^2(x-4)^2}{(x+4)^2-4x} \times \dfrac{64-x^3}{16-x^2} \div \dfrac{(x^2-4x)^3}{(x+4)^2}.$

30. $\dfrac{(p+q)^2-r^2}{(p+q+r)^2} \times \dfrac{p^2+pq+pr}{(p-r)^2-q^2} \div \dfrac{p^2-pq+pr}{(p-q)^2-r^2}.$

31. $\dfrac{a^4-x^4}{a^2-2ax+x^2} \div \left(\dfrac{a^2x+x^3}{a^3-x^3} \times \dfrac{a^4+a^2x^2+x^4}{a^2x-ax^2+x^3}\right).$

32. $\dfrac{a^3+8a^2b+15ab^2}{(64a^3-b^3)(a^3+b^3)} \times \dfrac{16a^4-17a^2b^2+b^4}{4a^2+21ab+5b^2} \div \dfrac{a^2+2ab-3b^2}{a^3-a^2b+ab^2}.$

CHAPTER XX.

Lowest Common Multiple.

[*The articles and examples marked with an asterisk must be omitted by those who adopt the suggestion printed at the top of page 130.*]

159. Definition. The **lowest common multiple** of two or more algebraical expressions is the expression of lowest dimensions, which is divisible by each of them without remainder.

In Chapter XI. we have explained how to write down by inspection the lowest common multiple of two or more *simple* expressions; the lowest common multiple of compound expressions which are given as the product of factors, or which can be easily resolved into factors, can be readily found by a similar method.

Example 1. The lowest common multiple of $6x^2(a-x)^2$, $8a^3(a-x)^3$ and $12ax(a-x)^5$ is $24a^3x^2(a-x)^5$.

For it consists of the product of

(1) the L.C.M. of the numerical coefficients;

(2) the lowest power of each factor which is divisible by every power of that factor occurring in the given expressions.

Example 2. Find the lowest common multiple of
$$3a^2+9ab, \quad 2a^3-18ab^2, \quad a^3+6a^2b+9ab^2.$$

$$3a^2+9ab = 3a(a+3b),$$
$$2a^3-18ab^2 = 2a(a+3b)(a-3b),$$
$$a^3+6a^2b+9ab^2 = a(a+3b)(a+3b)$$
$$= a(a+3b)^2.$$

Therefore the L.C.M. is $6a(a+3b)^2(a-3b)$.

EXAMPLES XX. a.

Find the lowest common multiple of

1. $x, \; x^2+x$.
2. $x^2, \; x^2-3x$.
3. $3x^2, \; 4x^2+8x$.
4. $21x^3, \; 7x^2(x+1)$.
5. $x^2-1, \; x^2+x$.
6. $a^2+ab, \; ab+b^2$.
7. $4x^2y-y, \; 2x^2+x$.
8. $6x^2-2x, \; 9x^2-3x$.
9. $x^2+2x, \; x^2+3x+2$.
10. $x^2-3x+2, \; x^2-1$.
11. $x^2+4x+4, \; x^2+5x+6$.
12. $x^2-5x+4, \; x^2-6x+8$.

13. $x^2 - x - 6$, $x^2 + x - 2$, $x^2 - 4x + 3$.
14. $x^2 + x - 20$, $x^2 - 10x + 24$, $x^2 - x - 30$.
15. $x^2 + x - 42$, $x^2 - 11x + 30$, $x^2 + 2x - 35$.
16. $2x^2 + 3x + 1$, $2x^2 + 5x + 2$, $x^2 + 3x + 2$.
17. $3x^2 + 11x + 6$, $3x^2 + 8x + 4$, $x^2 + 5x + 6$.
18. $5x^2 + 11x + 2$, $5x^2 + 16x + 3$, $x^2 + 5x + 6$.
19. $2x^2 + 3x - 2$, $2x^2 + 15x - 8$, $x^2 + 10x + 16$.
20. $3x^2 - x - 14$, $3x^2 - 13x + 14$, $x^2 - 4$.
21. $12x^2 + 3x - 42$, $12x^3 + 30x^2 + 12x$, $32x^2 - 40x - 28$.
22. $3x^4 + 26x^3 + 35x^2$, $6x^2 + 38x - 28$, $27x^3 + 27x^2 - 30x$.
23. $60x^4 + 5x^3 - 5x^2$, $60x^2y + 32xy + 4y$, $40x^3y - 2x^2y - 2xy$.
24. $8x^2 - 38xy + 35y^2$, $4x^2 - xy - 5y^2$, $2x^2 - 5xy - 7y^2$.
25. $12x^2 - 23xy + 10y^2$, $4x^2 - 9xy + 5y^2$, $3x^2 - 5xy + 2y^2$.
26. $6ax^3 + 7a^2x^2 - 3a^3x$, $3a^2x^2 + 14a^3x - 5a^4$, $6x^2 + 39ax + 45a^2$.
27. $4ax^2y^2 + 11axy^2 - 3ay^2$, $3x^3y^3 + 7x^2y^3 - 6xy^3$, $24ax^2 - 22ax + 4a$.
28. $(3x - 5x^2)^2$, $6 - 7x - 5x^2$, $4x + 4x^2 + x^3$.
29. $14a^4(a^3 - b^3)$, $21a^2b^2(a - b)^3$, $6a^3b(a - b)(a^2 - b^2)$.
30. $m^4 + m^2n^2 + n^4$, $m^3n + n^4$, $(m^2 - mn)^3$.
31. $(2c^2 - 3cd)^2$, $(4c - 6d)^3$, $8c^3 - 27d^3$.

*160. When the given expressions are such that their factors cannot be determined by inspection, they must be resolved by finding the highest common factor.

Example. Find the lowest common multiple of
$$2x^4 + x^3 - 20x^2 - 7x + 24 \text{ and } 2x^4 + 3x^3 - 13x^2 - 7x + 15.$$

The highest common factor is $x^2 + 2x - 3$.

By division, we obtain
$$2x^4 + x^3 - 20x^2 - 7x + 24 = (x^2 + 2x - 3)(2x^2 - 3x - 8).$$
$$2x^4 + 3x^3 - 13x^2 - 7x + 15 = (x^2 + 2x - 3)(2x^2 - x - 5).$$
Therefore the L.C.M. is $(x^2 + 2x - 3)(2x^2 - 3x - 8)(2x^2 - x - 5)$.

*161. We may now give the proof of the rule for finding the lowest common multiple of two compound algebraical expressions.

Let A and B be the two expressions, and F their highest common factor. Also suppose that a and b are the respective quotients when A and B are divided by F; then $A = aF$, $B = bF$. Therefore, since a and b have no common factor, the lowest common multiple of A and B is abF, by inspection.

*162. There is an important relation between the highest common factor and the lowest common multiple of two expressions which it is desirable to notice.

Let F be the highest common factor, and X the lowest common multiple of A and B. Then, as in the preceding article,
$$A = aF,\ B = bF,$$
and $$X = abF.$$
Therefore the product $$AB = aF \cdot bF$$
$$= F \cdot abF$$
$$= FX \quad \dots\dots\dots\dots\dots\dots(1).$$

Hence *the product of two expressions is equal to the product of their highest common factor and lowest common multiple.*

Again, from (1) $$X = \frac{AB}{F} = \frac{A}{F} \times B = \frac{B}{F} \times A\ ;$$

hence *the lowest common multiple of two expressions may be found by dividing their product by their highest common factor; or by dividing either of them by their highest common factor, and multiplying the quotient by the other.*

*163. The lowest common multiple of three expressions A, B, C may be obtained as follows.

First, find X the L.C.M. of A and B. Next find Y the L.C.M. of X and C; then Y will be the required L.C.M. of A, B, C.

For Y is the expression of lowest dimensions which is divisible by X and C, and X is the expression of lowest dimensions divisible by A and B. Therefore Y is the expression of lowest dimensions divisible by all three.

EXAMPLES XX. b.

1. Find the highest common factor and the lowest common multiple of $x^2 - 5x + 6$, $x^2 - 4$, $x^3 - 3x - 2$.

2. Find the lowest common multiple of
$$ab(x^2 + 1) + x(a^2 + b^2) \text{ and } ab(x^2 - 1) + x(a^2 - b^2).$$

3. Find the lowest common multiple of
$$xy - bx,\ xy - ay,\ y^2 - 3by + 2b^2,\ xy - 2bx - ay + 2ab,$$
$$xy - bx - ay + ab.$$

4. Find the highest common factor and the lowest common multiple of x^3+2x^2-3x, $2x^3+5x^2-3x$.

5. Find the lowest common multiple of
$$1-x, \quad (1-x^2)^2, \quad (1+x)^3.$$

6. Find the lowest common multiple of
$$x^2-10x+24, \quad x^2-8x+12, \quad x^2-6x+8.$$

7. Find the highest common factor and the lowest common multiple of $6x^3+x^2-5x-2$, $6x^3+5x^2-3x-2$.

8. Find the lowest common multiple of
$$(bc^2-abc)^2, \quad b^2(ac^2-a^3), \quad a^2c^2+2ac^3+c^4.$$

9. Find the lowest common multiple of
$$x^3-y^3, \quad x^3y-y^4, \quad y^2(x-y)^2, \quad x^2+xy+y^2.$$
Also find the highest common factor of the first three expressions.

10. Find the highest common factor of
$$6x^2-13x+6, \quad 2x^2+5x-12, \quad 6x^2-x-12.$$
Also shew that the lowest common multiple is the product of the three quantities divided by the square of the highest common factor.

11. Find the lowest common multiple of
$$x^4+ax^3+a^3x+a^4, \quad x^4+a^2x^2+a^4.$$

*12. Find the highest common factor and the lowest common multiple of $3x^3-7x^2y+5xy^2-y^3$, $x^2y+3xy^2-3x^3-y^3$,
$$3x^3+5x^2y+xy^2-y^3.$$

*13. Find the highest common factor of
$$4x^3-10x^2+4x+2, \quad 3x^4-2x^3-3x+2.$$

14. Find the lowest common multiple of
$$a^2-b^2, \quad a^3-b^3, \quad a^3-a^2b-ab^2-2b^3.$$

15. Find the highest common factor and the lowest common multiple of $(2x^2-3a^2)y+(2a^2-3y^2)x$, $(2a^2+3y^2)x+(2x^2+3a^2)y$.

*16. Find the highest common factor and the lowest common multiple of $x^3-9x^2+26x-24$, $x^3-12x^2+47x-60$.

*17. Find the highest common factor of
$$x^3-15ax^2+48a^2x+64a^3, \quad x^2-10ax+16a^2.$$

18. Find the lowest common multiple of
$$21x(xy-y^2)^2, \quad 35(x^4y^2-x^2y^4), \quad 15y(x^2+xy)^2.$$

CHAPTER XXI.

ADDITION AND SUBTRACTION OF FRACTIONS.

164. HAVING explained the rules for finding the lowest common multiple of any given expressions, we now proceed to shew how the addition and subtraction of fractions may be effected.

165. *To prove* $\quad \dfrac{a}{b} + \dfrac{c}{d} = \dfrac{ad+bc}{bd}.$

We have $\quad \dfrac{a}{b} = \dfrac{ad}{bd},$ and $\dfrac{c}{d} = \dfrac{bc}{bd}.$ [Art. 150].

Thus in each case we divide the unit into bd equal parts, and we take first ad of these parts, and then bc of them; that is, we take $ad+bc$ of the bd parts of the unit; and this is expressed by the fraction $\dfrac{ad+bc}{bd}.$

$$\therefore \quad \frac{a}{b} + \frac{c}{d} = \frac{ad+bc}{bd}.$$

Similarly, $\quad \dfrac{a}{b} - \dfrac{c}{d} = \dfrac{ad-bc}{bd}.$

166. Here the fractions have been both expressed with a common denominator bd. But if b and d have a common factor, the product bd is not the lowest common denominator, and the fraction $\dfrac{ad+bc}{bd}$ will not be in its lowest terms. To avoid working with fractions which are not in their lowest terms, some modification of the above will be necessary. In practice it will be found advisable to take the *lowest* common denominator, which is the lowest common multiple of the denominators of the given fractions.

XXI.] ADDITION AND SUBTRACTION OF FRACTIONS. 149

Rule. I. To reduce fractions to their lowest common denominator: *find the L.C.M. of the given denominators, and take it for the common denominator; divide it by the denominator of the first fraction, and multiply the numerator of this fraction by the quotient so obtained; and do the same with all the other given fractions.*

Example. Express with lowest common denominator
$$\frac{5x}{2a(x-a)} \text{ and } \frac{4a}{3x(x^2-a^2)}.$$

The lowest common denominator is $6ax(x-a)(x+a)$.

We must therefore multiply the numerators by $3x(x+a)$ and $2a$ respectively.

Hence the equivalent fractions are
$$\frac{15x^2(x+a)}{6ax(x-a)(x+a)} \text{ and } \frac{8a^2}{6ax(x-a)(x+a)}.$$

167. We may now enunciate the rule for the addition or subtraction of fractions.

Rule II. To add or subtract fractions: *reduce them to the lowest common denominator; find the algebraical sum of the numerators, and retain the common denominator.*

Example 1. Find the value of $\dfrac{2x+a}{3a} + \dfrac{5x-4a}{9a}$.

The lowest common denominator is $9a$.

Therefore the expression $= \dfrac{3(2x+a)+5x-4a}{9a}$

$$= \frac{6x+3a+5x-4a}{9a} = \frac{11x-a}{9a}.$$

Example 2. Find the value of $\dfrac{x-2y}{xy} + \dfrac{3y-a}{ay} - \dfrac{3x-2a}{ax}$.

The lowest common denominator is axy.

Thus the expression $= \dfrac{a(x-2y)+x(3y-a)-y(3x-2a)}{axy}$

$$= \frac{ax-2ay+3xy-ax-3xy+2ay}{axy}$$

$$= 0,$$

since the terms in the numerator destroy each other.

Note. To ensure accuracy the beginner is recommended to use brackets as in the first line of work above.

EXAMPLES XXI. a.

Find the value of

1. $\dfrac{x-1}{2} + \dfrac{x+3}{5} + \dfrac{x+7}{10}$.

2. $\dfrac{2x-1}{3} + \dfrac{x-5}{6} + \dfrac{x-4}{4}$.

3. $\dfrac{5x-1}{8} - \dfrac{3x-2}{7} + \dfrac{x-5}{4}$.

4. $\dfrac{2x-3}{9} - \dfrac{x+2}{6} + \dfrac{5x+8}{12}$.

5. $\dfrac{x-7}{15} + \dfrac{x-9}{25} - \dfrac{x+3}{45}$.

6. $\dfrac{2x+5}{x} - \dfrac{x+3}{2x} - \dfrac{27}{8x^2}$.

7. $\dfrac{a-b}{ab} + \dfrac{b-c}{bc} + \dfrac{c-a}{ca}$.

8. $\dfrac{a-2b}{2a} - \dfrac{a-5b}{4a} + \dfrac{a+7b}{8a}$.

9. $\dfrac{b+c}{2a} + \dfrac{c+a}{4b} - \dfrac{a-b}{3c}$.

10. $\dfrac{a-x}{x} + \dfrac{a+x}{a} - \dfrac{a^2-x^2}{2ax}$.

11. $\dfrac{x+2}{17x} - \dfrac{x-5}{34x} + \dfrac{x+2}{51x}$.

12. $\dfrac{2a^2-b^2}{a^2} - \dfrac{b^2-c^2}{b^2} - \dfrac{c^2-a^2}{c^2}$.

13. $\dfrac{x-3}{5x} + \dfrac{x^2-9}{10x^2} - \dfrac{8-x^3}{15x^3}$.

14. $\dfrac{2}{xy} - \dfrac{3y^2-x^2}{xy^3} + \dfrac{xy+y^2}{x^2y^2}$.

15. $\dfrac{2x-3y}{xy} + \dfrac{3x-2z}{xz} + \dfrac{5}{x}$.

16. $\dfrac{a^2-bc}{bc} - \dfrac{ac-b^2}{ac} - \dfrac{ab-c^2}{ab}$.

Example 3. Simplify $\dfrac{2x-3a}{x-2a} - \dfrac{2x-a}{x-a}$.

The lowest common denominator is $(x-2a)(x-a)$.

Hence, multiplying the numerators by $x-a$ and $x-2a$ respectively, we have

$$\text{the expression} = \dfrac{(2x-3a)(x-a) - (2x-a)(x-2a)}{(x-2a)(x-a)}$$

$$= \dfrac{2x^2 - 5ax + 3a^2 - (2x^2 - 5ax + 2a^2)}{(x-2a)(x-a)}$$

$$= \dfrac{2x^2 - 5ax + 3a^2 - 2x^2 + 5ax - 2a^2}{(x-2a)(x-a)}$$

$$= \dfrac{a^2}{(x-2a)(x-a)}.$$

Note. In finding the value of such an expression as
$$-(2x-a)(x-2a),$$
the beginner should first express the product in brackets, and then remove the brackets, as we have done. After a little practice he will be able to take both steps together.

The work will sometimes be shortened by first reducing the fractions to their lowest terms.

ADDITION AND SUBTRACTION OF FRACTIONS.

Example 4. Simplify $\dfrac{x^2+5xy-4y^2}{x^2-16y^2} - \dfrac{2xy}{2x^2+8xy}$.

The expression $= \dfrac{x^2+5xy-4y^2}{x^2-16y^2} - \dfrac{y}{x+4y}$

$= \dfrac{x^2+5xy-4y^2 - y(x-4y)}{x^2-16y^2}$

$= \dfrac{x^2+5xy-4y^2 - xy+4y^2}{x^2-16y^2}$

$= \dfrac{x^2+4xy}{x^2-16y^2} = \dfrac{x}{x-4y}$.

EXAMPLES XXI. b.

Find the value of

1. $\dfrac{1}{x+2} + \dfrac{1}{x+3}$.

2. $\dfrac{2}{x+3} - \dfrac{1}{x+4}$.

3. $\dfrac{1}{x-5} - \dfrac{1}{x-4}$.

4. $\dfrac{3}{x-6} - \dfrac{1}{x+2}$.

5. $\dfrac{a}{x+a} - \dfrac{b}{x+b}$.

6. $\dfrac{a}{x-a} + \dfrac{b}{x-b}$.

7. $\dfrac{x+3}{x+4} - \dfrac{x+1}{x+2}$.

8. $\dfrac{a+x}{a-x} - \dfrac{a-x}{a+x}$.

9. $\dfrac{x+2}{x-2} - \dfrac{x-2}{x+2}$.

10. $\dfrac{x-4}{x-2} - \dfrac{x-7}{x-5}$.

11. $\dfrac{a}{x-a} - \dfrac{a^2}{x^2-a^2}$.

12. $\dfrac{3}{x-3} + \dfrac{2x}{x^2-9}$.

13. $\dfrac{1}{2x-3y} - \dfrac{x+y}{4x^2-9y^2}$.

14. $\dfrac{x+a}{x-2a} - \dfrac{x^2+2a^2}{x^2-4a^2}$.

15. $\dfrac{4a^2+b^2}{4a^2-b^2} - \dfrac{2a-b}{2a+b}$.

16. $\dfrac{2x^2}{x^2-y^2} - \dfrac{2x^2}{x^2+xy}$.

17. $\dfrac{x^2}{x-x^3} - \dfrac{x}{1+x^2}$.

18. $\dfrac{1}{x(x-y)} + \dfrac{1}{y(x+y)}$.

19. $\dfrac{xy}{25x^2-y^2} + \dfrac{2x^2y}{10x^2y+2xy^2}$.

20. $\dfrac{y}{x(x^2-y^2)} + \dfrac{x}{y(x^2+y^2)}$.

21. $\dfrac{x^2-4a^2}{x^2-2ax} - \dfrac{x^2+2ax-8a^2}{x^2-4a^2}$.

22. $\dfrac{x^2+xy+y^2}{x+y} + \dfrac{x^2-xy+y^2}{x-y}$.

23. $\dfrac{1}{a-2x} - \dfrac{(a+2x)^2}{a^3-8x^3}$.

24. $\dfrac{a^3+b^3}{a^2-ab+b^2} - \dfrac{a^3-b^3}{a^2+ab+b^2}$.

25. $\dfrac{3}{x^2-4} + \dfrac{1}{(x-2)^2}$.

26. $\dfrac{1}{a(x^2-a^2)} - \dfrac{1}{x(x+a)^2}$.

E.A.

168. Some modification of the foregoing general methods may sometimes be used with advantage. The most useful artifices are explained in the examples which follow, but no general rules can be given which will apply to all cases.

Example 1. Simplify $\dfrac{a+3}{a-4} - \dfrac{a+4}{a-3} - \dfrac{8}{a^2-16}$.

Taking the first two fractions together, we have

$$\text{the expression} = \frac{a^2 - 9 - (a^2 - 16)}{(a-4)(a-3)} - \frac{8}{a^2 - 16}$$

$$= \frac{7}{(a-4)(a-3)} - \frac{8}{(a+4)(a-4)}$$

$$= \frac{7(a+4) - 8(a-3)}{(a+4)(a-4)(a-3)}$$

$$= \frac{52 - a}{(a+4)(a-4)(a-3)}.$$

Example 2. Simplify $\dfrac{1}{2x^2 + x - 1} + \dfrac{1}{3x^2 + 4x + 1}$.

$$\text{The expression} = \frac{1}{(2x-1)(x+1)} + \frac{1}{(3x+1)(x+1)}$$

$$= \frac{3x + 1 + 2x - 1}{(2x-1)(x+1)(3x+1)}$$

$$= \frac{5x}{(2x-1)(x+1)(3x+1)}.$$

Example 3. Simplify $\dfrac{1}{a-x} - \dfrac{1}{a+x} - \dfrac{2x}{a^2+x^2} - \dfrac{4x^3}{a^4+x^4}$.

Here it should be evident that the first two denominators give L.C.M. $a^2 - x^2$, which readily combines with $a^2 + x^2$ to give L.C.M. $a^4 - x^4$, which again combines with $a^4 + x^4$ to give L.C.M. $a^8 - x^8$. Hence it will be convenient to proceed as follows:

$$\text{The expression} = \frac{a + x - (a - x)}{a^2 - x^2} - \ldots\ldots - \ldots\ldots$$

$$= \frac{2x}{a^2 - x^2} - \frac{2x}{a^2 + x^2} - \ldots\ldots$$

$$= \frac{4x^3}{a^4 - x^4} - \frac{4x^3}{a^4 + x^4}$$

$$= \frac{8x^7}{a^8 - x^8}.$$

ADDITION AND SUBTRACTION OF FRACTIONS.

EXAMPLES XXI. c.

Find the value of

1. $\dfrac{1}{x+y} - \dfrac{1}{x-y} + \dfrac{2x}{x^2-y^2}.$

2. $\dfrac{1}{2x+y} + \dfrac{1}{2x-y} - \dfrac{3x}{4x^2-y^2}.$

3. $\dfrac{5}{1+2x} - \dfrac{3x}{1-2x} - \dfrac{4-13x}{1-4x^2}.$

4. $\dfrac{2a}{2a+3b} + \dfrac{3b}{2a-3b} - \dfrac{8b^2}{4a^2-9b^2}.$

5. $\dfrac{10}{9-a^2} - \dfrac{2}{3+a} - \dfrac{1}{3-a}.$

6. $\dfrac{5x}{6(x^2-1)} - \dfrac{1}{2(x-1)} + \dfrac{1}{3(x+1)}.$

7. $\dfrac{1}{2(a-b)} - \dfrac{1}{2(a+b)} - \dfrac{b}{a^2-b^2}.$

8. $\dfrac{2a}{2a-3} - \dfrac{5}{6a+9} - \dfrac{4(3a+2)}{3(4a^2-9)}.$

9. $\dfrac{3}{x-2} + \dfrac{2}{3x+6} + \dfrac{5x}{x^2-4}.$

10. $\dfrac{x}{x^3+y^3} - \dfrac{y}{x^3-y^3} + \dfrac{x^3y+xy^3}{x^6-y^6}.$

11. $\dfrac{1}{x^2-9x+20} + \dfrac{1}{x^2-11x+30}.$

12. $\dfrac{1}{x^2-7x+12} - \dfrac{1}{x^2-5x+6}.$

13. $\dfrac{1}{2x^2-x-1} - \dfrac{1}{2x^2+x-3}.$

14. $\dfrac{1}{2x^2-x-1} - \dfrac{3}{6x^2-x-2}.$

15. $\dfrac{4}{4-7a-2a^2} - \dfrac{3}{3-a-10a^2}.$

16. $\dfrac{5}{5+x-18x^2} - \dfrac{2}{2+5x+2x^2}.$

17. $\dfrac{1}{x+1} - \dfrac{1}{(x+1)(x+2)} + \dfrac{1}{(x+1)(x+2)(x+3)}.$

18. $\dfrac{5x}{2(x+1)(x-3)} - \dfrac{15(x-1)}{16(x-3)(x-2)} - \dfrac{9(x+3)}{16(x+1)(x-2)}.$

19. $\dfrac{a+3b}{4(a+b)(a+2b)} + \dfrac{a+2b}{(a+b)(a+3b)} - \dfrac{a+b}{4(a+2b)(a+3b)}.$

20. $\dfrac{2}{x^2-3x+2} + \dfrac{2}{x^2-x-2} - \dfrac{1}{x^2-1}.$

21. $\dfrac{x}{x^2+5x+6} + \dfrac{15}{x^2+9x+14} - \dfrac{12}{x^2+10x+21}.$

22. $\dfrac{3}{x^2-1} + \dfrac{4}{2x+1} + \dfrac{4x+2}{2x^2+3x+1}.$

23. $\dfrac{5(2x-3)}{11(6x^2+x-1)} + \dfrac{7x}{6x^2+7x-3} - \dfrac{12(3x+1)}{11(4x^2+8x+3)}.$

24. $\dfrac{x-3}{x+2} - \dfrac{x-2}{x+3} + \dfrac{1}{x-1}.$

25. $\dfrac{x-3}{x-4} - \dfrac{x+4}{x+3} - \dfrac{5}{x^2-16}.$

Find the value of

26. $\dfrac{1+2a}{1-2a} - \dfrac{1-2a}{1+2a} - \dfrac{8a}{(1-2a)^2}.$

27. $\dfrac{24x}{9-12x+4x^2} - \dfrac{3+2x}{3-2x} + \dfrac{3-2x}{3+2x}.$

28. $\dfrac{1}{3-x} - \dfrac{1}{3+x} - \dfrac{2x}{9+x^2}.$

29. $\dfrac{1}{2a+3} + \dfrac{1}{2a-3} - \dfrac{4a}{4a^2+9}.$

30. $\dfrac{1}{4(1+x)} + \dfrac{1}{4(1-x)} + \dfrac{1}{2(1+x^2)}.$

31. $\dfrac{3}{8(a-x)} + \dfrac{1}{8(a+x)} - \dfrac{a-x}{4(a^2+x^2)}.$

32. $\dfrac{2x}{4+x^2} + \dfrac{1}{2-x} - \dfrac{1}{2+x}.$

33. $\dfrac{5}{3-6x} - \dfrac{5}{3+6x} - \dfrac{x}{2+8x^2}.$

34. $\dfrac{1}{2a-8x} - \dfrac{a}{3a^2+48x^2} + \dfrac{1}{2a+8x}.$

35. $\dfrac{1}{6a^2+54} + \dfrac{1}{3a-9} - \dfrac{a}{3a^2-27}.$

36. $\dfrac{1}{8-8x} - \dfrac{1}{8+8x} + \dfrac{w}{4+4x^2} - \dfrac{r}{2+2x^4}.$

37. $\dfrac{1}{6a-18} - \dfrac{1}{6a+18} - \dfrac{1}{a^2+9} + \dfrac{18}{a^4+81}.$

38. $\dfrac{x+1}{2x^3-4x^2} + \dfrac{x-1}{2x^3+4x^2} - \dfrac{1}{x^2-4}.$

39. $\dfrac{1}{3x^2-4xy+y^2} + \dfrac{1}{x^2-4xy+3y^2} - \dfrac{3}{3x^2-10xy+3y^2}.$

40. $\dfrac{1}{x-1} + \dfrac{2}{x+1} - \dfrac{3x-2}{x^2-1} - \dfrac{1}{(x+1)^2}.$

41. $\dfrac{108-52x}{x(3-x)^2} - \dfrac{4}{3-x} - \dfrac{12}{x} + \left(\dfrac{1+x}{3-x}\right)^2.$

42. $\dfrac{(a+b)^2}{(x-a)(x+a+b)} - \dfrac{a+2b+x}{2(x-a)} + \dfrac{(a+b)x}{x^2+bx-a^2-ab} + \dfrac{1}{2}.$

43. $\dfrac{3(x^2+x-2)}{x^2-x-2} - \dfrac{3(x^2-x-2)}{x^2+x-2} - \dfrac{8x}{x^2-4}.$

169. We have thus far assumed both numerator and denominator to be positive integers, and have shewn in Art. 155 that a fraction itself is the quotient resulting from the division of the numerator by denominator. But in algebra division is a process not restricted to positive integers, and we shall extend this definition as follows:

The algebraic fraction $\dfrac{a}{b}$ *is the quotient resulting from the division of* a *by* b, *where* a *and* b *may have any values whatever.*

170. By the preceding article $\dfrac{-a}{-b}$ is the quotient resulting from the division of $-a$ by $-b$; and this is obtained by dividing a by b, and, by the rule of signs, prefixing $+$.

Again, $\dfrac{-a}{b}$ is the quotient resulting from the division of $-a$ by b; and this is obtained by dividing a by b, and, by the rule of signs, prefixing $-$.

Therefore
$$\dfrac{-a}{b} = -\dfrac{a}{b} \quad \dotfill (2).$$

Likewise $\dfrac{a}{-b}$ is the quotient resulting from the division of a by $-b$; and this is obtained by dividing a by b, and, by the rule of signs, prefixing $-$.

Therefore
$$\dfrac{a}{-b} = -\dfrac{a}{b} \quad \dotfill (3).$$

These results may be enunciated as follows:

1. *If the signs of* BOTH *numerator and denominator of a fraction be changed, the sign of the whole fraction will be unchanged.*

2. *If the sign of* EITHER *numerator or denominator alone be changed, the sign of the whole fraction will be changed.*

The principles here involved are so useful in certain cases of reduction of fractions that we quote them in another form, which will sometimes be found more easy of application.

1. *We may change the sign of every term in the numerator and denominator of a fraction without altering its value.*

2. *We may change the sign of a fraction by simply changing the sign of every term in* EITHER *the numerator or denominator.*

Example 1. $\quad \dfrac{b-a}{y-x} = \dfrac{-b+a}{-y+x} = \dfrac{a-b}{x-y}.$

Example 2. $\quad \dfrac{x-x^2}{2y} = -\dfrac{-x+x^2}{2y} = -\dfrac{x^2-x}{2y}.$

Example 3. $\quad \dfrac{3x}{4-x^2} = -\dfrac{3x}{-4+x^2} = -\dfrac{3x}{x^2-4}.$

The intermediate step may usually be omitted.

Example 4. Simplify $\dfrac{a}{x+a} + \dfrac{2x}{x-a} + \dfrac{a(3x-a)}{a^2-x^2}$.

Here it is evident that the lowest common denominator of the first two fractions is $x^2 - a^2$, therefore it will be convenient to alter the sign of the denominator in the third fraction.

Thus the expression $= \dfrac{a}{x+a} + \dfrac{2x}{x-a} - \dfrac{a(3x-a)}{x^2-a^2}$

$$= \dfrac{a(x-a) + 2x(x+a) - a(3x-a)}{x^2-a^2}$$

$$= \dfrac{ax - a^2 + 2x^2 + 2ax - 3ax + a^2}{x^2-a^2}$$

$$= \dfrac{2x^2}{x^2-a^2}.$$

Example 5. Simplify $\dfrac{5}{3x-3} + \dfrac{3x-1}{1-x^2} + \dfrac{1}{2x+2}$.

The expression $= \dfrac{5}{3(x-1)} - \dfrac{3x-1}{x^2-1} + \dfrac{1}{2(x+1)}$

$$= \dfrac{10(x+1) - 6(3x-1) + 3(x-1)}{6(x^2-1)}$$

$$= \dfrac{10x + 10 - 18x + 6 + 3x - 3}{6(x^2-1)}$$

$$= \dfrac{13 - 5x}{6(x^2-1)}.$$

EXAMPLES XXI. d.

Simplify

1. $\dfrac{1}{4x-4} - \dfrac{1}{5x+5} + \dfrac{1}{1-x^2}$.

2. $\dfrac{3}{1+a} - \dfrac{2}{1-a} - \dfrac{5a}{a^2-1}$.

3. $\dfrac{x-2a}{x+a} + \dfrac{2(a^2-4ax)}{a^2-x^2} - \dfrac{3a}{x-a}$.

4. $\dfrac{x-a}{x+a} + \dfrac{a^2+3ax}{a^2-x^2} + \dfrac{x+a}{x-a}$.

5. $\dfrac{1}{2x+1} + \dfrac{1}{2x-1} + \dfrac{4x}{1-4x^2}$.

6. $\dfrac{3x}{1-x^2} - \dfrac{2}{x-1} - \dfrac{2}{x+1}$.

7. $\dfrac{2-5x}{x+3} - \dfrac{3+x}{3-x} + \dfrac{2x(2x-11)}{x^2-9}$.

8. $\dfrac{3-2x}{2x+3} - \dfrac{2x+3}{3-2x} + \dfrac{12}{4x^2-9}$.

9. $\dfrac{5}{2b+2} - \dfrac{3}{4b-4} + \dfrac{11}{6-6b^2}$.

10. $\dfrac{1}{6a+6} + \dfrac{1}{6-6a} - \dfrac{1}{3a^2-3}$.

ADDITION AND SUBTRACTION OF FRACTIONS.

11. $\dfrac{y^2}{x^3-y^3}+\dfrac{x^3y^2}{y^6-x^6}.$

12. $\dfrac{x^2-y^2}{xy}-\dfrac{xy-y^2}{xy-x^2}.$

13. $\dfrac{x^2+y^2}{x^2-y^2}+\dfrac{x}{x+y}+\dfrac{y}{y-x}.$

14. $\dfrac{x^2+2x+4}{x+2}-\dfrac{x^2-2x+4}{2-x}.$

15. $\dfrac{1}{2a+5b}+\dfrac{3a}{25b^2-4a^2}+\dfrac{1}{2a-5b}.$

16. $\dfrac{2b-a}{x-b}-\dfrac{3x(a-b)}{b^2-x^2}+\dfrac{b-2a}{b+x}.$

17. $\dfrac{ax^2+b}{2x-1}+\dfrac{2(bx+ax^2)}{1-4x^2}-\dfrac{ax^2-b}{2x+1}.$

18. $\dfrac{a+c}{(a-b)(x-a)}+\dfrac{b+c}{(b-a)(x-b)}.$

19. $\dfrac{a-c}{(a-b)(x-a)}-\dfrac{b-c}{(b-a)(b-x)}.$

20. $\dfrac{2a+y}{(x-a)(a-b)}+\dfrac{a+b+y}{(x-b)(b-a)}-\dfrac{x+y-a}{(x-a)(x-b)}.$

21. $\dfrac{1}{(a^2-b^2)(x^2+b^2)}+\dfrac{1}{(b^2-a^2)(x^2+a^2)}-\dfrac{1}{(x^2+a^2)(x^2+b^2)}.$

22. $\dfrac{1}{x+a}+\dfrac{4a}{x^2-a^2}+\dfrac{1}{a-x}-\dfrac{2a}{x^2+a^2}.$

23. $\dfrac{3}{x+a}-\dfrac{1}{x+3a}+\dfrac{3}{a-x}+\dfrac{1}{x-3a}.$

24. $\dfrac{1}{4a^3(a+x)}-\dfrac{1}{4a^3(x-a)}+\dfrac{1}{2a^2(a^2+x^2)}-\dfrac{a^4}{a^8-x^8}.$

25. $\dfrac{x}{x^2-y^2}-\dfrac{y}{x^2+y^2}+\dfrac{x^3+y^3}{y^4-x^4}+\dfrac{xy}{(x+y)(x^2+y^2)}.$

26. $\dfrac{b}{a(a^2-b^2)}+\dfrac{a}{b(a^2+b^2)}+\dfrac{a^4+b^4}{ab(b^4-a^4)}-\dfrac{a^6}{b^8-a^8}.$

27. $\dfrac{a^2-2ax+x^2}{2(a^2-x^2)}-\dfrac{2ax(a+x)}{(a-x)(a^2+2ax+x^2)}-\dfrac{x^2-a^2}{2(x-a)^2}.$

28. $\dfrac{2}{a+b}-\dfrac{1}{a-b}-\dfrac{3b}{b^2-a^2}+\dfrac{ab}{a^3+b^3}.$

29. $\dfrac{3}{8(1-x)}+\dfrac{1}{8(1+x)}-\dfrac{1-x}{4(1+x^2)}-\dfrac{3}{4(x^2-1)}.$

30. $\dfrac{1}{x}+\dfrac{1}{x-1}+\dfrac{1}{x+1}+\dfrac{x}{1-x^2}+\dfrac{3}{x(x^2-1)}.$

31. $\dfrac{a^2+ac}{a^2c-c^3}-\dfrac{a^2-c^2}{a^2c+2ac^2+c^3}+\dfrac{2c}{c^2-a^2}-\dfrac{3}{a+c}.$

32. $\dfrac{4a+6b}{a+b}+\dfrac{6a-4b}{a-b}+\dfrac{4a^2+6b^2}{b^2-a^2}+\dfrac{4b^2-6a^2}{a^2+b^2}-\dfrac{20b^4}{b^4-a^4}.$

*171. Consider the expression
$$\frac{1}{(a-b)(a-c)}+\frac{1}{(b-c)(b-a)}+\frac{1}{(c-a)(c-b)}.$$

Here in finding the L.C.M. of the denominators it must be observed that there are not *six* different compound factors to be considered; for three of them differ from the other three only in sign.

Thus
$$(a-c) = -(c-a),$$
$$(b-a) = -(a-b),$$
$$(c-b) = -(b-c).$$

Hence, replacing the second factor in each denominator by its equivalent, we may write the expression in the form
$$-\frac{1}{(a-b)(c-a)}-\frac{1}{(b-c)(a-b)}-\frac{1}{(c-a)(b-c)}\ldots\ldots(1).$$

Now the L.C.M. is $(b-c)(c-a)(a-b)$,

and the expression
$$=\frac{-(b-c)-(c-a)-(a-b)}{(b-c)(c-a)(a-b)}$$
$$=\frac{-b+c-c+a-a+b}{(b-c)(c-a)(a-b)}$$
$$=0.$$

*172. There is a peculiarity in the arrangement of this example which it is desirable to notice. In the expression (1) the letters occur in what is known as **Cyclic Order**; that is, b follows a, a follows c, c follows b. Thus if a, b, c are arranged round the circumference of a circle, as in the annexed diagram, if we start from any letter and move round in the direction of the arrows, the other letters follow in cyclic order, namely, abc, bca, cab.

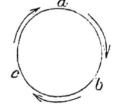

The observance of this principle is especially important in a large class of examples in which the differences of three letters are involved. Thus we are observing cyclic order when we write $b-c$, $c-a$, $a-b$; whereas we are violating cyclic order by the use of arrangements such as $b-c$, $a-c$, $a-b$, or $a-c$, $b-a$, $b-c$. It will always be found that the work is rendered shorter and easier by following cyclic order from the beginning, and adhering to it throughout the question.

In the present chapter we shall confine our attention to a few of the simpler cases, resuming the subject in Chapter XXIX.

*EXAMPLES XXI. e.

Find the value of

1. $\dfrac{a}{(a-b)(a-c)} + \dfrac{b}{(b-c)(b-a)} + \dfrac{c}{(c-a)(c-b)}.$

2. $\dfrac{b}{(a-b)(a-c)} + \dfrac{c}{(b-c)(b-a)} + \dfrac{a}{(c-a)(c-b)}.$

3. $\dfrac{z}{(x-y)(x-z)} + \dfrac{x}{(y-z)(y-x)} + \dfrac{y}{(z-x)(z-y)}.$

4. $\dfrac{y+z}{(x-y)(x-z)} + \dfrac{z+x}{(y-z)(y-x)} + \dfrac{x+y}{(z-x)(z-y)}.$

5. $\dfrac{b-c}{(a-b)(a-c)} + \dfrac{c-a}{(b-c)(b-a)} + \dfrac{a-b}{(c-a)(c-b)}.$

6. $\dfrac{x^2yz}{(x-y)(x-z)} + \dfrac{y^2zx}{(y-z)(y-x)} + \dfrac{z^2xy}{(z-x)(z-y)}.$

7. $\dfrac{1+a}{(a-b)(a-c)} + \dfrac{1+b}{(b-c)(b-a)} + \dfrac{1+c}{(c-a)(c-b)}.$

8. $\dfrac{p-a}{(p-q)(p-r)} + \dfrac{q-a}{(q-r)(q-p)} + \dfrac{r-a}{(r-p)(r-q)}.$

9. $\dfrac{p+q-r}{(p-q)(p-r)} + \dfrac{q+r-p}{(q-r)(q-p)} + \dfrac{r+p-q}{(r-p)(r-q)}.$

10. $\dfrac{a^2}{(a^2-b^2)(a^2-c^2)} + \dfrac{b^2}{(b^2-c^2)(b^2-a^2)} + \dfrac{c^2}{(c^2-a^2)(c^2-b^2)}.$

11. $\dfrac{x+y}{(p-q)(p-r)} + \dfrac{x+y}{(q-r)(q-p)} + \dfrac{x+y}{(r-p)(r-q)}.$

12. $\dfrac{q+r}{(x-y)(x-z)} + \dfrac{r+p}{(y-z)(y-x)} + \dfrac{p+q}{(z-x)(z-y)}.$

CHAPTER XXII.

Miscellaneous Fractions.

[Examples marked with an asterisk may be taken at a later stage.]

173. We now propose to consider some miscellaneous questions involving fractions of a more complicated kind than those already discussed.

In the previous chapters on Fractions, the numerator and denominator have been regarded as integers; but cases frequently occur in which the numerator or denominator of a fraction is itself fractional.

174. Definition. A fraction of which the numerator or denominator is itself a fraction is called a **Complex Fraction**.

Thus $\dfrac{a}{\dfrac{b}{c}}$, $\dfrac{\dfrac{a}{b}}{x}$, $\dfrac{\dfrac{a}{b}}{\dfrac{c}{d}}$ are Complex Fractions.

In the last of these types, the outside quantities, a and d, are sometimes referred to as the *extremes*, while the two middle quantities, b and c, are called the *means*.

175. Instead of using the horizontal line to separate numerator and denominator, it is sometimes convenient to write complex fractions in the forms

$$a \Big/ \frac{b}{c}, \quad \frac{a}{b} \Big/ x, \quad \frac{a}{b} \Big/ \frac{c}{d}.$$

176. By definition (Art. 169) $\dfrac{\dfrac{a}{b}}{\dfrac{c}{d}}$ is the quotient resulting from the division of $\dfrac{a}{b}$ by $\dfrac{c}{d}$; and this by Art. 158 is $\dfrac{ad}{bc}$;

$$\therefore \quad \frac{\dfrac{a}{b}}{\dfrac{c}{d}} = \frac{ad}{bc}.$$

[CHAP. XXII.] MISCELLANEOUS FRACTIONS. 161

Simplification of Complex Fractions.

177. From the preceding article we deduce an easy method of writing down the simplified form of a complex fraction.

Multiply the extremes for a new numerator, and the means for a new denominator.

Example. $\dfrac{\dfrac{a+x}{b}}{\dfrac{a^2-x^2}{ab}} = \dfrac{ab(a+x)}{b(a^2-x^2)} = \dfrac{a}{a-x},$

by cancelling common factors in numerator and denominator.

178. The student should especially notice the following cases, and should be able to write down the results readily.

$$\dfrac{1}{\dfrac{a}{b}} = 1 \div \dfrac{a}{b} = 1 \times \dfrac{b}{a} = \dfrac{b}{a},$$

$$\dfrac{a}{\dfrac{1}{b}} = a \div \dfrac{1}{b} = a \times b = ab.$$

$$\dfrac{\dfrac{1}{a}}{\dfrac{1}{b}} = \dfrac{1}{a} \div \dfrac{1}{b} = \dfrac{1}{a} \times \dfrac{b}{1} = \dfrac{b}{a}.$$

179. The following examples illustrate the simplification of complex fractions.

Example 1. $\dfrac{\dfrac{a}{b}+\dfrac{c}{d}}{\dfrac{a}{b}-\dfrac{c}{d}} = \left(\dfrac{a}{b}+\dfrac{c}{d}\right) \div \left(\dfrac{a}{b}-\dfrac{c}{d}\right) = \dfrac{ad+bc}{bd} \div \dfrac{ad-bc}{bd}$

$= \dfrac{ad+bc}{bd} \times \dfrac{bd}{ad-bc} = \dfrac{ad+bc}{ad-bc}.$

Or more simply thus :

Multiply the fractions above and below by bd which is the L.C.M. of their denominators.

Then the fraction becomes $\dfrac{ad+bc}{ad-bc}$, as before.

Example 2. Simplify $\dfrac{x = \dfrac{a^2}{x}}{x - \dfrac{a^4}{x^3}}$.

Here by multiplying above and below by x^3, we have

$$\text{the fraction} = \frac{x^4 + a^2 x^2}{x^4 - a^4} = \frac{x^2(x^2 + a^2)}{x^4 - a^4}$$

$$= \frac{x^2}{x^2 - a^2}.$$

Example 3. Simplify $\dfrac{\dfrac{3}{a} \div \dfrac{a}{3} - 2}{\dfrac{a}{6} + \dfrac{1}{2} - \dfrac{3}{a}}$.

Here the expression $= \dfrac{18 + 2a^2 - 12a}{a^2 + 3a - 18}$

$$= \frac{2(a^2 - 6a + 9)}{(a+6)(a-3)} = \frac{2(a-3)}{a+6}.$$

Example 4. Simplify $\dfrac{\dfrac{a^2 + b^2}{a^2 - b^2} - \dfrac{a^2 - b^2}{a^2 + b^2}}{\dfrac{a+b}{a-b} - \dfrac{a-b}{a+b}}$.

The numerator $= \dfrac{(a^2 + b^2)^2 - (a^2 - b^2)^2}{(a^2 + b^2)(a^2 - b^2)}$

$$= \frac{4a^2 b^2}{(a^2 + b^2)(a^2 - b^2)}.$$

Similarly the denominator $= \dfrac{4ab}{(a+b)(a-b)}$.

Hence the fraction $= \dfrac{4a^2 b^2}{(a^2 + b^2)(a^2 - b^2)} \div \dfrac{4ab}{(a+b)(a-b)}$

$$= \frac{4a^2 b^2}{(a^2 + b^2)(a^2 - b^2)} \times \frac{(a+b)(a-b)}{4ab}$$

$$= \frac{ab}{a^2 + b^2}.$$

Note. To ensure accuracy and neatness, when the numerator and denominator are somewhat complicated, the beginner is advised to simplify each separately as in the above example.

MISCELLANEOUS FRACTIONS.

180. In the case of fractions like the following, called **Continued Fractions**, we begin from the lowest fraction and simplify step by step.

Example. Simplify $\dfrac{9x^2-64}{x-1-\dfrac{1}{1-\dfrac{x}{4+x}}}$.

The expression $=\dfrac{9x^2-64}{x-1-\dfrac{1}{\dfrac{4+x-x}{4+x}}}=\dfrac{9x^2-64}{x-1-\dfrac{4+x}{4}}$

$=\dfrac{9x^2-64}{\dfrac{4x-4-(4+x)}{4}}=\dfrac{9x^2-64}{\dfrac{3x-8}{4}}$

$=\dfrac{4(9x^2-64)}{3x-8}=4(3x+8).$

EXAMPLES XXII. a.

Find the value of

1. $\dfrac{\dfrac{m}{n}-\dfrac{l}{m}}{\dfrac{a}{m}-\dfrac{b}{n}}.$

2. $\dfrac{\dfrac{1}{x}+\dfrac{1}{y}}{\dfrac{1}{x}-\dfrac{1}{y}}.$

3. $\dfrac{a+\dfrac{b}{d}}{x-\dfrac{y}{d}}.$

4. $\dfrac{1+\dfrac{c}{x}}{\dfrac{b}{x}-1}.$

5. $\dfrac{2+\dfrac{3a}{4b}}{a+\dfrac{8b}{3}}.$

6. $\dfrac{3a+\dfrac{7b}{8c}}{3c+\dfrac{7b}{8a}}.$

7. $\dfrac{1-\dfrac{y^2}{x^2}}{1+\dfrac{y^2}{x^2}}.$

8. $\dfrac{1}{a+\dfrac{b}{c}}.$

9. $\dfrac{a}{b+\dfrac{c}{d}}.$

10. $\dfrac{x}{x-\dfrac{m}{n}}.$

11. $\dfrac{\dfrac{a}{b}+\dfrac{c}{d}}{\dfrac{m}{n}+\dfrac{k}{p}}.$

12. $\dfrac{x-\dfrac{1}{x}}{1+\dfrac{1}{x}},$

13. $\dfrac{x+5+\dfrac{6}{x}}{1+\dfrac{6}{x}+\dfrac{8}{x^2}}.$

14. $\dfrac{\dfrac{1}{x}-\dfrac{2}{x^2}-\dfrac{3}{x^3}}{\dfrac{9}{x}-x}.$

15. $\dfrac{2x^2-x-6}{\dfrac{4}{x^2}-1}.$

Find the value of

16. $\dfrac{2}{1-x^2} \div \left(\dfrac{1}{1-x} - \dfrac{1}{1+x} \right).$

17. $\left(\dfrac{a^3-b^3}{a-b} - \dfrac{a^3+b^3}{a+b} \right) \div \dfrac{4ab}{a^2-b^2}.$

18. $\left(\dfrac{a^2-ax+x^2}{a-x} - \dfrac{a^2+ax+x^2}{a+x} \right) \div \dfrac{x^3}{a^2-x^2}.$

19. $\left(y + \dfrac{xy}{y-x} \right) \left(y - \dfrac{xy}{x+y} \right) \times \dfrac{y^2-x^2}{y^2+x^2}.$

20. $\left(\dfrac{x}{1+x} \div \dfrac{1-x}{x} \right) \div \left(\dfrac{x}{1+x} - \dfrac{1-x}{x} \right).$

21. $\dfrac{\dfrac{a+b}{a-b} - \dfrac{a-b}{a+b}}{1 - \dfrac{a^2+b^2}{(a+b)^2}}.$

22. $\dfrac{\dfrac{a}{x^2} + \dfrac{x}{a^2}}{\dfrac{1}{a^2} - \dfrac{1}{ax} + \dfrac{1}{x^2}}.$

23. $\dfrac{\dfrac{1}{3x-2} - \dfrac{1}{3x+2}}{9 - \dfrac{4}{x^2}}.$

24. $1 + \dfrac{x}{1+x+\dfrac{2x^2}{1-x}}.$

25. $\dfrac{1}{a - \dfrac{a^2-1}{a+\dfrac{1}{a-1}}}.$

26. $\dfrac{1}{4x + \dfrac{4x}{1+\dfrac{2(x+y)}{6-x}}}.$

27. $\dfrac{a}{x + \dfrac{m}{y+\dfrac{n}{z}}}.$

28. $\dfrac{1}{1 - \dfrac{1+x}{x-\dfrac{1}{x}}}.$

*29. $\dfrac{x-2}{x-2-\dfrac{x}{x-\dfrac{x-1}{x-2}}}.$

*30. $\dfrac{1}{x - \dfrac{1}{x+\dfrac{1}{x}}} - \dfrac{1}{x + \dfrac{1}{x-\dfrac{1}{x}}}.$

*31. $\dfrac{\dfrac{a-b}{1+ab} + \dfrac{b-c}{1+bc}}{1 - \dfrac{(a-b)(b-c)}{(1+ab)(1+bc)}}.$

*32. $\dfrac{1+\dfrac{a-b}{a+b}}{1-\dfrac{a-b}{a+b}} \div \dfrac{1+\dfrac{a^2-b^2}{a^2+b^2}}{1-\dfrac{a^2-b^2}{a^2+b^2}}.$

XXII.]   MISCELLANEOUS FRACTIONS.   165

*33. $\dfrac{a-x}{a^2 - ax - \dfrac{(a-x)^2}{1 - \dfrac{a}{x}}}$.

*34. $\dfrac{\dfrac{2x^2 + 2x}{x} - \dfrac{3}{2}}{\dfrac{x}{x-2} - 1} - \dfrac{\dfrac{3(x-1)}{x} + \dfrac{3}{4}}{\dfrac{x}{x-4} - 1}$.

*35. $\dfrac{2 - 4x}{4x - 2 - \dfrac{4x}{1 + \dfrac{1}{2x-1}{1 + \dfrac{1}{4x-1}}}}$.

*36. $\dfrac{x^2}{1 - \dfrac{1}{x^2 + \dfrac{\dfrac{1}{x}}{x + \dfrac{1}{x}}}} \div \dfrac{x^2 - 2}{1 - \dfrac{1}{x^2 - \dfrac{\dfrac{1}{x}}{x - \dfrac{1}{x}}}}$.

181. Sometimes it is convenient to express a single fraction as a group of fractions.

Example. $\dfrac{5x^2y - 10xy^2 + 15y^3}{10x^2y^2} = \dfrac{5x^2y}{10x^2y^2} - \dfrac{10xy^2}{10x^2y^2} + \dfrac{15y^3}{10x^2y^2}$

$= \dfrac{1}{2y} - \dfrac{1}{x} + \dfrac{3y}{2x^2}$.

182. Since a fraction represents the quotient of the numerator by the denominator, we may often express a fraction in an equivalent form, partly integral and partly fractional.

Example 1. $\dfrac{x+7}{x+2} = \dfrac{(x+2) + 5}{x+2} = 1 + \dfrac{5}{x+2}$.

Example 2. $\dfrac{3x-2}{x+5} = \dfrac{3(x+5) - 15 - 2}{x+5} = \dfrac{3(x+5) - 17}{x+5} = 3 - \dfrac{17}{x+5}$.

In some cases actual division may be advisable.

Example 3. Shew that $\dfrac{2x^2 - 7x - 1}{x - 3} = 2x - 1 - \dfrac{4}{x-3}$.

By division, $x - 3 \,)\, 2x^2 - 7x - 1 \,(\, 2x - 1$
$\underline{2x^2 - 6x}$
$\, -x - 1$
$\,\underline{-x + 3}$
$\,-4$

Thus the quotient is $2x - 1$, and the remainder -4.

Therefore $\dfrac{2x^2 - 7x - 1}{x-3} = 2x - 1 - \dfrac{4}{x-3}$.

183. If the numerator be of lower dimensions than the denominator, we may still perform the division, and express the result in a form which is partly integral and partly fractional.

Example. Prove that $\dfrac{2x}{1+3x^2} = 2x - 6x^3 + 18x^5 - \dfrac{54x^7}{1+3x^2}$.

By division
$$1+3x^2 \,)\, 2x \quad (\, 2x - 6x^3 + 18x^5$$
$$\underline{2x + 6x^3}$$
$$-6x^3$$
$$\underline{-6x^3 - 18x^5}$$
$$18x^5$$
$$\underline{18x^5 + 54x^7}$$
$$-54x^7$$

whence the result follows.

Here the division may be carried on to any number of terms in the quotient, and we can stop at any term we please by taking for our remainder the fraction whose numerator is the remainder last found, and whose denominator is the divisor.

Thus, if we carried on the quotient to four terms, we should have

$$\dfrac{2x}{1+3x^2} = 2x - 6x^3 + 18x^5 - 54x^7 + \dfrac{162x^9}{1+3x^2}.$$

The terms in the quotient may be fractional; thus if x^2 is divided by $x^3 - a^3$, the first four terms of the quotient are $\dfrac{1}{x} + \dfrac{a^3}{x^4} + \dfrac{a^6}{x^7} + \dfrac{a^9}{x^{10}}$, and the remainder is $\dfrac{a^{12}}{x^{10}}$.

184. Miscellaneous examples in multiplication and division occur which can be dealt with by the preceding rules for the reduction of fractions.

Example. Multiply $x + 2a - \dfrac{a^2}{2x+3a}$ by $2x - a - \dfrac{2a^2}{x+a}$.

The product $= \left(x + 2a - \dfrac{a^2}{2x+3a}\right) \times \left(2x - a - \dfrac{2a^2}{x+a}\right)$

$= \dfrac{2x^2 + 7ax + 6a^2 - a^2}{2x+3a} \times \dfrac{2x^2 + ax - a^2 - 2a^2}{x+a}$

$= \dfrac{2x^2 + 7ax + 5a^2}{2x+3a} \times \dfrac{2x^2 + ax - 3a^2}{x+a}$

$= \dfrac{(2x+5a)(x+a)}{2x+3a} \times \dfrac{(2x+3a)(x-a)}{x+a}$

$= (2x+5a)(x-a).$

EXAMPLES XXII. b.

Express each of the following fractions as a group of simple fractions in lowest terms:

1. $\dfrac{3x^2y + xy^2 - y^3}{9xy}$.

2. $\dfrac{3a^3x - 4a^2x^2 + 6ax^3}{12ax}$.

3. $\dfrac{a^3 - 3a^2b + 3ab^2 + b^3}{2ab}$.

4. $\dfrac{a+b+c}{abc}$.

5. $\dfrac{bc + ca + ab}{abc}$.

6. $\dfrac{a^3bc - 3ab^3c + 2abc}{6xbc}$.

Perform the following divisions, giving the remainder after four terms in the quotient:

7. $x \div (1+x)$.

8. $a \div (a-b)$.

9. $(1+x) \div (1-x)$.

10. $1 \div (1-x+x^2)$.

11. $x^2 \div (x+3)$.

12. $1 \div (1-x)^2$.

13. Shew that $\dfrac{a^3 - b^3}{(a-b)^2} = a + 2b + \dfrac{3b^2}{a-b}$.

14. Shew that $x^2 - xy + y^2 - \dfrac{2y^3}{x+y} = \dfrac{x^3 - y^3}{x+y}$.

15. Shew that $\dfrac{60x^3 - 17x^2 - 4x + 1}{5x^2 + 9x - 2} = 12x - 25 + \dfrac{49}{x+2}$.

16. Shew that $1 + \dfrac{a^2 + b^2 - c^2}{2ab} = \dfrac{(a+b+c)(a+b-c)}{2ab}$.

17. Divide $x + \dfrac{16x - 27}{x^2 - 16}$ by $x - 1 + \dfrac{13}{x+4}$.

18. Multiply $a^2 - 2ax + 4x^2 - \dfrac{16x^3}{a+2x}$ by $3 - \dfrac{6x(a+4x)}{a^2 + 2ax + 4x^2}$.

19. Divide $b^2 + 3b - 2 - \dfrac{12}{b-3}$ by $3b + 6 - \dfrac{2b^2}{b-3}$.

20. Divide $a^2 + 9b^2 + \dfrac{65b^4}{a^2 - 9b^2}$ by $a + 3b + \dfrac{13b^2}{a - 3b}$.

21. Multiply $4x^2 + 14x + \dfrac{98x - 27}{2x - 7}$ by $\dfrac{1}{6} - \dfrac{3x + 29}{12x^2 + 18x + 27}$.

E.A. M

185. The following exercise contains miscellaneous examples which illustrate most of the processes connected with fractions.

*EXAMPLES XXII. c.

Simplify the following fractions:

1. $\dfrac{4a(a^2-x^2)}{3b(c^2-x^2)} \div \left[\dfrac{a^2-ax}{bc+bx} \times \dfrac{a^2+2ax+x^2}{c^2-2cx+x^2}\right].$

2. $\dfrac{x(x+a)(x+2a)}{3a} - \dfrac{x(x+a)(2x+a)}{6a}.$

3. $\dfrac{1}{b}\left(\dfrac{1}{a-b} - \dfrac{1}{a+2b}\right) - \dfrac{2}{a^2+ab-2b^2}.$

4. $\left(\dfrac{x+y}{x-y}\right)^2 - \left(\dfrac{x-y}{x+y}\right)^2.$ 5. $\dfrac{2}{x-1} + \dfrac{2}{x+1} - \dfrac{4x}{x^2-x+1}.$

6. $\left(\dfrac{x^2}{1-x^4} + \dfrac{2x^4}{1-x^8}\right) \div \left(\dfrac{x^2+1}{x}\right)^2.$ 7. $\dfrac{1}{x} - \dfrac{1}{(x+1)^2} - \dfrac{2}{x+1} + \dfrac{x}{1+x+x^2}.$

8. $\dfrac{1+x^3}{1+2x+2x^2+x^3}.$ 9. $\dfrac{2x^3-9x^2+27}{3x^3-81x+162}.$

10. $\dfrac{a}{b} - \dfrac{(a^2-b^2)x}{b^2} + \dfrac{a(a^2-b^2)x^2}{b^2(b+ax)}.$

11. $\left\{\dfrac{x^4-a^4}{x^2-2ax+a^2} \div \dfrac{x^2+ax}{x-a}\right\} \times \dfrac{x^5-a^2x^3}{x^3+a^3} \div \left(\dfrac{x}{a} - \dfrac{a}{x}\right).$

12. $\dfrac{a^2-x^2}{a^2+ax+x^2} \div \dfrac{\left(1-\dfrac{x}{a}\right)^3 \left(1+\dfrac{x}{a}\right)^3}{a^3-x^3}.$

13. $\dfrac{x^4-2x^2+1}{3x^5-10x^3+15x-8}.$ 14. $\dfrac{a^3+a(1+a)y+y^2}{a^4-y^2}.$

15. $\dfrac{1}{a} + \dfrac{2}{a+1} + \dfrac{3}{a+2} - \dfrac{\dfrac{4}{a}}{1+\dfrac{1}{a}}.$

16. $\dfrac{x+3}{2x^2+9x+9} + \dfrac{1}{2} \cdot \dfrac{1}{2x-3} - \dfrac{1}{x-\dfrac{9}{4x}}.$

MISCELLANEOUS FRACTIONS.

17. $\dfrac{2}{x^3+x^2+x+1} - \dfrac{2}{x^3-x^2+x-1}.$

18. $\dfrac{1-a^2}{(1+ax)^2-(a+x)^2} \div \dfrac{1}{2}\left(\dfrac{1}{1-x}+\dfrac{1}{1+x}\right).$

19. $\dfrac{2x^3-x^2-2x+1}{x^3-3x+2}.$

20. $\dfrac{x^2-6x+8}{4x^3-21x^2+15x+20}.$

21. $\dfrac{2a}{(x-2a)^2} - \dfrac{x-a}{x^2-5ax+6a^2} + \dfrac{2}{x-3a}.$

22. $\dfrac{1}{2}\left(\dfrac{a^2+x^2}{a^2-x^2}\right) - \dfrac{1}{2}\cdot\dfrac{a+x}{a-x} - \left(\dfrac{a}{a+x}\right)^2.$

23. $\dfrac{x}{2}\left(\dfrac{1}{x-y}-\dfrac{1}{x+y}\right) \times \dfrac{x^2-y^2}{x^2y+xy^2} \div \dfrac{1}{x+y}.$

24. $\dfrac{1}{x+y} \div \left[\dfrac{y}{2}\left(\dfrac{1}{x+y}+\dfrac{1}{x-y}\right) \times \dfrac{x^2-y^2}{x^2y+xy^2}\right].$

25. $\left(3x-5-\dfrac{2}{x}\right)\left(3x+5-\dfrac{2}{x}\right) \div \left(x-\dfrac{4}{x}\right).$

26. $\left\{\dfrac{2}{x}-\dfrac{1}{a+x}+\dfrac{1}{a-x}\right\} \div \left(\dfrac{a+x}{a-x}-\dfrac{a-x}{a+x}\right).$

27. $\dfrac{1}{2x-1} - \dfrac{2x-\dfrac{1}{2x}}{4x^2-1}.$

28. $\left(b+\dfrac{ab}{b-a}\right)\left(b-\dfrac{ab}{a+b}\right)\left(\dfrac{b^2-a^2}{b^2+a^2}\right).$

29. $\left\{\dfrac{b+\dfrac{a-b}{1+ab}}{1-\dfrac{(a-b)b}{1+ab}} - \dfrac{a-\dfrac{a-b}{1-ab}}{1-\dfrac{a(a-b)}{1-ab}}\right\} \div \left(\dfrac{a}{b}-\dfrac{b}{a}\right).$

30. $\dfrac{\dfrac{x^2+y^2}{y}-x}{\dfrac{1}{y}-\dfrac{1}{x}} \times \dfrac{x^2-y^2}{x^3+y^3}.$

31. $\dfrac{2\left(x^2-\dfrac{1}{4}\right)}{2x+1} + \dfrac{1}{2}.$

32. $\dfrac{a+\dfrac{b-a}{1+ab}}{1-\dfrac{a(b-a)}{1+ab}} \times \dfrac{\dfrac{x+y}{1-xy}-y}{1+\dfrac{y(x+y)}{1-xy}}.$

33. $\dfrac{\dfrac{a+b}{a-b}+\dfrac{a-b}{a+b}}{\dfrac{a-b}{a+b}-\dfrac{a+b}{a-b}} \times \dfrac{ab^3-a^3b}{a^2+b^2}.$

Simplify the following fractions:

34. $\dfrac{(1-x^2)(1-x^3)}{x(1+x)(1-x)^2} - \dfrac{x^3+\dfrac{1}{x^3}}{x^2+\dfrac{1}{x^2}-1}.$

35. $\left\{x^3-\dfrac{1}{x^3}-3\left(x-\dfrac{1}{x}\right)\right\} \div \left(x-\dfrac{1}{x}\right).$

36. $\dfrac{1+\dfrac{1}{m}}{\dfrac{1}{m}} \times \dfrac{\dfrac{1}{m}}{m^2+\dfrac{1}{m}} \div \dfrac{\dfrac{1}{m}}{m-1+\dfrac{1}{m}}.$

37. $\dfrac{\dfrac{x}{y}+\dfrac{y}{x}-1}{\dfrac{x^2}{y^2}+\dfrac{x}{y}+1} \times \dfrac{1+\dfrac{y}{x}}{x-y} \div \dfrac{1+\dfrac{y^3}{x^3}}{\dfrac{x^2}{y}-\dfrac{y^2}{x}}.$

38. $\dfrac{\left(\dfrac{3x+x^3}{1+3x^2}\right)^2-1}{\dfrac{3x^2-1}{x^3-3x}+1} \div \dfrac{\dfrac{9}{x^2}-\dfrac{33-x^2}{3x^2+1}}{\dfrac{3}{x^2}-\dfrac{2(x^2+3)}{(x^3-x)^2}}.$

39. $\dfrac{1}{a^2-2}-\dfrac{2}{a^2-1}+\dfrac{2}{a^2+1}-\dfrac{1}{a^2+2}.$

40. $\dfrac{1}{6m-2n}+\dfrac{1}{3m+2n}-\dfrac{3}{6m+2n}.$

41. $\dfrac{3}{4(1-x)^2}+\dfrac{3}{8(1-x)}+\dfrac{1}{8(1+x)}+\dfrac{x-1}{4(1+x^2)}.$

42. $\dfrac{4}{9(x-2)}+\dfrac{5}{9(x+1)}-\dfrac{1}{3(x+1)^2}-\dfrac{1}{x+2+\dfrac{1}{x}}.$

43. $\left(\dfrac{x^2}{y}+\dfrac{y^2}{x}\right)\left(\dfrac{1}{y^2-x^2}\right)-\dfrac{y}{x^2+xy}+\dfrac{x}{xy-y^2}.$

44. $\dfrac{x^2-(y-z)^2}{(x+z)^2-y^2}+\dfrac{y^2-(z-x)^2}{(x+y)^2-z^2}+\dfrac{z^2-(x-y)^2}{(y+z)^2-x^2}.$

45. $\dfrac{x^2-(y-2z)^2}{(2z+x)^2-y^2}+\dfrac{y^2-(2z-x)^2}{(x+y)^2-4z^2}+\dfrac{4z^2-(x-y)^2}{(y+2z)^2-x^2}.$

46. $\dfrac{(x-y)(y-z)+(y-z)(z-x)+(z-x)(x-y)}{x(z-x)+y(x-y)+z(y-z)}.$

47. $\dfrac{a-b-c}{(a-b)(a-c)} + \dfrac{b-c-a}{(b-c)(b-a)} + \dfrac{c-a-b}{(c-a)(c-b)}$.

48. $\dfrac{c+a}{(a-b)(a-c)} + \dfrac{a+b}{(b-c)(b-a)} + \dfrac{b+c}{(c-a)(c-b)}$.

49. $\dfrac{x^2-(2y-3z)^2}{(3z+x)^2-4y^2} + \dfrac{4y^2-(3z-x)^2}{(x+2y)^2-9z^2} + \dfrac{9z^2-(x-2y)^2}{(2y+3z)^2-x^2}$.

50. $\dfrac{9y^2-(4z-2x)^2}{(2x+3y)^2-16z^2} + \dfrac{16z^2-(2x-3y)^2}{(3y+4z)^2-4x^2} + \dfrac{4x^2-(3y-4z)^2}{(4z+2x)^2-9y^2}$.

51. $\dfrac{\dfrac{1}{x} - \dfrac{x+a}{x^2+a^2}}{\dfrac{1}{a} - \dfrac{a+x}{a^2+x^2}} + \dfrac{\dfrac{1}{x} - \dfrac{x-a}{x^2+a^2}}{\dfrac{1}{a} - \dfrac{a-x}{a^2+x^2}}$.

52. $\dfrac{(x+a)(x+b)-(y+a)(y+b)}{x-y} - \dfrac{(x-a)(y-b)-(x-b)(y-a)}{a-b}$.

53. $\left(\dfrac{a+x}{a^2-ax+x^2} - \dfrac{a-x}{a^2+ax+x^2}\right) \div \left(\dfrac{a^2+x^2}{a^3-x^3} - \dfrac{a^2-x^2}{a^3+x^3}\right)$.

54. $\dfrac{\dfrac{x-1}{3} + \dfrac{x-1}{x-2}}{\dfrac{x+2}{4} + \dfrac{x+2}{x-3}} \div \dfrac{\dfrac{x+3}{7} - \dfrac{x+3}{x+4}}{\dfrac{x-2}{3} + \dfrac{x-2}{x-1}}$.   55. $\dfrac{x-2+\dfrac{6}{x+3}}{x-4+\dfrac{12}{x+3}} \times \dfrac{\dfrac{x+3}{4} - \dfrac{x+3}{x+1}}{\dfrac{x-3}{7} + \dfrac{x-3}{x-4}}$.

56. $(x+2)\left\{1 + \dfrac{6(x+2)}{x^2-x-6}\right\}\left\{1 - \dfrac{5(x+1)}{x^2+3x+2}\right\}$.

57. $(1+a)^2 \div \left\{1 + \dfrac{a}{1-a+\dfrac{a}{1+a+a^2}}\right\}$.

58. $\dfrac{\{ax^2+(b-c)x-f\}^2 - \{ax^2+(b+c)x-f\}^2}{\{ax^2+(b+e)x-f\}^2 - \{ax^2+(b-e)x-f\}^2}$.

MISCELLANEOUS EXAMPLES IV.

[The following Examples for revision are arranged in groups under different headings; each group illustrates one or more of the principal rules and processes already discussed, and for the most part the Examples present more variety and difficulty than those of the same type which have appeared in previous exercises.]

Substitutions and Brackets.

1. Find the value of $\dfrac{a+\sqrt{a^2+b^2}}{a^3-2b(a^2-b^2)}$ when $a=-4$, $b=-3$.

2. When $a=1$, $b=-1$, $c=2$ evaluate the expression
$$\sqrt{3a^3(b-c)+3b^3(c-a)+3c^3(a-b)}.$$

3. Simplify
$$a(b-c)^3 - a(b-c)(2b^2-bc+2c^2) + (ab+ac)(b^2-c^2);$$
and find its value when $a=1$, $b=2$, $c=3$.

4. Find the value of
$$\sqrt[3]{\{5(b^2-c^2)-a^2\}} + \sqrt[4]{3\{a(a^2-c^2)-1\}}$$
when $a=4$, $b=5$, $c=3$.

5. Find the value of
$$\sqrt{(x^2+y^3+z)(x-y-3z)} \div \sqrt[3]{xy^3z^2}$$
when $x=-1$, $y=-3$, $z=1$.

6. When $a=0$, $b=2$, $x=1$, $y=-3$, $z=5$, find the numerical value of
   (1) $(x-y)^3 - 3a(x-y)^2 + 3b(x^3-y^3)$;
   (2) $(x-a)^3 - b^2(x-y+z) + \sqrt{(bx^2-axy+y^2+z^2)}$.

7. If $x=6$, $y=7$, $z=8$, find the value of
   (1) $x\left\{\dfrac{x}{7} - \dfrac{2}{3}\left(\dfrac{y}{4}+1\right)\right\} - \dfrac{2}{7}\left\{y(x+1) + \dfrac{1}{2}(x^2-2y) - \dfrac{7x}{3}\right\}$;
   (2) $x\left\{y\left(z-\dfrac{1}{z}\right)\right\} - \dfrac{y}{z}\left\{y - 3\left(\dfrac{x}{3}-xz^2\right)\right\} - 2xz\left\{-y\left(1+\dfrac{y}{2xz^2}\right)\right\}$.

XXII.] MISCELLANEOUS EXAMPLES. 173

8. Evaluate $\dfrac{2a-b\{c-a(b-c)\}}{a^2(b+c)+b^2(c+a)} \cdot \sqrt{\dfrac{a+bc}{b^2+4ac}}$,

when $a=2$, $b=-1$, $c=1$.

9. Find the value of
$$\dfrac{a-[b-\{a-(b-c)\}]-\sqrt{2a^2+b^2+c^2}}{\sqrt{c^4+a^2b^2+2a^3c}},$$
when $a=-1$, $b=5$, $c=3$.

10. When $a=4$, $b=-2$, $c=\tfrac{3}{2}$, $d=-1$, find the value of

(1) $a^3-b^3-(a-b)^3-11(3b+2c)\left(2c^2-\dfrac{d^2}{2}\right)$;

(2) $\sqrt[3]{4c^2-a(a-2b-d)} - \sqrt[3]{b^4c+11b^3d^2}$.

11. Find the value of
$$l-m-3\left[\left\{l-m+lm\left(\dfrac{1}{l}+\dfrac{1}{m}\right)\right\}^2-4l^2\right]$$
when $l=\tfrac{2}{3}$, $m=-\tfrac{1}{3}$.

12. When $a=1$, $b=-\tfrac{1}{2}$, $c=0$, evaluate
$$\dfrac{a-[b-c-\{2a-2b-\tfrac{1}{2}(3c-b)\}]}{a-\dfrac{b}{c-\dfrac{1}{a}}}.$$

13. Simplify
$$42\left\{\dfrac{4x-3y}{6}-\dfrac{3}{7}\left(x-\dfrac{4}{3}y\right)\right\} - 56\left\{\dfrac{1}{7}(3x-2y)-\dfrac{3}{8}\left(\dfrac{16}{3}x-y\right)\right\};$$
and find its value when $x=\tfrac{1}{6}$, $y=-\tfrac{3}{2}$.

14. If $x=6$, $y=7$, $z=8$ find the value of

(1) $\dfrac{z}{x}\left\{y-\left(2z-\dfrac{x}{z}\right)\right\} - \dfrac{y}{z}\left[\dfrac{z^2}{x}-3\left\{1+\dfrac{2z^3}{3xy}\right\}+\dfrac{x}{2}\right]$;

(2) $3y\left\{\dfrac{2yz}{3}-\left(\dfrac{y}{5}-1\right)\right\} + \dfrac{y^2}{5}\left[3-2\left\{10-\dfrac{x}{2}(y-7)\right\}\right]$.

Resolution into Factors.

(On Arts. 128–132.)

Resolve into two or more factors:

15. $x^2 + 21x + 108$. 16. $a^2 + 6a - 91$. 17. $x^2 - 20xy + 96y^2$.
18. $a^2b^2 - 14ab - 51$. 19. $c^3 + c^2 - 156c$. 20. $m^2n - 6mn^2 + 9n^3$.
21. $p^4 - p^2q^2 - 56q^4$. 22. $d^4 - 4d^2c^2 - 45c^4$. 23. $x^3y - x^2y^2 - 42xy^3$.
24. $m^2 + 28m + 195$. 25. $210 - a - a^2$. 26. $57 + 16pq - p^2q^2$.
27. $x^4 + 27x^2 + 176$. 28. $a^4 + 7a^2 - 98$. 29. $c^2 + 54c + 729$.
30. $72 + xy - x^2y^2$. 31. $a^4 + 9a^2x^2 + 14x^4$. 32. $p^2 - 3pq - 108q^2$.
33. $2a^6 + 2a^3 - 264$. 34. $x^4 - 2x^3 - 63x^2$. 35. $b^2c^2 + 5bc - 84$.
36. $z^2 + 34z + 289$. 37. $a^2 - 22ac + 57c^2$. 38. $y^3z + 6y^2z - 91yz$.
39. $2 + x^3 - 3x^6$. 40. $2a^2b^2 + ab - 15$. 41. $9p^2 - 24p + 16$.
42. $35 + 12mn + m^2n^2$. 43. $119 - 10c - c^2$. 44. $6x^3 - 5x^4 + x^5$.
45. $6m^2 + 7m - 3$. 46. $4a^2 - 8ab - 5b^2$. 47. $6p^2 - 13pq + 2q^2$.
48. $20x^2 - 9xz - 20z^2$. 49. $8x^4 + 2x^2 - 15$. 50. $12y^2 - 30y + 12$.
51. $12(a^2b^2 - 1) + 7ab$. 52. $2(a^4b^2 + 5) - 9a^2b$.
53. $21x^2 + 2y(5x - 8y)$. 54. $3(6m^2 - 5n^2) + 17mn$.

(On Arts. 133–137.)

Resolve into two or more factors:

55. $c^2 - a^2 - b^2 + 2ab$. 56. $a^2 + 2bc - b^2 - c^2$.
57. $125x^3 + 27y^3$. 58. $a^3b^3 + 343$. 59. $512b^3 - a^6$.
60. $a^2 - 4(x - y)^2$. 61. $2mn + m^2 - 1 + n^2$.
62. $8c^4 - 2c^2(d + c)^2$. 63. $(a^2b^2 - 1)^2 - x^2 + 2xy - y^2$.
64. $1 - 64m^6$. 65. $p^3 + 1000p^3q^3$. 66. $6561 - a^4$.
67. $x^4 - 2x^2 - y^2 - z^2 + 2yz + 1$. 68. $a^2 - 16(b - c)^2$.
69. $c - d - 4(c - d)^3$. 70. $p^2 - 16q^2 + p - 4q$.
71. $2 + 128(a + b)^3$. 72. $x + 3y + x^3 + 27y^3$.

XXII.]   MISCELLANEOUS EXAMPLES.   175

(*Miscellaneous Factors.*)

Resolve into two or more factors:

73. $x^3 + x^2y + xy^2 + y^3$.   74. $acx^2 + bcx - adx - bd$.

75. $14 - 5a - a^2$.   76. $98x^4 - 7x^2y^2 - y^4$.   77. $51 - 14a - a^2$.

78. $1 - (m^2 + p^2) - 2mp$.   79. $ab(x^2 + 1) - x(a^2 + b^2)$.

80. $9b^2 - 6bc - 16 + c^2$.   81. $x^2c^3 - c^3 + x^2 - 1$.

82. $3x^2 - 2ab - x(b - 6a)$.   83. $m^3 - n^3 - (x^2 - mn)(m - n)$.

84. $a(b^2 + c^2 - a^2) + b(a^2 + c^2 - b^2)$.   85. $x^7 - x^3 + 8x^4 - 8$.

86. Express in factors the square root of
$$(x^2 + 8x + 7)(2x^2 - x - 3)(2x^2 + 11x - 21).$$

87. Find the expression whose square is
$$(2x^2 - xy - 15y^2)(4x^2 - 25y^2)(2x^2 - 11xy + 15y^2).$$

Highest Common Factor and Lowest Common Multiple.

88. Find the lowest common multiple of
$$13ab^2(x^3 - 3a^2x + 2a^3),\quad 65a^3b(x^2 + ax - 2a^2),\quad 25b^3(x^2 - a^2)^2.$$

89. Find the highest common factor of
$$2(x^4 + 9) - 5x^2(x + 1),\quad 2x^3(2x - 9) + 81(x - 1).$$

90. Find the expression of lowest dimensions which is divisible by each of the following expressions:
$$(2x^4 + 4x^3)(x^2 + 2x - 8),\quad (2x^3 - 4x^2)(x^2 - 2x - 8),$$
$$(x^2 - 4x)(x^2 + 2x - 8).$$

91. Find the H.C.F. and the L.C.M. of the three expressions
$$a(a + c) - b(b + c),\quad b(b + a) - c(c + a),\quad c(c + b) - a(a + b).$$

92. Find the divisor of highest dimensions of the expressions
$$(a + b)(a - b) + c(c - 2a),\quad (a + c)(a - c) + b(b + 2a).$$

93. Find the expression of lowest dimensions such that the L.C.M. of it and $2a^2 - 3ab + b^2$ is
$$2a^4 - 3a^3b - a^2b^2 + 3ab^3 - b^4.$$

94. Shew that $x^2 - 4y^2$ is the H.C.F. of the expressions
$$x^4 - 3x^2y^2 - 4y^4, \quad x^6 - 64y^6,$$
and $\quad x^5 + 32y^5 - 8x^3y^2 + 2x^4y + 16xy^4 - 16x^2y^3.$

95. Find the lowest common multiple of
$$(a-c)^2 - (b-c)^2, \quad a^2 + b^2 - 2ac - 2bc + 2ab, \quad a^4 - b^4.$$

96. Shew that the lowest common multiple of
$$a(a-b)^2 - ac^2, \quad a^2b - b(b-c)^2, \quad (a+c)^2c - b^2c$$
is $\quad abc(a^4 + b^4 + c^4 - 2b^2c^2 - 2c^2a^2 - 2a^2b^2).$

97. Prove that $\quad x^4 - 15x^3 + 75x^2 - 145x + 84$
and $\quad x^4 - 17x^3 + 101x^2 - 247x + 210$
have the same H.C.F and L.C.M. as
$$x^4 - 13x^3 + 53x^2 - 83x + 42$$
and $\quad x^4 - 19x^3 + 131x^2 - 389x + 420.$

Simplification of Fractions.

Simplify

98. $\dfrac{1+x+x^2}{1-x^3} + \dfrac{x-x^2}{(1-x)^3}.$

99. $\dfrac{2x-7}{(x-3)^2} - \dfrac{2(x+2)}{x^2-9}.$

100. $\dfrac{1}{2x^2 - \dfrac{1}{2}} + \dfrac{1}{(2x+1)^2}.$

101. $\dfrac{(x+1)^3 - (x-1)^3}{3x^3 + x}.$

102. $\dfrac{1}{(1-x)^2} + \dfrac{2}{1-x^2} + \dfrac{1}{(1+x)^2}.$

103. $\dfrac{(x^3 - 2x)^2 - (x^2 - 2)^2}{(x-1)(x+1)(x^2-2)^2}.$

104. $\dfrac{1}{6x-2} - \dfrac{1}{2\left(x - \dfrac{1}{3}\right)} - \dfrac{1}{1-3x}.$

105. $\dfrac{x}{9} + \dfrac{2}{3} + \dfrac{4}{x-6} - \dfrac{2}{3} \cdot \dfrac{1}{1 - \dfrac{6}{x}}.$

106. $\dfrac{x}{x^2-y^2} - \dfrac{1}{x-y} + \dfrac{1}{x+y} + \dfrac{1}{x} - \dfrac{1}{y} + \dfrac{x^2 - xy + y^2}{xy(x-y)}.$

107. $\left(x - y - \dfrac{4y^2}{x-y}\right)\left(x + y - \dfrac{4x^2}{x+y}\right) \div \left\{3(x+y) - \dfrac{8xy}{x-y}\right\}.$

XXII.]   MISCELLANEOUS EXAMPLES. IV.   177

108. $\dfrac{a^4+b^4+ab(a^2+b^2)}{(a+b)^2} - \dfrac{a^4+b^4-ab(a^2+b^2)}{(a-b)^2} + \dfrac{12a^2b^2}{(a+b)^2-(a-b)^2}$.

109. $\dfrac{(ac+bd)^3-(ad+bc)^3}{(a-b)(c-d)} - \dfrac{(ac+bd)^3+(ad+bc)^3}{(a+b)(c+d)}$.

110. $\left[\left(\dfrac{x}{a}\right)^2+\left(\dfrac{z-x}{b}\right)^2\right] \div \left[\dfrac{z^2}{a^2+b^2}+\dfrac{a^2+b^2}{a^2b^2}\left(x-\dfrac{za^2}{a^2+b^2}\right)^2\right]$.

111. $\dfrac{x^4-(x-1)^2}{(x^2+1)^2-x^2} + \dfrac{x^2-(x^2-1)^2}{x^2(x+1)^2-1} + \dfrac{x^2(x-1)^2-1}{x^4-(x+1)^2}$.

112. $\dfrac{\dfrac{1+x}{1-x}+\dfrac{4x}{1+x^2}+\dfrac{8x}{1-x^2}-\dfrac{1-x}{1+x}}{\dfrac{1+x^2}{1-x^2}+\dfrac{4x^2}{1+x^4}-\dfrac{1-x^2}{1+x^2}}$.   113. $\left\{\dfrac{f}{g-\dfrac{g^2}{f}}+\dfrac{g}{f-\dfrac{f^2}{g}}\right\} \times \dfrac{1}{\dfrac{f^2}{g}-\dfrac{g^2}{f}}$.

114. $\dfrac{1}{x-\dfrac{2}{x+\dfrac{1}{2}}} \times \dfrac{1}{2+\dfrac{1}{x}} \div \dfrac{x}{2x-\dfrac{x+4}{x+1}}$.

115. $\dfrac{x}{1+\dfrac{x}{1-x+\dfrac{x}{1+x}}} \div \dfrac{1+x+x^2}{1+3x+3x^2+2x^3}$.

116. $\dfrac{ab}{a+b}\left(3c+\dfrac{b}{a}\right)-\dfrac{b^2}{(a+b)^3}(a^2+b^2)-2a\left(\dfrac{b}{a+b}\right)^3$.

117. $\dfrac{\dfrac{(a+b)^2+(a-b)^2}{b-a}-(a+b)}{\dfrac{1}{b-a}-\dfrac{1}{a+b}} \div \dfrac{(a+b)^3+(b-a)^3}{(a+b)^2-(a-b)^2}$.

118. $\left\{\dfrac{c-b}{(a-b)(a-c)}-\dfrac{c-a}{(b-c)(b-a)}+\dfrac{b-a}{(c-a)(c-b)}\right\}$
$\qquad \div \dfrac{2(a^2+b^2+c^2-bc-ca-ab)}{(a-b)(b-c)(c-a)}$.

CHAPTER XXIII.

HARDER EQUATIONS.

186. In this chapter we propose to give a miscellaneous collection of equations. Some of these will serve as a useful exercise for revision of the methods already explained in previous chapters; but we also add others presenting more difficulty, the solution of which will often be facilitated by some special artifice.

The following examples worked in full will sufficiently illustrate the most useful methods.

Example 1. Solve $\dfrac{6x-3}{2x+7} = \dfrac{3x-2}{x+5}$.

Multiplying up, we have
$$(6x-3)(x+5) = (3x-2)(2x+7),$$
$$6x^2 + 27x - 15 = 6x^2 + 17x - 14;$$
$$\therefore\ 10x = 1;$$
$$\therefore\ x = \frac{1}{10}.$$

Note. By a simple reduction many equations can be brought to the form in which the above equation is given. When this is the case, the necessary simplification is readily completed by multiplying up, or "multiplying across," as it is sometimes called.

Example 2. Solve $\dfrac{8x+23}{20} - \dfrac{5x+2}{3x+4} = \dfrac{2x+3}{5} - 1$.

Multiply by 20, and we have
$$8x + 23 - \frac{20(5x+2)}{3x+4} = 8x + 12 - 20.$$

By transposition, $\qquad 31 = \dfrac{20(5x+2)}{3x+4}.$

Multiplying across, $93x + 124 = 20(5x+2),$
$$84 = 7x;$$
$$\therefore\ x = 12.$$

CHAP. XXIII.] HARDER EQUATIONS. 179

When two or more fractions have the same denominator they should be taken together and simplified.

Example 3. Solve $\dfrac{13-2x}{x+3}+\dfrac{23x+8\frac{1}{3}}{4x+5}=\dfrac{16-\frac{1}{4}x}{x+3}+4.$

By transposition, we have

$$\dfrac{23x+8\frac{1}{3}}{4x+5}-4=\dfrac{16-\frac{1}{4}x-13+2x}{x+3}.$$

$$\therefore \dfrac{7x-\frac{35}{3}}{4x+5}=\dfrac{3+\frac{7x}{4}}{x+3}.$$

Multiplying across, we have

$$7x^2-\dfrac{35x}{3}+21x-35=12x+7x^2+15+\dfrac{35x}{4}.$$

$$-\dfrac{137x}{12}=50;$$

$$\therefore x=-\dfrac{600}{137}.$$

Example 4. Solve $\dfrac{x-8}{x-10}+\dfrac{x-4}{x-6}=\dfrac{x-5}{x-7}+\dfrac{x-7}{x-9}.$

This equation might be solved by clearing of fractions, but the work would be very laborious. The solution will be much simplified by proceeding as follows:

Transposing, $\quad \dfrac{x-8}{x-10}-\dfrac{x-5}{x-7}=\dfrac{x-7}{x-9}-\dfrac{x-4}{x-6}.$

Simplifying each side *separately*, we have

$$\dfrac{(x-8)(x-7)-(x-5)(x-10)}{(x-10)(x-7)}=\dfrac{(x-7)(x-6)-(x-4)(x-9)}{(x-9)(x-6)};$$

$$\therefore \dfrac{x^2-15x+56-(x^2-15x+50)}{(x-10)(x-7)}=\dfrac{x^2-13x+42-(x^2-13x+36)}{(x-9)(x-6)};$$

$$\therefore \dfrac{6}{(x-10)(x-7)}=\dfrac{6}{(x-9)(x-6)}.$$

Hence, since the numerators are equal, the denominators must be equal;

that is, $\quad (x-10)(x-7)=(x-9)(x-6),$

$$x^2-17x+70=x^2-15x+54;$$

$$\therefore 16=2x;$$

$$\therefore x=8.$$

The above equation may also be solved very neatly by the following artifice.

The equation may be written in the form
$$\frac{(x-10)+2}{x-10}+\frac{(x-6)+2}{x-6}=\frac{(x-7)+2}{x-7}+\frac{(x-9)+2}{x-9};$$
whence we have
$$1+\frac{2}{x-10}+1+\frac{2}{x-6}=1+\frac{2}{x-7}+1+\frac{2}{x-9};$$
which gives
$$\frac{1}{x-10}+\frac{1}{x-6}=\frac{1}{x-7}+\frac{1}{x-9}.$$
Transposing,
$$\frac{1}{x-10}-\frac{1}{x-7}=\frac{1}{x-9}-\frac{1}{x-6};$$
$$\therefore \frac{3}{(x-10)(x-7)}=\frac{3}{(x-9)(x-6)};$$
and the solution may be completed as before.

Example 5. Solve $\dfrac{5x-64}{x-13}-\dfrac{2x-11}{x-6}=\dfrac{4x-55}{x-14}-\dfrac{x-6}{x-7}$.

We have $5+\dfrac{1}{x-13}-\left(2+\dfrac{1}{x-6}\right)=4+\dfrac{1}{x-14}-\left(1+\dfrac{1}{x-7}\right);$

$$\therefore \frac{1}{x-13}-\frac{1}{x-6}=\frac{1}{x-14}-\frac{1}{x-7}.$$

The solution may now be completed as before, and we obtain $x=10$.

EXAMPLES XXIII. a.

1. $\dfrac{x+4}{3x-8}=\dfrac{x+5}{3x-7}.$

2. $\dfrac{3x+1}{3(x-2)}=\dfrac{x-2}{x-1}.$

3. $\dfrac{7-5x}{1+x}=\dfrac{11-15x}{1+3x}.$

4. $\dfrac{3(7+6x)}{2+9x}=\dfrac{35+4x}{9+2x}.$

5. $\dfrac{6x+13}{15}-\dfrac{3x+5}{5x-25}=\dfrac{2x}{5}.$

6. $\dfrac{6x+8}{2x+1}-\dfrac{2x+38}{x+12}=1.$

7. $\dfrac{3x-1}{2x-1}-\dfrac{4x-2}{3x-1}=\dfrac{1}{6}.$

8. $\dfrac{x+25}{x-5}=\dfrac{2x+75}{2x-15}.$

9. $\dfrac{x}{x+2}+\dfrac{4}{x+6}=1.$

10. $\dfrac{6x+7}{9x+6}=\dfrac{1}{12}+\dfrac{5x-5}{12x+8}.$

HARDER EQUATIONS.

11. $\dfrac{2x-5}{5} + \dfrac{x-3}{2x-15} = \dfrac{4x-3}{10} - 1\dfrac{1}{10}.$

12. $\dfrac{4(x+3)}{9} = \dfrac{8x+37}{18} - \dfrac{7x-29}{5x-12}.$

13. $\dfrac{(2x-1)(3x+8)}{6x(x+4)} - 1 = 0.$ 14. $\dfrac{2x+5}{5x+3} - \dfrac{2x+1}{5x+2} = 0.$

15. $\dfrac{4}{x+3} - \dfrac{2}{x+1} = \dfrac{5}{2x+6} - \dfrac{2\frac{1}{2}}{2x+2}.$

16. $\dfrac{7}{x-4} - \dfrac{60}{5x-30} = \dfrac{10\frac{1}{2}}{3x-12} - \dfrac{8}{x-6}.$

17. $\dfrac{3}{4-2x} + \dfrac{30}{8(1-x)} = \dfrac{3}{2-x} + \dfrac{5}{2-2x}.$

18. $\dfrac{25-\frac{x}{3}}{x+1} + \dfrac{16x+4\frac{1}{5}}{3x+2} = 5 + \dfrac{23}{x+1}.$ 19. $\dfrac{30+6x}{x+1} + \dfrac{60+8x}{x+3} = 14 + \dfrac{48}{x+1}.$

20. $\dfrac{x}{x-2} - \dfrac{x+1}{x-1} = \dfrac{x-8}{x-6} - \dfrac{x-9}{x-7}.$ 21. $\dfrac{x+5}{x+4} - \dfrac{x-6}{x-7} = \dfrac{x-4}{x-5} - \dfrac{x-15}{x-16}.$

22. $\dfrac{x-7}{x-9} - \dfrac{x-9}{x-11} = \dfrac{x-13}{x-15} - \dfrac{x-15}{x-17}.$

23. $\dfrac{x+3}{x+6} - \dfrac{x+6}{x+9} = \dfrac{x+2}{x+5} - \dfrac{x+5}{x+8}.$ 24. $\dfrac{x+2}{x} + \dfrac{x-7}{x-5} - \dfrac{x+3}{x+1} = \dfrac{x-6}{x-4}.$

25. $\dfrac{4x-17}{x-4} + \dfrac{10x-13}{2x-3} = \dfrac{8x-30}{2x-7} + \dfrac{5x-4}{x-1}.$

26. $\dfrac{5x-8}{x-2} + \dfrac{6x-44}{x-7} - \dfrac{10x-8}{x-1} = \dfrac{x-8}{x-6}.$

27. $\dfrac{2x-3}{\cdot 3x - \cdot 4} = \dfrac{\cdot 4x - \cdot 6}{\cdot 06x - \cdot 07}.$ 28. $\dfrac{x-2}{\cdot 05} - \dfrac{x-4}{\cdot 0625} = 56.$

29. $\cdot 08\dot{3}(x - \cdot 625) = \cdot 0\dot{9}(x - \cdot 59375).$

30. $(2x + 1\cdot 5)(3x - 2\cdot 25) = (2x - 1\cdot 125)(3x + 1\cdot 25).$

31. $\dfrac{\cdot 3x - 1}{\cdot 5x - \cdot 4} = \dfrac{\cdot 5 + 1\cdot 2x}{2x - \cdot 1}.$ 32. $\dfrac{1 - 1\cdot 4x}{\cdot 2 + x} = \dfrac{\cdot 7(x-1)}{\cdot 1 - \cdot 5x}.$

33. $\dfrac{(\cdot 3x - 2)(\cdot 3x - 1)}{\cdot 2x - 1} - \dfrac{1}{6}(\cdot 3x - 2) = \cdot 4x - 2.$

Literal Equations.

187. In the equations we have discussed hitherto the coefficients have been numerical quantities, but equations often involve *literal* coefficients. [Art. 6.] These are supposed to be known, and will appear in the solution.

Example 1. Solve $(x+a)(x+b)-c(a+c)=(x-c)(x+c)+ab$.

Multiplying out, we have
$$x^2+ax+bx+ab-ac-c^2=x^2-c^2+ab;$$
whence
$$ax+bx=ac,$$
$$(a+b)x=ac;$$
$$\therefore\ x=\frac{ac}{a+b}.$$

Example 2. Solve $\dfrac{a}{x-a}-\dfrac{b}{x-b}=\dfrac{a-b}{x-c}$.

Simplifying the left side, we have
$$\frac{a(x-b)-b(x-a)}{(x-a)(x-b)}=\frac{a-b}{x-c}.$$
$$\frac{(a-b)x}{(x-a)(x-b)}=\frac{a-b}{x-c};$$
$$\therefore\ \frac{x}{(x-a)(x-b)}=\frac{1}{x-c}.$$

Multiplying across,
$$x^2-cx=x^2-ax-bx+ab,$$
$$ax+bx-cx=ab,$$
$$(a+b-c)x=ab;$$
$$\therefore\ x=\frac{ab}{a+b-c}.$$

EXAMPLES XXIII. b.

Solve the equations:

1. $ax-2b=5bx-3a$.
2. $a^2(x-a)+b^2(x-b)=abx$.
3. $x^2+a^2=(b-x)^2$.
4. $(x-a)(x+b)=(x-a+b)^2$.
5. $a(x-2)+2x=6+a$.
6. $m^2(m-x)-mnx=n^2(n+x)$.
7. $(a+x)(b+x)=x(x-c)$.
8. $(a-b)(x-a)=(a-c)(x-b)$.
9. $\dfrac{2x+3a}{x+a}=\dfrac{2(3x+2a)}{3x+a}$.
10. $\dfrac{2(x-b)}{3x-c}=\dfrac{2x+b}{3(x-c)}$.

11. $\dfrac{1}{a} - \dfrac{1}{x} = \dfrac{1}{x} - \dfrac{1}{b}$.

12. $\dfrac{2}{3}\left(\dfrac{x}{a} + 1\right) = \dfrac{3}{4}\left(\dfrac{x}{a} - 1\right)$.

13. $\dfrac{a}{x} = c(a-b) + \dfrac{b}{x}$.

14. $\dfrac{9a}{b} - \dfrac{3x}{b} = \dfrac{4b}{a} - \dfrac{2x}{a}$.

15. $\dfrac{x-a}{b-x} = \dfrac{x-b}{a-x}$.

16. $\dfrac{x-a}{2} = \dfrac{(x-b)^2}{2x-a}$.

17. $\dfrac{1}{4}x(x-a) - \left(\dfrac{x+a}{2}\right)^2 = \dfrac{2a}{3}\left(x - \dfrac{a}{2}\right)$.

18. $(a+b)x^2 - a(bx + a^2) = bx(x-a) + ax(x-b)$.

19. $b(a+x) - (a+x)(b-x) = x^2 + \dfrac{bc^2}{a}$.

20. $b(a-x) - \dfrac{a}{b}(b+x)^2 + ab\left(\dfrac{x}{b} + 1\right)^2 = 0$.

21. $x^2 + a(2a - x) - \dfrac{3b^2}{4} = \left(x - \dfrac{b}{2}\right)^2 + a^2$.

22. $(2x - a)\left(x + \dfrac{2a}{3}\right) = 4x\left(\dfrac{a}{3} - x\right) - \dfrac{1}{2}(a - 4x)(2a + 3x)$.

23. $\dfrac{x - a + b}{x - a} + \dfrac{x - b}{x - 2b} = \dfrac{x}{x - b} + \dfrac{x - a}{x - a - b}$.

24. $\left(\dfrac{x}{a} - 3\right)\left(\dfrac{3x}{a} - 1\right) - \dfrac{1}{a^2}(x - 2a)(2x - a) = \left(\dfrac{x}{a} - 1\right)^2 - 1$.

25. $\dfrac{b(x+a)}{x^2 - b^2} + \dfrac{2x + 3b - a}{x + b} = \dfrac{2(x^2 + bx - b^2)}{x^2 - b^2}$.

Example 3. Solve $\quad ax + by = c$ ....(1),

$\qquad\qquad\qquad\qquad\quad a'x + b'y = c'$ ....(2).

The notation here first used is one that the student will frequently meet with in the course of his reading. In the first equation we choose certain letters as the coefficients of x and y, and we choose *corresponding letters with accents* to denote corresponding quantities in the second equation. There is no necessary connection between the values of a and a', and they are as different as a and b; but it is often convenient to use the same letter thus slightly varied to mark some common meaning of such letters, and thereby assist the memory. Thus a, a' have a common property as being coefficients of x; b, b' as being coefficients of y.

Sometimes instead of accents letters are used with a *suffix*, such as a_1, a_2, a_3; b_1, b_2, b_3, etc.

To return to the equations $ax+by=c$ ....(1),
$$a'x+b'y=c' \quad\quad\quad\quad\quad\quad\quad (2).$$

Multiply (1) by b' and (2) by b. Thus
$$ab'x+bb'y=b'c,$$
$$a'bx+bb'y=bc';$$
by subtraction, $(ab'-a'b)x=b'c-bc';$
$$\therefore\ x=\frac{b'c-bc'}{ab'-a'b}\ \ldots\ldots\ldots\ldots\ldots(3).$$

As previously explained in Art. 104, we might obtain y by substituting this value of x in *either* of the equations (1) or (2); but y is more conveniently found by eliminating x, as follows:

Multiplying (1) by a' and (2) by a, we have
$$aa'x+a'by=a'c,$$
$$aa'x+ab'y=ac';$$
by subtraction, $(a'b-ab')y=a'c-ac';$
$$\therefore\ y=\frac{a'c-ac'}{a'b-ab'},$$

or, changing signs in the terms of the denominator so as to have the same denominator as in (3),
$$y=\frac{ac'-a'c}{ab'-a'b},\ \text{and}\ \ x=\frac{b'c-bc'}{ab'-a'b}.$$

Example 4. Solve $\dfrac{x-a}{c-a}+\dfrac{y-b}{c-b}=1$ ....(1),

$$\frac{x+a}{c}+\frac{y-a}{a-b}=\frac{a}{c} \ldots\ldots\ldots\ldots\ldots(2).$$

From (1) by clearing of fractions, we have
$$x(c-b)-a(c-b)+y(c-a)-b(c-a)=(c-a)(c-b),$$
$$x(c-b)+y(c-a)=ac-ab+bc-ab+c^2-ac-bc+ab,$$
$$x(c-b)+y(c-a)=c^2-ab\ldots\ldots\ldots\ldots\ldots(3).$$

Again, from (2), we have
$$x(a-b)+a(a-b)+cy-ca=a(a-b)$$
$$x(a-b)+cy=ac\ \ldots\ldots\ldots\ldots\ldots(4).$$

Multiply (3) by c and (4) by $c-a$ and subtract,
$$x\{c(c-b)-(c-a)(a-b)\}=c^3-abc-ac(c-a),$$
$$x(c^2-ac+a^2-ab)=c(c^2-ab-ac+a^2);$$
$$\therefore\ x=c;$$
and therefore from (4) $y=b.$

EXAMPLES XXIII. c.

Solve the equations :

1. $ax + by = l,$
   $bx + ay = m.$

2. $lx + my = n,$
   $px + qy = r.$

3. $ax = by,$
   $bx + ay = c.$

4. $ax + by = a^2,$
   $bx + ay = b^2.$

5. $x + ay = a',$
   $ax + a'y = 1.$

6. $px - qy = r,$
   $rx - py = q.$

7. $\dfrac{x}{a} + \dfrac{y}{b} = \dfrac{1}{ab},$
   $\dfrac{x}{a'} - \dfrac{y}{b'} = \dfrac{1}{a'b'}.$

8. $\dfrac{x}{a} - \dfrac{y}{b} = 0,$
   $bx + ay = 4ab$

9. $\dfrac{3x}{a} + \dfrac{2y}{b} = 3,$
   $\dfrac{9x}{a} - \dfrac{6y}{b} = 3.$

10. $qx - rb = p(a - y),$
    $\dfrac{qx}{a} + r = p\left(1 + \dfrac{y}{b}\right).$

11. $\dfrac{x}{m} + \dfrac{y}{m'} = 1,$
    $\dfrac{x}{m'} - \dfrac{y}{m} = 1.$

12. $px + qy = 0,$
    $lx + my = n.$

13. $(a - b)x = (a + b)y,$
    $x + y = c.$

14. $(a - b)x + (a + b)y = 2a^2 - 2b^2.$
    $(a + b)x - (a - b)y = 4ab.$

15. $\dfrac{x}{a} + \dfrac{y}{b} = 1,$
    $\dfrac{x}{3a} + \dfrac{y}{6b} = \dfrac{2}{3}.$

16. $\dfrac{x}{a} + \dfrac{y}{b} = 2,$
    $\dfrac{x}{a'} = \dfrac{y}{b'}.$

17. $\dfrac{x}{a} - \dfrac{y}{b} = 1,$
    $\dfrac{x}{b} + \dfrac{y}{a} = \dfrac{a}{b}.$

18. $\dfrac{m}{l}x + \dfrac{l}{m}y = \left(\dfrac{1}{l} + \dfrac{1}{m}\right)(m^2 + l^2),$
    $(x + y)(m^2 + l^2) = 2(m^3 + l^3) + ml(x + y).$

19. $bx + cy = a + b,$
    $ax\left(\dfrac{1}{a-b} - \dfrac{1}{a+b}\right) + cy\left(\dfrac{1}{b-a} - \dfrac{1}{b+a}\right) = \dfrac{2a}{a+b}.$

20. $(a - b)x + (a + b)y = 2(a^2 - b^2), \quad ax - by = a^2 + b^2.$

21. $x\left(a - b + \dfrac{ab}{a-b}\right) = y\left(a + b - \dfrac{ab}{a+b}\right), \quad x + y = 2a^3.$

CHAPTER XXIV.

Harder Problems.

188. In previous chapters we have given collections of problems which lead to simple equations. We add here a few examples of somewhat greater difficulty.

Example 1. A grocer buys 15 lbs. of figs and 28 lbs. of currants for £1. 1s. 8d.; by selling the figs at a loss of 10 per cent., and the currants at a gain of 30 per cent., he clears 2s. 6d. on his outlay; how much per pound did he pay for each?

Let x, y denote the number of pence in the price of a pound of figs and currants respectively; then the outlay is

$$15x + 28y \text{ pence.}$$

Therefore $\qquad 15x + 28y = 260 \ldots\ldots\ldots\ldots\ldots\ldots\ldots\ldots\ldots(1).$

The loss upon the figs is $\frac{1}{10} \times 15x$ pence, and the gain upon the currants is $\frac{3}{10} \times 28y$ pence; therefore the total gain is

$$\frac{42y}{5} - \frac{3x}{2} \text{ pence;}$$

$$\therefore \quad \frac{42y}{5} - \frac{3x}{2} = 30 \ldots\ldots\ldots\ldots\ldots\ldots\ldots\ldots\ldots(2).$$

From (1) and (2) we find that $x = 8$, and $y = 5$; that is the figs cost 8d. a pound, and the currants cost 5d. a pound.

Example 2. At what time between 4 and 5 o'clock will the minute-hand of a watch be 13 minutes in advance of the hour-hand?

Let x denote the required number of minutes after 4 o'clock; then, as the minute-hand travels twelve times as fast as the hour-hand, the hour-hand will move over $\frac{x}{12}$ minute divisions in x minutes. At 4 o'clock the minute-hand is 20 divisions behind the hour-hand, and finally the minute-hand is 13 divisions in advance; therefore the minute-hand moves over $20 + 13$, or 33 divisions more than the hour-hand.

Hence
$$x = \frac{x}{12} + 33,$$
$$\frac{11}{12}x = 33;$$
$$\therefore x = 36.$$

Thus the time is 36 minutes past 4.

If the question be asked as follows: "At what *times* between 4 and 5 o'clock will there be 13 minutes between the two hands?" we must also take into consideration the case when the minute-hand is 13 divisions *behind* the hour-hand. In this case the minute-hand gains $20-13$, or 7 divisions.

Hence
$$x = \frac{x}{12} + 7,$$
which gives
$$x = 7\frac{7}{11}.$$

Therefore the *times* are $7\frac{7'}{11}$ past 4, and 36' past 4.

Example 3. Two persons A and B start simultaneously from two places, c miles apart, and walk in the same direction. A travels at the rate of p miles an hour, and B at the rate of q miles; how far will A have walked before he overtakes B?

Suppose A has walked x miles, then B has walked $x-c$ miles.

A walking at the rate of p miles an hour will travel x miles in $\frac{x}{p}$ hours; and B will travel $x-c$ miles in $\frac{x-c}{q}$ hours: these two times being equal, we have
$$\frac{x}{p} = \frac{x-c}{q},$$
$$qx = px - pc;$$
whence
$$x = \frac{pc}{p-q}.$$

Therefore A has travelled $\frac{pc}{p-q}$ miles.

Example 4. A train travelled a certain distance at a uniform rate. Had the speed been 6 miles an hour more, the journey would have occupied 4 hours less; and had the speed been 6 miles an hour less, the journey would have occupied 6 hours more. Find the distance.

Let the speed of the train be x miles per hour, and let the time occupied be y hours; then the distance traversed will be represented by xy miles.

On the first supposition the speed per hour is $x+6$ miles, and the time taken is $y-4$ hours. In this case the distance traversed will be represented by $(x+6)(y-4)$ miles.

On the second supposition the distance traversed will be represented by $(x-6)(y+6)$ miles.

All these expressions for the distance must be equal;

$$\therefore\ xy=(x+6)(y-4)=(x-6)(y+6).$$

From these equations we have

$$xy = xy + 6y - 4x - 24,$$

or $\qquad\qquad 6y - 4x = 24$ ....(1);

and $\qquad\qquad xy = xy - 6y + 6x - 36,$

or $\qquad\qquad 6x - 6y = 36$ ....(2).

From (1) and (2) we obtain $x=30$, $y=24$.

Hence the distance is 720 miles.

Example 5. A person invests £3770, partly in 3 per cent. Stock at £102, and partly in Railway Stock at £84 which pays a dividend of $4\frac{1}{2}$ per cent.: if his income from these investments is £136. 5s. per annum, what sum does he invest in each?

Let x denote the number of pounds invested in 3 per cent., y the number of pounds invested in Railway Stock; then

$$x+y=3770\ \dots\dots\dots\dots\dots\dots(1).$$

The income from 3 per cent. Stock is £$\dfrac{3x}{102}$, or £$\dfrac{x}{34}$; and that from Railway Stock is £$\dfrac{4\frac{1}{2}y}{84}$, or £$\dfrac{3y}{56}$.

Therefore $\qquad\dfrac{x}{34}+\dfrac{3y}{56}=136\frac{1}{4}$ ....(2).

From (2) $\qquad\qquad x+\dfrac{51}{28}y=4632\frac{1}{2}$;

therefore by subtracting (1)

$$\dfrac{23}{28}y=862\tfrac{1}{2};$$

whence $\qquad\qquad y=28\times 37\tfrac{1}{2}=1050$;

and from (1) $\qquad x=2720.$

Therefore he invests £2720 in 3 per cent Stock, and £1050 in Railway Stock.

EXAMPLES XXIV.

1. A sum of £10 is divided among a number of persons: if the number had been increased by one-fourth each would have received a shilling less: find the number of persons.

HARDER PROBLEMS.

2. I bought a certain number of eggs at four a penny; I kept one-fifth of them, and sold the rest at three a penny, and gained a penny: how many did I buy?

3. I bought a certain number of articles at five for sixpence; if they had been eleven for one shilling, I should have spent sixpence less: how many did I buy?

4. A man at whist wins twice as much as he had to begin with, and then loses 16s.; he then loses four-fifths of what remained, and afterwards wins as much as he had at first: how much had he originally, if he leaves off with £4?

5. I spend £14. 5s. in buying 20 yards of calico and 30 yards of silk; the silk costs as many shillings per yard as the calico costs pence per yard: find the price of each.

6. A number of two digits exceeds five times the sum of its digits by 9, and its ten-digit exceeds its unit-digit by 1: find the number.

7. The sum of the digits of a number less than 100 is 6; if the digits be reversed the resulting number will be less by 18 than the original number: find it.

8. A man being asked his age replied, "If you take 2 years from my present age the result will be double my wife's age, and 3 years ago her age was one-third of what mine will be in 12 years." What were their ages?

9. At what time between one and two o'clock are the hands of a watch first at right angles?

10. At what time between 3 and 4 o'clock is the minute-hand one minute ahead of the hour hand?

11. When are the hands of a clock together between the hours of 6 and 7?

12. It is between 2 and 3 o'clock, and in 10 minutes the minute-hand will be as much before the hour-hand as it is now behind it: what is the time?

13. At an election the majority was 162, which was three-elevenths of the whole numbers of voters: what was the number of the votes on each side?

14. A certain number of persons paid a bill; if there had been ten more each would have paid 2s. less; if there had been 5 less each would have paid 2s. 6d. more: find the number of persons, and what each had to pay.

15. A man spends £5 in buying two kinds of silk at 4s. 6d. and 4s. a yard; by selling it all at 4s. 3d. per yard he gains 2 per cent.: how much of each did he buy?

16. Ten years ago the sum of the ages of two sons was one-third of their father's age: one is two years older than the other, and the present sum of their ages is fourteen years less than their father's age: how old are they?

17. A and B start from the same place walking at different rates; when A has walked 15 miles B doubles his pace, and 6 hours later passes A: if A walks at the rate of 5 miles an hour, what is B's rate at first?

18. A basket of oranges is emptied by one person taking half of them and one more, a second person taking half of the remainder and one more, and a third person taking half of the remainder and six more. How many did the basket contain at first?

19. A person swimming in a stream which runs $1\frac{1}{2}$ miles per hour, finds that it takes him four times as long to swim a mile up the stream as it does to swim the same distance down: at what rate does he swim?

20. At what *times* between 7 and 8 o'clock will the hands of a watch be at right angles to each other? When will they be in the same straight line?

21. The denominator of a fraction exceeds the numerator by 4; and if 5 is taken from each, the sum of the reciprocal of the new fraction and four times the original fraction is 5: find the original fraction.

22. Two persons start at noon from towns 60 miles apart. One walks at the rate of four miles an hour, but stops $2\frac{1}{2}$ hours on the way; the other walks at the rate of 3 miles an hour without stopping: when and where will they meet?

23. A, B, and C travel from the same place at the rates of 4, 5, and 6 miles an hour respectively; and B starts 2 hours after A. How long after B must C start in order that they may overtake A at the same instant?

24. A dealer bought a horse, expecting to sell it again at a price that would have given him 10 per cent. profit on his purchase; but he had to sell it for £50 less than he expected, and he then found that he had lost 15 per cent. on what it cost him: what did he pay for the horse?

25. A man walking from a town, A, to another, B, at the rate of 4 miles an hour, starts one hour before a coach travelling 12 miles an hour, and is picked up by the coach. On arriving at B, he finds that his coach journey has lasted 2 hours: find the distance between A and B.

26. What is the property of a person whose income is £1140, when one-twelfth of it is invested at 2 per cent., one-half at 3 per cent., one-third at $4\frac{1}{2}$ per cent., and the remainder pays him no dividend?

HARDER PROBLEMS.

27. A person spends one-third of his income, saves one-fourth, and pays away 5 per cent. on the whole as interest at $7\frac{1}{2}$ per cent. on debts previously incurred, and then has £110 remaining: what was the amount of his debts?

28. Two vessels contain mixtures of wine and water; in one there is three times as much wine as water, in the other, five times as much water as wine. Find how much must be drawn off from each to fill a third vessel which holds seven gallons, in order that its contents may be half wine and half water.

29. There are two mixtures of wine and water, one of which contains twice as much water as wine, and the other three times as much wine as water. How much must there be taken from each to fill a pint cup, in which the water and the wine shall be equally mixed?

30. Two men set out at the same time to walk, one from A to B, and the other from B to A, a distance of a miles. The former walks at the rate of p miles, and the latter at the rate of q miles an hour: at what distance from A will they meet?

31. A train on the North Western line passes from London to Birmingham in 3 hours; a train on the Great Western line which is 15 miles longer, travelling at a speed which is less by 1 mile per hour, passes from one place to the other in $3\frac{1}{2}$ hours: find the length of each line.

32. Coffee is bought at 1s. and chicory at 3d. per lb.; in what proportion must they be mixed that 10 per cent. may be gained by selling the mixture at 11d. per lb.?

33. A man has one kind of coffee at a pence per pound, and another at b pence per pound. How much of each must he take to form a mixture of $a-b$ lbs., which he can sell at c pence a pound without loss?

34. A man spends c half-crowns in buying two kinds of silk at a shillings and b shillings a yard respectively; he could have bought 3 times as much of the first and half as much of the second for the same money: how many yards of each did he buy?

35. A man rides one-third of the distance from A to B at the rate of a miles an hour, and the remainder at the rate of $2b$ miles an hour. If he had travelled at a uniform rate of $3c$ miles an hour, he could have ridden from A to B and back again in the same time. Prove that
$$\frac{2}{c} = \frac{1}{a} + \frac{1}{b}.$$

36. A, B, C are three towns forming a triangle. A man has to walk from one to the next, ride thence to the next, and drive thence to his starting point. He can walk, ride, and drive a mile in a, b, c minutes respectively. If he starts from B he takes $a+c-b$ hours, if he starts from C he takes $b+a-c$ hours, and if he starts from A he takes $c+b-a$ hours. Find the length of the circuit.

CHAPTER XXV.

Quadratic Equations.

189. Suppose the following problem were proposed for solution:

A dealer bought a number of horses for £280. If he had bought four less each would have cost £8 more: how many did he buy?

We should proceed thus:

Let $x=$ the number of horses; then $\dfrac{280}{x}=$ the number of pounds each cost.

If he had bought 4 less he would have had $x-4$ horses, and each would have cost $\dfrac{280}{x-4}$ pounds.

$$\therefore\ 8+\frac{280}{x}=\frac{280}{x-4};$$

whence $\qquad x(x-4)+35(x-4)=35x;$

$$\therefore\ x^2-4x+35x-140=35x;$$

$$\therefore\ x^2-4x=140.$$

Here we have an equation which involves the *square* of the unknown quantity; and in order to complete the solution of the problem we must discover a method of solving such equations.

190. Definition. An equation which contains the square of the unknown quantity, but no higher power, is called a **quadratic equation**, or an **equation of the second degree**.

If the equation contains both the square and the first power of the unknown it is called an *adfected* quadratic; if it contains only the square of the unknown it is said to be a *pure* quadratic.

Thus $2x^2-5x=3$ is an adfected quadratic,

and $\qquad\qquad 5x^2=20$ is a pure quadratic.

QUADRATIC EQUATIONS.

191. A pure quadratic may be considered as a simple equation in which the *square* of the unknown quantity is to be found.

Example. Solve $\dfrac{9}{x^2-27}=\dfrac{25}{x^2-11}$.

Multiplying up, $9x^2-99=25x^2-675$;

$\therefore\ 16x^2=576$;

$\therefore\ x^2=36$;

and taking the square root of these equals, we have
$$x=\pm 6.$$

Note. We prefix the double sign to the number on the right-hand side for the reason given in Art. 117.

192. In extracting the square root of the two sides of the equation $x^2=36$, it might seem that we ought to prefix the double sign to the quantities on both sides, and write $\pm x=\pm 6$. But an examination of the various cases shews this to be unnecessary. For $\pm x=\pm 6$ gives the four cases:
$$+x=+6,\ \ +x=-6,\ \ -x=+6,\ \ -x=-6,$$
and these are all included in the two already given, namely, $x=+6$, $x=-6$. Hence, when we extract the square root of the two sides of an equation, it is sufficient to put the double sign before the square root of *one* side.

193. The equation $x^2=36$ is an instance of the simplest form of quadratic equations. The equation $(x-3)^2=25$ may be solved in a similar way; for taking the square root of both sides, we have two *simple* equations,
$$x-3=\pm 5.$$

Taking the upper sign, $x-3=+5$, whence $x=8$;

taking the lower sign, $x-3=-5$, whence $x=-2$.

$\therefore$ the solution is $x=8$, or -2.

Now the given equation, $(x-3)^2=25$

may be written $x^2-6x+(3)^2=25$,

or $x^2-6x=16$.

Hence, by retracing our steps, we learn that the equation
$$x^2-6x=16.$$
can be solved by first adding $(3)^2$ or 9 to each side, and then extracting the square root; and the reason why we add 9 to each side is that this quantity added to the left side makes it a *perfect square*.

Now whatever the quantity a may be,
$$x^2+2ax+a^2=(x+a)^2,$$
and
$$x^2-2ax+a^2=(x-a)^2;$$
so that if a trinomial is a perfect square, and *its highest power*, x^2, *has unity for its coefficient*, we must always have the term without x equal to the square of half the coefficient of x. If, therefore, the terms in x^2 and x are given, the square may be completed by adding the square of half the coefficient of x.

Note. When an expression is a perfect square, the *square terms* are always *positive*. [Art. 114, Note.] Hence, if necessary, the coefficient of x^2 must be made equal to $+1$ before completing the square.

Example 1. Solve $\quad x^2+14x=32.$

The square of half 14 is $(7)^2$.
$$\therefore\ x^2+14x+(7)^2=32+49;$$
that is,
$$(x+7)^2=81;$$
$$\therefore\ x+7=\pm 9;$$
$$\therefore\quad x=-7+9,\ \text{or}\ -7-9;$$
$$\therefore\quad x=2,\quad\text{or}\ -16.$$

Example 2. Solve $\quad 7x=x^2-8.$

Transpose so as to have the terms involving x on one side, and the square term positive.

Thus $\quad x^2-7x=8.$

Completing the square, $x^2-7x+\left(\dfrac{7}{2}\right)^2=8+\dfrac{49}{4};$

that is,
$$\left(x-\dfrac{7}{2}\right)^2=\dfrac{81}{4};$$
$$\therefore\ x-\dfrac{7}{2}=\pm\dfrac{9}{2};$$
$$\therefore\quad x=\dfrac{7}{2}\pm\dfrac{9}{2};$$
$$\therefore\quad x=8,\ \text{or}\ -1.$$

Note. We do not work out $\left(\dfrac{7}{2}\right)^2$ on the left-hand side.

EXAMPLES XXV. a.

1. $5(x^2+5)=6x^2.$
2. $3x^2=4(x^2-4).$
3. $x^2+22x=75.$
4. $x^2+24x=25.$
5. $x^2=10x-21.$
6. $(9+x)(9-x)=17.$
7. $x^2+3x=18.$
8. $x^2+5x=14.$
9. $x^2-5x-36=0.$

10. $x^2 = x + 72$. 11. $x^2 - 341 = 20x$. 12. $9x - x^2 + 220 = 0$.
13. $68 - x^2 = 13x$. 14. $x + 156 = x^2$. 15. $187 = x^2 + 6x$.
16. $23x = 120 + x^2$. 17. $42 + x^2 = 13x$. 18. $22x + 23 - x^2 = 0$.
19. $x^2 - \dfrac{2}{3}x = 32$. 20. $x^2 + \dfrac{4}{15}x = \dfrac{1}{5}$. 21. $x^2 - \dfrac{7}{6}x - \dfrac{1}{2} = 0$.
22. $\dfrac{19}{5}x = \dfrac{4}{5} - x^2$. 23. $\dfrac{3}{5}(x+6)(x-2) = \dfrac{2}{3}\left(62\dfrac{1}{10} + \dfrac{18x}{5}\right)$.

194. We have shewn that the square may readily be completed when the coefficient of x^2 is unity. All cases may be reduced to this by dividing the equation throughout by the coefficient of x^2.

Example 1. Solve $\qquad 32 - 3x^2 = 10x$.

Transposing, $\qquad 3x^2 + 10x = 32$.

Divide throughout by 3, so as to make the coefficient of x^2 unity.

Thus $$x^2 + \dfrac{10}{3}x = \dfrac{32}{3};$$

completing the square, $x^2 + \dfrac{10}{3}x + \left(\dfrac{5}{3}\right)^2 = \dfrac{32}{3} + \dfrac{25}{9}$;

that is, $$\left(x + \dfrac{5}{3}\right)^2 = \dfrac{121}{9};$$

$$\therefore\ x + \dfrac{5}{3} = \pm\dfrac{11}{3};$$

$$\therefore\ x = -\dfrac{5}{3} \pm \dfrac{11}{3} = 2,\ \text{or}\ -5\dfrac{1}{3}.$$

Note. We do not add $\left(\dfrac{10}{6}\right)^2$ but $\left(\dfrac{5}{3}\right)^2$ to the left-hand side.

Example 2. Solve $\qquad 5x^2 + 11x = 12$.

Dividing by 5, $\qquad x^2 + \dfrac{11}{5}x = \dfrac{12}{5}$;

completing the square, $x^2 + \dfrac{11}{5}x + \left(\dfrac{11}{10}\right)^2 = \dfrac{12}{5} + \dfrac{121}{100}$;

that is, $$\left(x + \dfrac{11}{10}\right)^2 = \dfrac{361}{100};$$

$$\therefore\ x + \dfrac{11}{10} = \pm\dfrac{19}{10};$$

$$\therefore\ x = -\dfrac{11}{10} \pm \dfrac{19}{10} = \dfrac{4}{5},\ \text{or}\ -3.$$

195. We see then that the following are the steps required for solving an adfected quadratic equation :

(1) If necessary, simplify the equation so that the terms in x^2 and x are on one side of the equation, and the term without x on the other.

(2) Make the coefficient of x^2 unity and positive by dividing throughout by the coefficient of x^2.

(3) Add to each side of the equation the square of half the coefficient of x.

(4) Take the square root of each side.

(5) Solve the resulting simple equations.

196. In the examples which follow some preliminary reduction and simplification may be necessary.

Example 1. Solve $\quad \dfrac{3x-2}{2x-3} = \dfrac{5x}{x+4} - 2.$

Simplifying, $\quad \dfrac{3x-2}{2x-3} = \dfrac{3x-8}{x+4};$

multiplying across, $\quad 3x^2 + 10x - 8 = 6x^2 - 25x + 24;$

that is, $\quad -3x^2 + 35x = 32.$

Dividing by -3, $\quad x^2 - \dfrac{35}{3}x = -\dfrac{32}{3};$

completing the square, $x^2 - \dfrac{35}{3}x + \left(\dfrac{35}{6}\right)^2 = \dfrac{1225}{36} - \dfrac{32}{3};$

that is, $\quad \left(x - \dfrac{35}{6}\right)^2 = \dfrac{841}{36};$

$\therefore \quad x - \dfrac{35}{6} = \pm \dfrac{29}{6};$

$\therefore \quad x = 10\tfrac{2}{3},$ or $1.$

Example 2. Solve $\quad 7(x+2a)^2 + 3a^2 = 5a(7x+23a).$

Simplifying, $7x^2 + 28ax + 28a^2 + 3a^2 = 35ax + 115a^2,$

that is, $\quad 7x^2 - 7ax = 84a^2.$

Whence $\quad x^2 - ax = 12a^2;$

completing the square, $x^2 - ax + \left(\dfrac{a}{2}\right)^2 = 12a^2 + \dfrac{a^2}{4};$

that is, $\quad \left(x - \dfrac{a}{2}\right)^2 = \dfrac{49a^2}{4};$

$\therefore \quad x - \dfrac{a}{2} = \pm \dfrac{7a}{2};$

$\therefore \quad x = 4a,$ or $-3a.$

197. Sometimes there is *only one solution*. Thus if
$$x^2 - 2x + 1 = 0, \text{ then } (x-1)^2 = 0,$$
whence $x=1$ is the only solution. Nevertheless, in this and similar cases we find it convenient to say that the quadratic has *two equal roots*.

EXAMPLES XXV. b.

1. $5x^2 + 14x = 55$.
2. $3x^2 + 121 = 44x$.
3. $25x = 6x^2 + 21$.
4. $8x^2 + x = 30$.
5. $3x^2 + 35 = 22x$.
6. $x + 22 - 6x^2 = 0$.
7. $15 = 17x + 4x^2$.
8. $21 + x = 2x^2$.
9. $9x^2 - 143 - 6x = 0$.
10. $12x^2 = 29x - 14$.
11. $20x^2 = 12 - x$.
12. $19x = 15 - 8x^2$.
13. $21x^2 + 22x + 5 = 0$.
14. $50x^2 - 15x = 27$.
15. $18x^2 - 27x - 26 = 0$.
16. $5x^2 = 8x + 21$.
17. $15x^2 - 2ax = a^2$.
18. $21x^2 = 2ax + 3a^2$.
19. $6x^2 = 11kx + 7k^2$.
20. $12x^2 + 23kx + 10k^2 = 0$.
21. $12x^2 - cx - 20c^2 = 0$.
22. $2(x-3) = 3(x+2)(x-3)$.
23. $(x+1)(2x+3) = 4x^2 - 22$.
24. $(3x-5)(2x-5) = x^2 + 2x - 3$.
25. $\dfrac{5x+7}{x-1} = 3x + 2$.
26. $\dfrac{5x-1}{x+1} = \dfrac{3x}{2}$.
27. $\dfrac{3x-8}{x-2} = \dfrac{5x-2}{x+5}$.
28. $\dfrac{3x-1}{4x+7} = 1 - \dfrac{6}{x+7}$.
29. $\dfrac{5x-7}{7x-5} = \dfrac{x-5}{2x-13}$.
30. $\dfrac{x+3}{2x-7} - \dfrac{2x-1}{x-3} = 0$.
31. $\dfrac{1}{1+x} - \dfrac{1}{3-x} = \dfrac{6}{35}$.
32. $\dfrac{x+4}{x-4} + \dfrac{x-2}{x-3} = 6\tfrac{1}{3}$.
33. $\dfrac{1}{3-x} - \dfrac{4}{5} = \dfrac{1}{9-2x}$.
34. $\dfrac{4}{x-1} - \dfrac{5}{x+2} = \dfrac{3}{x}$.
35. $\dfrac{5}{x-2} - \dfrac{4}{x} = \dfrac{3}{x+6}$.
36. $\dfrac{x-2}{x-3} + \dfrac{3x-11}{x-4} = \dfrac{4x+13}{x+1}$.
37. $\dfrac{1}{2x-5a} + \dfrac{5}{2x-a} = \dfrac{2}{a}$.
38. $\dfrac{2}{3x-2c} + \dfrac{3}{2x-3c} = \dfrac{7}{2c}$.
39. $\dfrac{a^2b}{x^2} + \left(1 + \dfrac{b}{x}\right)a = 2b + \dfrac{a^2}{x}$.

198. From the preceding examples it appears that after suitable reduction and transposition every quadratic equation can be written in the form
$$ax^2 + bx + c = 0,$$
where a, b, c may have any numerical values whatever. If therefore we can solve this quadratic we can solve any.

Transposing, $ax^2 + bx = -c$;

dividing by a, $x^2 + \dfrac{b}{a}x = -\dfrac{c}{a}$.

Completing the square by adding to each side $\left(\dfrac{b}{2a}\right)^2$,

$$x^2 + \dfrac{b}{a}x + \left(\dfrac{b}{2a}\right)^2 = \dfrac{b^2}{4a^2} - \dfrac{c}{a};$$

that is, $\left(x + \dfrac{b}{2a}\right)^2 = \dfrac{b^2 - 4ac}{4a^2};$

extracting the square root,

$$x + \dfrac{b}{2a} = \dfrac{\pm\sqrt{(b^2 - 4ac)}}{2a};$$

$$x = \dfrac{-b \pm \sqrt{(b^2 - 4ac)}}{2a}.$$

199. Instead of going through the process of completing the square in each particular example, we may now make use of this general formula, adapting it to the case in question by substituting the values of a, b, c.

Example. Solve $5x^2 + 11x - 12 = 0$.

Here $a = 5$, $b = 11$, $c = -12$.

$$\therefore x = \dfrac{-11 \pm \sqrt{(11)^2 - 4 \cdot 5(-12)}}{10}$$

$$= \dfrac{-11 \pm \sqrt{361}}{10} = \dfrac{-11 \pm 19}{10} = \dfrac{4}{5}, \text{ or } -3,$$

which agrees with the solution of Example 2, Art. 194.

200. In the result $x = \dfrac{-b \pm \sqrt{(b^2 - 4ac)}}{2a}$,

it must be remembered that the expression $\sqrt{(b^2 - 4ac)}$ is the square root of the compound quantity $b^2 - 4ac$, *taken as a whole.* We cannot simplify the solution unless we know the numerical values of a, b, c. It may sometimes happen that these values do not make $b^2 - 4ac$ a perfect square. In such a case the exact numerical solution of the equation cannot be determined.

QUADRATIC EQUATIONS.

Example 1. Solve $5x^2 - 15x + 11 = 0$.

We have
$$x = \frac{15 \pm \sqrt{(-15)^2 - 4.5.11}}{2.5}$$
$$= \frac{15 \pm \sqrt{5}}{10} \quad\quad\quad\quad\quad\quad\quad\quad\quad\quad\quad\quad\quad\quad\quad (1)$$

Now $\sqrt{5} = 2\cdot236$ approximately.

$$\therefore \quad x = \frac{15 \pm 2\cdot236}{10} = 1\cdot7236, \text{ or } 1\cdot2764.$$

These solutions are correct only to four places of decimals, and neither of them will be found to *exactly* satisfy the equation.

Unless the *numerical* values of the unknown quantity are required it is usual to leave the roots in the form (1).

Example 2. Solve $x^2 - 3x + 5 = 0$.

We have
$$x = \frac{3 \pm \sqrt{(-3)^2 - 4.1.5}}{2}$$
$$= \frac{3 \pm \sqrt{9 - 20}}{2}$$
$$= \frac{3 \pm \sqrt{-11}}{2}.$$

Now there is no quantity, positive or negative, whose square is negative (Art. 110). Therefore it is impossible to find any quantity exactly or approximately to represent the square root of -11. Thus there is no real value of x which satisfies the equation. In such a case the roots are said to be *imaginary* or *impossible*. A reference to the general formula of Art. 198 will shew that the roots of a quadratic $ax^2 + bx + c = 0$ are always imaginary when $b^2 - 4ac$ is negative.

Note. If the equation $x^2 - 3x + 5 = 0$ is treated graphically, as explained in Art. 427, it will be found that the graph never meets the axis of x. In other words there is no numerical value of x which makes the expression $x^2 - 3x + 5$ equal to zero.

[Chap. XLIV., Arts. 425-427 *may be read here.*]

201. Solution by Factors. The following method of solution will sometimes be found shorter than either of the methods given.

Consider the equation $x^2 + \dfrac{7}{3}x = 2$.

Clearing of fractions, $3x^2 + 7x - 6 = 0$ ....(1);
by resolving the left-hand side into factors we have
$$(3x - 2)(x + 3) = 0.$$

E.A.

Now if *either* of the factors $3x-2$, $x+3$ is zero their product is zero. Hence the quadratic equation is satisfied by either of the suppositions
$$3x-2=0, \text{ or } x+3=0.$$
Thus the roots are $\dfrac{2}{3}$, -3.

It appears from this that *when a quadratic equation has been simplified and brought to the form of equation* (1), its solution can always be readily obtained if the expression on the left-hand side can be resolved into factors. Each of these factors equated to zero gives a simple equation, and a corresponding root of the quadratic.

Example 1. Solve $\quad 2x^2 - ax + 2bx = ab.$

Transposing, *so as to have all the terms on one side of the equation,* we have
$$2x^2 - ax + 2bx - ab = 0.$$
Now $\quad 2x^2 - ax + 2bx - ab = x(2x-a) + b(2x-a)$
$$= (2x-a)(x+b).$$
Therefore $\quad (2x-a)(x+b) = 0\,;$
whence $\quad 2x-a=0, \text{ or } x+b=0.$
$$\therefore \ x = \frac{a}{2}, \text{ or } -b.$$

Example 2. Solve $\quad 2(x^2 - 6) = 3(x-4).$

We have $\quad 2x^2 - 12 = 3x - 12\,;$
that is, $\quad 2x^2 = 3x \ \dotfill (1).$
Transposing, $\quad 2x^2 - 3x = 0,$
$$x(2x-3) = 0.$$
$$\therefore \ x = 0, \text{ or } 2x - 3 = 0.$$
Thus the roots are 0, $\dfrac{3}{2}.$

Note. In equation (1) above we might have divided both sides by x and obtained the simple equation $2x = 3$, whence $x = \dfrac{3}{2}$, which is *one* of the solutions of the given equation. But the student must be particularly careful to notice that whenever an x, or a factor containing x, is removed by division from every term of an equation it must not be neglected, since the equation is satisfied by $x=0$, which is therefore one of the roots.

202. There are some equations which are not really quadratics, but which may be solved by the methods explained in this chapter.

Example 1. Solve $x^4 - 13x^2 + 36 = 0$.

By resolution into factors, $(x^2 - 9)(x^2 - 4) = 0$;
$$\therefore x^2 - 9 = 0, \text{ or } x^2 - 4 = 0;$$
that is, $x^2 = 9$, or 4,
and $x = \pm 3$, or ± 2.

Example 2. Solve $x^2 + 3x - \dfrac{20}{x^2 + 3x} = 8$.

Write y for $x^2 + 3x$, then we have
$$y - \frac{20}{y} = 8,$$
or $y^2 - 8y - 20 = 0$.

From this quadratic $y = 10$, or -2;
$$\therefore x^2 + 3x = 10, \text{ or } -2.$$

Thus we have *two* quadratics to solve, and finally we obtain $x = -5, 2$; or $-1, -2$.

EXAMPLES XXV. c.

Solve by Art. 200, and verify graphically by Art. 427:

1. $3x^2 = 15 - 4x$.
2. $2x^2 + 7x = 15$.
3. $2x^2 + 7 - 9x = 0$.
4. $x^2 = 3x + 5$.
5. $5x^2 + 4 + 21x = 0$.
6. $x^2 + 11 = 7x$.
7. $8x^2 = x + 7$.
8. $5x^2 = 17x - 10$.
9. $35 + 9x - 2x^2 = 0$.
10. $3x^2 = x + 1$.
11. $3x^2 + 5x = 2$.
12. $2x^2 + 5x - 33 = 0$.

Solve by resolution into factors:

13. $6x^2 = 7 + x$.
14. $21 + 8x^2 = 26x$.
15. $26x - 21 + 11x^2 = 0$.
16. $5x^2 + 26x + 24 = 0$.
17. $4x^2 = \dfrac{4}{15}x + 3$.
18. $x^2 - 2 = \dfrac{23}{12}x$.
19. $7x^2 = 28 - 96x$.
20. $96x^2 = 4x + 15$.
21. $25x^2 = 5x + 6$.
22. $35 - 4x = 4x^2$.
23. $12x^2 - 11ax = 36a^2$.
24. $12x^2 + 36a^2 = 43ax$.
25. $35b^2 = 9x^2 + 6bx$.
26. $36x^2 - 35b^2 = 12bx$.
27. $x^2 - 2ax + 4ab = 2bx$.
28. $x^2 - 2ax + 8x = 16a$.
29. $3x^2 - 2ax - bx = 0$.
30. $ax^2 + 2x = bx$.

Solve as explained in Art. 202 :

31. $4 = 5x^2 - x^4$.
32. $x^4 + 36 = 13x^2$.
33. $x^6 + 7x^3 = 8$.
34. $x^6 - 19x^3 = 216$.
35. $16\left(x^2 + \dfrac{1}{x^2}\right) = 257$.
36. $x^2 + \dfrac{a^2 b^2}{x^2} = a^2 + b^2$.
37. $x^3(19 + x^3) = 216$.
38. $(x^2+2)^2 + 198 = 29(x^2+2)$.
39. $x^2 - x + \dfrac{72}{x^2 - x} = 18$.
40. $x(x-2a) = \dfrac{8a^4}{x^2 - 2ax} + 7a^2$.

202$_A$. The method of solution by factors is applicable to equations of higher degree than the second.

For example, if
$$(x-2)(x+1)(x+2) = 0,$$
the equation must be satisfied by each of the values which satisfy the equations
$$x - 2 = 0, \quad x + 1 = 0, \quad x + 2 = 0.$$
Thus the roots are $x = 2, -1, -2$.

Example. Solve the equation $3x^3 + 5x^2 = 3x + 5$.

Putting the equation in the form
$$3x^3 + 5x^2 - 3x - 5 = 0,$$
we have $\qquad x^2(3x+5) - (3x+5) = 0$,*
or $\qquad (x^2 - 1)(3x+5) = 0$;
that is, $\qquad (x+1)(x-1)(3x+5) = 0$;
whence $\qquad x+1 = 0$, or $x-1 = 0$, or $3x+5 = 0$.

Thus the roots are $-1, 1, -\dfrac{5}{3}$.

Note. At the stage marked with an asterisk we might have divided throughout by $3x+5$, but in so doing the factor must be equated to zero to furnish one root of the equation.

202$_B$. If one root of an equation is known, or can be obtained by trial, a corresponding factor of the first degree can be removed. When this is done we have left an equation of lower degree than the original equation.

Example. Solve the equation
$$x^3 - 3x^2 - 6x - 16 = 0.$$

By trial it will be found that the left-hand side vanishes when $x = 2$.

Hence $x=2$ is one root of the equation and corresponding to this root we have a factor $x-2$; the equation may now be written
$$x^2(x-2) - x(x-2) - 8(x-2) = 0;$$
or
$$(x^2 - x - 8)(x-2) = 0.$$

Removing the factor $x-2$, we have
$$x^2 - x - 8 = 0;$$
whence
$$x = \frac{1 \pm \sqrt{33}}{2}.$$

Thus the three roots are $2, \dfrac{1+\sqrt{33}}{2}, \dfrac{1-\sqrt{33}}{2}$.

EXAMPLES XXV. d.

Solve the following equations by the method of factors.

1. $x^3 + x^2 - x - 1 = 0$. 2. $x^3 - 2x^2 - x + 2 = 0$.
3. $x^3 - 4x = x^2 - 4$. 4. $x^3 + 7x^2 + 7x - 15 = 0$.
5. $x^3 - 3x - 2 = 0$. 6. $x^4 + 2x = 3x^2$. 7. $x^3 + 30 = 19x$.

Solve the following equations having given one root in each case.

8. $x^3 - 39x + 70 = 0$. $[x=5.]$ 9. $x^3 - 37x - 84 = 0$. $[x=-3.]$
10. $x^3 - 12a^2x = 16a^3$. $[x=4a.]$ 11. $x^4 + 432a^3x = 108a^2x^2$. $[x=6a.]$

Solve the following equations by the method of Art. 200, giving the roots to two places of decimals.

12. $x^2 + 2x = 3\cdot 2$. 13. $x^2 - 3x - 3\cdot 51 = 0$.
14. $x^2 + x = 1\cdot 0956$. 15. $x^2 - 36x + 323\cdot 7 = 0$.
16. $x^2 - 7x + 6\cdot 035 = 0$. 17. $x^2 - 5\cdot 5x + 7\cdot 3776 = 0$.

18. Find two values of x which will make $x(3x-1)$ equal to $\cdot 362$, giving each value to the nearest hundredth.

19. Find to the nearest tenth the values of x which will make $2x(2-x)$ equal to $1\cdot 73$.

20. Solve the equation $x^2 + ax - a^2 = 0$. If $a=12$, give the numerical values of the roots to three decimal places.

21. Solve the equation $x(a-x) = c^2$. Give the numerical values of the roots to three decimal places, when $a=16$, $c=6$.

CHAPTER XXVI.

SIMULTANEOUS QUADRATIC EQUATIONS.

203. We shall now consider some of the most useful methods of solving simultaneous equations, one or more of which may be of a degree higher than the first; but no fixed rules can be laid down which are applicable to all cases.

Example 1. Solve
$$x+y=15 \quad \ldots\ldots\ldots\ldots\ldots(1),$$
$$xy=36 \quad \ldots\ldots\ldots\ldots\ldots(2).$$

From (1) by squaring, $x^2+2xy+y^2=225$;
from (2) $\quad 4xy=144$;
by subtraction, $\quad x^2-2xy+y^2=81$;
by taking the square root, $\quad x-y=\pm 9.$

Combining this with (1) we have to consider the two cases,

$$\left.\begin{array}{l}x+y=15,\\x-y=9.\end{array}\right\} \quad \left.\begin{array}{l}x+y=15,\\x-y=-9.\end{array}\right\}$$

from which we find $\left.\begin{array}{l}x=12,\\y=3.\end{array}\right\} \quad \left.\begin{array}{l}x=3,\\y=12.\end{array}\right\}$

Example 2. Solve
$$x-y=12 \quad \ldots\ldots\ldots\ldots\ldots(1),$$
$$xy=85 \quad \ldots\ldots\ldots\ldots\ldots(2).$$

From (1) $\quad x^2-2xy+y^2=144$;
from (2) $\quad 4xy=340$;
by addition, $\quad x^2+2xy+y^2=484$;
by taking the square root, $\quad x+y=\pm 22.$

Combining this with (1) we have the two cases,

$$\left.\begin{array}{l}x+y=22,\\x-y=12.\end{array}\right\} \quad \left.\begin{array}{l}x+y=-22,\\x-y=12.\end{array}\right\}$$

Whence $\left.\begin{array}{l}x=17,\\y=5.\end{array}\right\} \quad \left.\begin{array}{l}x=-5,\\y=-17.\end{array}\right\}$ [See Art. 441.]

204. These are the simplest cases that arise, but they are specially important since the solution in a large number of other cases is dependent upon them.

As a rule our object is to solve the proposed equations *symmetrically*, by finding the values of $x+y$ and $x-y$. From the foregoing examples it will be seen that we can always do this as soon as we have obtained the product of the unknowns, and either their sum or their difference.

Example 1. Solve
$$x^2+y^2=74 \quad\quad\quad\quad\quad (1),$$
$$xy=35 \quad\quad\quad\quad\quad (2).$$

Multiply (2) by 2; then by addition and subtraction we have
$$x^2+2xy+y^2=144,$$
$$x^2-2xy+y^2=4;$$
Whence
$$x+y=\pm 12,$$
$$x-y=\pm 2.$$

We have now four cases to consider; namely,

$$\left.\begin{array}{l}x+y=12,\\x-y=2.\end{array}\right\} \quad \left.\begin{array}{l}x+y=12,\\x-y=-2.\end{array}\right\} \quad \left.\begin{array}{l}x+y=-12,\\x-y=2.\end{array}\right\} \quad \left.\begin{array}{l}x+y=-12,\\x-y=-2.\end{array}\right\}$$

From which the values of x are $\quad 7,\ 5,\ -5,\ -7$; [Compare
and the corresponding values of y are $5,\ 7,\ -7,\ -5.\quad$ Art. 441.]

Example 2. Solve
$$x^2+y^2=185 \quad\quad\quad\quad\quad (1),$$
$$x+y=17 \quad\quad\quad\quad\quad (2).$$

By subtracting (1) from the square of (2) we have
$$2xy=104;$$
$$\therefore\ xy=52 \quad\quad\quad\quad\quad (3).$$

Equations (1) and (3) can now be solved by the method of Example 1; and the solution is
$$\left.\begin{array}{l}x=13,\ \text{or}\ 4,\\y=4,\ \text{or}\ 13.\end{array}\right\}$$

EXAMPLES XXVI. a.

Solve the following equations:

1. $x+y=28,$
   $xy=187.$

2. $x+y=51,$
   $xy=518.$

3. $x+y=74,$
   $xy=1113.$

4. $x-y=5,$
   $xy=126.$

5. $x-y=8,$
   $xy=513.$

6. $xy=1075,$
   $x-y=18.$

7. $xy=923,$
   $x+y=84.$

8. $x-y=-8,$
   $xy=1353.$

9. $x-y=-22,$
   $xy=3848.$

Solve the following equations:

10. $xy = -2193,$
    $x + y = -8.$

11. $x - y = -18,$
    $xy = 1363.$

12. $xy = -1914,$
    $x + y = -65.$

13. $x^2 + y^2 = 89,$
    $xy = 40.$

14. $x^2 + y^2 = 170,$
    $xy = 13.$

15. $x^2 + y^2 = 65,$
    $xy = 28.$

16. $x^2 + y^2 = 178,$
    $x + y = 16.$

17. $x + y = 15,$
    $x^2 + y^2 = 125.$

18. $x - y = 4,$
    $x^2 + y^2 = 106.$

19. $x^2 + y^2 = 180,$
    $x - y = 6.$

20. $x^2 + y^2 = 185,$
    $x - y = 3.$

21. $x + y = 13,$
    $x^2 + y^2 = 97.$

22. $x + y = 9,$
    $x^2 + xy + y^2 = 61.$

23. $x - y = 3,$
    $x^2 - 3xy + y^2 = -19.$

24. $x^2 - xy + y^2 = 76,$
    $x + y = 14.$

25. $\frac{1}{10}(x-y) = 1,$
    $x^2 - 4xy + y^2 = 52.$

26. $\frac{1}{x} + \frac{1}{y} = 2,$
    $x + y = 2.$

27. $\frac{1}{x} + \frac{1}{y} = \frac{7}{12},$
    $xy = 12.$

28. $ax + by = 2,$
    $abxy = 1.$

29. $x^2 + pxy + y^2 = p + 2,$
    $qx^2 + xy + qy^2 = 2q + 1.$

205. Any pair of equations of the form

$$x^2 \pm pxy + y^2 = a^2 \quad \ldots\ldots\ldots\ldots\ldots (1),$$
$$x \pm y = b \quad \ldots\ldots\ldots\ldots\ldots (2),$$

where p is any numerical quantity, can be reduced to one of the cases already considered; for by squaring (2) and combining with (1), an equation to find xy is obtained; the solution can then be completed by the aid of equation (2).

Example 1. Solve $\quad x^3 - y^3 = 999 \quad \ldots\ldots\ldots\ldots\ldots (1),$
$\quad\quad\quad\quad\quad\quad\quad\quad x - y = 3 \quad \ldots\ldots\ldots\ldots\ldots (2),$

By division, $\quad x^2 + xy + y^2 = 333 \quad \ldots\ldots\ldots\ldots\ldots (3);$
from (2) $\quad\quad x^2 - 2xy + y^2 = 9;$
by subtraction, $\quad 3xy = 324,$
$\quad\quad\quad\quad\quad\quad xy = 108 \quad \ldots\ldots\ldots\ldots\ldots (4).$

From (2) and (4) $\quad x = 12, \text{ or } -9,\}$
$\quad\quad\quad\quad\quad\quad y = 9, \text{ or } -12.\}$

Example 2. Solve $\quad x^4 + x^2y^2 + y^4 = 2613 \quad \ldots\ldots\ldots (1),$
$\quad\quad\quad\quad\quad\quad\quad x^2 + xy + y^2 = 67 \quad \ldots\ldots\ldots (2).$

Dividing (1) by (2) $\quad x^2 - xy + y^2 = 39 \quad \ldots\ldots\ldots (3).$
From (2) and (3) by addition, $x^2 + y^2 = 53;$
by subtraction, $\quad\quad\quad\quad xy = 14;$
whence $\quad\quad\quad x = \pm 7, \pm 2,\}$
$\quad\quad\quad\quad\quad y = \pm 2, \pm 7.\}$ [Art. 204, Ex. 1.]

SIMULTANEOUS QUADRATIC EQUATIONS.

Example 3. Solve

$$\frac{1}{x}-\frac{1}{y}=\frac{1}{3}\quad\ldots\ldots\ldots\ldots(1),$$

$$\frac{1}{x^2}+\frac{1}{y^2}=\frac{5}{9}\quad\ldots\ldots\ldots\ldots(2).$$

From (1) by squaring, $\dfrac{1}{x^2}-\dfrac{2}{xy}+\dfrac{1}{y^2}=\dfrac{1}{9}$;

by subtraction, $\dfrac{2}{xy}=\dfrac{4}{9}$;

adding to (2), $\dfrac{1}{x^2}+\dfrac{2}{xy}+\dfrac{1}{y^2}=1$;

$$\therefore\ \frac{1}{x}+\frac{1}{y}=\pm 1.$$

Combining with (1), $\dfrac{1}{x}=\dfrac{2}{3}$, or $-\dfrac{1}{3}$,

$\dfrac{1}{y}=\dfrac{1}{3}$, or $-\dfrac{2}{3}$;

$$\therefore\ x=\frac{3}{2},\ \text{or}\ -3,\ \text{and}\ y=3,\ \text{or}\ -\frac{3}{2}.$$

EXAMPLES XXVI. b.

Solve the equations:

1. $x^3+y^3=407$,
   $x+y=11$.

2. $x^3+y^3=637$,
   $x+y=13$.

3. $x+y=23$,
   $x^3+y^3=3473$.

4. $x^3-y^3=218$,
   $x-y=2$.

5. $x-y=4$,
   $x^3-y^3=988$.

6. $x^3-y^3=2197$,
   $x-y=13$.

7. $x^4+x^2y^2+y^4=2128$,
   $x^2+xy+y^2=76$.

8. $x^4+x^2y^2+y^4=2923$,
   $x^2-xy+y^2=37$.

9. $x^4+x^2y^2+y^4=9211$,
   $x^2-xy+y^2=61$.

10. $x^4+x^2y^2+y^4=7371$,
    $x^2-xy+y^2=63$.

11. $\dfrac{1}{x^2}+\dfrac{1}{y^2}=\dfrac{481}{576}$,
    $\dfrac{1}{x}+\dfrac{1}{y}=\dfrac{29}{24}$,

12. $\dfrac{1}{x^2}+\dfrac{1}{y^2}=\dfrac{61}{900}$,
    $xy=30$.

13. $\dfrac{x}{y}+\dfrac{y}{x}=2\dfrac{1}{2}$,
    $x+y=6$.

14. $\dfrac{x}{y}+\dfrac{y}{x}=2\dfrac{16}{21}$,
    $x-y=4$.

15. $\dfrac{34}{x^2+y^2}=\dfrac{15}{xy}$,
    $x+y=8$.

16. $x^3-y^3=56$,
    $x^2+xy+y^2=28$.

17. $4(x^2+y^2)=17xy$,
    $x-y=6$.

18. $x^3+y^3=126$,
    $x^2-xy+y^2=21$.

Solve the equations:

19. $\dfrac{1}{x^3}+\dfrac{1}{y^3}=1\dfrac{1}{125}$,

    $\dfrac{1}{x}+\dfrac{1}{y}=1\dfrac{1}{5}$.

20. $\dfrac{1}{x^3}-\dfrac{1}{y^3}=91$,

    $\dfrac{1}{x}-\dfrac{1}{y}=1$.

206. The following method of solution may always be used when the equations are *of the same degree and homogeneous.*

[See Art. 24.]

Example. Solve $\quad x^2+xy+2y^2=74 \quad\ldots\ldots\ldots\ldots\ldots\ldots(1)$,

$\quad\quad\quad\quad\quad\quad\quad 2x^2+2xy+y^2=73 \quad\ldots\ldots\ldots\ldots\ldots\ldots(2)$.

Put $y=mx$, and substitute in both equations. Thus

$$x^2(1+m+2m^2)=74 \quad\ldots\ldots\ldots\ldots\ldots\ldots(3).$$

and $\quad\quad\quad\quad x^2(2+2m+m^2)=73 \quad\ldots\ldots\ldots\ldots\ldots\ldots(4).$

By division, $\quad\dfrac{1+m+2m^2}{2+2m+m^2}=\dfrac{74}{73}$;

$\therefore\ 73+73m+146m^2=148+148m+74m^2$;

$\therefore\quad 72m^2-75m-75=0$,

or $\quad\quad\quad 24m^2-25m-25=0$;

$\therefore\quad (8m+5)(3m-5)=0$;

$\therefore\quad m=-\dfrac{5}{8},\ \text{or}\ \dfrac{5}{3}.$

(i) Take $m=-\dfrac{5}{8}$, and substitute in either (3) or (4).

From (3) $\quad\quad x^2\left(1-\dfrac{5}{8}+\dfrac{50}{64}\right)=74$;

$\therefore\ x^2=\dfrac{64\times 74}{74}=64$;

$\therefore\ x=\pm 8$;

$\therefore\ y=mx=-\dfrac{5}{8}x=\mp 5.$

(ii) Take $m=\dfrac{5}{3}$; then from (3)

$\quad\quad\quad x^2\left(1+\dfrac{5}{3}+\dfrac{50}{9}\right)=74.$

$\quad\quad\quad x^2=\dfrac{74\times 9}{74}=9$;

$\therefore\ x=\pm 3$;

$\therefore\ y=mx=\dfrac{5}{3}x=\pm 5.$

SIMULTANEOUS QUADRATIC EQUATIONS.

207. When one of the equations is of the first degree and the other of a higher degree, we may from the simple equation find the value of one of the unknowns in terms of the other, and substitute in the second equation.

Example. Solve
$$3x - 4y = 5 \quad\quad\quad\quad\quad\quad (1),$$
$$3x^2 - xy - 3y^2 = 21 \quad\quad\quad\quad (2).$$

From (1) we have $x = \dfrac{5+4y}{3}$;

and substituting in (2), $\dfrac{3(5+4y)^2}{9} - \dfrac{y(5+4y)}{3} - 3y^2 = 21$;

$$\therefore\ 75 + 120y + 48y^2 - 15y - 12y^2 - 27y^2 = 189;$$
$$9y^2 + 105y - 114 = 0,$$
$$3y^2 + 35y - 38 = 0;$$
$$\therefore\ (y-1)(3y+38) = 0;$$
$$\therefore\ y = 1,\ \text{or}\ -\dfrac{38}{3};$$

and by substituting in (1), $x = 3,\ \text{or}\ -\dfrac{137}{9}$;

208. The examples we have given will be sufficient as a general explanation of the methods to be employed; but in some cases special artifices are necessary.

Example 1. Solve $x^2 + 4xy + 3x = 40 - 6y - 4y^2$ ....(1),
$$2xy - x^2 = 3 \quad\quad\quad\quad\quad\quad (2).$$

From (1) we have $x^2 + 4xy + 4y^2 + 3x + 6y = 40$;

that is, $(x+2y)^2 + 3(x+2y) - 40 = 0,$

or $(x+2y+8)(x+2y-5) = 0;$

whence $x + 2y = -8,\ \text{or}\ 5.$

(i) Combining $x + 2y = 5$ with (2) we obtain
$$2x^2 - 5x + 3 = 0;$$

whence $x = 1,\ \text{or}\ \dfrac{3}{2}$;

and by substituting in $x + 2y = 5,$ $y = 2,\ \text{or}\ \dfrac{7}{4}.$

(ii) Combining $x + 2y = -8$ with (2) we obtain
$$2x^2 + 8x + 3 = 0;$$

whence $x = \dfrac{-4 \pm \sqrt{10}}{2}$; and $y = \dfrac{-12 \mp \sqrt{10}}{4}$.

Example 2. Solve $\quad x^2y^2 - 6x = 34 - 3y$ ....(1),
$$3xy + y = 2(9+x) \quad\ldots\ldots\ldots\ldots\ldots\ldots(2).$$

From (1) $\quad x^2y^2 - 6x + 3y = 34;$

from (2) $\quad 9xy - 6x + 3y = 54;$

by subtraction, $\quad x^2y^2 - 9xy + 20 = 0,$

$$(xy - 5)(xy - 4) = 0;$$

$$\therefore xy = 5, \text{ or } 4.$$

(i) Substituting $xy = 5$ in (2) gives $y - 2x = 3$.

From these equations we obtain $\left.\begin{array}{l} x = 1, \text{ or } -\dfrac{5}{2}, \\ y = 5, \text{ or } -2. \end{array}\right\}$

(ii) Substituting $xy = 4$ in (2) gives $y - 2x = 6$.

From these equations we obtain $\left.\begin{array}{l} x = \dfrac{-3 \pm \sqrt{17}}{2}, \\ y = 3 \pm \sqrt{17}. \end{array}\right\}$

and

EXAMPLES XXVI. c.

Solve the equations:

1. $5x - y = 17,$
$xy = 12.$

2. $x^2 + xy = 15,$
$y^2 + xy = 10.$

3. $x - y = 10,$
$x^2 - 2xy - 3y^2 = 84.$

4. $3x + 2y = 16,$
$xy = 10.$

5. $3x - y = 11,$
$3x^2 - y^2 = 47.$

6. $x - 3y = 1,$
$x^2 - 2xy + 9y^2 = 17.$

7. $x + 2y = 9,$
$3y^2 - 5x^2 = 43.$

8. $x^2 + y^2 = 5,$
$2xy - y^2 = 3.$

9. $5x + y = 3,$
$2x^2 - 3xy - y^2 = 1.$

10. $3x^2 - 5y^2 = 28,$
$3xy - 4y^2 = 8.$

11. $3x^2 - y^2 = 23,$
$2x^2 - xy = 12.$

12. $x^2 + xy + y^2 = 3\tfrac{1}{4},$
$2x^2 - 3xy + 2y^2 = 2\tfrac{3}{4}.$

13. $x^2 - 3xy + y^2 + 1 = 0,$
$3x^2 - xy + 3y^2 = 13.$

14. $7xy - 8x^2 = 10,$
$8y^2 - 9xy = 18.$

15. $x^2 - 2xy = 21,$
$xy + y^2 = 18.$

16. $x^2 + 3xy = 54,$
$xy + 4y^2 = 115.$

17. $x^3 + y^3 = 152,$
$x^2y + xy^2 = 120.$

18. $x^3 - y^3 = 127,$
$x^2y - xy^2 = 42.$

19. $x^3 - y^3 = 208,$
$xy(x - y) = 48.$

20. $x^2y^2 + 5xy = 84,$
$x + y = 8.$

21. $x^2 + 4y^2 + 80 = 15x + 30y,$
$xy = 6.$

22. $9x^2 + y^2 - 63x - 21y + 128 = 0,$
$xy = 4.$

CHAPTER XXVII.

Problems leading to Quadratic Equations.

209. We shall now discuss some problems which give rise to quadratic equations.

Example 1. A train travels 300 miles at a uniform rate; if the rate had been 5 miles an hour more, the journey would have taken two hours less: find the rate of the train.

Suppose the train travels at the rate of x miles per hour, then the time occupied is $\dfrac{300}{x}$ hours.

On the other supposition the time is $\dfrac{300}{x+5}$ hours;

$$\therefore \quad \frac{300}{x+5} = \frac{300}{x} - 2 \quad \ldots\ldots\ldots\ldots\ldots\ldots\ldots(1);$$

whence $\quad x^2 + 5x - 750 = 0,$
or $\quad (x+30)(x-25) = 0,$
$$\therefore \quad x = 25, \text{ or } -30.$$

Hence the train travels 25 miles per hour, the negative value being inadmissible.

It will frequently happen that the algebraical statement of the question leads to a result which does not apply to the actual problem we are discussing. But such results can sometimes be explained by a suitable modification of the conditions of the question. In the present case we may explain the negative solution as follows.

Since the values $x=25$ and -30 satisfy the equation (1), if we write $-x$ for x the resulting equation,

$$\frac{300}{-x+5} = \frac{300}{-x} - 2 \quad \ldots\ldots\ldots\ldots\ldots\ldots\ldots(2),$$

will be satisfied by the values $x = -25$ and 30. Now, by changing signs throughout, equation (2) becomes $\dfrac{300}{x-5} = \dfrac{300}{x} + 2$;

and this is the algebraical statement of the following question:

A train travels 300 miles at a uniform rate; if the rate had been 5 miles an hour *less*, the journey would have taken two hours *more*: find the rate of the train. The rate is 30 miles an hour.

Example 2. A person selling a horse for £72 finds that his loss per cent. is one-eighth of the number of pounds that he paid for the horse: what was the cost price?

Suppose that the cost price of the horse is x pounds; then the loss on £100 is £$\frac{x}{8}$.

Hence the loss on £x is $x \times \frac{x}{800}$, or $\frac{x^2}{800}$ pounds;

$\therefore$ the selling price is $x - \frac{x^2}{800}$ pounds.

Hence $$x - \frac{x^2}{800} = 72,$$
or $$x^2 - 800x + 57600 = 0;$$
that is, $$(x-80)(x-720) = 0;$$
$$\therefore x = 80, \text{ or } 720;$$

and each of these values will be found to satisfy the conditions of the problem. Thus the cost is either £80, or £720.

Example 3. A cistern can be filled by two pipes in $33\frac{1}{3}$ minutes; if the larger pipe takes 15 minutes less than the smaller to fill the cistern, find in what time it will be filled by each pipe singly.

Suppose that the two pipes running singly would fill the cistern in x and $x - 15$ minutes. When running together they will fill $\left(\frac{1}{x} + \frac{1}{x-15}\right)$ of the cistern in one minute. But they fill $\frac{1}{33\frac{1}{3}}$, or $\frac{3}{100}$ of the cistern in one minute.

Hence $$\frac{1}{x} + \frac{1}{x-15} = \frac{3}{100},$$
$$100(2x-15) = 3x(x-15),$$
$$3x^2 - 245x + 1500 = 0,$$
$$(x-75)(3x-20) = 0;$$
$$\therefore x = 75, \text{ or } 6\frac{2}{3}.$$

Thus the smaller pipe takes 75 minutes, the larger 60 minutes. The other solution $6\frac{2}{3}$ is inadmissible.

Example 4. By rowing half the distance and walking the other half, a man can travel 24 miles on a river in 5 hours with the stream, and in 7 hours against the stream. If there were no current, the journey would take $5\frac{2}{7}$ hours: find the rate of his walking, and rowing, and the rate of the stream.

PROBLEMS LEADING TO QUADRATIC EQUATIONS.

Suppose that the man walks x miles per hour, rows y miles per hour, and that the stream flows at the rate of z miles per hour.

With the current the man rows $y+z$ miles, and against the current $y-z$ miles per hour.

Hence we have the following equations:

$$\frac{12}{x}+\frac{12}{y+z}=5\ldots\ldots\ldots\ldots\ldots\ldots\ldots(1),$$

$$\frac{12}{x}+\frac{12}{y-z}=7\ldots\ldots\ldots\ldots\ldots\ldots\ldots(2),$$

$$\frac{12}{x}+\frac{12}{y}=5\frac{2}{3}\ \ldots\ldots\ldots\ldots\ldots\ldots\ldots(3).$$

From (1) and (3) by subtraction, $\dfrac{1}{y}-\dfrac{1}{y+z}=\dfrac{1}{18}\ldots\ldots\ldots\ldots\ldots(4)$.

Similarly, from (2) and (3) $\dfrac{1}{y-z}-\dfrac{1}{y}=\dfrac{1}{9}\ \ldots\ldots\ldots\ldots\ldots\ldots(5)$.

From (4) $\qquad\qquad\qquad 18z=y(y+z)\ldots\ldots\ldots\ldots\ldots(6)$;

and from (5) $\qquad\qquad\quad 9z=y(y-z)\ \ldots\ldots\ldots\ldots\ldots\ldots(7)$.

From (6) and (7) by division, $\quad 2=\dfrac{y+z}{y-z}$;

whence $\qquad\qquad\qquad\qquad y=3z$;

$\therefore$ from (4) $z=1\frac{1}{2}$; and hence $y=4\frac{1}{2}$, $x=4$.

Thus the rates of walking and rowing are 4 miles and $4\frac{1}{2}$ miles per hour respectively; and the stream flows at the rate of $1\frac{1}{2}$ miles per hour.

EXAMPLES XXVII.

1. Find a number whose square diminished by 119 is equal to ten times the excess of the number over 8.

2. A man is five times as old as his son, and the sum of the squares of their ages is equal to 2106: find their ages.

3. The sum of the reciprocals of two consecutive numbers is $\dfrac{15}{56}$: find them.

4. Find a number which when increased by 17 is equal to 60 times the reciprocal of the number.

5. Find two numbers whose sum is 9 times their difference, and the difference of whose squares is 81.

6. The sum of a number and its square is nine times the next highest number: find it.

7. If a train travelled 5 miles an hour faster it would take one hour less to travel 210 miles: what time does it take?

8. Find two numbers the sum of whose squares is 74, and whose sum is 12.

9. The perimeter of a rectangular field is 500 yards, and its area is 14400 square yards: find the length of the sides.

10. The perimeter of one square exceeds that of another by 100 feet; and the area of the larger square exceeds three times the area of the smaller by 325 square feet: find the length of their sides.

11. A cistern can be filled by two pipes running together in $22\frac{1}{2}$ minutes; the larger pipe would fill the cistern in 24 minutes less than the smaller one: find the time taken by each.

12. A man travels 108 miles, and finds that he could have made the journey in $4\frac{1}{2}$ hours less had he travelled 2 miles an hour faster; at what rate did he travel?

13. I buy a number of cricket balls for £5; had they cost a shilling apiece less, I should have had five more for the money: find the cost of each.

14. A boy was sent out for a shilling's worth of eggs. He broke 3 on his way home, and his master therefore had to pay at the rate of a penny more than the market price for 5. How many did the master get for a shilling?

15. What are eggs a dozen when two more in a shilling's worth lowers the price a penny per dozen?

16. A lawn 50 feet long and 34 feet broad has a path of uniform width round it; if the area of the path is 540 square feet, find its width.

17. A hall can be paved with 200 square tiles of a certain size; if each tile were one inch longer each way it would take 128 tiles: find the length of each tile.

18. In the centre of a square garden is a square lawn; outside this is a gravel walk 4 feet wide, and then a flower border 6 feet wide. If the flower border and lawn together contain 721 square feet, find the area of the lawn.

19. By lowering the price of apples and selling them one penny a dozen cheaper, an applewoman finds that she can sell 60 more than she used to do for 5s. At what price per dozen did she sell them at first?

20. Two rectangles contain the same area, 480 square yards. The difference of their lengths is 10 yards, and of their breadths 4 yards: find their sides.

XXVII.] PROBLEMS LEADING TO QUADRATIC EQUATIONS. 213

21. There is a number between 10 and 100; when multiplied by the digit on the left the product is 280; if the sum of the digits be multiplied by the same digit the product is 55: required the number.

22. A farmer having sold at 75s. a head, a flock of sheep which cost him x shillings a head, finds that he has realised x per cent. profit on his outlay: find x.

23. A tradesman bought a number of yards of cloth for £5; he kept 5 yards and sold the rest at 2s. per yard more than he gave, and got £1 more than he originally spent: how many yards did he buy?

24. If a carriage wheel $14\frac{2}{3}$ ft. in circumference takes one second more to revolve, the rate of the carriage per hour will be $2\frac{2}{3}$ miles less: how fast is the carriage travelling?

25. A broker bought as many railway shares as cost him £1875; he reserved 15, and sold the remainder for £1740, gaining £4 a share on their cost price. How many shares did he buy?

26. A and B are two stations 300 miles apart. Two trains start simultaneously from A and B, each to the opposite station. The train from A reaches B nine hours, the train from B reaches A four hours after they meet: find the rate at which each train travels.

27. A train A starts to go from P to Q, two stations 240 miles apart, and travels uniformly. An hour later another train B starts from P, and after travelling for 2 hours, comes to a point that A had passed 45 minutes previously. The pace of B is now increased by 5 miles an hour, and it overtakes A just on entering Q. Find the rates at which they started.

28. A cask P is filled with 50 gallons of water, and a cask Q with 40 gallons of brandy; x gallons are drawn from each cask, mixed and replaced; and the same operation is repeated. Find x when there are $8\frac{7}{8}$ gallons of brandy in P after the second replacement.

29. Two farmers A and B have 30 cows between them; they sell at different prices, but each receives the same sum. If A had sold his at B's price, he would have received £320: and if B had sold his at A's price, he would have received £245. How many had each?

30. A man arrives at the railway station nearest to his house $1\frac{1}{2}$ hours before the time at which he had ordered his carriage to meet him. He sets out at once to walk at the rate of 4 miles an hour, and, meeting his carriage when it had travelled 8 miles, reaches home exactly 1 hour earlier than he had originally expected. How far is his house from the station, and at what rate was his carriage driven?

31. P is a point in a line AB of length a. Find AP when $AB \cdot BP = AP^2$. Explain both solutions.

32. If a straight line 6 cm. in length is divided internally so that the rectangle contained by the whole and one part is equal to the square on the other part, find the segments of the line to the nearest millimetre.

33. A line AB is produced to P so that $AB \cdot AP = BP^2$. If $AB = 8$ cm., find the lengths of AP and BP to the nearest millimetre.

34. If a line AB of any length is divided externally as in the last Example, shew that

   (i) $AB^2 + AP^2 = 3BP^2$;   (ii) $(AB + AP)^2 = 5BP^2$.

35. A line AB is produced to P so that $BP^2 = 2AB^2$. If $AB = 3\cdot5$ cm., find AP to the nearest millimetre.

36. Find a point P in a straight line AB so that
$$AP(AP - BP) = BP^2.$$
If $AB = 4\cdot2$ cm., find AP and BP to the nearest millimetre. By substituting these values verify the truth of the given relation.

37. Divide a straight line 13 centimetres long into two parts so that the rectangle contained by them may be equal to 36 square centimetres.

38. Justify the following graphical solution of the previous Example:

On AB, a line 13 cm. in length, describe a semicircle. At A draw AP perpendicular to AB and 6 cm. in length; through P draw a line PQR to cut the semicircle in Q and R; draw QX, RY perpendicular to AB. Then AB is divided as required either at X or Y. Verify the algebraical solution of Example 37 by actual measurement.

39. Solve the following equations graphically, taking a centimetre as unit and giving the roots to the nearest millimetre.

   (i) $x(7 - x) = 12$;   (ii) $x^2 - 11x + 30 = 0$;
   (iii) $x^2 - 6x + 4 = 0$;   (iv) $x^2 + 13 = 8x$.

CHAPTER XXVIII.

Harder Factors.

210. In Chapter XVII. we have explained several rules for resolving algebraical expressions into factors; in the present chapter we shall continue the subject by discussing cases of greater difficulty.

211. By a slight modification some expressions admit of being written in the form of the difference of two squares, and may then be resolved into factors by the method of Art. 133.

Example 1. Resolve into factors $x^4 + x^2y^2 + y^4$.
$$x^4 + x^2y^2 + y^4 = (x^4 + 2x^2y^2 + y^4) - x^2y^2$$
$$= (x^2 + y^2)^2 - (xy)^2$$
$$= (x^2 + y^2 + xy)(x^2 + y^2 - xy)$$
$$= (x^2 + xy + y^2)(x^2 - xy + y^2).$$

Example 2. Resolve into factors $x^4 - 15x^2y^2 + 9y^4$.
$$x^4 - 15x^2y^2 + 9y^4 = (x^4 - 6x^2y^2 + 9y^4) - 9x^2y^2$$
$$= (x^2 - 3y^2)^2 - (3xy)^2$$
$$= (x^2 - 3y^2 + 3xy)(x^2 - 3y^2 - 3xy).$$

212. Expressions which can be put into the form $x^3 \pm \dfrac{1}{y^3}$ may be separated into factors by the rules for resolving the sum or the difference of two cubes. [Art. 136.]

Example 1. $\quad \dfrac{8}{a^3} - 27b^6 = \left(\dfrac{2}{a}\right)^3 - (3b^2)^3$
$$= \left(\dfrac{2}{a} - 3b^2\right)\left(\dfrac{4}{a^2} + \dfrac{6b^2}{a} + 9b^4\right).$$

Example 2. Resolve $a^2x^3 - \dfrac{8a^2}{y^3} - x^3 + \dfrac{8}{y^3}$ into four factors.
$$a^2x^3 - \dfrac{8a^2}{y^3} - x^3 + \dfrac{8}{y^3} = x^3(a^2 - 1) - \dfrac{8}{y^3}(a^2 - 1)$$
$$= (a^2 - 1)\left(x^3 - \dfrac{8}{y^3}\right)$$
$$= (a+1)(a-1)\left(x - \dfrac{2}{y}\right)\left(x^2 + \dfrac{2x}{y} + \dfrac{4}{y^2}\right).$$

Example 3. Resolve $a^9 - 64a^3 - a^6 + 64$ into six factors.

The expression $= a^3(a^6 - 64) - (a^6 - 64)$
$= (a^6 - 64)(a^3 - 1)$
$= (a^3 + 8)(a^3 - 8)(a^3 - 1)$
$= (a+2)(a^2-2a+4)(a-2)(a^2+2a+4)(a-1)(a^2+a+1).$

Example 4.

$a(a-1)x^2 - (a-b-1)xy - b(b+1)y^2 = \{ax - (b+1)y\}\{(a-1)x + by\}.$

Note. In examples of this kind the coefficients of x and y in the binomial factors can usually be guessed at once, and it only remains to verify the coefficient of the middle term.

213. From Example 2, Art. 52, we see that the quotient of

$a^3 + b^3 + c^3 - 3abc$ by $a + b + c$ is $a^2 + b^2 + c^2 - bc - ca - ab.$

Thus $a^3 + b^3 + c^3 - 3abc = (a+b+c)(a^2+b^2+c^2-bc-ca-ab)\ldots(1).$

This result is important and should be carefully remembered. We may note that the expression on the left consists of the sum of the cubes of three quantities a, b, c, diminished by 3 times the product abc. Whenever an expression admits of a similar arrangement, the above formula will enable us to resolve it into factors.

Example 1. Resolve into factors $a^3 - b^3 + c^3 + 3abc.$

$a^3 - b^3 + c^3 + 3abc = a^3 + (-b)^3 + c^3 - 3a(-b)c$
$= (a-b+c)(a^2+b^2+c^2+bc-ca+ab),$

$-b$ taking the place of b in formula (1).

Example 2.

$x^3 - 8y^3 - 27 - 18xy = x^3 + (-2y)^3 + (-3)^3 - 3x(-2y)(-3)$
$= (x-2y-3)(x^2+4y^2+9-6y+3x+2xy).$

EXAMPLES XXVIII. a.

Resolve into factors:

1. $x^4 + 16x^2 + 256.$
2. $81a^4 + 9a^2b^2 + b^4.$
3. $x^4 + y^4 - 7x^2y^2.$
4. $m^4 + n^4 - 18m^2n^2.$
5. $x^4 - 6x^2y^2 + y^4.$
6. $4x^4 + 9y^4 - 93x^2y^2.$
7. $4m^4 + 9n^4 - 24m^2n^2.$
8. $9x^4 + 4y^4 + 11x^2y^2.$
9. $x^4 - 19x^2y^2 + 25y^4.$
10. $16a^4 + b^4 - 28a^2b^2.$

HARDER FACTORS.

11. $\dfrac{27}{a^3b^3} - 1$. 12. $216a^3 - \dfrac{b^3}{8}$. 13. $\dfrac{x^3}{125} + y^3$.

14. $\dfrac{m^3n^3}{729} - 1$. 15. $\dfrac{a^3b^3}{125} + 1000$. 16. $\dfrac{x^3}{512} - \dfrac{64}{x^3}$.

Resolve into two or more factors :

17. $x^2y + 3xy^2 - 3x^3 - y^3$.
18. $4mn^2 - 20n^3 + 45nm^2 - 9m^3$.
19. $ab(x^2+1) + x(a^2+b^2)$.
20. $y^2z^2(x^4-1) + x^2(y^4-z^4)$.
21. $a^3 + (a+b)ax + bx^2$.
22. $pn(m^2+1) - m(p^2+n^2)$.
23. $6bx(a^2+1) - a(4x^2+9b^2)$.
24. $(2a^2+3y^2)x + (2x^2+3a^2)y$.
25. $(2x^2 - 3a^2)y + (2a^2 - 3y^2)x$.
26. $a(a-1)x^2 + (2a^2-1)x + a(a+1)$.
27. $3x^2 - (4a+2b)x + a^2 + 2ab$.
28. $2a^2x^2 - 2(3b-4c)(b-c)y^2 + abxy$.
29. $(a^2 - 3a + 2)x^2 + (2a^2 - 4a + 1)x + a(a-1)$.
30. $a(a+1)x^2 + (a+b)xy - b(b-1)y^2$.
31. $b^3 + c^3 - 1 + 3bc$. 32. $a^3 + 8c^3 + 1 - 6ac$.
33. $a^3 + b^3 + 8c^3 - 6abc$. 34. $a^3 - 27b^3 + c^3 + 9abc$.
35. $a^3 - b^3 - c^3 - 3abc$. 36. $8a^3 + 27b^3 + c^3 - 18abc$.
37. Resolve $x^8 + 81x^4 + 6561$ into three factors.
38. Resolve $(a^4 - 2a^2b^2 - b^4)^2 - 4a^4b^4$ into four factors.
39. Resolve $4(ab+cd)^2 - (a^2+b^2-c^2-d^2)^2$ into four factors.
40. Resolve $x^8 - \dfrac{1}{256}$ into four factors.
41. Resolve $x^{16} - y^{16}$ into five factors.
42. Resolve $x^{18} - y^{18}$ into six factors.

Resolve into four factors :

43. $\dfrac{a^3}{x^2} - 8x - a^3 + 8x^3$. 44. $x^9 + x^3y^6 - 8x^6y^3 - 8y^9$.

45. $x^9 + x^6 + 64x^3 + 64$. 46. $4a - 9b + \dfrac{4b^3}{a^2} - \dfrac{9a^3}{b^2}$.

47. $\dfrac{xy^3}{72} - \dfrac{x^3y^5}{32} - \dfrac{1}{9x^2} + \dfrac{y^2}{4}$. 48. $x^6 - 25x^2 + 6\tfrac{1}{4} - \tfrac{1}{4}x^4$.

Resolve into five factors :

49. $x^7 + x^4 - 16x^3 - 16$. 50. $16x^7 - 81x^3 - 16x^4 + 81$.

214. The actual processes of multiplication and division can often be partially or wholly avoided by a skilful use of factors.

It should be observed that the formulæ which the student has seen exemplified in the preceding pages are just as useful in their converse as in their direct application. Thus the formula for resolving into factors the difference of two squares is equally useful as enabling us to write down at once the product of the sum and the difference of two quantities.

Example 1.   Multiply $2a+3b-c$ by $2a-3b+c$.

These expressions may be arranged thus:
$$2a+(3b-c) \text{ and } 2a-(3b-c).$$
Hence the product $=\{2a+(3b-c)\}\{2a-(3b-c)\}$
$$=(2a)^2-(3b-c)^2 \quad\quad\quad\quad \text{[Art. 133.]}$$
$$=4a^2-(9b^2-6bc+c^2)$$
$$=4a^2-9b^2+6bc-c^2.$$

Example 2.   Multiply $(a^2+a+1)x-a-1$ by $(a-1)x-a^2+a-1$.

The product $=\{(a^2+a+1)x-(a+1)\}\{(a-1)x-(a^2-a+1)\}$
$$=(a^3-1)x^2-\{(a^4+a^2+1)+(a^2-1)\}x+(a^3+1)$$
$$=(a^3-1)x^2-(a^4+2a^2)x+a^3+1$$
$$=(a^3-1)x^2-a^2(a^2+2)x+a^3+1.$$

Note. The product of a^2+a+1 and a^2-a+1 is a^4+a^2+1 and should be written down without actual multiplication.

Example 3.   Multiply $(3+x-2x^2)^2-(3-x+2x^2)^2$ ....(1),

by $\quad\quad\quad\quad\quad\quad (3+x+2x^2)^2-(3-x-2x^2)^2$ ....(2).

The expression (1)
$$=(3+x-2x^2+3-x+2x^2)(3+x-2x^2-3+x-2x^2)$$
$$=6(2x-4x^2)$$
$$=12x(1-2x).$$
The expression (2)
$$=(3+x+2x^2+3-x-2x^2)(3+x+2x^2-3+x+2x^2)$$
$$=6(2x+4x^2)$$
$$=12x(1+2x).$$
Therefore the product $=12x(1-2x) \times 12x(1+2x)$
$$=144x^2(1-4x^2).$$

XXVIII.]   HARDER FACTORS.   219

Example 4. Divide the product of $2x^2+x-6$, and $6x^2-5x+1$ by $3x^2+5x-2$.

Denoting the division by means of a fraction, the required quotient

$$= \frac{(2x^2+x-6)(6x^2-5x+1)}{3x^2+5x-2}$$

$$= \frac{(2x-3)(x+2)(3x-1)(2x-1)}{(3x-1)(x+2)}$$

$$= (2x-3)(2x-1).$$

Example 5. Shew that $(2x+3y-z)^3+(3x+7y+z)^3$ is divisible by $5(x+2y)$.

The given expression is of the form A^3+B^3, and therefore has a divisor of the form $A+B$.

Therefore $(2x+3y-z)^3+(3x+7y+z)^3$
is divisible by $(2x+3y-z)+(3x+7y+z)$,
that is, by $5x+10y$,
or by $5(x+2y)$.

Example 6. Find the quotient when $a^3+8-5b(25b^2-6a)$ is divided by $a-5b+2$.

The expression $= a^3+8-125b^3+30ab$

$$= a^3+(-5b)^3+(2)^3-3 \cdot a(-5b)(2)$$

$$= (a-5b+2)(a^2+25b^2+4+10b-2a+5ab).$$

[Art. 213.]

∴ the quotient is $a^2+25b^2+4+10b-2a+5ab$.

Example 7. If $x+y=a$, and $x-y=b$ shew that
$$4(x^4-6x^2y^2+y^4)=6a^2b^2-a^4-b^4.$$

$$x^4-6x^2y^2+y^4 = (x^4-2x^2y^2+y^4)-4x^2y^2$$

$$= (x^2-y^2)^2 - \frac{1}{4}(4xy)^2$$

$$= \{(x+y)(x-y)\}^2 - \frac{1}{4}\{(x+y)^2-(x-y)^2\}^2$$

$$= (ab)^2 - \frac{1}{4}(a^2-b^2)^2;$$

∴ $4(x^4-6x^2y^2+y^4) = 4a^2b^2-(a^2-b^2)^2$
$$= 6a^2b^2-a^4-b^4.$$

EXAMPLES XXVIII. b.

Find the product of

1. $2x-7y+3z$ and $2x+7y-3z$.
2. $3x^2-4xy+7y^2$ and $3x^2+4xy+7y^2$.
3. $5x^2+5xy-9y^2$ and $5x^2-5xy-9y^2$.
4. $7x^2-8xy+3y^2$ and $7x^2+8xy-3y^2$.
5. $x^3+2x^2y+2xy^2+y^3$ and $x^3-2x^2y+2xy^2-y^3$.
6. $(x+y)^2+2(x+y)+4$ and $(x+y)^2-2(x+y)+4$.
7. $(1+x+2x^2)^2-(1-x-2x^2)^2$ and $(1+x-2x^2)^2-(1-x+2x^2)^2$.
8. $(a^2+3a-1)^2-(a^2-3a-1)^2$ and $(a^2+a+1)^2-(a^2-a+1)^2$.
9. x^3-4x^2+8x-8 and x^3+4x^2+8x+8.
10. $x^3-6ax^2+18a^2x-27a^3$ and $x^3+6ax^2+18a^2x+27a^3$.
11. $x-a-\dfrac{x^2}{a}-\dfrac{a^2}{x}$ and $x+a+\dfrac{x^2}{a}-\dfrac{a^2}{x}$.
12. $(2x^2+3x+1)^2-(2x^2-3x-1)^2$ and $(x^2+6x-2)^2-(x^2-6x+2)^2$.

Find the continued product of

13. x^2+ax+a^2, x^2-ax+a^2, $x^4-a^2x^2+a^4$.
14. $1-x+x^2$, $1+x+x^2$, $1-x^2+x^4$, $1-x^4+x^8$.
15. $(a-x)^3$, $(a+x)^3$, $(a^2+x^2)^3$.
16. $(1-x)^2$, $(1+x)^2$, $(1+x^2)^2$, $(1+x^4)^2$.
17. x^2+4x+3, x^2+x-2, x^2-5x+6.
18. x^2+2x-3, x^2-5x+6, x^2+3x+2.
19. $x+2$, x^2+2x+4, $x-2$, x^2-2x+4.
20. Multiply the square of $a+3b$ by $a^2-6ab+9b^2$.
21. Multiply $\dfrac{1}{2}(a-b)^2+\dfrac{1}{2}(b-c)^2+\dfrac{1}{2}(c-a)^2$ by $a+b+c$.
22. Divide $(4x+3y-2z)^2-(3x-2y+3z)^2$ by $x+5y-5z$.
23. Divide $x^8+16a^4x^4+256a^8$ by $x^2+2ax+4a^2$.
24. Divide $(3x+4y-2z)^2-(2x+3y-4z)^2$ by $x+y+2z$.
25. Divide the product of $x^2+7x+10$ and $x+3$ by x^2+5x+6.
26. Divide $2x(x^2-1)(x+2)$ by x^2+x-2.
27. Divide $5x(x-11)(x^2-x-156)$ by x^3+x^2-132x.

HARDER FACTORS.

28. Divide $x^6 + 19x^3 - 216$ by $(x^2 - 3x + 9)(x - 2)$.
29. Divide $(5x^2 - 3x - 6)^2 - (2x^2 - 7x + 9)^2$ by the product of $3x - 5$ and $x + 3$.
30. Divide $a^9 - b^9$ by the product of $a^2 + ab + b^2$ and $a^6 + a^3b^3 + b^6$.
31. Divide $(x^3 - 3x^2y)^2 - (3xy^2 - y^3)^2$ by $(x - y)^3$.
32. Divide $(x^2 - yz)^3 + 8y^3z^3$ by $x^2 + yz$.
33. Divide $18xy + 1 + 27x^3 - 8y^3$ by $1 + 3x - 2y$.
34. Divide $(2x^2 + 3x - 1)^2 - (x^2 + 4x + 5)^2$ by the product of $3x + 4$ and $x + 2$.
35. Divide the product of $6a^2 - 23a + 20$ and $22a^2 - 81a + 14$ by $33a^2 - 50a + 8$.
36. Divide the product of $x^2 + (a - b)x - ab$ and $x^2 - (a - b)x - ab$ by $x^2 + (a + b)x + ab$.
37. Divide $a^3 - 8y^3 - 9x(3x^2 + 2ay)$ by $a - 3x - 2y$.
38. Divide $27 - 8x^3 - 64y^3 - 72xy$ by $3 - 2(x + 2y)$.
39. Shew that $(2x - 3y + 1)^3 - (1 - 3x + 2y)^3$ is divisible by $5(x - y)$.
40. Shew that the square of $x + 1$ exactly divides
$$(x^3 + x^2 + 4)^3 - (x^3 - 2x + 3)^3.$$
41. Shew that $2b + 2d$ is a factor of the expression
$$(a + b + c + d)^3 - (a - b + c - d)^3.$$
42. Shew that $(3x^2 - 7x + 2)^3 - (x^2 - 8x + 8)^3$ is divisible by $2x - 3$ and by $x + 2$.
43. Shew that $(7x^2 + 3x - 3)^3 + (5x^2 - 4x - 3)^3$ is divisible by $4x - 3$ and by $3x + 2$.
44. Shew that the sum of the cubes of $2x^2 - 5x - 9$ and $x^2 + 6x - 5$ is divisible by the product of $3x + 7$ and $x - 2$.
45. If $x + y = m$ and $x - y = n$, express $x^3 + y^3$ in terms of m and n.
46. If $x + y = m$ and $x - y = n$, shew that
$$16(x^4 - 7x^2y^2 + y^4) = (5m^2 - n^2)(5n^2 - m^2).$$
47. Find the value of $x^4 + x^2y^2 + y^4$ when $x + y = 2a$, $x - y = 2b$.
48. If $x + y = 2a$ and $x - y = 2b$ prove that
$$x^4 - 23x^2y^2 + y^4 = (7a^2 - 3b^2)(7b^2 - 3a^2).$$
49. Find the value of $x^4 - 47x^2y^2 + y^4$ in terms of p and q when $x + y = p$ and $x - y = q$.
50. Find the value of $x^4 - 2x^3y + 2xy^3 - y^4$ when $x = a + b$ and $y = a - b$.

CHAPTER XXIX.

Miscellaneous Theorems and Examples.

215. Examples upon the simple rules, *e.g.* Division, Highest Common Factor, Evolution, etc., frequently occur which cannot be neatly and concisely worked without a ready use of factors and compound expressions. These we have hitherto excluded as unsuitable for the student until he has gained confidence and power by practice. We propose in the present chapter to bring together a miscellaneous collection of examples, for the most part not new in principle, but requiring some skill for their solution. The chapter will be found useful as a revision of the earlier chapters.

Example. Divide
$ax^4 - (ap-b)x^3 + (aq-bp-c)x^2 + (bq+cp)x - cq$ by $ax^2 + bx - c$.

$$ax^2+bx-c \,\big)\, ax^4 - (ap-b)x^3 + (aq-bp-c)x^2 + (bq+cp)x - cq \,\big(\, x^2 - px + q$$

$$\underline{ax^4 + bx^3 - cx^2}$$

$$-apx^3 + (aq-bp)x^2 + (bq+cp)x$$
$$\underline{-apx^3 - bpx^2 \quad\quad + cpx}$$

$$aqx^2 \quad\quad + bqx - cq$$
$$\underline{aqx^2 \quad\quad + bqx - cq}$$

Note. When the coefficients in divisor or dividend are compound quantities it is best to retain them in brackets throughout the work.

216. In the process of finding the highest common factor, by the rules explained in Chap. XVIII., every remainder that occurs in the course of the work contains the factor we are seeking. Hence when any one of the remainders admits of being resolved into factors, we may often shorten the work.

Example 1. Find the H.C.F. of $2x^3 - (4a-3c)x^2 + 6(b-ac)x + 9bc$ and $2x^3 + (2a+3c)x^2 + (3ac-4b)x - 6bc$.

$$2x^3 - (4a-3c)x^2 + 6(b-ac)x + 9bc \,\big)\, 2x^3 + (2a+3c)x^2 + (3ac-4b)x - 6bc \,\big(\, 1$$
$$\underline{2x^3 - (4a-3c)x^2 + (6b-6ac)x + 9bc}$$
$$6ax^2 + (9ac-10b)x - 15bc$$

Now the remainder $= 6ax^2 + 9acx - 10bx - 15bc$
$$= 3ax(2x+3c) - 5b(2x+3c)$$
$$= (2x+3c)(3ax-5b).$$

CHAP. XXIX.] THEOREMS AND EXAMPLES. 223

Of these factors, $3ax-5b$ may clearly be rejected; therefore if there is a common factor it must be $2x+3c$. And by division, or by the method explained in Art. 152, we find that $2x+3c$ is a factor of each expression.

Hence the H.C.F. is $2x+3c$.

Example 2. Find the H.C.F. of $(a^2-2a)x^2+2(2a-1)x-a^2+1$ and $(a^2-a-2)x^2+(4a+1)x-a^2-a$.

Each of these expressions can be resolved into factors as explained in Art. 212, Ex. 4. Thus
$(a^2-2a)x^2+2(2a-1)x-a^2+1 = a(a-2)x^2+2(2a-1)x-(a+1)(a-1)$
$\qquad\qquad = \{(a-2)x+(a+1)\}\{ax-(a-1)\}.$
$(a^2-a-2)x^2+(4a+1)x-a^2-a = (a-2)(a+1)x^2+(4a+1)x-a(a+1)$
$\qquad\qquad = \{(a-2)x+(a+1)\}\{(a+1)x-a\}.$

Hence the H.C.F. is $\quad (a-2)x+a+1$.

EXAMPLES XXIX. a.

Divide

1. $x^3+(a+b+c)x^2+(bc+ca+ab)x+abc$ by $x^2+(a+b)x+ab$.
2. $x^4-(5+a)x^3+(4+5a+b)x^2-(4a+5b)x+4b$ by x^2-5x+4.
3. $x^3-(a-b)x^2-(ab+2b^2)x+2ab^2$ by $x-b$.
4. $x^3-(p^2+3q^2)x+2p^2q-2q^3$ by $x+p+q$.
5. $x^3-(3mn+n^2)x+m(m^2-n^2)$ by $x+m+n$.
6. $a(a-1)x^2+(2a^2-1)x+a(a+1)$ by $(a-1)x+a$.
7. $x^4+(a+b)x^3+(a^2+ab+b^2)x^2+(a^3+b^3)x+a^2b^2$ by x^2+ax+b^2.
8. $2l^2x^2-2(3m-4n)(m-n)y^2+lmxy$ by $lx+2(m-n)y$.
9. $(a^2+a-2)x^2-(2a+1)xy-(a^2+a)y^2$ by $(a-1)x-ay$.
10. $x^3-(a-b-2)x^2-(ab+2a-2b)x-2ab$ by $(x-a)(x+2)$.
11. $(x+1)^8+4(x+1)^6+6(x+1)^4+4(x+1)^2+1$ by x^2+2x+2.
12. $(m+1)(bx+an)b^2x^2-(n+1)(mbx+a)a^2$ by $bx-a$.

Find the H.C.F. of

13. $(m^2-3m+2)x^2+(2m^2-4m+1)x+m(m-1)$ and
    $m(m-1)x^2+(2m^2-1)x+m(m+1)$.
14. $mpx^3+(mq-np)x^2-(mr+nq)x+nr$ and
    $max^3-(mc+na)x^2-(mb-nc)x+nb$.
15. $2ap^3+(3a-2b)p^2q+(a-3b)pq^2-bq^3$ and
    $3ap^3-(a+3b)p^2q+(2a+b)pq^2-2bq^3$.
16. $acx^3+(bc+ad)x^2+(bd+ac)x+bc$ and
    $2acx^3+(2bc-ad)x^2-(3ac+bd)x-3bc$.

Find the H.C.F. of

17. $2a^2x^3 - (4b+3)ax^2 + 2(3b-ac)x + 3c$ and
     $2a^2x^3 + (2b-3)ax^2 - (4ac+3b)x + 6c$.

18. $2ax^3 + (4a^2-1)bx^2 - (2ab^2+3c)x - 6abc$ and
     $ax^3 - (3-2a^2)bx^2 + (2c-6ab^2)x + 4abc$.

Find the L.C.M. of

19. $x^4 - px^3 + (q-1)x^2 + px - q$ and $x^4 - qx^3 + (p-1)x^2 + qx - p$.
20. $p(p+1)x^2 + x - p(p-1)$ and $p(p+2)x^2 + 2x - p^2 + 1$.
21. $(a^2 - 5a + 6)x^2 + 2(a-1)x - a(a+1)$ and
     $a(a-3)x^2 + 12x - (a+1)(a+4)$.

217. We add some miscellaneous questions in Evolution.

The *fourth* root of an expression is obtained by extracting the square root of the square root of the expression.

Similarly by successive applications of the rule for finding the square root, we may find the *eighth, sixteenth* ... root. The *sixth* root of an expression is found by taking the cube root of the square root, or the square root of the cube root.

Similarly by combining the two processes for extraction of cube and square roots, certain other higher roots may be obtained.

Example 1. Find the fourth root of
$$81x^4 - 216x^3y + 216x^2y^2 - 96xy^3 + 16y^4.$$

Extracting the square root by the rule we obtain $9x^2 - 12xy + 4y^2$; and *by inspection*, the square root of this is $3x - 2y$, which is the required fourth root.

Example 2. Find the sixth root of
$$\left(x^3 - \frac{1}{x^3}\right)^2 - 6\left(x - \frac{1}{x}\right)\left(x^3 - \frac{1}{x^3}\right) + 9\left(x - \frac{1}{x}\right)^2.$$

By inspection, the square root of this is
$$\left(x^3 - \frac{1}{x^3}\right) - 3\left(x - \frac{1}{x}\right),$$

which may be written $\quad x^3 - 3x + \dfrac{3}{x} - \dfrac{1}{x^3}$;

and the cube root of this is $\quad x - \dfrac{1}{x}$,

which is the required sixth root.

218. In Chap. VI. we have given examples of inexact division. In a similar manner when an expression is not an exact square

XXIX.] MISCELLANEOUS THEOREMS AND EXAMPLES. 225

or cube, we may perform the process of evolution, and obtain as many terms of the root as we please.

Example. To find four terms of the square root of $1+2x-2x^2$.

$$
\begin{array}{r|l}
& 1+2x-2x^2 \ \Big(\ 1+x-\dfrac{3}{2}x^2+\dfrac{3}{2}x^3 \\
& 1 \\
\hline
2+x & 2x-2x^2 \\
& 2x+\ x^2 \\
\hline
2+2x-\dfrac{3}{2}x^2 & -3x^2 \\
& -3x^2-3x^3+\dfrac{9}{4}x^4 \\
\hline
2+2x-3x^2+\dfrac{3}{2}x^3 & 3x^3-\dfrac{9}{4}x^4 \\
& 3x^3+3x^4-\dfrac{9}{2}x^5+\dfrac{9}{4}x^6 \\
\hline
& -\dfrac{21}{4}x^4+\dfrac{9}{2}x^5-\dfrac{9}{4}x^6.
\end{array}
$$

Thus the required result is $1+x-\dfrac{3}{2}x^2+\dfrac{3}{2}x^3$.

*219. In Art. 124 we pointed out the similarity between the arithmetical and algebraical methods of extracting square and cube roots. We shall now shew that in extracting either the square or the cube root of any number, when a certain number of figures have been obtained by the common rule, that number may be nearly doubled by ordinary division.

*220. *If the square root of a number consists of $2n+1$ figures, when the first $n+1$ of these have been obtained by the ordinary method, the remaining n may be obtained by division.*

Let N denote the given number; a the part of the square root already found, that is the first $n+1$ figures found by the common rule, with n ciphers annexed; x the remaining part of the root.

Then $\sqrt{N}=a+x$;

$$\therefore\ N=a^2+2ax+x^2;$$

$$\therefore\ \dfrac{N-a^2}{2a}=x+\dfrac{x^2}{2a} \quad \dots\dots\dots\dots\dots\dots\dots\dots(1).$$

Now $N-a^2$ is the remainder after $n+1$ figures of the root, represented by a, have been found; and $2a$ is the divisor at the same stage of the work. We see from (1) that $N - a^2$ divided by $2a$ gives x, the rest of the quotient required, increased by $\dfrac{x^2}{2a}$. We shall shew that $\dfrac{x^2}{2a}$ is a *proper fraction*, so that by neglecting the remainder arising from the division, we obtain x, the rest of the root.

For x contains n figures, and therefore x^2 contains $2n$ figures at most; also a is a number of $2n+1$ figures (the last n of which are ciphers) and thus $2a$ contains $2n+1$ figures at least; and therefore $\dfrac{x^2}{2a}$ is a proper fraction.

From the above investigation, by putting $n=1$, we see that *two* at least of the figures of a square root must have been obtained in order that the method of division, which is employed to obtain the next figure of the square root, may give that figure correctly.

Example. Find the square root of 290 to five places of decimals.

```
           290 ( 17·02
           1
       27 | 190
          | 189
     3402 | 10000
          |  6804
            3196
```

Here we have obtained four figures in the square root by the ordinary method. Three more may be obtained by division only, using 2×1702, that is 3404, for divisor, and 3196 as remainder. Thus

```
   3404 ) 31960 ( 938
          30636
          13240
          10212
          30280
          27232
           3048
```

And therefore to five places of decimals $\sqrt{290} = 17\cdot 02938$.

When the divisor consists of several digits, the method of contracted division may be employed with advantage.

Again, it may be noticed that in obtaining the second figure of the root, the division of 190 by 20 gives 9 for the next figure; this is too great, and the figure 7 has to be obtained tentatively. This is one of the modifications of the algebraical rule to which we referred in Art. 124.

*221. *If the cube root of a number consists of $2n+2$ figures, when the first $n+2$ of these have been obtained by the ordinary method, the remaining n may be obtained by division.*

Let N denote the given number; a the part of the cube root already found, that is the first $n+2$ figures found by the common rule, with n ciphers annexed; x the remaining part of the root.

Then
$$\sqrt[3]{N} = a+x;$$
$$\therefore\quad N = a^3 + 3a^2x + 3ax^2 + x^3;$$
$$\therefore\quad \frac{N-a^3}{3a^2} = x + \frac{x^2}{a} + \frac{x^3}{3a^2} \quad\ldots\ldots\ldots\ldots(1)$$

Now $N-a^3$ is the remainder after $n+2$ figures of the root, represented by a, have been found; and $3a^2$ is the divisor at the same stage of the work. We see from (1) that $N-a^3$ divided by $3a^2$ gives x, the rest of the quotient required, increased by $\frac{x^2}{a} + \frac{x^3}{3a^2}$. We shall shew that this expression is a *proper fraction*, so that by neglecting the remainder arising from the division, we obtain x, the rest of the root.

By supposition, x is $< 10^n$, and a is $> 10^{2n+1}$;
$$\therefore\quad \frac{x^2}{a} \text{ is } < \frac{10^{2n}}{10^{2n+1}}; \text{ that is, } < \frac{1}{10};$$
and
$$\frac{x^3}{3a^2} \text{ is } < \frac{10^{3n}}{3 \times 10^{4n+2}}; \text{ that is, } < \frac{1}{3 \times 10^{n+1}};$$
hence
$$\frac{x^2}{a} + \frac{x^3}{3a^2} \text{ is } < \frac{1}{10} + \frac{1}{3 \times 10^{n+1}},$$
and is therefore a proper fraction.

EXAMPLES XXIX. b.

Find the fourth roots of the following expressions:

1. $x^4 - 28x^3 + 294x^2 - 1372x + 2401$.
2. $16 - \frac{32}{m} + \frac{24}{m^2} - \frac{8}{m^3} + \frac{1}{m^4}$.
3. $a^4 + 8a^3x + 16x^4 + 32ax^3 + 24a^2x^2$.
4. $1 + 4x + 2x^2 - 8x^3 - 5x^4 + 8x^5 + 2x^6 - 4x^7 + x^8$.
5. $1 + 8x + 20x^2 + 8x^3 - 26x^4 - 8x^5 + 20x^6 - 8x^7 + x^8$.

Find the sixth roots of the following expressions:

6. $1+6x+15x^2+20x^3+15x^4+6x^5+x^6$.
7. $x^6-12ax^5+240a^4x^2-192a^5x+60a^2x^4-160a^3x^3+64a^6$.
8. $a^6-18a^5x+135a^4x^2-540a^3x^3+1215a^2x^4-1458ax^5+729x^6$.

Find the eighth roots of the following expressions:

9. $x^8-8x^7y+28x^6y^2-56x^5y^3+70x^4y^4-56x^3y^5+28x^2y^6-8xy^7+y^8$.
10. $\{x^4+2(p-1)x^3+(p^2-2p-1)x^2-2(p-1)x+1\}^4$.

Find to four terms the square root of

11. $1+x$. 12. $1-2x$. 13. $4+2x$. 14. $1-x-x^2$.
15. a^2-x. 16. x^2+a^2. 17. a^4-3x^2. 18. $9a^2+12ax$.

Find to three terms the cube root of

19. x^3-a^3. 20. $8+x$. 21. $\dfrac{1}{a^3}+9x$.
22. $1-6x+21x^2$. 23. $27x^6-27x^5-18x^4$. 24. $64-48x+9x^2$.

Identities and Transformations.

*222. DEFINITION. An **identity** is an algebraical statement which is true for all values of the letters involved in it.

Examples. $\quad a^3+b^3=(a+b)(a^2-ab+b^2)$.
$x^3+y^3+z^3-3xyz=(x+y+z)(x^2+y^2+z^2-yz-zx-xy)$.

*223. An identity asserts that two expressions are always equal; and the proof of this equality is called "proving the identity." The method of procedure is to choose one of the expressions given, and to shew by successive transformations that it can be made to assume the form of the other.

Example 1. To prove that
$$bc(b-c)+ca(c-a)+ab(a-b)=-(b-c)(c-a)(a-b).$$
The first side $=bc(b-c)+c^2a-ca^2+a^2b-ab^2$
$\qquad =bc(b-c)+a^2(b-c)-a(b^2-c^2)$
$\qquad =(b-c)\{bc+a^2-a(b+c)\}$
$\qquad =(b-c)\{bc+a^2-ab-ac\}$
$\qquad =(b-c)\{a(a-b)-c(a-b)\}$
$\qquad =(b-c)(a-b)(a-c)$
$\qquad =-(b-c)(c-a)(a-b)$,
changing the signs of the factor $a-c$, so as to preserve cyclic order.
[Compare Art. 229, Example 3.]

XXIX.] IDENTITIES AND TRANSFORMATIONS. 229

The expression on the left-hand side can be readily put in the following forms:
$$a^2(b-c)+b^2(c-a)+c^2(a-b);$$
$$-\{a(b^2-c^2)+b(c^2-a^2)+c(a^2-b^2)\}.$$

Hence we have the following results:
$$bc(b-c)+ca(c-a)+ab(a-b)=-(b-c)(c-a)(a-b);$$
$$a^2(b-c)+b^2(c-a)+c^2(a-b)=-(b-c)(c-a)(a-b);$$
$$a(b^2-c^2)+b(c^2-a^2)+c(a^2-b^2)=(b-c)(c-a)(a-b).$$

These identities are of such frequent occurrence that they should be carefully noticed and remembered.

Example 2. If $2s=a+b+c$ prove that
$$\frac{1}{s-a}+\frac{1}{s-b}+\frac{1}{s-c}-\frac{1}{s}=\frac{abc}{s(s-a)(s-b)(s-c)}.$$

The first side $=\left(\dfrac{1}{s-a}+\dfrac{1}{s-b}\right)+\left(\dfrac{1}{s-c}-\dfrac{1}{s}\right)$

$$=\frac{s-b+s-a}{(s-a)(s-b)}+\frac{s-s+c}{s(s-c)}$$

$$=\frac{2s-a-b}{(s-a)(s-b)}+\frac{c}{s(s-c)}$$

$$=\frac{c}{(s-a)(s-b)}+\frac{c}{s(s-c)}$$

$$=c\left\{\frac{s(s-c)+(s-a)(s-b)}{s(s-a)(s-b)(s-c)}\right\}$$

$$=\frac{c\{s^2-cs+s^2-as-bs+ab\}}{s(s-a)(s-b)(s-c)}$$

$$=\frac{c\{2s^2-s(a+b+c)+ab\}}{s(s-a)(s-b)(s-c)}$$

$$=\frac{abc}{s(s-a)(s-b)(s-c)}, \text{ for } s(a+b+c)=s\cdot 2s=2s^2.$$

Note. Here $2s$ is a convenient abbreviation of $a+b+c$; and the reduction is much simplified by working in terms of s instead of substituting its value at once. In examples of this kind, as a rule, the student should avoid substituting as long as the work can be carried on in terms of the symbol of abbreviation.

Example 3. If $x^2+u^2=2(xy+yz+zu-y^2-z^2)$
prove that $x=y=z=u.$

By transposing, we have
$$x^2-2xy+y^2+y^2-2yz+z^2+z^2-2zu+u^2=0,$$
or $$(x-y)^2+(y-z)^2+(z-u)^2=0.$$

Now since the square of any quantity is always positive, each of the expressions $(x-y)^2$, $(y-z)^2$, $(z-u)^2$ is positive. Hence their sum cannot be zero unless each of them be separately equal to zero.

$$\therefore \quad x-y=0, \quad y-z=0, \quad z-u=0;$$
or
$$x=y=z=u.$$

Note. The student should be careful to notice the difference between the conclusions to be drawn from the two statements

$$(x-a)^2 + (y-b)^2 = 0 \quad \ldots\ldots\ldots\ldots\ldots\ldots\ldots (1),$$
and
$$(x-a)(y-b) = 0 \quad \ldots\ldots\ldots\ldots\ldots\ldots\ldots (2).$$

From (1) we infer that *both* $x-a=0$ and $y-b=0$ *simultaneously*, while from (2) we infer that *either* $x-a=0$ or $y-b=0$.

*EXAMPLES XXIX. c.

Prove the following identities:

1. $b(x^3+a^3) + ax(x^2-a^2) + a^3(x+a) = (a+b)(x+a)(x^2-ax+a^2)$.
2. $(ax+by)^2 + (ay-bx)^2 + c^2x^2 + c^2y^2 = (x^2+y^2)(a^2+b^2+c^2)$.
3. $(x+y)^3 + 3(x+y)^2z + 3(x+y)z^2 + z^3$
$\qquad = (x+z)^3 + 3(x+z)^2y + 3(x+z)y^2 + y^3$.
4. $(a+b+c)(ab+bc+ca) - abc = (a+b)(b+c)(c+a)$.
5. $(a+b+c)^2 - a(b+c-a) - b(a+c-b) - c(a+b-c) = 2(a^2+b^2+c^2)$.
6. $(x-y)^3 + (x+y)^3 + 3(x-y)^2(x+y) + 3(x+y)^2(x-y) = 8x^3$.
7. $x^2(y-z) + y^2(z-x) + z^2(x-y) + (y-z)(z-x)(x-y) = 0$.
8. $a^3(b-c) + b^3(c-a) + c^3(a-b) = -(b-c)(c-a)(a-b)(a+b+c)$.
9. If $x+y+z=0$, prove that $x^3+y^3+z^3 = 3xyz$.
10. Prove that $(b-c)^3 + (c-a)^3 + (a-b)^3 = 3(b-c)(c-a)(a-b)$.

If $2s = a+b+c$, shew that

11. $(s-a)^2 + (s-b)^2 + (s-c)^2 + s^2 = a^2+b^2+c^2$.
12. $(s-a)^3 + (s-b)^3 + (s-c)^3 + 3abc = s^3$.
13. $16s(s-a)(s-b)(s-c) = 2b^2c^2 + 2c^2a^2 + 2a^2b^2 - a^4 - b^4 - c^4$.
14. $2(s-a)(s-b)(s-c) + a(s-b)(s-c) + b(s-c)(s-a)$
$\qquad\qquad + c(s-a)(s-b) = abc$.

If $a+b+c=0$, shew that

15. $(2a-b)^3 + (2b-c)^3 + (2c-a)^3 = 3(2a-b)(2b-c)(2c-a)$.
16. $\dfrac{a^2}{2a^2+bc} + \dfrac{b^2}{2b^2+ca} + \dfrac{c^2}{2c^2+ab} = 1$.
17. Prove that
$(x+y+z)^3 + (x+y-z)^3 + (x-y+z)^3 + (x-y-z)^3 = 4x(x^2+3y^2+3z^2)$.

XXIX.] IDENTITIES AND TRANSFORMATIONS. 231

18. If $a+b+c=s$, prove that
$$(s-3a)^3+(s-3b)^3+(s-3c)^3-3(s-3a)(s-3b)(s-3c)=0.$$

19. If $X=b+c-2a$, $Y=c+a-2b$, $Z=a+b-2c$, find the value of $X^3+Y^3+Z^3-3XYZ$.

20. Find the value of $a(a^2+bc)+b(b^2+ac)-c(c^2-ab)$ when $a=\cdot 7$, $b=\cdot 08$, $c=\cdot 78$.

21. Prove that $(a-b)^2+(b-c)^2+(c-a)^2$
$$=2(c-b)(c-a)+2(b-a)(b-c)+2(a-b)(a-c).$$

22. Prove that $a^2(b^3-c^3)+b^2(c^3-a^3)+c^2(a^3-b^3)$
$$=(a-b)(b-c)(c-a)(ab+bc+ca)$$
$$=a^2(b-c)^3+b^2(c-a)^3+c^2(a-b)^3$$
$$=-[a^2b^2(a-b)+b^2c^2(b-c)+c^2a^2(c-a)].$$

23. If $(a+b)^2+(b+c)^2+(c+d)^2=4(ab+bc+cd)$, prove that
$$a=b=c=d.$$

24. If $x=a+d$, $y=b+d$, $z=c+d$, prove that
$$x^2+y^2+z^2-yz-zx-xy=a^2+b^2+c^2-bc-ca-ab.$$

25. If $a+b+c=0$, prove that
$$\frac{1}{b^2+c^2-a^2}+\frac{1}{c^2+a^2-b^2}+\frac{1}{a^2+b^2-c^2}=0.$$

26. If $a+b+c=0$, simplify
$$\frac{b+c}{bc}(b^2+c^2-a^2)+\frac{c+a}{ca}(c^2+a^2-b^2)+\frac{a+b}{ab}(a^2+b^2-c^2).$$

27. Prove that the equation
$$(x-a)^2+(y-b)^2+(a^2+b^2-1)(x^2+y^2-1)=0,$$
is equivalent to the equation
$$(ax+by-1)^2+(bx-ay)^2=0;$$
hence shew that the only possible values of x and y are
$$\frac{a}{a^2+b^2}, \frac{b}{a^2+b^2}.$$

28. If $2(x^2+a^2-ax)(y^2+b^2-by)=x^2y^2+a^2b^2$, shew that
$$(x-a)^2(y-b)^2+(bx-ay)^2=0,$$
and therefore that $x=a$, $y=b$ are the only possible solutions.

*224. We shall now give some further examples of fractions to illustrate the advantage of arranging expressions with regard to cyclic order. [Art. 172.]

Example. Find the value of
$$\frac{a}{(a-b)(a-c)(x-a)}+\frac{b}{(b-c)(b-a)(x-b)}+\frac{c}{(c-a)(c-b)(x-c)}.$$

Changing the sign of one factor in each denominator, so as to preserve cyclic order, we get for the lowest common denominator,
$$(a-b)(b-c)(c-a)(x-a)(x-b)(x-c).$$

The whole expression has for its numerator
$$-[a(b-c)(x-b)(x-c)+\ldots\ldots+\ldots\ldots]$$
or $\qquad -[a(b-c)\{x^2-(b+c)x+bc\}+\ldots\ldots+\ldots\ldots].$

Arrange it according to powers of x; thus

coefficient of $x^2 = -\{a(b-c)+b(c-a)+c(a-b)\}$
$\qquad\qquad\qquad = 0;$

coefficient of $x = \{a(b^2-c^2)+b(c^2-a^2)+c(a^2-b^2)\}$
$\qquad\qquad\qquad = (b-c)(c-a)(a-b);$ [Art. 223.]

terms which do not contain x
$\qquad = -\{abc(b-c)+abc(c-a)+abc(a-b)\}$
$\qquad = -abc\{b-c+c-a+a-b\}$
$\qquad = 0.$

Hence the expression $= \dfrac{(b-c)(c-a)(a-b)x}{(b-c)(c-a)(a-b)(x-a)(x-b)(x-c)}$

$\qquad\qquad\qquad\qquad = \dfrac{x}{(x-a)(x-b)(x-c)}.$

Note. In examples of this kind the work will be much facilitated if the student accustoms himself to readily writing down the following equivalents:
$$(b-c)+(c-a)+(a-b)=0.$$
$$a(b-c)+b(c-a)+c(a-b)=0.$$
$$a^2(b-c)+b^2(c-a)+c^2(a-b)=-(a-b)(b-c)(c-a).$$
$$bc(b-c)+ca(c-a)+ab(a-b)=-(a-b)(b-c)(c-a).$$
$$a(b^2-c^2)+b(c^2-a^2)+c(a^2-b^2)=(a-b)(b-c)(c-a).$$

Some of the identities in XXIX. c. may also be remembered with advantage.

*EXAMPLES XXIX. d.

1. $\dfrac{a}{(a-b)(a-c)}+\dfrac{b}{(b-c)(b-a)}+\dfrac{c}{(c-a)(c-b)}.$

2. $\dfrac{bc}{(a-b)(a-c)}+\dfrac{ca}{(b-c)(b-a)}+\dfrac{ab}{(c-a)(c-b)}.$

CYCLIC ORDER.

3. $\dfrac{a^2}{(a-b)(a-c)} + \dfrac{b^2}{(b-c)(b-a)} + \dfrac{c^2}{(c-a)(c-b)}$.

4. $\dfrac{a^3}{(a-b)(a-c)} + \dfrac{b^3}{(b-c)(b-a)} + \dfrac{c^3}{(c-a)(c-b)}$.

5. $\dfrac{a(b+c)}{(a-b)(c-a)} + \dfrac{b(a+c)}{(a-b)(b-c)} + \dfrac{c(a+b)}{(c-a)(b-c)}$.

6. $\dfrac{1}{a(a-b)(a-c)} + \dfrac{1}{b(b-c)(b-a)} + \dfrac{1}{c(c-a)(c-b)}$.

7. $\dfrac{bc}{a(a^2-b^2)(a^2-c^2)} + \dfrac{ca}{b(b^2-c^2)(b^2-a^2)} + \dfrac{ab}{c(c^2-a^2)(c^2-b^2)}$.

8. $\dfrac{(x-b)(x-c)}{(a-b)(a-c)} + \dfrac{(x-c)(x-a)}{(b-c)(b-a)} + \dfrac{(x-a)(x-b)}{(c-a)(c-b)}$.

9. $\dfrac{bc(a+d)}{(a-b)(a-c)} + \dfrac{ca(b+d)}{(b-c)(b-a)} + \dfrac{ab(c+d)}{(c-a)(c-b)}$.

10. $\dfrac{1}{(a-b)(a-c)(x-a)} + \dfrac{1}{(b-c)(b-a)(x-b)} + \dfrac{1}{(c-a)(c-b)(x-c)}$.

11. $\dfrac{a^2}{(a-b)(a-c)(x+a)} + \dfrac{b^2}{(b-c)(b-a)(x+b)} + \dfrac{c^2}{(c-a)(c-b)(x+c)}$

12. $a^2\dfrac{(a+b)(a+c)}{(a-b)(a-c)} + b^2\dfrac{(b+c)(b+a)}{(b-c)(b-a)} + c^2\dfrac{(c+a)(c+b)}{(c-a)(c-b)}$.

13. $\dfrac{a^3(b-c)+b^3(c-a)+c^3(a-b)}{(b-c)^3+(c-a)^3+(a-b)^3}$.

14. $\dfrac{a^2(b-c)+b^2(c-a)+c^2(a-b)+2(a-b)(b-c)(c-a)}{(b-c)^3+(c-a)^3+(a-b)^3}$.

15. $\dfrac{a^3(b-c)+b^3(c-a)+c^3(a-b)}{a^2(b-c)+b^2(c-a)+c^2(a-b)}$.

16. $\dfrac{a^2(b-c)^3+b^2(c-a)^3+c^2(a-b)^3}{(a-b)(b-c)(c-a)}$.

17. $\dfrac{\dfrac{1}{a}(b-c)+\dfrac{1}{b}(c-a)+\dfrac{1}{c}(a-b)}{\dfrac{1}{a}\left(\dfrac{1}{b^2}-\dfrac{1}{c^2}\right)+\dfrac{1}{b}\left(\dfrac{1}{c^2}-\dfrac{1}{a^2}\right)+\dfrac{1}{c}\left(\dfrac{1}{a^2}-\dfrac{1}{b^2}\right)}$.

18. $\dfrac{a^2\left(\dfrac{1}{c^2}-\dfrac{1}{b^2}\right)+b^2\left(\dfrac{1}{a^2}-\dfrac{1}{c^2}\right)+c^2\left(\dfrac{1}{b^2}-\dfrac{1}{a^2}\right)}{\dfrac{1}{bc}\left(\dfrac{1}{c}-\dfrac{1}{b}\right)+\dfrac{1}{ca}\left(\dfrac{1}{a}-\dfrac{1}{c}\right)+\dfrac{1}{ab}\left(\dfrac{1}{b}-\dfrac{1}{a}\right)}$.

***225.** *To find when* $\quad x^3+px^2+qx+r$ ....(1),
is divisible by $\quad x^2+ax+b$ .... (2).

Divide (1) by (2) in the ordinary way; thus

$$x^2+ax+b \,\big)\, \underline{x^3+px^2+qx+r} \,\big(\, x+(p-a)$$
$$\underline{x^3+ax^2+bx}$$
$$(p-a)x^2+\ (q-b)\ x+r$$
$$\underline{(p-a)x^2+a(p-a)\ x+b(p-a)}$$
$$\{(q-b)-a(p-a)\}x+r-b(p-a) \ldots\ldots\ldots (3)$$

Now if the remainder is zero the division is exact. This is the case when

$$\{(q-b)-a(p-a)\}x+r-b(p-a)=0,$$

or
$$x=\frac{b(p-a)-r}{q-b-a(p-a)}$$

Hence when x has this value, (1) is divisible by (2).

But if in (3), $\quad q-b-a(p-a)=0,$
and also $\quad r-b(p-a)=0,$

the remainder is equal to zero *whatever value* x *may have*. Thus x^3+px^2+qx+b is divisible by x^2+ax+b for *all* values of x, provided that

$$q-b-a(p-a)=0,$$
and $\quad r-b(p-a)=0.$

*226. *To find the condition that* x^2+px+q *may be a perfect square*.

Using the ordinary rule for square root, we have

$$x^2+px+q\,\big(\,x+\frac{p}{2}$$
$$\underline{x^2}$$
$$2x+\frac{p}{2}\,\big)\, px+q$$
$$\underline{px+\frac{p^2}{4}}$$
$$q-\frac{p^2}{4}.$$

If therefore x^2+px+q be a perfect square, the remainder, $q-\frac{p^2}{4}$, must be zero.

Hence $q-\frac{p^2}{4}=0$, or $p^2=4q$, is the condition required.

***227.** *To prove that* $x^4+px^3+qx^2+rx+s$ *is a perfect square if* $\left(q-\dfrac{p^2}{4}\right)^2=4s$ *and* $r^2=p^2s$.

The square root must clearly be a trinomial expression of the form x^2+lx+m; if therefore we put

$$x^4+px^3+qx^2+rx+s=(x^2+lx+m)^2,$$

we have, on expanding the right-hand side

$$x^4+px^3+qx^2+rx+s=x^4+2lx^3+x^2(l^2+2m)+2lmx+m^2.$$

Since this is to be true for all values of x, *we may assume that the coefficients of the like powers of* x *are the same;* hence

$$2l=p, \qquad l^2+2m=q,$$
$$2lm=r, \qquad m^2=s.$$

From these equations, by eliminating the unknown quantities l and m, we shall obtain the necessary relations between p, q, r, and s.

Thus we have
$$q-\frac{p^2}{4}=2m=2\sqrt{s},$$
$$r=2lm=p\sqrt{s};$$
$$\therefore \left(q-\frac{p^2}{4}\right)^2=4s \text{ and } r^2=p^2s.$$

Note. The method of Art. 226 might have been used here. Also the method of the present article may be used to establish the results of Arts. 225 and 226.

*228. The proposition in the preceding article has been given to illustrate a useful method, which admits of very wide application. In the course of the proof we assume the truth of an important principle; namely,

If two rational integral expressions involving x *are identically equal, the coefficients of like powers of* x *in the two expressions are equal.*

[An expression is said to be *rational* when no term contains a square or other root, and it is said to be *integral with respect to* x when the powers of x are all positive integers.]

The demonstration of this principle belongs to a more advanced part of the subject, and could not be discussed completely here. [See *Higher Algebra.* Art. 311.]

The Remainder Theorem.

*229. *If a rational integral algebraical expression*

$$x^n + p_1 x^{n-1} + p_2 x^{n-2} + p_3 x^{n-3} + \ldots + p_{n-1} x + p_n$$

be divided by $x - a$, *the remainder will be*

$$a^n + p_1 a^{n-1} + p_2 a^{n-2} + p_3 a^{n-3} + \ldots + p_{n-1} a + p_n.$$

Divide the given expression by $x - a$ till a remainder is obtained which does not involve x. Let Q be the quotient, and R the remainder; then

$$x^n + p_1 x^{n-1} + p_2 x^{n-2} + \ldots + p_{n-1} x + p_n = Q(x - a) + R.$$

Since R does not contain x, it will remain unaltered whatever value we give to x.

Put $x = a$, then

$$a^n + p_1 a^{n-1} + p_2 a^{n-2} + \ldots + p_{n-1} a + p_n = Q \times 0 + R,$$

$$\therefore R = a^n + p_1 a^{n-1} + p_2 a^{n-2} + \ldots + p_{n-1} a + p_n;$$

which proves the proposition.

From this it appears that when an algebraical expression is divided by $x - a$, the remainder can be obtained at once by writing a in the place of x in the given expression.

Again, the remainder is zero when the given expression is exactly divisible by $x - a$; hence we deduce another important proposition, known as the **Factor Theorem**.

If a rational integral expression involving x *become equal to* 0 *when* a *is written for* x, *it will contain* $x - a$ *as a factor*.

Example 1. Resolve into factors $x^3 + 3x^2 - 13x - 15$.

By trial we find that this expression vanishes when $x = 3$; hence $x - 3$ is a factor.

$$\therefore \quad x^3 + 3x^2 - 13x - 15 = x^2(x - 3) + 6x(x - 3) + 5(x - 3)$$
$$= (x - 3)(x^2 + 6x + 5)$$
$$= (x - 3)(x + 1)(x + 5).$$

Note. The only numerical values that need be substituted for x are the factors of the last term of the expression. Thus, in the present case, by making trial of -5, we should have detected the factor $x + 5$.

Example 2. The remainder when $x^4 - 2x^3 + x - 7$ is divided by $x+2$ is
$$(-2)^4 - 2(-2)^3 + (-2) - 7;$$
that is, $16 + 16 - 2 - 7$, or 23.

Or the remainder may be found more shortly by substituting $x = -2$ in $[\{(x-2)x\}x+1]x-7$.

Example 3. Find the factors of $bc(b-c) + ca(c-a) + ab(a-b)$.

On trial, this expression vanishes when $b=c$; therefore $b-c$ is a factor. Similarly $c-a$, $a-b$ may be shewn to be factors.

$$\therefore bc(b-c) + ca(c-a) + ab(a-b) = M(b-c)(c-a)(a-b)\ldots\ldots(1);$$

and since the left-hand member of this identity is only of three dimensions in a, b, c, the factor M must be some numerical quantity independent of a, b, c; its value can therefore be found by giving particular values to a, b, c, or by equating the coefficients of like terms on each side.

Let $a=0$, $b=1$, $c=2$, then (1) becomes
$$2(-1) + 0 + 0 = M(-1) \times 2 \times (-1);$$
whence $M = -1$.

$$\therefore bc(b-c) + ca(c-a) + ab(a-b) = -(b-c)(c-a)(a-b).$$

*230. We shall now give general proofs of the statements made in Art. 55. We suppose n to be positive and integral.

I. *To prove that* $x^n - y^n$ *is always divisible by* $x-y$.

By the remainder theorem when $x^n - y^n$ is divided by $x-y$ the remainder is
$$y^n - y^n, \text{ or } 0,$$
that is, $x^n - y^n$ is always divisible by $x-y$.

II. *To prove that* $x^n + y^n$ *is divisible by* $x+y$ *when* n *is odd, but not when* n *is even.*

By the remainder theorem when $x^n + y^n$ is divided by $x+y$ the remainder is
$$(-y)^n + y^n.$$
(1) if n is odd, $(-y)^n + y^n = -y^n + y^n = 0$;
(2) if n is even, $(-y)^n + y^n = y^n + y^n = 2y^n$;

hence there is a remainder when n is even, but none when n is odd; which proves the proposition.

In like manner it may be proved that $x^n - y^n$ is divisible by $x+y$ when n is even; and $x^n + y^n$ is *never* divisible by $x-y$.

By going through a few steps of the division, the form of the quotient in each case is easily determined. The results of the present article may be conveniently stated as follows:

(i) For all values of n,
$$x^n - y^n = (x-y)(x^{n-1} + x^{n-2}y + x^{n-3}y^2 + \ldots + y^{n-1}).$$

(ii) When n is odd,
$$x^n + y^n = (x+y)(x^{n-1} - x^{n-2}y + x^{n-3}y^2 - \ldots + y^{n-1}).$$

(iii) When n is even,
$$x^n - y^n = (x+y)(x^{n-1} - x^{n-2}y + x^{n-3}y^2 - \ldots - y^{n-1}).$$

*EXAMPLES XXIX. e.

Find the values of x which will make each of the following expressions a perfect square:

1. $x^4 + 6x^3 + 13x^2 + 13x - 1$.
2. $x^4 + 6x^3 + 11x^2 + 3x + 31$.
3. $x^4 - 2ax^3 + (a^2 + 2b)x^2 - 3abx + 2b^2$.
4. $4p^2x^4 - 4pqx^3 + (q^2 + 2p^2)x^2 - 5pqx + \dfrac{p^2}{2}$.
5. $\dfrac{a^2x^6}{9} - \dfrac{abx^4}{2} + \dfrac{2acx^3}{3} + \dfrac{9b^2x^2}{16} - \dfrac{5bcx}{2} + 6c^2$.
6. $x^4 + 2ax^3 + 3a^2x^2 + cx + d$.
7. Find the conditions that $x^4 - ax^3 + bx^2 - cx + 1$ may be a perfect square for all values of x.

Find the values of x which will make each of the following expressions a perfect cube:

8. $8x^3 - 36x^2 + 56x - 39$.
9. $\dfrac{x^6}{27} - \dfrac{a^2x^4}{3} + 4a^4x^2 - 28a^6$.
10. $m^3x^6 - 9m^2nx^4 + 39mn^2x^2 - 51n^3$.
11. Find the relation between b and c in order that
$$x^3 + 3ax^2 + bx + c$$
may be a perfect cube for all values of x.
12. Find the conditions that
$$x^6 + 3ax^5 + 3bx^4 + a(6b - 5a^2)x^3 + 3b(b - a^2)x^2 + 3cx + d$$
may be a perfect cube for all values of x.
13. What number must be added to $x^3 + 2x^2$ in order that the expression may be divisible by $x + 4$?
14. If $x + a$ be a common factor of $x^2 + px + q$ and $x^2 + lx + m$, shew that $a = \dfrac{m - q}{l - p}$.

MISCELLANEOUS THEOREMS.

Resolve into factors:

15. $x^3 - 6x^2 + 11x - 6$.
16. $x^3 - 5x^2 - 2x + 24$.
17. $x^3 + 9x^2 + 26x + 24$.
18. $x^3 - x^2 - 41x + 105$.
19. $x^3 - 39x + 70$.
20. $x^3 - 8x^2 - 31x - 22$.
21. $6x^3 + 7x^2 - x - 2$.
22. $6x^3 + x^2 - 19x + 6$.

Write down the quotient in the following cases:

23. $\dfrac{x^7 + y^7}{x + y}$.
24. $\dfrac{x^8 - y^8}{x + y}$.
25. $\dfrac{x^6 - y^6}{x - y}$.
26. $\dfrac{x^9 - y^9}{x - y}$.

Find the square root of

27. $x^4 + (2a - 4)x^3 + (a^2 - 2a + 4)x^2 + (2a^2 - 4a)x + a^2$.
28. $(a+1)^2 x^4 + (2a^2 + 2a)x^3 + (3a^2 - 4a - 6)x^2 + (2a^2 - 6a)x + a^2 - 6a + 9$.
29. Find what values of m make $3mx^2 + (6m - 12)x + 8$ a perfect square.
30. If $4x^4 + 12x^3 y + Px^2 y^2 + 6xy^3 + y^4$ is a perfect square, find P.

Without actual division shew that

31. $32x^{10} - 33x^5 + 1$ is divisible by $x - 1$.
32. $3x^4 + 5x^3 - 13x^2 - 20x + 4$ .... $x^2 - 4$.
33. $x^4 + 4x^3 - 5x^2 - 36x - 36$ .... $x^2 - x - 6$.

Without actual division find the remainder when

34. $x^5 - 5x^2 + 5$ is divided by $x - 5$.
35. $x^3 - 7x^2 a + 8xa^2 + 15a^3$ .... $x + 2a$.
36. If $ax^2 - bx + c$ and $dx^3 - bx + c$ have a common factor, then
$$a^3 - abd + cd^2 = 0.$$
37. If n be any positive integer, prove that $5^{2n} - 1$ is always divisible by 24.
38. Shew that $1 - x - x^n + x^{n+1}$ is exactly divisible by
$$1 - 2x + x^2.$$
39. If $x^3 + px + r$ and $3x^2 + p$ have a common factor, prove that
$$\dfrac{p^3}{27} + \dfrac{r^2}{4} = 0.$$
40. Shew that if $x^n + py^n + qz^n$ is exactly divisible by
$$x^2 - (ay + bz)x + abyz,$$
then $\quad\dfrac{p}{a^n} + \dfrac{q}{b^n} + 1 = 0.$

CHAPTER XXX.

The Theory of Indices.

[*Logarithms* (Chap. XXXIX.) *may be taken in connection with this chapter after* Arts. 231–242 *have been read. The articles marked with an asterisk may be postponed on a first reading.*]

231. Hitherto all the definitions and rules with regard to indices have been based upon the supposition that they were positive integers; for instance

(1) $\quad a^{14} = a \cdot a \cdot a \ldots$ to fourteen factors.
(2) $\quad a^{14} \times a^{3} = a^{14+3} = a^{17}$.
(3) $\quad a^{14} \div a^{3} = a^{14-3} = a^{11}$.
(4) $\quad (a^{14})^{3} = a^{14 \times 3} = a^{42}$.

The object of the present chapter is twofold: first, to give *general* proofs which shall establish the laws of combination in the case of all positive integral indices; secondly, to explain how, in strict accordance with these laws, intelligible meanings may be given to symbols whose indices are fractional, zero, or negative.

We shall begin by proving, directly from the definition of a positive integral index, three important propositions.

232. Definition. When m is a *positive integer*, a^{m} stands for the product of m factors each equal to a.

233. Prop. I. *To prove that* $a^{m} \times a^{n} = a^{m+n}$, *when* m *and* n *are positive integers.*

By definition, $a^{m} = a \cdot a \cdot a \ldots$ to m factors;
$\qquad a^{n} = a \cdot a \cdot a \ldots$ to n factors;
$\therefore \ a^{m} \times a^{n} = (a \cdot a \cdot a \ldots \text{ to } m \text{ factors}) \times (a \cdot a \cdot a \ldots \text{ to } n \text{ factors})$
$\qquad = a \cdot a \cdot a \ldots$ to $m+n$ factors
$\qquad = a^{m+n}$, by definition.

Cor. If p is also a positive integer, then
$$a^{m} \times a^{n} \times a^{p} = a^{m+n+p};$$
and so for any number of factors.

234. Prop. II. *To prove that* $a^m \div a^n = a^{m-n}$, *when* m *and* n *are positive integers, and* m > n.

$$a^m \div a^n = \frac{a^m}{a^n} = \frac{a \cdot a \cdot a \ldots \text{ to } m \text{ factors}}{a \cdot a \cdot a \ldots \text{ to } n \text{ factors}}$$
$$= a \cdot a \cdot a \ldots \text{ to } m-n \text{ factors}$$
$$= a^{m-n}.$$

235. Prop. III. *To prove that* $(a^m)^n = a^{mn}$, *when* m *and* n *are positive integers.*

$(a^m)^n = a^m \cdot a^m \cdot a^m \ldots$ to n factors
$= (a \cdot a \cdot a \ldots \text{ to } m \text{ factors})(a \cdot a \cdot a \ldots \text{ to } m \text{ factors}) \ldots$
the bracket being repeated n times,
$$= a \cdot a \cdot a \ldots \text{ to } mn \text{ factors}$$
$$= a^{mn}.$$

236. These are the fundamental laws of combination of indices, and they are proved directly from a definition which is intelligible only on the supposition that the indices are *positive* and *integral*.

But it is found convenient to use fractional and negative indices, such as $a^{\frac{4}{5}}$, a^{-7}, or, more generally, $a^{\frac{p}{q}}$, a^{-n}; and these have at present no intelligible meaning. For it is plain that the definition of a^m, [Art. 232], upon which we based the three propositions just proved, is no longer applicable when m is *fractional*, or *negative*.

Now it is important that all indices, whether positive or negative, integral or fractional, should be governed by the same laws. We therefore determine meanings for symbols such as $a^{\frac{p}{q}}$, a^{-n}, in the following way: we assume that they conform to the fundamental law, $a^m \times a^n = a^{m+n}$, and accept the meaning to which this assumption leads us. It will be found that the symbols so interpreted will also obey the other laws enunciated in Props. II. and III.

237. *To find a meaning for* $a^{\frac{p}{q}}$, p *and* q *being positive integers.*

Since $a^m \times a^n = a^{m+n}$ is to be true for *all* values of m and n, by replacing each of the indices m and n by $\frac{p}{q}$, we have

$$a^{\frac{p}{q}} \times a^{\frac{p}{q}} = a^{\frac{p}{q} + \frac{p}{q}} = a^{\frac{2p}{q}}.$$

Similarly, $a^{\frac{p}{q}} \times a^{\frac{p}{q}} \times a^{\frac{p}{q}} = a^{\frac{2p}{q}} \times a^{\frac{p}{q}} = a^{\frac{2p}{q}+\frac{p}{q}} = a^{\frac{3p}{q}}$.

Proceeding in this way for 4, 5, ... q factors, we have
$$a^{\frac{p}{q}} \times a^{\frac{p}{q}} \times a^{\frac{p}{q}} \ldots\ldots \text{to } q \text{ factors} = a^{\frac{qp}{q}};$$

that is, $(a^{\frac{p}{q}})^q = a^p.$

Therefore, by taking the qth root,
$$a^{\frac{p}{q}} = \sqrt[q]{a^p}.$$

or, in words, $a^{\frac{p}{q}}$ is equal to "the qth root of a^p."

Examples. (1) $x^{\frac{5}{7}} = \sqrt[7]{x^5}$,

(2) $a^{\frac{1}{3}} = \sqrt[3]{a}.$

(3) $4^{\frac{3}{2}} = \sqrt{4^3} = \sqrt{64} = 8.$

(4) $a^{\frac{2}{3}} \times a^{\frac{5}{6}} = a^{\frac{2}{3}+\frac{5}{6}} = a^{\frac{3}{2}}.$

(5) $k^{\frac{a}{2}} \times k^{\frac{2}{3}} = k^{\frac{a}{2}+\frac{2}{3}} = k^{\frac{3a+4}{6}}.$

(6) $3a^{\frac{2}{3}}b^{\frac{1}{2}} \times 4a^{\frac{1}{6}}b^{\frac{5}{6}} = 12a^{\frac{2}{3}+\frac{1}{6}}b^{\frac{1}{2}+\frac{5}{6}} = 12a^{\frac{5}{6}}b^{\frac{4}{3}}.$

238. *To find a meaning for* a^0.

Since $a^m \times a^n = a^{m+n}$ is to be true for *all* values of m and n, by replacing the index m by 0, we have
$$a^0 \times a^n = a^{0+n}$$
$$= a^n;$$
$$\therefore a^0 = \frac{a^n}{a^n}$$
$$= 1.$$

Hence *any quantity* with zero index is equivalent to 1.

Example. $x^{b-c} \times x^{c-b} = x^{b-c+c-b} = x^0 = 1.$

239. *To find a meaning for* a^{-n}.

Since $a^m \times a^n = a^{m+n}$ is to be true for *all* values of m and n, by replacing the index m by $-n$, we have
$$a^{-n} \times a^n = a^{-n+n} = a^0.$$
But $a^0 = 1;$

THE THEORY OF INDICES.

hence
$$a^{-n} = \frac{1}{a^n},$$

and
$$a^n = \frac{1}{a^{-n}}.$$

From this it follows that any *factor* may be transferred from the numerator to the denominator of an expression, or vice-versâ, by merely changing the sign of the index.

Examples. (1) $x^{-3} = \frac{1}{x^3}$.

(2) $\frac{1}{y^{-\frac{1}{2}}} = y^{\frac{1}{2}} = \sqrt{y}$.

(3) $27^{-\frac{2}{3}} = \frac{1}{27^{\frac{2}{3}}} = \frac{1}{\sqrt[3]{(27)^2}} = \frac{1}{\sqrt[3]{3^6}} = \frac{1}{3^2} = \frac{1}{9}$.

240. *To prove that* $a^m \div a^n = a^{m-n}$ *for all values of* m *and* n.

$$a^m \div a^n = a^m \times \frac{1}{a^n}$$
$$= a^m \times a^{-n}$$
$$= a^{m-n}, \text{ by the fundamental law.}$$

Examples. (1) $a^3 \div a^5 = a^{3-5} = a^{-2} = \frac{1}{a^2}$.

(2) $c \div c^{-\frac{3}{6}} = c^{1+\frac{3}{6}} = c^{\frac{13}{6}}$.

(3) $x^{a-b} \div x^{a-c} = x^{a-b-(a-c)} = x^{c-b}$.

241. The method of finding a meaning for a symbol, as explained in the preceding articles, deserves careful attention. The usual algebraical process is to make choice of symbols, give them meanings, and then prove the rules for their combination. Here the process is reversed; the symbols are given, and the law to which they are to conform, and from this the meanings of the symbols are determined.

242. The following examples will illustrate the different principles we have established.

Examples. (1) $\frac{3a^{-2}}{5x^{-1}y} = \frac{3x}{5a^2y}$.

(2) $\frac{2a^{\frac{1}{2}} \times a^{\frac{2}{3}} \times 6a^{-\frac{7}{3}}}{9a^{-\frac{5}{3}} \times a^{\frac{3}{2}}} = \frac{4}{3}a^{\frac{1}{2}+\frac{2}{3}-\frac{7}{3}+\frac{5}{3}-\frac{3}{2}} = \frac{4}{3}a^{-1} = \frac{4}{3a}$.

(3) $\dfrac{\sqrt{x^3} \times \sqrt[3]{y^2}}{\sqrt[6]{y^{-2}} \times \sqrt[4]{x^6}} = \dfrac{x^{\frac{3}{2}} \times y^{\frac{2}{3}}}{y^{-\frac{1}{3}} \times x^{\frac{3}{2}}} = x^{\frac{3}{2}-\frac{3}{2}} y^{\frac{2}{3}+\frac{1}{3}} = x^0 y = y.$

(4) $2\sqrt{a} + \dfrac{3}{a^{-\frac{1}{2}}} + a^{\frac{5}{2}} = 2a^{\frac{1}{2}} + 3a^{\frac{1}{2}} + a^{\frac{5}{2}}$

$\qquad = 5a^{\frac{1}{2}} + a^{\frac{5}{2}} = a^{\frac{1}{2}}(5 + a^2).$

EXAMPLES XXX. a.

Express with positive indices:

1. $2x^{-\frac{1}{4}}.$ 2. $3a^{-\frac{2}{3}}.$ 3. $4x^{-2}a^3.$ 4. $3 \div a^{-2}.$

5. $\dfrac{1}{4a^{-2}}.$ 6. $\dfrac{1}{5x^{-\frac{1}{2}}}.$ 7. $\dfrac{3a^{-3}x^2}{5y^2c^{-4}}.$ 8. $\dfrac{x^a y^{-b}}{b^{-a}}.$

9. $2x^{\frac{1}{2}} \times 3x^{-1}.$ 10. $1 \div 2a^{-\frac{1}{2}}.$ 11. $xy^2 \times x^{-1}.$ 12. $a^{-2}x^{-1} \div 3x.$

13. $\dfrac{1}{\sqrt{x^3}}.$ 14. $\dfrac{1}{4\sqrt[5]{x^{-3}}}.$ 15. $\dfrac{2}{\sqrt{y^{-3}}}.$ 16. $\dfrac{\sqrt[4]{x^3}}{\sqrt{x^{-1}}}.$

17. $a^{-2}x^{-\frac{1}{2}} \div a^{-3}.$ 18. $\sqrt[3]{a^{-1}} \div \sqrt[3]{a}.$ 19. $\sqrt[5]{a^{-3}} \div \sqrt[5]{a^7}.$

Express with radical signs and positive indices:

20. $x^{\frac{3}{5}}.$ 21. $a^{-\frac{1}{2}}.$ 22. $5x^{-\frac{1}{2}}.$ 23. $2a^{-\frac{1}{x}}.$

24. $\dfrac{1}{2a^{\frac{1}{5}}}.$ 25. $\dfrac{2}{b^{-\frac{3}{4}}}.$ 26. $\dfrac{c^{-\frac{1}{3}}}{2}.$ 27. $\dfrac{1}{x^{-\frac{1}{x}}}.$

28. $a^{-\frac{1}{5}} \times 2a^{-\frac{1}{2}}.$ 29. $x^{-\frac{2}{3}} \div 2a^{-\frac{1}{2}}.$ 30. $7a^{-\frac{1}{2}} \times 3a^{-1}.$

31. $\dfrac{2a^{-2}}{a^{-\frac{2}{3}}}.$ 32. $\dfrac{a^{-\frac{1}{2}}}{3a}.$ 33. $\dfrac{4x^{-1}}{x^{-\frac{1}{3}}}.$ 34. $\dfrac{\sqrt[3]{x^{-a}}}{\sqrt[3]{x^2}}.$

35. $\sqrt[3]{a^2} \times \sqrt[2]{a^3}.$ 36. $\sqrt[5]{a^{-x}} \div \sqrt[5]{a^{-2x}}.$ 37. $\sqrt[2a]{x} \times \sqrt[a]{x^2}.$

38. $\sqrt[a]{x} \div \sqrt[2a]{x^3}.$ 39. $\sqrt[3x]{a^3} \div \sqrt[x]{a^2}.$ 40. $\sqrt[4]{a^n} \times \sqrt[3]{a^n} \div \sqrt[12]{a^{5n}}.$

Find the value of

41. $16^{\frac{3}{4}}$. 42. $4^{-\frac{5}{2}}$. 43. $125^{\frac{2}{3}}$. 44. $8^{-\frac{2}{3}}$. 45. $36^{-\frac{3}{2}}$.

46. $\dfrac{1}{25^{-2}}$. 47. $243^{\frac{2}{5}}$. 48. $\left(\dfrac{8}{27}\right)^{-\frac{1}{3}}$. 49. $\left(\dfrac{81}{16}\right)^{\frac{3}{4}}$. 50. $\left(\dfrac{32}{243}\right)^{-\frac{7}{5}}$.

*243. *To prove that* $(a^m)^n = a^{mn}$ *is universally true for all values of* m *and* n.

CASE I. Let n be a *positive integer*.

Now, *whatever be the value of* m

$$(a^m)^n = a^m \cdot a^m \cdot a^m \ldots \text{ to } n \text{ factors}$$
$$= a^{m+m+m+\ldots \text{ to } n \text{ terms}}$$
$$= a^{mn}.$$

CASE II. Let m be unrestricted as before, and let n be a *positive fraction*. Replacing n by $\dfrac{p}{q}$, where p and q are *positive integers*, we have $(a^m)^n = (a^m)^{\frac{p}{q}}$.

Now the qth power of $(a^m)^{\frac{p}{q}} = \{(a^m)^{\frac{p}{q}}\}^q$

$$= (a^m)^{\frac{p}{q} \cdot q}, \qquad \text{[Case I.]}$$
$$= (a^m)^p$$
$$= a^{mp}. \qquad \text{[Case I.]}$$

Hence by taking the qth root of these equals,

$$(a^m)^{\frac{p}{q}} = \sqrt[q]{a^{mp}}$$
$$= a^{\frac{mp}{q}}. \qquad \text{[Art. 237.]}$$

CASE III. Let m be unrestricted as before, and let n be *any negative quantity*. Replacing n by $-r$, where r is *positive*, we have

$$(a^m)^n = (a^m)^{-r} = \dfrac{1}{(a^m)^r}, \qquad \text{[Art. 239.]}$$
$$= \dfrac{1}{a^{mr}}, \qquad \text{[Case II.]}$$
$$= a^{-mr} = a^{mn}.$$

Hence Prop. III., Art. 235, $(a^m)^n = a^{mn}$ has been shewn to be universally true.

E. A. R

Examples. (1) $(b^{\frac{2}{3}})^{\frac{6}{7}} = b^{\frac{2}{3} \times \frac{6}{7}} = b^{\frac{4}{7}}$.

(2) $\{(x^{-2})^3\}^{-4} = (x^{-6})^{-4} = x^{24}$.

(3) $\left(x^{\frac{1}{a-c}}\right)^{a^2-c^2} = x^{\frac{1}{a-c} \times (a^2-c^2)} = x^{a+c}$.

*244. *To prove that* $(ab)^n = a^n b^n$, *whatever be the value of* n; a *and* b *being any quantities whatever.*

CASE I. Let n be a *positive integer*.

Now $(ab)^n = ab \cdot ab \cdot ab \ldots\ldots$ to n factors

$\qquad = (a \cdot a \cdot a \ldots$ to n factors$)(b \cdot b \cdot b \ldots$ to n factors$)$

$\qquad = a^n b^n$.

CASE II. Let n be a *positive fraction*. Replacing n by $\dfrac{p}{q}$, where p and q are positive integers, we have $(ab)^n = (ab)^{\frac{p}{q}}$.

Now the qth power of $(ab)^{\frac{p}{q}} = \{(ab)^{\frac{p}{q}}\}^q$

$\qquad\qquad = (ab)^p$, [Art. 243.]

$\qquad\qquad = a^p b^p$

$\qquad\qquad = (a^{\frac{p}{q}} b^{\frac{p}{q}})^q$. [Case I.]

Taking the qth root, $(ab)^{\frac{p}{q}} = a^{\frac{p}{q}} b^{\frac{p}{q}}$.

CASE III. Let n have *any negative value*. Replacing n by $-r$, where r is positive,

$$(ab)^n = (ab)^{-r} = \frac{1}{(ab)^r}$$

$$= \frac{1}{a^r b^r} = a^{-r} b^{-r}$$

$$= a^n b^n.$$

Hence the proposition is proved universally.

The result we have just proved may be expressed in a verbal form by saying that the index of a product may be *distributed* over its *factors*.

Note. An index is not distributive over the *terms* of an expression. Thus $(a^{\frac{1}{2}} + b^{\frac{1}{2}})^2$ is not equal to $a + b$. Again $(a^2 + b^2)^{\frac{1}{2}}$ is equal to $\sqrt{a^2 + b^2}$, and cannot be further simplified.

XXX.] THE THEORY OF INDICES. 247

Examples. (1) $(yz)^{a-c}(zx)^c(xy)^{-c} = y^{a-c}z^{a-c}z^c x^c x^{-c} y^{-c}$
$$= y^{a-2c} z^a.$$

(2) $\{(a-b)^k\}^{-l} \times \{(a+b)^{-k}\}^l = (a-b)^{-kl} \times (a+b)^{-kl}$
$$= \{(a-b)(a+b)\}^{-kl}$$
$$= (a^2 - b^2)^{-kl}.$$

*245. It should be observed that in the proof of Art. 244 the quantities a and b are *wholly unrestricted*, and may themselves involve indices.

Examples. (1) $(x^{\frac{1}{2}}y^{-\frac{1}{2}})^{\frac{4}{3}} \div (x^2 y^{-1})^{-\frac{1}{3}} = x^{\frac{2}{3}} y^{-\frac{2}{3}} \div x^{-\frac{2}{3}} y^{\frac{1}{3}}$
$$= x^{\frac{4}{3}} y^{-1}.$$

(2) $\left(\dfrac{a^{\frac{2}{3}}\sqrt{b^{-1}}}{b\sqrt[3]{a^{-2}}} \div \sqrt{\dfrac{a\sqrt{b^{-4}}}{b\sqrt{a^{-2}}}}\right)^6 = \left(\dfrac{a^{\frac{2}{3}}b^{-\frac{1}{2}}}{ba^{-\frac{2}{3}}} \div \sqrt{\dfrac{ab^{-2}}{ba^{-1}}}\right)^6$
$$= (a^{\frac{4}{3}} b^{-\frac{3}{2}} \div \sqrt{a^2 b^{-3}})^6$$
$$= (a^{\frac{4}{3}} b^{-\frac{3}{2}} \div ab^{-\frac{3}{2}})^6$$
$$= (a^{\frac{1}{3}})^6 = a^2.$$

EXAMPLES XXX. b.

Simplify and express with positive indices:

1. $(\sqrt{a^2 b^3})^6$.
2. $(\sqrt[9]{x^{-4} y^3})^{-3}$.
3. $(x^a y^{-b})^3 \times (x^3 y^2)^{-a}$.
4. $\left(\dfrac{16 x^2}{y^{-2}}\right)^{-\frac{1}{4}}$.
5. $\left(\dfrac{27 x^3}{8 a^{-3}}\right)^{-\frac{2}{3}}$.
6. $\left(\dfrac{a^{-\frac{1}{2}}}{4 c^2}\right)^{-2}$.
7. $\left\{\sqrt[4]{(x^{-\frac{2}{3}} y^{\frac{1}{2}})^3}\right\}^{-\frac{2}{3}}$.
8. $\sqrt[4]{x \sqrt[3]{x^{-1}}}$.
9. $(4a^{-2} \div 9x^2)^{-\frac{1}{2}}$.
10. $(x \div \sqrt[n]{x})^n$.
11. $\left(x \times \sqrt[n]{x^{-\frac{1}{n}}}\right)^{\frac{n^2}{1-n}}$.
12. $(\sqrt[b]{x^b} \div \sqrt[a]{x})^{\frac{1}{1-a}}$.
13. $\sqrt{a^{-2}b} \times \sqrt[3]{ab^{-3}}$.
14. $\sqrt[3]{ab^{-1}c^{-2}} \times (a^{-1}b^{-2}c^{-4})^{-\frac{1}{6}}$.
15. $\sqrt[6]{a^{4b}x^6} \times (a^{\frac{2}{3}} x^{-1})^{-b}$.
16. $\sqrt[3]{x^{-1}\sqrt{y^3}} \div \sqrt{y\sqrt[3]{x}}$.

Simplify and express with positive indices:

17. $(a^{-\frac{1}{2}}\sqrt[3]{x})^{-3} \times \sqrt{x^{-2}\sqrt{a^{-6}}}$.

18. $\sqrt[n]{a^{n+k}b^{2n-k}} \div (a^{\frac{1}{n}}b^{-\frac{1}{n}})^k$.

19. $\sqrt[3]{(a+b)^5} \times (a+b)^{-\frac{2}{3}}$.

20. $\{(x-y)^{-3}\}^n \div \{(x+y)^n\}^3$.

21. $\left(\dfrac{a^{-2}b}{a^3b^{-4}}\right)^{-3} \div \left(\dfrac{ab^{-1}}{a^{-3}b^2}\right)^5$.

22. $\left\{\dfrac{\sqrt[3]{a}}{\sqrt[4]{b^{-1}}} \cdot \left(\dfrac{b^{\frac{1}{4}}}{a^{\frac{1}{3}}}\right)^2 \div \dfrac{a^{-\frac{1}{3}}}{b^{-\frac{1}{2}}}\right\}^6$.

23. $\left(a^{-\frac{1}{2}}x^{\frac{1}{3}}\sqrt{ax^{-\frac{1}{3}}\sqrt[4]{x^{\frac{4}{3}}}}\right)^{\frac{1}{3}}$.

24. $\sqrt[4]{(a+b)^6} \times (a^2 - b^2)^{-\frac{1}{2}}$.

25. $\left(\dfrac{a^{-3}}{b^{-\frac{2}{3}}c}\right)^{-\frac{3}{2}} \div \left(\dfrac{\sqrt{a^{-\frac{1}{2}}}\sqrt[6]{b^3}}{a^2c^{-1}}\right)^{-2}$.

26. $\left(\dfrac{a^{-\frac{2}{3}}x^{\frac{1}{2}}}{x^{-1}a}\right)^2 \div \sqrt[3]{\dfrac{a^{-1}}{x^{-3}}}$.

27. $\left(\sqrt[5]{\dfrac{a^{\frac{1}{2}}x^{-2}}{x^{\frac{1}{2}}a^{-2}}} \times \sqrt[3]{\dfrac{a\sqrt{x}}{x^{-1}\sqrt{a}}}\right)^{-4}$.

28. $\dfrac{\sqrt[3]{(a^3b^3 + a^6)}}{\sqrt[3]{(b^6 - a^3b^3)^{-1}}}$.

29. $(a^{n^2-1})^{\frac{n}{n+1}} + \dfrac{\sqrt[n]{a^{2n}}}{a}$.

30. $\left(x^{\frac{n}{n+1}}\right)^{n^2-1} + \dfrac{\sqrt{x^{2n}}}{x}$.

31. $\left\{\dfrac{a^{p-q}}{\sqrt[q]{a^{q^2-pq}}} \times a^{2(p-q)}\right\}^n$.

32. $(x^{\frac{a}{b}}y^{-1})^b \div \left(\dfrac{x^{a^2-b^2}}{y^{ab+b^2}}\right)^{\frac{1}{a+b}}$.

33. $\left(\dfrac{x^{-2}y^3}{x^3y^{-2}}\right)^{-\frac{1}{5}} \times \left(\dfrac{y^3x^{-3}}{x^3y^{-3}}\right)^{-1}$.

34. $\left(\dfrac{y^{-3}}{x^{\frac{2}{7}}z^{-1}}\right)^{-\frac{3}{2}} \times \left(\dfrac{y^{\frac{14}{3}}x^{-1}}{z^{-\frac{21}{4}}}\right)^{\frac{2}{7}}$.

35. $\dfrac{2^n \times (2^{n-1})^n}{2^{n+1} \times 2^{n-1}} \times \dfrac{1}{4^{-n}}$.

36. $\dfrac{2^{n+1}}{(2^n)^{n-1}} \div \dfrac{4^{n+1}}{(2^{n-1})^{n+1}}$.

37. $\dfrac{3 \cdot 2^n - 4 \cdot 2^{n-2}}{2^n - 2^{n-1}}$.

38. $\dfrac{3^{n+4} - 6 \cdot 3^{n+1}}{3^{n+2} \times 7}$.

246. Since the index-laws are universally true, all the ordinary operations of multiplication, division, involution and evolution are applicable to expressions which contain fractional and negative indices.

247. In Art. 121, we pointed out that the descending powers of x are

$$\ldots\ldots\ x^3,\ x^2,\ x,\ 1,\ \dfrac{1}{x},\ \dfrac{1}{x^2},\ \dfrac{1}{x^3},\ \ldots\ldots$$

A reason for this may be seen if we write these terms in the form

$$\ldots\ldots\ x^3,\ x^2,\ x^1,\ x^0,\ x^{-1},\ x^{-2},\ x^{-3},\ \ldots\ldots$$

Example 1. Multiply $3x^{-\frac{1}{3}} + x + 2x^{\frac{2}{3}}$ by $x^{\frac{1}{3}} - 2$.

Arrange in descending powers of x.

$$x + 2x^{\frac{2}{3}} + 3x^{-\frac{1}{3}}$$
$$x^{\frac{1}{3}} - 2$$
$$\overline{x^{\frac{4}{3}} + 2x\ \ + 3}$$
$$-2x\ \ -4x^{\frac{2}{3}} - 6x^{-\frac{1}{3}}$$
$$\overline{x^{\frac{4}{3}} - 4x^{\frac{2}{3}} + 3\ \ -6x^{-\frac{1}{3}}.}$$

Example 2. Divide $16a^{-3} - 6a^{-2} + 5a^{-1} + 6$ by $1 + 2a^{-1}$.

$$2a^{-1} + 1\ )\ 16a^{-3} - 6a^{-2} + 5a^{-1} + 6\ (\ 8a^{-2} - 7a^{-1} + 6$$
$$\underline{16a^{-3} + 8a^{-2}}$$
$$-14a^{-2} + 5a^{-1}$$
$$\underline{-14a^{-2} - 7a^{-1}}$$
$$12a^{-1} + 6$$
$$\underline{12a^{-1} + 6}$$

Example 3. Find the square root of

$$\frac{4x^2}{y} + \frac{\sqrt{x^3}}{y^{-\frac{1}{2}}} - 2x + \frac{y}{4} + x^3 - 4\sqrt{(x^5 y^{-1})}.$$

Getting rid of the radical signs, and arranging in descending powers of x, we have

$$\begin{array}{r}x^3 - 4x^{\frac{5}{2}}y^{-\frac{1}{2}} + 4x^2y^{-1} + x^{\frac{3}{2}}y^{\frac{1}{2}} - 2x + \frac{y}{4}\ \Big(\ x^{\frac{3}{2}} - 2xy^{-\frac{1}{2}} + \frac{y^{\frac{1}{2}}}{2}\\ x^3\end{array}$$

$$2x^{\frac{3}{2}} - 2xy^{-\frac{1}{2}}\ \Big|\ -4x^{\frac{5}{2}}y^{-\frac{1}{2}} + 4x^2y^{-1}$$
$$\phantom{2x^{\frac{3}{2}} - 2xy^{-\frac{1}{2}}\ \Big|\ }-4x^{\frac{5}{2}}y^{-\frac{1}{2}} + 4x^2y^{-1}$$

$$2x^{\frac{3}{2}} - 4xy^{-\frac{1}{2}} + \frac{y^{\frac{1}{2}}}{2}\ \Big|\ x^{\frac{3}{2}}y^{\frac{1}{2}} - 2x + \frac{y}{4}$$
$$\phantom{2x^{\frac{3}{2}} - 4xy^{-\frac{1}{2}} + \frac{y^{\frac{1}{2}}}{2}\ \Big|\ }x^{\frac{3}{2}}y^{\frac{1}{2}} - 2x + \frac{y}{4}$$

Note. In this example it should be observed that the introduction of negative indices enables us to avoid the use of algebraical fractions.

EXAMPLES XXX. c.

1. Multiply $3x^{\frac{1}{3}} - 5 + 8x^{-\frac{1}{3}}$ by $4x^{\frac{1}{3}} + 3x^{-\frac{1}{3}}$.

2. Multiply $3a^{\frac{2}{5}} - 4a^{\frac{1}{5}} - a^{-\frac{1}{5}}$ by $3a^{\frac{1}{5}} + a^{-\frac{1}{5}} - 6a^{-\frac{2}{5}}$.

3. Find the product of $c^x + 2c^{-x} - 7$ and $5 - 3c^{-x} + 2c^x$.

4. Find the product of $5 + 2x^{2a} + 3x^{-2a}$ and $4x^a - 3x^{-a}$.

5. Divide $21x + x^{\frac{2}{3}} + x^{\frac{1}{3}} + 1$ by $3x^{\frac{1}{3}} + 1$.

6. Divide $15a - 3a^{\frac{1}{3}} - 2a^{-\frac{1}{3}} + 8a^{-1}$ by $5a^{\frac{2}{3}} + 4$.

7. Divide $16a^{-3} + 6a^{-2} + 5a^{-1} - 6$ by $2a^{-1} - 1$.

8. Divide $5h^{\frac{2}{3}} - 6h^{\frac{1}{3}} - 4h^{-\frac{2}{3}} - 4h^{-\frac{1}{3}} - 5$ by $h^{\frac{1}{3}} - 2h^{-\frac{1}{3}}$.

9. Divide $21a^{3x} + 20 - 27a^x - 26a^{2x}$ by $3a^x - 5$.

10. Divide $8c^{-n} - 8c^n + 5c^{3n} - 3c^{-3n}$ by $5c^n - 3c^{-n}$.

Find the square root of

11. $9x - 12x^{\frac{1}{2}} + 10 - 4x^{-\frac{1}{2}} + x^{-1}$.

12. $25a^{\frac{4}{3}} + 16 - 30a - 24a^{\frac{1}{3}} + 49a^{\frac{2}{3}}$.

13. $4x^n + 9x^{-n} + 28 - 24x^{-\frac{n}{2}} - 16x^{\frac{n}{2}}$.

14. $12a^x + 4 - 6a^{3x} + a^{4x} + 5a^{2x}$.

15. Multiply $a^{\frac{3}{2}} - 8a^{-\frac{3}{2}} + 4a^{-\frac{1}{2}} - 2a^{\frac{1}{2}}$ by $4a^{-\frac{3}{2}} + a^{\frac{1}{2}} + 4a^{-\frac{1}{2}}$.

16. Multiply $1 - 2\sqrt[3]{x} - 2x^{\frac{1}{2}}$ by $1 - \sqrt[6]{x}$.

17. Multiply $2\sqrt[3]{a^5} - a^{\frac{1}{3}} - \dfrac{3}{a}$ by $2a - 3\sqrt[3]{\dfrac{1}{a}} - a^{-\frac{5}{3}}$.

18. Divide $\sqrt[3]{x^2} + 2x^{\frac{1}{3}} - 16x^{-\frac{2}{3}} - \dfrac{32}{x}$ by $x^{\frac{1}{3}} + 4x^{-\frac{1}{3}} + \dfrac{4}{\sqrt{x}}$.

19. Divide $1 - \sqrt{a} - \dfrac{2}{a^{-1}} + 2a^2$ by $1 - a^{\frac{1}{2}}$.

20. Divide $4\sqrt[3]{x^2} - 8x^{\frac{1}{3}} - 5 + \dfrac{10}{\sqrt[3]{x}} + 3x^{-\frac{2}{3}}$ by $2x^{\frac{5}{12}} - \sqrt[12]{x} - \dfrac{3}{\sqrt[4]{x}}$.

THE THEORY OF INDICES.

Find the square root of

21. $9x^{-4} - 18x^{-3}\sqrt{y} + \dfrac{15y}{x^2} - 6\sqrt{\left(\dfrac{y^3}{x^2}\right)} + y^2$.

22. $4\sqrt{x^3} - 12\sqrt[4]{(x^3y)} + 25\sqrt{y} - 24\sqrt[4]{\left(\dfrac{y^3}{x^3}\right)} + 16x^{-\frac{3}{2}}y$.

23. $81\left(\dfrac{\sqrt[3]{x^4}}{y^2} + 1\right) + 36\dfrac{x^{\frac{1}{3}}}{\sqrt{y}}(x^{\frac{2}{3}}y^{-1} - 1) - 158\dfrac{\sqrt[3]{x^2}}{y}$.

24. $\dfrac{x^{-2}}{16} + 1 + \dfrac{9}{\sqrt[3]{y^{-2}}} + \dfrac{1 - 3\sqrt[3]{y}}{2x} - 6\sqrt[3]{y}$.

248. The following examples will illustrate the formulæ of earlier chapters when applied to expressions involving fractional and negative indices.

Example 1. $(a^{\frac{h}{k}} - b^{\frac{p}{q}})(a^{-\frac{h}{k}} + b^{-\frac{p}{q}}) = a^{\frac{h}{k} - \frac{h}{k}} - a^{-\frac{h}{k}}b^{\frac{p}{q}} + a^{\frac{h}{k}}b^{-\frac{p}{q}} - b^{\frac{p}{q} - \frac{p}{q}}$

$$= 1 - a^{-\frac{h}{k}}b^{\frac{p}{q}} + a^{\frac{h}{k}}b^{-\frac{p}{q}} - 1$$

$$= a^{\frac{h}{k}}b^{-\frac{p}{q}} - a^{-\frac{h}{k}}b^{\frac{p}{q}}.$$

Example 2. Multiply $2x^{2p} - x^p + 3$ by $2x^{2p} + x^p - 3$.

The product $= \{2x^{2p} - (x^p - 3)\}\{2x^{2p} + (x^p - 3)\}$

$$= (2x^{2p})^2 - (x^p - 3)^2$$

$$= 4x^{4p} - x^{2p} + 6x^p - 9.$$

Example 3. The square of $3x^{\frac{1}{2}} - 2 - x^{-\frac{1}{2}}$

$$= 9x + 4 + x^{-1} - 2 \cdot 3x^{\frac{1}{2}} \cdot 2 - 2 \cdot 3x^{\frac{1}{2}} \cdot x^{-\frac{1}{2}} + 2 \cdot 2 \cdot x^{-\frac{1}{2}}$$

$$= 9x + 4 + x^{-1} - 12x^{\frac{1}{2}} - 6 + 4x^{-\frac{1}{2}}$$

$$= 9x - 12x^{\frac{1}{2}} - 2 + 4x^{-\frac{1}{2}} + x^{-1},$$

by collecting like terms and rearranging.

Example 4. Divide $a^{\frac{3n}{2}} + a^{-\frac{3n}{2}}$ by $a^{\frac{n}{2}} + a^{-\frac{n}{2}}$.

The quotient $= (a^{\frac{3n}{2}} + a^{-\frac{3n}{2}}) \div (a^{\frac{n}{2}} + a^{-\frac{n}{2}})$

$$= \{(a^{\frac{n}{2}})^3 + (a^{-\frac{n}{2}})^3\} \div (a^{\frac{n}{2}} + a^{-\frac{n}{2}})$$

$$= (a^{\frac{n}{2}})^2 - a^{\frac{n}{2}} \cdot a^{-\frac{n}{2}} + (a^{-\frac{n}{2}})^2$$

$$= a^n - 1 + a^{-n}.$$

EXAMPLES XXX. d.

Write down the value of

1. $(x^{\frac{1}{2}} - 7)(x^{\frac{1}{2}} + 3)$.
2. $(4x - 5x^{-1})(4x + 3x^{-1})$.
3. $(7x - 9y^{-1})(7x + 9y^{-1})$.
4. $(x^m - y^n)(x^{-m} + y^{-n})$.
5. $(a^x - 2a^{-x})^2$.
6. $(a^x + a^{\frac{1}{x}})^2$.
7. $\left(x^{\frac{a}{2}} - \dfrac{1}{2}x^{-a}\right)^2$.
8. $(5x^a y^b - 3x^{-a} y^{-b})(4x^a y^b + 5x^{-a} y^{-b})$.
9. $\left(\dfrac{1}{3}a^{\frac{1}{3}} - a^{-\frac{1}{3}}\right)^2$.
10. $(3x^a y^{-b} + 5x^{-a} y^b)(3x^a y^b - 5x^{-a} y^{-b})$.
11. $\left(a^x - \dfrac{1}{2} - a^{-x}\right)^2$.
12. $(x^{\frac{1}{a}} - x^{-\frac{1}{a}} + x)^2$.
13. $\{(a+b)^{\frac{1}{2}} + (a-b)^{\frac{1}{2}}\}^2$.
14. $\{(a+b)^{\frac{1}{2}} - (a-b)^{-\frac{1}{2}}\}^2$.

Write down the quotient of

15. $x - 9a$ by $x^{\frac{1}{2}} + 3a^{\frac{1}{2}}$.
16. $x^{\frac{3}{2}} - 27$ by $x^{\frac{1}{2}} - 3$.
17. $a^{2x} - 16$ by $a^x - 4$.
18. $x^{3a} + 8$ by $x^a + 2$.
19. $c^{2x} - c^{-x}$ by $c^x - c^{-\frac{x}{2}}$.
20. $1 - 8a^{-3}$ by $1 - 2a^{-1}$.
21. $a^{4x} - x^6$ by $a^{2x} + x^3$.
22. $x^{-4} - 1$ by $x^{-1} + 1$.
23. $x^{\frac{5}{3}} - 1$ by $x^{\frac{1}{3}} - 1$.
24. $x^{5n} + 32$ by $x^n + 2$.

Find the value of

25. $(x + x^{\frac{1}{2}} - 4)(x + x^{\frac{1}{2}} + 4)$.
26. $(2x^{\frac{1}{3}} + 4 + 3x^{-\frac{1}{3}})(2x^{\frac{1}{3}} + 4 - 3x^{-\frac{1}{3}})$.
27. $(2 - x^{\frac{1}{3}} + x)(2 + x^{\frac{1}{3}} + x)$.
28. $(a^x + 7 + 3a^{-x})(a^x - 7 - 3a^{-x})$.
29. $\dfrac{a^{\frac{4}{3}} - 8a^{\frac{1}{3}}b}{a^{\frac{2}{3}} + 2\sqrt[3]{ab} + 4b^{\frac{2}{3}}}$.
30. $\dfrac{x - 7x^{\frac{1}{2}}}{x - 5\sqrt{x} - 14} \div \left(1 + \dfrac{2}{\sqrt{x}}\right)^{-1}$.
31. $\dfrac{x^{\frac{2}{3}} - 4\sqrt[3]{x^{-2}}}{\sqrt[3]{x^2} + 4 + 4x^{-\frac{2}{3}}}$.
32. $\dfrac{a^{\frac{3}{2}} + ab}{ab - b^3} - \dfrac{\sqrt{a}}{\sqrt{a - b}}$.

CHAPTER XXXI.

ELEMENTARY SURDS.

249. DEFINITION. If the root of a quantity cannot be exactly obtained the root is called a **surd**.

Thus $\sqrt{2}$, $\sqrt[3]{5}$, $\sqrt[5]{a^3}$, $\sqrt{a^2+b^2}$ are surds.

By reference to the preceding chapter it will be seen that these are only cases of fractional indices; for the above quantities might be written

$$2^{\frac{1}{2}},\ 5^{\frac{1}{3}},\ a^{\frac{3}{5}},\ (a^2+b^2)^{\frac{1}{2}}.$$

Since surds may always be expressed as quantities with fractional indices they are subject to the same laws of combination as other algebraical symbols.

250. A quantity may be expressed in a surd form without really being a surd. Thus $\sqrt[3]{x^6}$ or $x^{\frac{6}{3}}$, though apparently a surd, can be expressed in the equivalent form x^2.

251. A surd is sometimes called an **irrational quantity**: and quantities which are not surds are, for the sake of distinction, termed **rational quantities**.

252. In the case of numerical surds such as $\sqrt{2}$, $\sqrt[3]{5}$, ..., although the *exact* value can never be found, it can be determined to any degree of accuracy by carrying the process of evolution far enough.

Thus $\sqrt{5} = 2\cdot 236068\ldots\ldots\ldots$;

that is $\sqrt{5}$ lies between $2\cdot 23606$ and $2\cdot 23607$; and therefore the error in using either of these quantities instead of $\sqrt{5}$ is less than $\cdot 00001$. By taking the root to a greater number of decimal places we can approximate still nearer to the true value.

It thus appears that it will never be *absolutely necessary* to introduce surds into numerical work, which can always be carried on to a certain degree of accuracy; but we shall in the present chapter prove laws for combination of surd quantities which will enable us to work with symbols such as $\sqrt{2}$, $\sqrt[3]{5}$, $\sqrt[4]{a}$, ... with absolute accuracy so long as the symbols are kept in their surd form. Moreover it will be found that even where approximate numerical results are required, the work is considerably simplified and shortened by operating with surd symbols, and afterwards substituting numerical values, if necessary.

253. The *order* of a surd is indicated by the root symbol, or surd index. Thus $\sqrt[3]{x}$, $\sqrt[n]{a}$ are respectively surds of the third and n^{th} orders.

The surds of the most frequent occurrence are those of the second order; they are sometimes called **quadratic surds**. Thus $\sqrt{3}$, $\sqrt{a}$, $\sqrt{x+y}$ are quadratic surds.

254. It will frequently be found convenient to express a rational quantity in a surd form.

A rational quantity may be expressed in the form of a surd of *any required order* by raising it to the power whose root the surd expresses, and prefixing the radical sign. Thus
$$5 = \sqrt{25} = \sqrt[3]{125} = \sqrt[4]{625} = \sqrt[n]{5^n};$$
$$a + x = \sqrt{(a+x)^2} = \sqrt[6]{(a+x)^6} = \sqrt[n]{(a+x)^n}.$$

255. A surd of any order may be transformed into a surd of a different order.

Examples. (1) $\sqrt[3]{2} = 2^{\frac{1}{3}} = 2^{\frac{4}{12}} = \sqrt[12]{2^4}$.

(2) $\sqrt[p]{a} = a^{\frac{1}{p}} = a^{\frac{q}{pq}} = \sqrt[pq]{a^q}$.

256. Surds of different orders may be transformed into surds of the same order. This order may be *any* common multiple of each of the given orders, but it is usually most convenient to choose the *least* common multiple.

Example. Express $\sqrt[4]{a^3}$, $\sqrt[3]{b^2}$, $\sqrt[6]{a^5}$ as surds of the same lowest order.

The least common multiple of 4, 3, 6 is 12; and expressing the given surds as surds of the twelfth order they become $\sqrt[12]{a^9}$, $\sqrt[12]{b^8}$, $\sqrt[12]{a^{10}}$.

257. Surds of different orders may be arranged according to magnitude by transforming them into surds of the same order.

Example. Arrange $\sqrt{3}$, $\sqrt[3]{6}$, $\sqrt[4]{10}$ according to magnitude.

The least common multiple of 2, 3, 4 is 12; and, expressing the given surds as surds of the twelfth order, we have
$$\sqrt{3} = \sqrt[12]{3^6} = \sqrt[12]{729},$$
$$\sqrt[3]{6} = \sqrt[12]{6^4} = \sqrt[12]{1296},$$
$$\sqrt[4]{10} = \sqrt[12]{10^3} = \sqrt[12]{1000}.$$

Hence arranged in ascending order of magnitude the surds are
$$\sqrt{3}, \ \sqrt[4]{10}, \ \sqrt[3]{6}.$$

EXAMPLES XXXI. a.

Express as surds of the twelfth order with positive indices:

1. $x^{\frac{1}{3}}$.
2. $a^{-1} \div a^{-\frac{1}{2}}$.
3. $\sqrt[4]{ax^3} \times \sqrt[3]{a^{-1}x^{-2}}$.
4. $\dfrac{1}{a^{-\frac{3}{4}}}$.
5. $\dfrac{1}{\sqrt[8]{a^{-14}}}$.
6. $\sqrt[6]{\dfrac{1}{a^{-2}}}$.

Express as surds of the nth order with positive indices:

7. $\sqrt[3]{x^2}$.
8. x^a.
9. $a^{\frac{1}{2}}$.
10. $\sqrt{a^{-\frac{1}{n}}}$.
11. $\sqrt[3]{\dfrac{1}{x^n y^n}}$.
12. $\dfrac{1}{a^{-1}}$.
13. $\dfrac{x^{-\frac{1}{2}}}{y^2}$.
14. $\dfrac{a^{\frac{1}{2}}}{x^{-n}}$.

Express as surds of the same lowest order:

15. $\sqrt{a},\ \sqrt[9]{a^5}$.
16. $\sqrt[5]{a^3},\ \sqrt{a}$.
17. $\sqrt[8]{x^3},\ \sqrt[9]{x^6},\ \sqrt[20]{x^5}$.
18. $\sqrt[16]{x^4},\ \sqrt[12]{x^{10}}$.
19. $\sqrt[21]{a^3 b^4},\ \sqrt[7]{ab}$.
20. $\sqrt{ax^2},\ \sqrt[39]{a^9 x^6}$.
21. $\sqrt{5},\ \sqrt[3]{11},\ \sqrt[6]{13}$.
22. $\sqrt[4]{8},\ \sqrt{3},\ \sqrt[8]{6}$.
23. $\sqrt[3]{2},\ \sqrt[9]{8},\ \sqrt[6]{4}$.

258. The root of any expression is equal to the product of the roots of the separate factors of the expression.

For
$$\sqrt[n]{ab} = (ab)^{\frac{1}{n}}$$
$$= a^{\frac{1}{n}} b^{\frac{1}{n}}, \qquad \text{[Art. 244}$$
$$= \sqrt[n]{a} \cdot \sqrt[n]{b}.$$

Similarly, $\sqrt[n]{abc} = \sqrt[n]{a} \cdot \sqrt[n]{b} \cdot \sqrt[n]{c}$;
and so for any number of factors.

Examples. (1) $\sqrt[4]{15} = \sqrt[4]{3} \cdot \sqrt[4]{5}$.

(2) $\sqrt[3]{a^6 b} = \sqrt[3]{a^6} \cdot \sqrt[3]{b} = a^2 \sqrt[3]{b}$.

(3) $\sqrt{50} = \sqrt{25} \cdot \sqrt{2} = 5\sqrt{2}$.

Hence it appears that a surd may sometimes be expressed as the product of a rational quantity and a surd; when so reduced the surd is said to be in its *simplest form*.

Thus the simplest form of $\sqrt{128}$ is $8\sqrt{2}$.

Conversely, the coefficient of a surd may be brought under the radical sign by first reducing it to the form of a surd, and then multiplying the surds together.

Examples. (1) $7\sqrt{5} = \sqrt{49} \cdot \sqrt{5} = \sqrt{245}$.

(2) $a\sqrt[8]{b} = \sqrt[8]{a^8} \cdot \sqrt[8]{b} = \sqrt[8]{a^8 b}$.

When so reduced a surd is said to be an *entire surd*.

259. When surds have, or can be reduced to have, the same irrational factor, they are said to be *like*; otherwise, they are said to be *unlike*. Thus

$$5\sqrt{3},\ 2\sqrt{3},\ \frac{1}{5}\sqrt{3} \text{ are like surds.}$$

But $3\sqrt{2}$ and $2\sqrt{3}$ are unlike surds.

Again, $3\sqrt{20},\ 4\sqrt{5},\ \sqrt{\dfrac{1}{5}}$ are like surds;

for $3\sqrt{20} = 3\sqrt{4} \cdot \sqrt{5} = 3 \cdot 2\sqrt{5} = 6\sqrt{5}$;

and $\sqrt{\dfrac{1}{5}} = \sqrt{\dfrac{5}{25}} = \dfrac{1}{5}\sqrt{5}$.

260. In finding the sum of a number of like surds we reduce them to their simplest form, and prefix to their common irrational part the sum of the coefficients.

Example 1. The sum of $3\sqrt{20},\ 4\sqrt{5},\ \dfrac{1}{\sqrt{5}}$

$$= 6\sqrt{5} + 4\sqrt{5} + \frac{1}{5}\sqrt{5}$$

$$= \frac{51}{5}\sqrt{5}.$$

Example 2. The sum of $x\sqrt[3]{8x^3 a} + y\sqrt[3]{-y^3 a} - z\sqrt[3]{z^3 a}$

$$= x \cdot 2x\sqrt[3]{a} + y(-y)\sqrt[3]{a} - z \cdot z\sqrt[3]{a}$$

$$= (2x^2 - y^2 - z^2)\sqrt[3]{a}.$$

261. Unlike surds cannot be collected.

Thus the sum of $5\sqrt{2},\ -2\sqrt{3}$ and $\sqrt{6}$ is $5\sqrt{2} - 2\sqrt{3} + \sqrt{6}$ and cannot be further simplified.

EXAMPLES XXXI. b.

Express in the simplest form:

1. $\sqrt{288}$.
2. $\sqrt{147}$.
3. $\sqrt[3]{256}$.
4. $\sqrt[3]{432}$.
5. $3\sqrt{150}$.
6. $2\sqrt{720}$.
7. $5\sqrt{245}$.
8. $\sqrt[3]{1029}$.
9. $\sqrt[4]{3125}$.
10. $\sqrt[3]{-2187}$.
11. $\sqrt{36a^3}$.
12. $\sqrt{27a^3 b^5}$.

ELEMENTARY SURDS.

13. $\sqrt[5]{-108x^4y^3}$. 14. $\sqrt[n]{x^{3n}y^{2n+5}}$. 15. $\sqrt[p]{x^{a+p}y^{2p}}$.

16. $\sqrt{a^3 + 2a^2b + ab^2}$. 17. $\sqrt[3]{8x^4y - 24x^3y^2 + 24x^2y^3 - 8xy^4}$.

Express as entire surds:

18. $11\sqrt{2}$. 19. $14\sqrt{5}$. 20. $6\sqrt[3]{4}$. 21. $5\sqrt[3]{6}$.

22. $\dfrac{4}{11}\sqrt{\dfrac{77}{8}}$. 23. $\dfrac{3ab}{2c}\sqrt{\dfrac{20c^2}{9a^2b}}$. 24. $\dfrac{3x}{y}\sqrt{\dfrac{a^2y^3}{x^2}}$.

25. $\dfrac{a}{x^2}\sqrt{\dfrac{3x^3}{a}}$. 26. $\dfrac{2a}{3x}\sqrt[3]{\dfrac{27x^4}{a^2}}$. 27. $\dfrac{2a}{b}\sqrt[4]{\dfrac{b^4}{8a^3}}$.

28. $a\sqrt[n]{\dfrac{b^2}{a^{n-2}}}$. 29. $\dfrac{a}{b}\sqrt[p]{\dfrac{b^{p+1}}{a^{p-1}}}$. 30. $\dfrac{y}{x^n}\sqrt{\dfrac{x^{2n+1}}{y^3}}$.

31. $(x+y)\sqrt{\dfrac{x-y}{x+y}}$. 32. $\dfrac{ax}{a-x}\sqrt{\dfrac{a^2-x^2}{a^2x^2}}$.

Find the value of

33. $3\sqrt{45} - \sqrt{20} + 7\sqrt{5}$. 34. $4\sqrt{63} + 5\sqrt{7} - 8\sqrt{28}$.

35. $\sqrt{44} - 5\sqrt{176} + 2\sqrt{99}$. 36. $2\sqrt{363} - 5\sqrt{243} + \sqrt{192}$.

37. $2\sqrt[3]{189} + 3\sqrt[3]{875} - 7\sqrt[3]{56}$. 38. $5\sqrt[3]{81} - 7\sqrt[3]{192} + 4\sqrt[3]{648}$.

39. $3\sqrt[4]{162} - 7\sqrt[4]{32} + \sqrt[4]{1250}$. 40. $5\sqrt[3]{-54} - 2\sqrt[3]{-16} + 4\sqrt[3]{686}$.

41. $4\sqrt{128} + 4\sqrt{75} - 5\sqrt{162}$. 42. $5\sqrt{24} - 2\sqrt{54} - \sqrt{6}$.

43. $\sqrt{252} - \sqrt{294} - 48\sqrt{\dfrac{1}{6}}$. 44. $3\sqrt{147} - \dfrac{7}{3}\sqrt{\dfrac{1}{3}} - \sqrt{\dfrac{1}{27}}$.

262. *To multiply two surds of the same order: multiply separately the rational factors and the irrational factors.*

For
$$a\sqrt[n]{x} \times b\sqrt[n]{y} = ax^{\frac{1}{n}} \times by^{\frac{1}{n}}$$
$$= ab x^{\frac{1}{n}} y^{\frac{1}{n}}$$
$$= ab(xy)^{\frac{1}{n}}$$
$$= ab\sqrt[n]{xy}.$$

Examples. (1) $5\sqrt{3} \times 3\sqrt{7} = 15\sqrt{21}$.

(2) $2\sqrt{x} \times 3\sqrt{x} = 6x$.

(3) $\sqrt[4]{a+b} \times \sqrt[4]{a-b} = \sqrt[4]{(a+b)(a-b)} = \sqrt[4]{a^2 - b^2}$.

258   ALGEBRA.   [CHAP.

263. If the surds are not in their simplest form, it will save labour to reduce them to this form before multiplication.

Example. The product of $5\sqrt{32}$, $\sqrt{48}$, $2\sqrt{54}$
$$= 5 \cdot 4\sqrt{2} \times 4\sqrt{3} \times 2 \cdot 3\sqrt{6} = 480 \cdot \sqrt{2} \cdot \sqrt{3} \cdot \sqrt{6} = 480 \times 6 = 2880.$$

264. *To multiply surds which are not of the same order: reduce them to equivalent surds of the same order, and proceed as before.*

Example. Multiply $5\sqrt[3]{2}$ by $2\sqrt{5}$.

The product $= 5\sqrt[6]{2^2} \times 2\sqrt[6]{5^3} = 10\sqrt[6]{2^2 \times 5^3} = 10\sqrt[6]{500}$.

265. Suppose it is required to find the numerical value of the quotient when $\sqrt{5}$ is divided by $\sqrt{7}$.

At first sight it would seem that we must find the square root of 5, which is 2·236..., and then the square root of 7, which is 2·645..., and finally divide 2·236... by 2·645...; three troublesome operations.

But we may avoid much of this labour by multiplying both numerator and denominator by $\sqrt{7}$, so as to make the denominator a rational quantity. Thus

$$\frac{\sqrt{5}}{\sqrt{7}} = \frac{\sqrt{5}}{\sqrt{7}} \times \frac{\sqrt{7}}{\sqrt{7}} = \frac{\sqrt{5 \times 7}}{7} = \frac{\sqrt{35}}{7}.$$

Now $\sqrt{35} = 5\cdot916...$

$$\therefore \frac{\sqrt{5}}{\sqrt{7}} = \frac{5\cdot916...}{7} = \cdot 845....$$

266. The great utility of this artifice in calculating the numerical value of surd fractions suggests its convenience in the case of *all* surd fractions, even where numerical values are not required. Thus it is usual to simplify $\dfrac{a\sqrt{b}}{\sqrt{c}}$ as follows:

$$\frac{a\sqrt{b}}{\sqrt{c}} = \frac{a\sqrt{b} \times \sqrt{c}}{\sqrt{c} \times \sqrt{c}} = \frac{a\sqrt{bc}}{c}.$$

The process by which surds are removed from the denominator of any fraction is known as **rationalising the denominator**. It is effected by multiplying both numerator and denominator by any factor which renders the denominator rational. We shall return to this point in Art. 270.

267. The quotient of one surd by another may be found by expressing the result as a fraction, and rationalising the denominator.

Example 1. Divide $4\sqrt{75}$ by $25\sqrt{56}$.

The quotient $= \dfrac{4\sqrt{75}}{25\sqrt{56}} = \dfrac{4 \times 5\sqrt{3}}{25 \times 2\sqrt{14}} = \dfrac{2\sqrt{3}}{5\sqrt{14}}$

$= \dfrac{2\sqrt{3} \times \sqrt{14}}{5\sqrt{14} \times \sqrt{14}} = \dfrac{2\sqrt{42}}{5 \times 14} = \dfrac{\sqrt{42}}{35}.$

Example 2. $\dfrac{\sqrt[3]{b}}{\sqrt[3]{c^2}} = \dfrac{\sqrt[3]{b} \times \sqrt[3]{c}}{\sqrt[3]{c^2} \times \sqrt[3]{c}} = \dfrac{\sqrt[3]{bc}}{\sqrt[3]{c^3}} = \dfrac{\sqrt[3]{bc}}{c}.$

EXAMPLES XXXI. c.

Find the value of

1. $2\sqrt{14} \times \sqrt{21}.$
2. $3\sqrt{8} \times \sqrt{6}.$
3. $5\sqrt{a} \times 2\sqrt{3}.$
4. $2\sqrt{15} \times 3\sqrt{5}.$
5. $8\sqrt{12} \times 3\sqrt{24}.$
6. $\sqrt[3]{x+2} \times \sqrt[3]{x-2}.$
7. $21\sqrt{384} \div 8\sqrt{98}.$
8. $5\sqrt{27} \div 3\sqrt{24}.$
9. $-13\sqrt{125} \div 5\sqrt{65}.$
10. $\sqrt[3]{168} \times \sqrt[3]{147}.$
11. $5\sqrt[3]{128} \times 2\sqrt[3]{432}.$
12. $6\sqrt{14} \div 2\sqrt{21}.$
13. $a\sqrt{b^3} \times b^2\sqrt{a}.$
14. $\dfrac{3\sqrt{11}}{2\sqrt{98}} \div \dfrac{5}{7\sqrt{22}}.$
15. $\dfrac{3\sqrt{48}}{5\sqrt{112}} \div \dfrac{6\sqrt{84}}{\sqrt{392}}.$
16. $\dfrac{3}{x}\sqrt{\dfrac{a^2}{x}} \times \dfrac{4}{3}\sqrt{\dfrac{x^3}{2a^4}}.$
17. $\dfrac{3}{a-b}\sqrt{\dfrac{2x}{a-b}} \div \sqrt{\dfrac{18x^3}{(a-b)^5}}.$

Given $\sqrt{2}=1{\cdot}41421$, $\sqrt{3}=1{\cdot}73205$, $\sqrt{5}=2{\cdot}23607$, $\sqrt{6}=2{\cdot}44949$, $\sqrt{7}=2{\cdot}64575$: find to four places of decimals the numerical value of

18. $\dfrac{14}{\sqrt{2}}.$
19. $\dfrac{25}{\sqrt{5}}.$
20. $\dfrac{10}{\sqrt{7}}.$
21. $\dfrac{48}{\sqrt{6}}.$
22. $\dfrac{60}{\sqrt{5}}.$
23. $144 \div \sqrt{6}.$
24. $\sqrt{2} \div \sqrt{3}.$
25. $\dfrac{1}{2\sqrt{3}}.$
26. $\dfrac{1}{\sqrt{500}}.$
27. $\dfrac{4}{\sqrt{243}}.$
28. $\dfrac{25}{\sqrt{252}}.$
29. $\sqrt{\dfrac{256}{1575}}.$

268. Hitherto we have confined our attention to **simple surds**, such as $\sqrt[4]{5}$, $\sqrt[3]{a}$, $\sqrt{x+y}$. An expression involving two or more simple surds is called a **compound surd**; thus $2\sqrt{a}-3\sqrt{b}$; $\sqrt[3]{a}+\sqrt[4]{b}$ are compound surds.

269. The multiplication of compound surds is performed like the multiplication of compound algebraical expressions.

Example 1. Multiply $2\sqrt{x}-5$ by $3\sqrt{x}$.

The product $= 3\sqrt{x}(2\sqrt{x}-5)$
$= 6x - 15\sqrt{x}.$

Example 2. Multiply $2\sqrt{5}+3\sqrt{x}$ by $\sqrt{5}-\sqrt{x}$.

The product $= (2\sqrt{5}+3\sqrt{x})(\sqrt{5}-\sqrt{x})$
$= 2\sqrt{5}.\sqrt{5}+3\sqrt{5}.\sqrt{x}-2\sqrt{5}.\sqrt{x}-3\sqrt{x}.\sqrt{x}$
$= 10 - 3x + \sqrt{5x}.$

Example 3. Find the square of $2\sqrt{x}+\sqrt{7-4x}$.

$(2\sqrt{x}+\sqrt{7-4x})^2 = (2\sqrt{x})^2 + (\sqrt{7-4x})^2 + 4\sqrt{x}.\sqrt{7-4x}$
$= 4x + 7 - 4x + 4\sqrt{7x-4x^2}$
$= 7 + 4\sqrt{7x - 4x^2}.$

EXAMPLES XXXI. d.

Find the value of

1. $(3\sqrt{x}-5) \times 2\sqrt{x}.$
2. $(\sqrt{x}-\sqrt{a}) \times 2\sqrt{x}.$
3. $(\sqrt{a}+\sqrt{b}) \times \sqrt{ab}.$
4. $(\sqrt{x+y}-1) \times \sqrt{x+y}.$
5. $(2\sqrt{3}+3\sqrt{2})^2.$
6. $(\sqrt{7}+5\sqrt{3})(2\sqrt{7}-4\sqrt{3}).$
7. $(3\sqrt{5}-4\sqrt{2})(2\sqrt{5}+3\sqrt{2}).$
8. $(3\sqrt{a}-2\sqrt{x})(2\sqrt{a}+3\sqrt{x}).$
9. $(\sqrt{x}+\sqrt{x-1}) \times \sqrt{x-1}.$
10. $(\sqrt{x+a}-\sqrt{x-a}) \times \sqrt{x+a}.$
11. $(\sqrt{a+x}-2\sqrt{a})^2.$
12. $(2\sqrt{a}-\sqrt{1+4a})^2.$
13. $(\sqrt{a+x}-\sqrt{a-x})^2.$
14. $(\sqrt{a+x}-2)(\sqrt{a+x}-1).$
15. $(\sqrt{2}+\sqrt{3}-\sqrt{5})(\sqrt{2}+\sqrt{3}+\sqrt{5}).$
16. $(\sqrt{5}+3\sqrt{2}+\sqrt{7})(\sqrt{5}+3\sqrt{2}-\sqrt{7}).$

Write down the square of

17. $\sqrt{2x+a}-\sqrt{2x-a}.$
18. $\sqrt{x^2-2y^2}+\sqrt{x^2+2y^2}.$
19. $\sqrt{m+n}+\sqrt{m-n}.$
20. $3\sqrt{a^2+b^2}-2\sqrt{a^2-b^2}.$
21. $3x\sqrt{2}-3\sqrt{7-2x^2}.$
22. $\sqrt{4x^2+1}-\sqrt{4x^2-1}.$

270. One case of the multiplication of compound surds deserves careful attention. For if we multiply together the sum and the difference of any two quadratic surds we obtain a rational product.

Examples. (1) $(\sqrt{a}+\sqrt{b})(\sqrt{a}-\sqrt{b})=(\sqrt{a})^2-(\sqrt{b})^2=a-b$.

(2) $(3\sqrt{5}+4\sqrt{3})(3\sqrt{5}-4\sqrt{3})=(3\sqrt{5})^2-(4\sqrt{3})^2=45-48=-3$.

Similarly, $(4-\sqrt{a+b})(4+\sqrt{a+b})=(4)^2-(\sqrt{a+b})^2=16-a-b$.

271. DEFINITION. When two binomial quadratic surds differ only in the sign which connects their terms they are said to be *conjugate*.

Thus $3\sqrt{7}+5\sqrt{11}$ is conjugate to $3\sqrt{7}-5\sqrt{11}$.

Similarly, $a-\sqrt{a^2-x^2}$ is conjugate to $a+\sqrt{a^2-x^2}$.

The product of two conjugate surds is rational. [Art. 270.]

Example. $(3\sqrt{a}+\sqrt{x-9a})(3\sqrt{a}-\sqrt{x-9a})$
$=(3\sqrt{a})^2-(\sqrt{x-9a})^2=9a-(x-9a)=18a-x$.

272. The only case of the division of compound surds which we shall here consider is that in which the divisor is a binomial quadratic surd. If we express the division by means of a fraction, we can always rationalise the denominator by multiplying numerator and denominator by the surd which is conjugate to the divisor.

Example 1. Divide $4+3\sqrt{2}$ by $5-3\sqrt{2}$.

The quotient $=\dfrac{4+3\sqrt{2}}{5-3\sqrt{2}}=\dfrac{4+3\sqrt{2}}{5-3\sqrt{2}}\times\dfrac{5+3\sqrt{2}}{5+3\sqrt{2}}$

$=\dfrac{20+18+12\sqrt{2}+15\sqrt{2}}{25-18}=\dfrac{38+27\sqrt{2}}{7}$.

Example 2. Rationalise the denominator of $\dfrac{b^2}{\sqrt{a^2+b^2}+a}$.

The expression $=\dfrac{b^2}{\sqrt{a^2+b^2}+a}\times\dfrac{\sqrt{a^2+b^2}-a}{\sqrt{a^2+b^2}-a}$

$=\dfrac{b^2\{\sqrt{a^2+b^2}-a\}}{(a^2+b^2)-a^2}$

$=\sqrt{a^2+b^2}-a$.

E.A. S

Example 3. Divide $\dfrac{\sqrt{3}+\sqrt{2}}{2-\sqrt{3}}$ by $\dfrac{7+4\sqrt{3}}{\sqrt{3}-\sqrt{2}}$.

The quotient $=\dfrac{\sqrt{3}+\sqrt{2}}{2-\sqrt{3}} \times \dfrac{\sqrt{3}-\sqrt{2}}{7+4\sqrt{3}} = \dfrac{(\sqrt{3})^2-(\sqrt{2})^2}{14-12+8\sqrt{3}-7\sqrt{3}}$

$= \dfrac{1}{2+\sqrt{3}} = 2-\sqrt{3}$, on rationalising.

Example 4. Given $\sqrt{5}=2\cdot236068$, find the value of $\dfrac{87}{7-2\sqrt{5}}$.

Rationalising the denominator,

$$\dfrac{87}{7-2\sqrt{5}} = \dfrac{87(7+2\sqrt{5})}{49-20} = 3(7+2\sqrt{5}) = 34\cdot416408.$$

It will be seen that by rationalising the denominator we have avoided the use of a divisor consisting of 7 figures.

EXAMPLES XXXI. e.

Find the value of

1. $(9\sqrt{2}-7)(9\sqrt{2}+7)$.
2. $(3+5\sqrt{7})(3-5\sqrt{7})$.
3. $(5\sqrt{8}-2\sqrt{7})(5\sqrt{8}+2\sqrt{7})$.
4. $(2\sqrt{11}+5\sqrt{2})(2\sqrt{11}-5\sqrt{2})$.
5. $(\sqrt{a}+2\sqrt{b})(\sqrt{a}-2\sqrt{b})$.
6. $(3c-2\sqrt{x})(3c+2\sqrt{x})$.
7. $(\sqrt{a+x}-\sqrt{a})(\sqrt{a+x}+\sqrt{a})$.
8. $(\sqrt{2p+3q}-2\sqrt{q})(\sqrt{2p+3q}+2\sqrt{q})$.
9. $(\sqrt{a+x}+\sqrt{a-x})(\sqrt{a+x}-\sqrt{a-x})$.
10. $(5\sqrt{x^2-3y^2}+7a)(5\sqrt{x^2-3y^2}-7a)$.
11. $29 \div (11+3\sqrt{7})$.
12. $17 \div (3\sqrt{7}+2\sqrt{3})$.
13. $(3\sqrt{2}-1) \div (3\sqrt{2}+1)$.
14. $(2\sqrt{3}+7\sqrt{2}) \div (5\sqrt{3}-4\sqrt{2})$.
15. $(2x-\sqrt{xy}) \div (2\sqrt{xy}-y)$.
16. $(3+\sqrt{5})(\sqrt{5}-2) \div (5-\sqrt{5})$.
17. $\dfrac{\sqrt{a}}{\sqrt{a}-\sqrt{x}} \div \dfrac{\sqrt{a}+\sqrt{x}}{\sqrt{x}}$.
18. $\dfrac{2\sqrt{15}+8}{5+\sqrt{15}} \div \dfrac{8\sqrt{3}-6\sqrt{5}}{5\sqrt{3}-3\sqrt{5}}$.

Rationalise the denominator of

19. $\dfrac{25\sqrt{3}-4\sqrt{2}}{7\sqrt{3}-5\sqrt{2}}$.
20. $\dfrac{10\sqrt{6}-2\sqrt{7}}{3\sqrt{6}+2\sqrt{7}}$.
21. $\dfrac{\sqrt{7}+\sqrt{2}}{9+2\sqrt{14}}$.
22. $\dfrac{2\sqrt{3}+3\sqrt{2}}{5+2\sqrt{6}}$.
23. $\dfrac{y^2}{x+\sqrt{x^2-y^2}}$.
24. $\dfrac{x^2}{\sqrt{x^2+a^2}+a}$.
25. $\dfrac{\sqrt{1+x^2}-\sqrt{1-x^2}}{\sqrt{1+x^2}+\sqrt{1-x^2}}$.
26. $\dfrac{2\sqrt{a+b}+3\sqrt{a-b}}{2\sqrt{a+b}-\sqrt{a-b}}$.

27. $\dfrac{\sqrt{9+x^2}-3}{\sqrt{9+x^2}+3}$.   28. $\dfrac{3+\sqrt{6}}{5\sqrt{3}-2\sqrt{12}-\sqrt{32}+\sqrt{50}}$.

Given $\sqrt{2}=1\cdot41421$, $\sqrt{3}=1\cdot73205$, $\sqrt{5}=2\cdot23607$: find to four places of decimals the value of

29. $\dfrac{1}{2+\sqrt{3}}$.   30. $\dfrac{3+\sqrt{5}}{\sqrt{5}-2}$.   31. $\dfrac{\sqrt{5}+\sqrt{3}}{4+\sqrt{15}}$.   32. $\dfrac{\sqrt{5}-2}{9-4\sqrt{5}}$.

33. $\dfrac{7\sqrt{5}+15}{\sqrt{5}-1} \times \dfrac{\sqrt{5}-2}{3+\sqrt{5}}$.   34. $(2-\sqrt{3})(7-4\sqrt{3}) \div (3\sqrt{3}-5)$.

273. *The square root of a rational quantity cannot be partly rational and partly a quadratic surd.*

If possible let $\qquad \sqrt{n}=a+\sqrt{m}$;

then by squaring, $\qquad n=a^2+m+2a\sqrt{m}$;

$$\therefore \sqrt{m}=\dfrac{n-a^2-m}{2a} ;$$

that is a surd is equal to a rational quantity ; which is impossible.

274. *If* $x+\sqrt{y}=a+\sqrt{b}$, *where* x *and* a *are both rational and* $\sqrt{y}$ *and* $\sqrt{b}$ *are both irrational, then will* x=a *and* y=b.

For if x is not equal to a, let $x=a+m$; then

$$a+m+\sqrt{y}=a+\sqrt{b} ;$$

that is, $\qquad \sqrt{b}=m+\sqrt{y}$;

which is impossible. [Art. 273.]

Therefore $\qquad x=a,$

and consequently, $\qquad y=b.$

If therefore $\qquad x+\sqrt{y}=a+\sqrt{b},$

we must also have $\qquad x-\sqrt{y}=a-\sqrt{b}.$

275. It appears from the preceding article that in any equation of the form

$$X+\sqrt{Y}=A+\sqrt{B} \dots\dots\dots\dots(1),$$

we may equate the rational parts on each side, and also the irrational parts ; so that the equation (1) is really equivalent to *two* independent equations, $X=A$ and $Y=B$. But this is only true when $\sqrt{Y}$ and $\sqrt{B}$ are irrational.

276. *If* $\sqrt{a+\sqrt{b}}=\sqrt{x}+\sqrt{y}$ *then will* $\sqrt{a-\sqrt{b}}=\sqrt{x}-\sqrt{y}$.

For by squaring, we obtain
$$a+\sqrt{b}=x+2\sqrt{xy}+y;$$
$$\therefore\ a=x+y,\ \sqrt{b}=2\sqrt{xy}. \qquad [\text{Art. 275.}]$$

Hence $\qquad a-\sqrt{b}=x-2\sqrt{xy}+y,$

and $\qquad \sqrt{a-\sqrt{b}}=\sqrt{x}-\sqrt{y}.$

277. *To find the square root of* $a+\sqrt{b}$.

Suppose $\qquad \sqrt{a+\sqrt{b}}=\sqrt{x}+\sqrt{y};$

then as in the last article,
$$x+y=a \ \dots\dots\dots\dots\dots\dots\dots\dots (1),$$
$$2\sqrt{xy}=\sqrt{b} \ \dots\dots\dots\dots\dots\dots\dots\dots (2).$$
$$\therefore\ (x-y)^2=(x+y)^2-4xy$$
$$=a^2-b, \qquad \text{from (1) and (2)}.$$
$$\therefore\ x-y=\sqrt{a^2-b}.$$

Combining this with (1) we find
$$x=\frac{a+\sqrt{a^2-b}}{2},\ \text{and}\ y=\frac{a-\sqrt{a^2-b}}{2}$$
$$\therefore\ \sqrt{a+\sqrt{b}}=\sqrt{\frac{a+\sqrt{(a^2-b)}}{2}}+\sqrt{\frac{a-\sqrt{(a^2-b)}}{2}}.$$

278. From the values just found for x and y, it appears that each of them is itself a compound surd unless a^2-b is a perfect square. Hence the method of Art. 277 for finding the square root of $a+\sqrt{b}$ is of no practical utility except when a^2-b is a perfect square.

Example. Find the square root of $16+2\sqrt{55}$.

Assume $\qquad \sqrt{16+2\sqrt{55}}=\sqrt{x}+\sqrt{y}.$

Then $\qquad 16+2\sqrt{55}=x+2\sqrt{xy}+y.$
$$\therefore\ x+y=16 \ \dots\dots\dots\dots\dots\dots\dots\dots (1),$$
$$2\sqrt{xy}=2\sqrt{55} \ \dots\dots\dots\dots\dots\dots (2).$$
$$\therefore\ (x-y)^2=(x+y)^2-4xy$$
$$=16^2-4\times 55, \qquad \text{by (1) and (2)}.$$
$$=4\times 9.$$
$$\therefore\ x-y=\pm 6 \ \dots\dots\dots\dots\dots\dots\dots\dots (3).$$

From (1) and (3) we obtain
$$x=11,\ \text{or}\ 5,\ \text{and}\ y=5,\ \text{or}\ 11.$$

That is, the required square root is $\sqrt{11}+\sqrt{5}$.

XXXI.] ELEMENTARY SURDS. 265

In the same way we may shew that
$$\sqrt{16-2\sqrt{55}}=\sqrt{11}-\sqrt{5}.$$

Note. Since every quantity has two square roots equal in magnitude but opposite in sign, strictly speaking we should have

the square root of $16+2\sqrt{55}=\pm(\sqrt{11}+\sqrt{5})$,
.... $16-2\sqrt{55}=\pm(\sqrt{11}-\sqrt{5})$.

However it is usually sufficient to take the positive value of the square root, so that in assuming $\sqrt{a-\sqrt{b}}=\sqrt{x}-\sqrt{y}$ it is understood that x is greater than y. With this proviso it will be unnecessary in any numerical example to use the double sign at the stage of work corresponding to equation (3) of the last example.

279. When the binomial whose square root we are seeking consists of *two* quadratic surds, we proceed as explained in the following example.

Example. Find the square root of $\sqrt{175}-\sqrt{147}$.

Since $\sqrt{175}-\sqrt{147}=\sqrt{7}(\sqrt{25}-\sqrt{21})=\sqrt{7}(5-\sqrt{21})$.

$\therefore \sqrt{\sqrt{175}-\sqrt{147}}=\sqrt[4]{7}\cdot\sqrt{5-\sqrt{21}}$.

And, proceeding as in the last article,
$$\sqrt{5-\sqrt{21}}=\sqrt{\tfrac{7}{2}}-\sqrt{\tfrac{3}{2}};$$
$$\therefore \sqrt{\sqrt{175}-\sqrt{147}}=\sqrt[4]{7}\left(\sqrt{\tfrac{7}{2}}-\sqrt{\tfrac{3}{2}}\right).$$

280. The square root of a binomial surd may often be found by inspection.

Example 1. Find the square root of $11+2\sqrt{30}$.

We have only to find two quantities whose sum is 11, and whose product is 30; thus
$$11+2\sqrt{30}=6+5+2\sqrt{6\times 5}$$
$$=(\sqrt{6}+\sqrt{5})^2.$$
$$\therefore \sqrt{11+2\sqrt{30}}=\sqrt{6}+\sqrt{5}.$$

Example 2. Find the square root of $53-12\sqrt{10}$.

First write the binomial so that the surd part has a coefficient 2; thus $53-12\sqrt{10}=53-2\sqrt{360}$.

We have now to find two quantities whose sum is 53 and whose product is 360; these are 45 and 8;

hence
$$53 - 12\sqrt{10} = 45 + 8 - 2\sqrt{45 \times 8}$$
$$= (\sqrt{45} - \sqrt{8})^2;$$
$$\therefore \sqrt{53 - 12\sqrt{10}} = \sqrt{45} - \sqrt{8}$$
$$= 3\sqrt{5} - 2\sqrt{2}.$$

EXAMPLES XXXI. f.

Find the square root of each of the following binomial surds:

1. $7 - 2\sqrt{10}$.
2. $13 + 2\sqrt{30}$.
3. $8 - 2\sqrt{7}$.
4. $5 + 2\sqrt{6}$.
5. $75 + 12\sqrt{21}$.
6. $18 - 8\sqrt{5}$.
7. $41 - 24\sqrt{2}$.
8. $83 + 12\sqrt{35}$.
9. $47 - 4\sqrt{33}$.
10. $2\tfrac{1}{4} + \sqrt{5}$.
11. $4\tfrac{1}{3} - \tfrac{4}{3}\sqrt{3}$.
12. $16 + 5\sqrt{7}$.
13. $\sqrt{27} + 2\sqrt{6}$.
14. $\sqrt{32} - \sqrt{24}$.
15. $3\sqrt{5} + \sqrt{40}$.

Find the fourth roots of the following binomial surds:

16. $17 + 12\sqrt{2}$.
17. $56 + 24\sqrt{5}$.
18. $\tfrac{3}{2}\sqrt{5} + 3\tfrac{1}{2}$.
19. $14 + 8\sqrt{3}$.
20. $49 - 20\sqrt{6}$.
21. $248 + 32\sqrt{60}$.

Find, by inspection, the value of

22. $\sqrt{3 - 2\sqrt{2}}$.
23. $\sqrt{4 + 2\sqrt{3}}$.
24. $\sqrt{6 - 2\sqrt{5}}$.
25. $\sqrt{19 + 8\sqrt{3}}$.
26. $\sqrt{8 + 2\sqrt{15}}$.
27. $\sqrt{9 - 2\sqrt{14}}$.
28. $\sqrt{11 + 4\sqrt{6}}$.
29. $\sqrt{15 - 4\sqrt{14}}$.
30. $\sqrt{29 + 6\sqrt{22}}$.

Equations involving Surds.

281. Sometimes equations are proposed in which the unknown quantity appears under the radical sign. Such equations are very varied in character and often require special artifices for their solution. Here we shall only consider a few of the simpler cases, which can generally be solved by the following method. Bring to one side of the equation a single radical term by itself: on squaring both sides this radical will disappear. By repeating this process any remaining radicals can in turn be removed.

Example 1. Solve $\quad 2\sqrt{x} - \sqrt{4x-11} = 1$.

Transposing $\quad\quad\quad\quad 2\sqrt{x} - 1 = \sqrt{4x-11}$.

Square both sides; then $\quad 4x - 4\sqrt{x} + 1 = 4x - 11$,
$$4\sqrt{x} = 12,$$
$$\sqrt{x} = 3;$$
$$\therefore\ x = 9.$$

Example 2. Solve $\quad 2 + \sqrt[3]{x-5} = 13$.

Transposing $\quad\quad\quad\quad \sqrt[3]{x-5} = 11$.

Here we must *cube* both sides; thus $x - 5 = 1331$; whence $\quad\quad\quad\quad\quad\quad\quad x = 1336$.

Example 3. Solve $\quad \sqrt{x+5} + \sqrt{3x+4} = \sqrt{12x+1}$.

Squaring both sides,
$$x + 5 + 3x + 4 + 2\sqrt{(x+5)(3x+4)} = 12x + 1.$$

Transposing and dividing by 2,
$$\sqrt{(x+5)(3x+4)} = 4x - 4 \quad\ldots\ldots\ldots\ldots\ldots(1).$$

Squaring, $\quad\quad (x+5)(3x+4) = 16x^2 - 32x + 16$,

or $\quad\quad\quad 13x^2 - 51x - 4 = 0$,
$$(x-4)(13x+1) = 0;$$
$$\therefore\ x = 4,\ \text{or}\ -\frac{1}{13}.$$

If we proceed to verify the solution by substituting these values in the original equation, it will be found that it is satisfied by $x = 4$, but not by $x = -\frac{1}{13}$. But this latter value will be found on trial to satisfy the given equation if we alter the sign of the second radical; thus
$$\sqrt{x+5} - \sqrt{3x+4} = \sqrt{12x+1}.$$

On squaring this and reducing, we obtain
$$-\sqrt{(x+5)(3x+4)} = 4x - 4 \quad\ldots\ldots\ldots\ldots\ldots(2);$$

and a comparison of (1) and (2) shews that in the next stage of the work *the same quadratic equation is obtained* in each case, the roots of which are 4 and $-\frac{1}{13}$, as already found.

From this it appears that when the solution of an equation requires that both sides should be squared, we cannot be certain without trial which of the values found for the unknown quantity will satisfy the original equation.

In order that all the values found by the solution of the equation may be applicable it will be necessary to take into account both signs of the radicals in the given equation.

EXAMPLES XXXI. g.

Solve the equations :

1. $\sqrt{x-5}=3$.
2. $\sqrt[3]{4x-7}=5$.
3. $7-\sqrt{x-4}=3$.
4. $13-\sqrt[3]{5x-4}=7$.
5. $\sqrt{5x-1}=2\sqrt{x+3}$.
6. $2\sqrt{3-7x}-3\sqrt{8x-12}=0$.
7. $2\sqrt[3]{5x-35}=5\sqrt[3]{2x-7}$.
8. $\sqrt{9x^2-11x-5}=3x-2$.
9. $\sqrt[4]{2x+11}=\sqrt{5}$.
10. $\sqrt{4x^2-7x+1}=2x-1\frac{4}{5}$.
11. $\sqrt{x+25}=1+\sqrt{x}$.
12. $\sqrt{8x+33}-3=2\sqrt{2x}$.
13. $\sqrt{x+3}+\sqrt{x}=5$.
14. $10-\sqrt{25+9x}=3\sqrt{x}$.
15. $\sqrt{x-4}+3=\sqrt{x+11}$.
16. $\sqrt{9x-8}=3\sqrt{x+4}-2$.
17. $\sqrt{4x+5}-\sqrt{x}=\sqrt{x+3}$.
18. $\sqrt{25x-29}-\sqrt{4x-11}=3\sqrt{x}$.
19. $\sqrt{8x+17}-\sqrt{2x}=\sqrt{2x+9}$.
20. $\sqrt{3x-11}+\sqrt{3x}=\sqrt{12x-23}$.
21. $\sqrt{12x-5}-\sqrt{3x-1}=\sqrt{27x-2}$.
22. $\sqrt{x+3}+\sqrt{x+8}-\sqrt{4x+21}=0$.
23. $\sqrt{x+2}+\sqrt{4x+1}-\sqrt{9x+7}=0$.
24. $\sqrt{x+4ab}=2a+\sqrt{x}$.
25. $\sqrt{x+\sqrt{4a+x}}=2\sqrt{b+x}$.
26. $\sqrt{a-x}+\sqrt{b+x}=\sqrt{a}+\sqrt{b}$.
27. $5\sqrt[3]{70x+29}=9\sqrt[3]{14x-15}$.
28. $\sqrt[3]{x^3-3x^2+7x-11}=x-1$.
29. $\sqrt[3]{8x^3+12x^2+12x-11}=2x+1$.
30. $\sqrt[3]{1+x}+\sqrt[3]{1-x}=\sqrt[3]{2}$.

282. When radicals appear in a fractional form in an equation, we must clear of fractions in the ordinary way, combining the irrational factors by the rules already explained in this chapter.

Example 1. Solve $\dfrac{6\sqrt{x}-11}{3\sqrt{x}}=\dfrac{2\sqrt{x}+1}{\sqrt{x}+6}$.

Multiplying across, we have
$$6x+25\sqrt{x}-66=6x+3\sqrt{x},$$
that is,
$$25\sqrt{x}-3\sqrt{x}=66,$$
$$22\sqrt{x}=66,$$
$$\sqrt{x}=3,$$
$$x=9.$$

XXXI.] EQUATIONS INVOLVING SURDS. 269

Example 2. Solve $\sqrt{9+2x} - \sqrt{2x} = \dfrac{5}{\sqrt{9+2x}}$.

Clearing of fractions, $9 + 2x - \sqrt{2x(9+2x)} = 5$,

$$4 + 2x = \sqrt{2x(9+2x)}.$$

Squaring, $16 + 16x + 4x^2 = 18x + 4x^2$,

$$16 = 2x,$$
$$x = 8.$$

EXAMPLES XXXI. h.

Solve the equations:

1. $\dfrac{6\sqrt{x} - 21}{3\sqrt{x} - 14} = \dfrac{8\sqrt{x} - 11}{4\sqrt{x} - 13}$.

2. $\dfrac{9\sqrt{x} - 23}{3\sqrt{x} - 8} = \dfrac{6\sqrt{x} - 17}{2\sqrt{x} - 6}$.

3. $\dfrac{\sqrt{x} + 3}{\sqrt{x} - 2} = \dfrac{3\sqrt{x} - 5}{3\sqrt{x} - 13}$.

4. $2 - \dfrac{\sqrt{x} + 3}{\sqrt{x} + 2} = \dfrac{\sqrt{x} + 9}{\sqrt{x} + 7}$.

5. $\dfrac{2\sqrt{x} - 1}{2\sqrt{x} + \frac{4}{3}} = \dfrac{\sqrt{x} - 2}{\sqrt{x} - \frac{4}{3}}$.

6. $\dfrac{6\sqrt{x} - 7}{\sqrt{x} - 1} - 5 = \dfrac{7\sqrt{x} - 26}{7\sqrt{x} - 21}$.

7. $\dfrac{12\sqrt{x} - 11}{4\sqrt{x} - 4\frac{2}{3}} = \dfrac{6\sqrt{x} + 5}{2\sqrt{x} + \frac{2}{3}}$.

8. $\sqrt{1+x} + \sqrt{x} = \dfrac{2}{\sqrt{1+x}}$.

9. $\sqrt{x-1} + \sqrt{x} = \dfrac{2}{\sqrt{x}}$.

10. $\sqrt{x} - \sqrt{x-8} = \dfrac{2}{\sqrt{x-8}}$.

11. $\sqrt{x+5} + \sqrt{x} = \dfrac{10}{\sqrt{x}}$.

12. $2\sqrt{x} - \sqrt{4x-3} = \dfrac{1}{\sqrt{4x-3}}$.

13. $3\sqrt{x} = \dfrac{8}{\sqrt{9x-32}} + \sqrt{9x-32}$.

14. $\sqrt{x} - 7 = \dfrac{1}{\sqrt{x+7}}$.

15. $(\sqrt{x}+11)(\sqrt{x}-11) + 110 = 0$.

16. $2\sqrt{x} = \dfrac{12 - 6\sqrt{x}}{2\sqrt{x} - 3}$.

17. $3\sqrt{x} - 1 = \dfrac{5}{3\sqrt{x} + 7} + 6$.

18. $\dfrac{x-1}{\sqrt{x}-1} = 3 + \dfrac{\sqrt{x}+1}{2}$.

19. $\dfrac{1}{1-x} + \dfrac{1}{\sqrt{x}+1} + \dfrac{1}{\sqrt{x}-1} = 0$.

20. $2 = \dfrac{\sqrt{2+x} + \sqrt{2-x}}{\sqrt{2+x} - \sqrt{2-x}}$.

21. $\dfrac{2x-3}{\sqrt{x-2}+1} = 2\sqrt{x-2} - 1$.

22. $\dfrac{2}{x-6+\sqrt{x}} + \dfrac{3}{\sqrt{x}-2} = \dfrac{4}{\sqrt{x}+3}$.

[*Further practice in surd equations will be found on pages* 362, 363.]

CHAPTER XXXII.

Ratio, Proportion, and Variation.

Ratio.

283. DEFINITION. **Ratio** is the relation which one quantity bears to another of the *same* kind, the comparison being made by considering what multiple, part, or parts, one quantity is of the other.

The ratio of A to B is usually written $A:B$. The quantities A and B are called the *terms* of the ratio. The first term is called the **antecedent**, the second term the **consequent**.

284. To find what multiple or part A is of B we divide A by B; hence the ratio $A:B$ may be measured by the fraction $\frac{A}{B}$, and we shall usually find it convenient to adopt this notation.

In order to compare two quantities they must be expressed in terms of the same unit. Thus the ratio of £2 to 15s. is measured by the fraction $\frac{2 \times 20}{15}$ or $\frac{8}{3}$.

Note. Since a ratio expresses the *number* of times that one quantity contains another, *every ratio is an abstract quantity*.

285. By Art. 151, $$\frac{a}{b} = \frac{ma}{mb};$$

and thus the ratio $a:b$ is equal to the ratio $ma:mb$; that is, *the value of a ratio remains unaltered if the antecedent and the consequent are multiplied or divided by the same quantity*.

286. Two or more ratios may be compared by reducing their equivalent fractions to a common denominator. Thus suppose $a:b$ and $x:y$ are two ratios. Now $\frac{a}{b} = \frac{ay}{by}$, and $\frac{x}{y} = \frac{bx}{by}$; hence the ratio $a:b$ is greater than, equal to, or less than the ratio $x:y$ according as ay is greater than, equal to, or less than bx.

287. The ratio of two fractions can be expressed as a ratio of two integers. Thus the ratio $\frac{a}{b} : \frac{c}{d}$ is measured by the fraction $\dfrac{\frac{a}{b}}{\frac{c}{d}}$, or $\frac{ad}{bc}$; and is therefore equivalent to the ratio $ad : bc$.

288. If either, or both, of the terms of a ratio be a surd quantity, then no two integers can be found which will *exactly* measure their ratio. Thus the ratio $\sqrt{2} : 1$ cannot be exactly expressed by any two integers.

289. DEFINITION. If the ratio of any two quantities can be expressed exactly by the ratio of two integers the quantities are said to be **commensurable**; otherwise, they are said to be **incommensurable**.

Although we cannot find two integers which will exactly measure the ratio of two incommensurable quantities, we can always find two integers whose ratio differs from that required by as small a quantity as we please.

Thus $$\frac{\sqrt{5}}{4} = \frac{2\cdot 236067\ldots}{4} = \cdot 559016\ldots$$

and therefore $\dfrac{\sqrt{5}}{4} > \dfrac{559016}{1000000}$ and $< \dfrac{559017}{1000000}$,

and it is evident that by carrying the decimals further, any degree of approximation may be arrived at.

290. DEFINITION. Ratios are *compounded* by multiplying together the fractions which denote them; or by multiplying together the antecedents for a new antecedent, and the consequents for a new consequent.

Example. Find the ratio compounded of the three ratios

$$2a : 3b,\ 6ab : 5c^2,\ c : a.$$

The required ratio $= \dfrac{2a}{3b} \times \dfrac{6ab}{5c^2} \times \dfrac{c}{a} = \dfrac{4a}{5c}$.

291. DEFINITION. When the ratio $a : b$ is compounded with itself the resulting ratio is $a^2 : b^2$, and is called the **duplicate ratio** of $a : b$. Similarly $a^3 : b^3$ is called the **triplicate ratio** of $a : b$. Also $a^{\frac{1}{2}} : b^{\frac{1}{2}}$ is called the **subduplicate ratio** of $a : b$.

Examples. (1) The duplicate ratio of $2a : 3b$ is $4a^2 : 9b^2$.

(2) The subduplicate ratio of $49 : 25$ is $7 : 5$.

(3) The triplicate ratio of $2x : 1$ is $8x^3 : 1$.

292. DEFINITION. A ratio is said to be a ratio of *greater inequality*, of *less inequality*, or of *equality*, according as the antecedent is *greater than, less than,* or *equal to* the consequent.

293. If to each term of the ratio $8 : 3$ we add 4, a new ratio $12 : 7$ is obtained, and we see that it is less than the former because $\dfrac{12}{7}$ is clearly less than $\dfrac{8}{3}$.

This is a particular case of a more general proposition which we shall now prove.

A ratio of greater inequality is diminished, and a ratio of less inequality is increased, by adding the same quantity to both its terms.

Let $\dfrac{a}{b}$ be the ratio, and let $\dfrac{a+x}{b+x}$ be the new ratio formed by adding x to both its terms.

Now
$$\frac{a}{b} - \frac{a+x}{b+x} = \frac{ax-bx}{b(b+x)}$$
$$= \frac{x(a-b)}{b(b+x)};$$

and $a-b$ is positive or negative according as a is greater or less than b.

Hence if $a > b$, $\qquad \dfrac{a}{b} > \dfrac{a+x}{b+x}$;

and if $\quad a < b$, $\qquad \dfrac{a}{b} < \dfrac{a+x}{b+x}$,

which proves the proposition.

Similarly it can be proved that *a ratio of greater inequality is increased, and a ratio of less inequality is diminished, by taking the same quantity from both its terms.*

294. When two or more ratios are equal, many useful propositions may be proved by introducing a single symbol to denote each of the equal ratios.

The proof of the following important theorem will illustrate the method of procedure.

If
$$\frac{a}{b}=\frac{c}{d}=\frac{e}{f}=\ldots\ldots,$$

each of these ratios $=\left(\dfrac{pa^n+qc^n+re^n+\ldots}{pb^n+qd^n+rf^n+\ldots}\right)^{\frac{1}{n}},$

where p, q, r, n *are any quantities whatever.*

Let $\dfrac{a}{b}=\dfrac{c}{d}=\dfrac{e}{f}=\ldots =k;$

then $a=bk,\ c=dk,\ e=fk,\ \ldots;$

whence $pa^n=pb^nk^n,\ qc^n=qd^nk^n,\ re^n=rf^nk^n,\ \ldots;$

$$\therefore \frac{pa^n+qc^n+re^n+\ldots}{pb^n+qd^n+rf^n+\ldots}=\frac{pb^nk^n+qd^nk^n+rf^nk^n+\ldots}{pb^n+qd^n+rf^n+\ldots}$$
$$=k^n;$$

$$\therefore \left(\frac{pa^n+qc^n+re^n+\ldots}{pb^n+qd^n+rf^n+\ldots}\right)^{\frac{1}{n}}=k=\frac{a}{b}=\frac{c}{d}=\ldots.$$

By giving different values to p, q, r, n many particular cases of this general proposition may be deduced; or they may be proved independently by using the same method. For instance,

if $\dfrac{a}{b}=\dfrac{c}{d}=\dfrac{e}{f},$ each of these ratios $=\dfrac{a+c+e}{b+d+f};$

a result which will frequently be found useful.

Example 1. If $\dfrac{x}{y}=\dfrac{3}{4}$ find the value of $\dfrac{5x-3y}{7x+2y},$

$$\frac{5x-3y}{7x+2y}=\frac{\dfrac{5x}{y}-3}{\dfrac{7x}{y}+2}=\frac{\dfrac{15}{4}-3}{\dfrac{21}{4}+2}=\frac{3}{29}.$$

Example 2. Two numbers are in the ratio of $5:8$. If 9 be added to each they are in the ratio of $8:11$. Find the numbers.

Let the numbers be denoted by $5x$ and $8x$.

Then $\dfrac{5x+9}{8x+9}=\dfrac{8}{11};$

$$\therefore\ x=3.$$

Hence the numbers are 15 and 24.

Example 3. If $A : B$ be in the duplicate ratio of $A+x : B+x$, prove that $x^2 = AB$.

By the given condition, $\left(\dfrac{A+x}{B+x}\right)^2 = \dfrac{A}{B}$;

$$\therefore B(A+x)^2 = A(B+x)^2,$$
$$A^2B + 2ABx + Bx^2 = AB^2 + 2ABx + Ax^2,$$
$$x^2(A-B) = AB(A-B);$$
$$\therefore x^2 = AB,$$

since $A - B$ is, by supposition, not zero.

EXAMPLES XXXII. a.

Find the ratio compounded of

1. The duplicate ratio of $4 : 3$, and the ratio $27 : 8$.
2. The ratio $32 : 27$, and the triplicate ratio of $3 : 4$.
3. The subduplicate ratio of $25 : 36$, and the ratio $6 : 25$.
4. The ratio $169 : 200$, and the duplicate ratio of $15 : 26$.
5. The triplicate ratio of $x : y$, and the ratio $2y^2 : 3x^2$.
6. The ratio $3a : 4b$, and the subduplicate ratio of $b^4 : a^4$.
7. If $x : y = 5 : 7$, find the value of $x+y : y-x$.
8. If $\dfrac{x}{y} = 3\frac{1}{3}$, find the value of $\dfrac{x-3y}{2x-5y}$.
9. If $b : a = 2 : 5$, find the value of $2a - 3b : 3b - a$.
10. If $\dfrac{a}{b} = \dfrac{3}{4}$, and $\dfrac{x}{y} = \dfrac{5}{7}$, find the value of $\dfrac{3ax - by}{4by - 7ax}$.
11. If $7x - 4y : 3x + y = 5 : 13$, find the ratio $x : y$.
12. If $\dfrac{2a^2 - 3b^2}{a^2 + b^2} = \dfrac{2}{41}$, find the ratio $a : b$.
13. If $2x : 3y$ be in the duplicate ratio of $2x - m : 3y - m$, prove that $m^2 = 6xy$.
14. If $P : Q$ be the subduplicate ratio of $P - x : Q - x$, prove that $x = \dfrac{PQ}{P+Q}$.
15. If $\dfrac{a}{b} = \dfrac{c}{d} = \dfrac{e}{f}$, prove that each of these ratios is equal to

$$\sqrt[3]{\dfrac{2a^2c + 3c^3e + 4e^2c}{2b^2d + 3d^3e + 4f^2d}}$$

16. Two numbers are in the ratio of 3 : 4, and if 7 be subtracted from each the remainders are in the ratio of 2 : 3. Find them.

17. What number must be taken from each term of the ratio 27 : 35 that it may become 2 : 3?

18. What number must be added to each term of the ratio 37 : 29 that it may become 8 : 7?

19. If $\dfrac{p}{b-c} = \dfrac{q}{c-a} = \dfrac{r}{a-b}$, shew that $p+q+r=0$.

20. If $\dfrac{x}{b+c} = \dfrac{y}{c+a} = \dfrac{z}{a-b}$, shew that $x-y+z=0$.

21. If $\dfrac{a}{b} = \dfrac{c}{d} = \dfrac{e}{f}$, shew that the square root of
$$\dfrac{a^6 b - 2c^5 e + 3a^4 c^3 e^2}{b^7 - 2d^5 f + 3b^4 cd^2 e^2} \text{ is equal to } \dfrac{ace}{bdf}.$$

22. Prove that the ratio $la + mc + ne : lb + md + nf$ will be equal to each of the ratios $a : b$, $c : d$, $e : f$, if these be all equal; and that it will be intermediate in value between the greatest and least of these ratios if they be not all equal.

23. If $\dfrac{bx - ay}{cy - az} = \dfrac{cx - az}{by - ax} = \dfrac{z+y}{x+z}$, then will each of these fractions be equal to $\dfrac{x}{y}$, unless $b+c=0$.

24. If $\dfrac{2x - 3y}{3z + y} = \dfrac{z - y}{z - x} = \dfrac{x + 3z}{2y - 3x}$, prove that each of these ratios is equal to $\dfrac{x}{y}$; hence shew that either $x=y$, or $z=x+y$.

Proportion.

295. Definition. When two ratios are equal, the four quantities composing them are said to be **proportionals**. Thus if $\dfrac{a}{b} = \dfrac{c}{d}$ then a, b, c, d are proportionals. This is expressed by saying that a is to b as c is to d, and the proportion is written
$$a : b :: c : d\,;$$
or $\qquad\qquad a : b = c : d.$

The terms a and d are called the *extremes*, b and c the *means*.

296. *If four quantities are in proportion, the product of the extremes is equal to the product of the means.*

Let a, b, c, d be the proportionals.

Then by definition
$$\frac{a}{b} = \frac{c}{d};$$

whence
$$ad = bc.$$

Hence if any three terms of a proportion are given, the fourth may be found. Thus if a, c, d are given, then $b = \dfrac{ad}{c}$.

Conversely, if there are any four quantities, a, b, c, d, such that $ad = bc$, then a, b, c, d are proportionals; a and d being the extremes, b and c the means; or vice versâ.

297. DEFINITION. Quantities are said to be in **continued proportion** when the first is to the second, as the second is to the third, as the third to the fourth; and so on. Thus $a, b, c, d, \ldots\ldots$ are in continued proportion when

$$\frac{a}{b} = \frac{b}{c} = \frac{c}{d} = \ldots\ldots\ldots$$

If three quantities a, b, c are in continued proportion, then
$$a : b = b : c;$$
$$\therefore \quad ac = b^2. \qquad \text{[Art. 296.]}$$

In this case b is said to be a **mean proportional** between a and c; and c is said to be a **third proportional** to a and b.

298. *If three quantities are proportionals the first is to the third in the duplicate ratio of the first to the second.*

Let the three quantities be a, b, c; then $\dfrac{a}{b} = \dfrac{b}{c}$.

Now
$$\frac{a}{c} = \frac{a}{b} \times \frac{b}{c} = \frac{a}{b} \times \frac{a}{b} = \frac{a^2}{b^2};$$

that is,
$$a : c = a^2 : b^2.$$

299. *If* $a : b = c : d$ *and* $e : f = g : h$ *then will*
$$ae : bf = cg : dh.$$

For
$$\frac{a}{b} = \frac{c}{d} \text{ and } \frac{e}{f} = \frac{g}{h},$$

$$\therefore \quad \frac{ae}{bf} = \frac{cg}{dh}, \text{ or } ae : bf = cg : dh.$$

Cor. If $a : b = c : d$,
and $b : x = d : y$,
then $a : x = c : y$.

This is the theorem known as *ex æquali* in Geometry.

300. If four quantities, a, b, c, d form a proportion, many other proportions may be deduced by the properties of fractions. The results of these operations are very useful, and some of them are often quoted by the annexed names borrowed from Geometry.

(1) If $a : b = c : d$, then $b : a = d : c$. [*Invertendo.*]

For $\dfrac{a}{b} = \dfrac{c}{d}$; therefore $1 \div \dfrac{a}{b} = 1 \div \dfrac{c}{d}$;

that is, $\dfrac{b}{a} = \dfrac{d}{c}$;

or $b : a = d : c$.

(2) If $a : b = c : d$, then $a : c = b : d$. [*Alternando.*]

For $ad = bc$; therefore $\dfrac{ad}{cd} = \dfrac{bc}{cd}$;

that is, $\dfrac{a}{c} = \dfrac{b}{d}$;

or $a : c = b : d$.

(3) If $a : b = c : d$, then $a+b : b = c+d : d$. [*Componendo.*]

For $\dfrac{a}{b} = \dfrac{c}{d}$; therefore $\dfrac{a}{b} + 1 = \dfrac{c}{d} + 1$;

that is, $\dfrac{a+b}{b} = \dfrac{c+d}{d}$;

or $a+b : b = c+d : d$.

(4) If $a : b = c : d$, then $a-b : b = c-d : d$. [*Dividendo.*]

For $\dfrac{a}{b} = \dfrac{c}{d}$; therefore $\dfrac{a}{b} - 1 = \dfrac{c}{d} - 1$;

that is, $\dfrac{a-b}{b} = \dfrac{c-d}{d}$;

or $a-b : b = c-d : d$.

(5) If $a:b=c:d$, then $a+b:a-b=c+d:c-d$.

For by (3) $$\frac{a+b}{b}=\frac{c+d}{d};$$

and by (4) $$\frac{a-b}{b}=\frac{c-d}{d};$$

∴ by division, $$\frac{a+b}{a-b}=\frac{c+d}{c-d};$$

or $\qquad a+b:a-b=c+d:c-d.$

Several other proportions may be proved in a similar way.

301. The results of the preceding article are the algebraical equivalents of some of the propositions in the fifth book of Euclid, and the student is advised to make himself familiar with them in their verbal form. For example, *dividendo* may be quoted as follows:

When there are four proportionals, the excess of the first above the second is to the second, as the excess of the third above the fourth is to the fourth.

302. We shall now compare the algebraical definition of proportion with that given in Euclid.

In algebraical symbols the definition of Euclid may be stated as follows:

Four magnitudes, a, b, c, d are in proportion when $pc \gtreqless qd$ according as $pa \gtreqless qb$, p and q being *any positive integers whatever*.

I. To deduce the geometrical definition of proportion from the algebraical definition.

Since $\frac{a}{b}=\frac{c}{d}$, by multiplying both sides by $\frac{p}{q}$, we obtain

$$\frac{pa}{qb}=\frac{pc}{qd};$$

hence, from the properties of fractions,

$$pc \gtreqless qd \text{ according as } pa \gtreqless qb,$$

which proves the proposition.

II. To deduce the algebraical definition of proportion from the geometrical definition.

Given that $pc \gtreqless qd$ according as $pa \gtreqless qb$, to prove
$$\frac{a}{b} = \frac{c}{d}$$

If $\frac{a}{b}$ is not equal to $\frac{c}{d}$ one of them must be the greater. Suppose $\frac{a}{b} > \frac{c}{d}$; then it will be possible to find some fraction $\frac{q}{p}$ which lies between them.

Hence $\qquad \frac{a}{b} > \frac{q}{p}$ ....(1),

and $\qquad \frac{c}{d} < \frac{q}{p}$ ....(2).

From (1) $\qquad pa > qb$;
from (2) $\qquad pc < qd$;
and these contradict the hypothesis.

Therefore $\frac{a}{b}$ and $\frac{c}{d}$ are not unequal; that is $\frac{a}{b} = \frac{d}{c}$; which proves the proposition.

Example 1. If $\qquad a:b=c:d=e:f$,
shew that $\qquad 2a^2 + 3c^2 - 5e^2 : 2b^2 + 3d^2 - 5f^2 = ae:bf$.

Let $\frac{a}{b} = \frac{c}{d} = \frac{e}{f} = k$; then $a = bk$, $c = dk$, $e = fk$;

$$\therefore \frac{2a^2 + 3c^2 - 5e^2}{2b^2 + 3d^2 - 5f^2} = \frac{2b^2k^2 + 3d^2k^2 - 5f^2k^2}{2b^2 + 3d^2 - 5f^2}$$

$$= k^2 = \frac{a}{b} \times \frac{e}{f} = \frac{ae}{bf},$$

or $\qquad 2a^2 + 3c^2 - 5e^2 : 2b^2 + 3d^2 - 5f^2 = ae:bf$.

Example 2. Solve the equation $\dfrac{x^2 + x - 2}{x - 2} = \dfrac{4x^2 + 5x - 6}{5x - 6}$.

Dividendo, $\qquad \dfrac{x^2}{x-2} = \dfrac{4x^2}{5x-6}$;

whence, dividing by x^2, which gives a solution $x = 0$, [Art. 201.]

$$\frac{1}{x-2} = \frac{4}{5x-6};$$

whence $\qquad x = -2$,
and therefore the roots are 0, -2.

Example 3. If
$$(3a+6b+c+2d)(3a-6b-c+2d)=(3a-6b+c-2d)(3a+6b-c-2d),$$
prove that a, b, c, d are in proportion.

We have $\qquad \dfrac{3a+6b+c+2d}{3a-6b+c-2d}=\dfrac{3a+6b-c-2d}{3a-6b-c+2d}.$ [Art. 296.]

Componendo and Dividendo,
$$\frac{2(3a+c)}{2(6b+2d)}=\frac{2(3a-c)}{2(6b-2d)}.$$

Alternando, $\qquad \dfrac{3a+c}{3a-c}=\dfrac{6b+2d}{6b-2d}.$

Again, Componendo and Dividendo,
$$\frac{6a}{2c}=\frac{12b}{4d};$$
whence $\qquad a:b=c:d.$

EXAMPLES XXXII. b.

Find a fourth proportional to

1. a, ab, c.   2. a^2, $2ab$, $3b^2$.   3. x^3, xy, $5x^2y$.

Find a third proportional to

4. a^2b, ab.   5. x^3, $2x^2$.   6. $3x$, $6xy$.   7. 1, x.

Find a mean proportional between

8. a^2, b^2.   9. $2x^3$, $8x$.   10. $12ax^2$, $3a^3$.   11. $27a^2b^3$, $3b$.

If a, b, c are three proportionals, shew that

12. $a:a+b=a-b:a-c.$
13. $(b^2+bc+c^2)(ac-bc+c^2)=b^4+ac^3+c^4.$

If $a:b=c:d$, prove that

14. $ab+cd:ab-cd=a^2+c^2:a^2-c^2.$
15. $a^2+ac+c^2:a^2-ac+c^2=b^2+bd+d^2:b^2-bd+d^2.$
16. $a:b=\sqrt{3a^2+5c^2}:\sqrt{3b^2+5d^2}.$
17. $\dfrac{a}{p}+\dfrac{b}{q}:a=\dfrac{c}{p}+\dfrac{d}{q}:c.$
18. $\dfrac{b}{a}+\dfrac{a}{b}:\dfrac{ab}{a^2+b^2}=\dfrac{d}{c}+\dfrac{c}{d}:\dfrac{cd}{c^2+d^2}.$

Solve the equations:

19. $3x-1 : 6x-7 = 7x-10 : 9x+10$.
20. $x-12 : y+3 = 2x-19 : 5y-13 = 5 : 14$.
21. $\dfrac{x^2-2x+3}{2x-3} = \dfrac{x^2-3x+5}{3x-5}$.
22. $\dfrac{2x-1}{x^2+2x-1} = \dfrac{x+4}{x^2+x+4}$.
23. If $(a+b-3c-3d)(2a-2b-c+d)$
$= (2a+2b-c-d)(a-b-3c+3d)$
prove that a, b, c, d are proportionals.
24. If a, b, c, d are in continued proportion, prove that
$$a : d = a^3+b^3+c^3 : b^3+c^3+d^3.$$
25. If b is a mean proportional between a and c, shew that $4a^2-9b^2$ is to $4b^2-9c^2$ in the duplicate ratio of a to b.
26. If a, b, c, d are in continued proportion, prove that $b+c$ is a mean proportional between $a+b$ and $c+d$.
27. If $a+b : b+c = c+d : d+a$,
prove that $a = c$, or $a+b+c+d = 0$.
28. If $a : b = c : d = e : f$, prove that
   (i) $5a-7c+3e : 5b-7d+3f = c : d$.
   (ii) $4a^2-5ace+6e^2f : 4b^2-5bde+6f^3 = ae : bf$.
   (iii) $a^2ce : b^2df = 2a^4b^2+3a^2e^2-5e^4f : 2b^6+3b^2f^2-5f^5$.
29. If $a : b = x : y$, prove that
   (i) $al+xm : bl+ym = ap+xq : bp+yq$.
   (ii) $pa^2+qax+rx^2 : pb^2+qby+ry^2 = a^2+x^2 : b^2+y^2$.

[*Examples for revision will be found on page* 307.]

Variation.

303. DEFINITION. One quantity A is said to **vary directly** as another B, when the two quantities depend upon each other in such a manner that if B is changed, A is changed *in the same ratio*.

Note. The word *directly* is often omitted, and A is said to vary as B.

304. For instance: if a train moving at a uniform rate travels 40 miles in 60 minutes, it will travel 20 miles in 30 minutes, 80 miles in 120 minutes, and so on; the distance in each case being increased or diminished in the same ratio as the time. This is expressed by saying that when the velocity is uniform *the distance is proportional to the time*, or more briefly, *the distance varies as the time*.

Again, if we refer to the general formula of Art. 84, we find that $\frac{s}{t} = v$ is a relation connecting the space described by a body which moves for a time t with *uniform* velocity v. That is, if $s_1, s_2, s_3 \ldots$ be spaces described in times $t_1, t_2, t_3 \ldots$ respectively,

we have $$\frac{s_1}{t_1} = \frac{s_2}{t_2} = \frac{s_3}{t_3} = \ldots\ldots = v.$$

From this it appears that the ratio of any value of s to the corresponding value of t is *constant*, that is, remains the same whatever numerical values s and t may have.

This is an instance of *direct variation*, and s is said to *vary as t*.

305. The symbol $\propto$ is used to denote variation; so that $A \propto B$ is read "*A* varies as *B*."

306. *If A varies as B, then A is equal to B multiplied by some constant quantity.*

For suppose that $a_1, a_2, a_3 \ldots, b_1, b_2, b_3 \ldots$ are corresponding values of A and B.

Then, by definition, $\dfrac{A}{a_1} = \dfrac{B}{b_1}$; $\dfrac{A}{a_2} = \dfrac{B}{b_2}$; $\dfrac{A}{a_3} = \dfrac{B}{b_3}$; and so on.

$$\therefore \frac{a_1}{b_1} = \frac{a_2}{b_2} = \frac{a_3}{b_3} = \ldots,$$ each being equal to $\dfrac{A}{B}$.

Hence $\dfrac{\text{any value of } A}{\text{the corresponding value of } B}$ is always the same;

that is, $\qquad \dfrac{A}{B} = m$, where m is constant.

$$\therefore A = mB.$$

307. DEFINITION. One quantity A is said to **vary inversely** as another B when A varies *directly* as the reciprocal of B. [See Art. 107].

Thus if A varies inversely as B, $A = \dfrac{m}{B}$, where m is constant.

The following is an illustration of inverse variation: If 6 men do a certain work in 8 hours, 12 men would do the same work in 4 hours, 2 men in 24 hours; and so on. Thus it appears that when the number of men is increased the time is proportionately decreased; and vice versâ.

308. Definition. One quantity is said to **vary jointly** as a number of others when it varies directly as their product.

Thus A varies jointly as B and C when $A = mBC$, where m is constant. For instance, the interest on a sum of money varies jointly as the principal, the time, and the rate per cent.

309. Definition. A is said to vary directly as B and inversely as C when A varies as $\dfrac{B}{C}$.

310. *If* A *varies as* B *when* C *is constant, and* A *varies as* C *when* B *is constant, then will* A *vary as* BC *when both* B *and* C *vary.*

The variation of A depends partly on that of B and partly on that of C. Suppose these latter variations to take place separately, each in its turn producing its own effect on A; also let a, b, c be certain simultaneous values of A, B, C.

1. *Let C be constant* while B changes to b; then A must undergo a partial change and will assume some intermediate value a', where

$$\frac{A}{a'} = \frac{B}{b} \quad \ldots\ldots\ldots\ldots\ldots\ldots\ldots\ldots\ldots (1).$$

2. *Let B be constant*, that is, let it retain its value b, while C changes to c; then A must complete its change and pass from its intermediate value a' to its final value a, where

$$\frac{a'}{a} = \frac{C}{c} \quad \ldots\ldots\ldots\ldots\ldots\ldots\ldots\ldots\ldots (2).$$

From (1) and (2) $\quad \dfrac{A}{a'} \times \dfrac{a'}{a} = \dfrac{B}{b} \times \dfrac{C}{c};$

that is, $\quad A = \dfrac{a}{bc} \cdot BC.$

or $\quad A$ varies as BC.

311. The following are illustrations of the theorem proved in the last article.

The amount of work done by *a given number of men* varies directly as the number of days they work, and the amount of work done *in a given time* varies directly as the number of men; therefore when the number of days and the number of men are both variable, the amount of work will vary as the product of the number of men and the number of days.

Again, in Geometry the area of a triangle varies directly as its base when the height is constant, and directly as the height when the base is constant; and when both the height and base are variable, the area varies as the product of the numbers representing the height and the base.

Example 1. If $A \propto B$, and $C \propto D$, then will $AC \propto BD$.

For, by supposition, $A = mB$, $C = nD$, where m and n are constants.

Therefore $AC = mnBD$; and as mn is constant, $AC \propto BD$.

Example 2. If x varies inversely as $y^2 - 1$, and is equal to 24 when $y = 10$; find x when $y = 5$.

By supposition, $x = \dfrac{m}{y^2 - 1}$, where m is constant.

Putting $x = 24$, $y = 10$, we obtain $24 = \dfrac{m}{99}$,

whence $m = 24 \times 99$

$$\therefore x = \frac{24 \times 99}{y^2 - 1};$$

hence, putting $y = 5$, we obtain $x = 99$.

Example 3. The volume of a pyramid varies jointly as its height and the area of its base; and when the area of the base is 60 square feet and the height 14 feet the volume is 280 cubic feet. What is the area of the base of a pyramid whose volume is 390 cubic feet and whose height is 26 feet?

Let V denote the volume, A the area of the base, and h the height;

then $V = mAh$, where m is constant.

Substituting the given values of V, A, h we have

$$280 = m \times 60 \times 14;$$

$$\therefore m = \frac{280}{60 \times 14} = \frac{1}{3}.$$

$$\therefore V = \frac{1}{3} Ah.$$

Also when $V = 390$, $h = 26$;

$$\therefore 390 = \frac{1}{3} A \times 26;$$

$$\therefore A = 45.$$

Hence the area of the base is 45 square feet.

EXAMPLES XXXII. c.

1. If $x \propto y$, and $y=7$ when $x=18$, find x when $y=21$.

2. If $x \propto y$, and $y=3$ when $x=2$, find y when $x=18$.

3. A varies jointly as B and C; and $A=6$ when $B=3$, $C=2$ find A when $B=5$, $C=7$.

4. A varies jointly as B and C; and $A=9$ when $B=5$, $C=7$: find B when $A=54$, $C=10$.

5. If $x \propto \dfrac{1}{y}$, and $y=4$ when $x=15$, find y when $x=6$.

6. If $y \propto \dfrac{1}{x}$, and $y=1$ when $x=1$, find x when $y=5$.

7. A varies as B directly, and as C inversely; and $A=10$ when $B=15$, $C=6$; find A when $B=8$, $C=2$.

8. If x varies as y directly, and as z inversely, and $x=14$ when $y=10$, $z=14$; find z when $x=49$, $y=45$.

9. If $x \propto \dfrac{1}{y}$, and $y \propto \dfrac{1}{z}$, prove that $z \propto x$.

10. If $a \propto b$, prove that $a^n \propto b^n$.

11. If $x \propto z$ and $y \propto z$, prove that $x^2 - y^2 \propto z^2$.

12. If $3a+7b \propto 3a+13b$, and when $a=5$, $b=3$, find the equation between a and b.

13. If $5x-y \propto 10x-11y$, and when $x=7$, $y=5$, find the equation between x and y.

14. If the cube of x varies as the square of y, and if $x=3$ when $y=5$, find the equation between x and y.

15. If the square root of a varies as the cube root of b, and if $a=4$ when $b=8$, find the equation between a and b.

16. If y varies inversely as the square of x, and if $y=8$ when $x=3$, find x when $y=2$.

17. If $x \propto y+a$, where a is constant, and $x=15$ when $y=1$, and $x=35$ when $y=5$; find x when $y=2$.

18. If $a+b \propto a-b$, prove that $a^2+b^2 \propto ab$; and if $a \propto b$, prove that $a^2-b^2 \propto ab$.

19. If y be the sum of three quantities which vary as x, x^2, x^3 respectively, and when $x=1$, $y=4$, when $x=2$, $y=8$, and when $x=3$, $y=18$, express y in terms of x.

20. Given that the area of a circle varies as the square of its radius, and that the area of a circle is 154 square feet when the radius is 7 feet; find the area of a circle whose radius is 10 feet 6 inches.

21. The area of a circle varies as the square of its diameter; prove that the area of a circle whose diameter is $2\frac{1}{2}$ inches is equal to the sum of the areas of two circles whose diameters are $1\frac{1}{2}$, and 2 inches respectively.

22. The pressure of wind on a plane surface varies jointly as the area of the surface, and the square of the wind's velocity. The pressure on a square foot is 1 lb. when the wind is moving at the rate of 15 miles per hour; find the velocity of the wind when the pressure on a square yard is 16 lbs.

23. The value of a silver coin varies directly as the square of its diameter, while its thickness remains the same; it also varies directly as its thickness while its diameter remains the same. Two silver coins have their diameters in the ratio of 4 : 3. Find the ratio of their thicknesses if the value of the first be four times that of the second.

24. The volume of a circular cylinder varies as the square of the radius of the base when the height is the same, and as the height when the base is the same. The volume is 88 cubic feet when the height is 7 feet, and the radius of the base is 2 feet; what will be the height of a cylinder on a base of radius 9 feet, when the volume is 396 cubic feet?

25. The altitude of a triangle varies directly as its area and inversely as its base. A triangle, 2 square yards in area, standing on a base of $13\frac{1}{2}$ feet, has an altitude of $2\frac{2}{3}$ feet: find the altitude of a triangle whose base is 1 foot 4 inches, and whose area is 2 square feet 96 inches.

26. The expenses of a school are partly constant and partly vary as the number of boys. The expenses were £1000 for 150 boys and £840 for 120 boys; what will the expenses be when there are 330 boys? [Compare Art. 442, Ex. 2.]

CHAPTER XXXIII.

ARITHMETICAL PROGRESSION.

312. DEFINITION. Quantities are said to be in **Arithmetical Progression** when they increase or decrease by a *common difference*.

Thus each of the following series forms an Arithmetical Progression:

$$3, 7, 11, 15, \ldots\ldots\ldots\ldots\ldots$$
$$8, 2, -4, -10, \ldots\ldots\ldots\ldots$$
$$a, a+d, a+2d, a+3d, \ldots\ldots$$

The common difference is found by subtracting *any* term of the series from that which *follows* it. In the first of the above examples the common difference is 4; in the second it is -6; in the third it is d.

313. If we examine the series

$$a, a+d, a+2d, a+3d, \ldots\ldots$$

we notice that *in any term the coefficient of* d *is always less by one than the number of the term in the series*.

Thus the   3rd term is $a+2d$;
             6th term is $a+5d$;
             20th term is $a+19d$;
and, generally, the pth term is $a+(p-1)d$.

If n be the number of terms, and if l denote the last, or nth term, we have $l = a + (n-1)d$.

314. *To find the sum of a number of terms in Arithmetical Progression.*

Let a denote the first term, d the common difference, and n the number of terms. Also let l denote the last term, and s the required sum; then

$$s = a + (a+d) + (a+2d) + \ldots\ldots + (l-2d) + (l-d) + l;$$

and, by writing the series in the reverse order,

$$s = l + (l-d) + (l-2d) + \ldots\ldots + (a+2d) + (a+d) + a.$$

Adding together these two series,

$$2s = (a+l) + (a+l) + (a+l) + \ldots \text{ to } n \text{ terms}$$
$$= n(a+l),$$
$$\therefore \quad s = \frac{n}{2}(a+l) \quad \ldots\ldots\ldots\ldots\ldots\ldots (1);$$

and
$$l = a + (n-1)d \quad \ldots\ldots\ldots\ldots\ldots\ldots (2),$$
$$\therefore \quad s = \frac{n}{2}\{2a + (n-1)d\} \quad \ldots\ldots\ldots\ldots\ldots\ldots (3).$$

315. In the last article we have three useful formulæ (1), (2), (3); in each of these any one of the letters may denote the unknown quantity when the three others are known. [See Art. 82, Chap. ix.] For instance, in (1) if we substitute given values for s, n, l, we obtain an equation for finding a; and similarly in the other formulæ. But it is necessary to guard against a too mechanical use of these general formulæ, and it will often be found better to solve simple questions by a mental rather than by an actual reference to the requisite formula.

Example 1. Find the 20th and 35th terms of the series

$$38, 36, 34, \ldots\ldots.$$

Here the common difference is $36 - 38$, or -2.

$$\therefore \text{ the 20th term} = 38 + 19(-2)$$
$$= 0;$$
$$\text{and the 35th term} = 38 + 34(-2)$$
$$= -30.$$

Example 2. Find the sum of the series $5\frac{1}{2}$, $6\frac{3}{4}$, 8, $\ldots\ldots$ to 17 terms.

Here the common difference is $1\frac{1}{4}$; hence from (3)

$$\text{The sum} = \frac{17}{2}\left\{2 \times \frac{11}{2} + 16 \times 1\frac{1}{4}\right\}$$
$$= \frac{17}{2}(11 + 20)$$
$$= \frac{17 \times 31}{2}$$
$$= 263\frac{1}{2}.$$

Example 3. The first term of a series is 5, the last 45, and the sum 400: find the number of terms, and the common difference.

If n be the number of terms, then from (1)
$$400 = \frac{n}{2}(5+45);$$
whence $\qquad n=16.$

If d be the common difference,
$$45 = \text{the } 16^{\text{th}} \text{ term}$$
$$= 5 + 15d;$$
whence $\qquad d = 2\frac{2}{3}.$

EXAMPLES XXXIII. a.

1. Find the 27th and 41st terms in the series 5, 11, 17, ...
2. Find the 13th and 109th terms in the series 71, 70, 69, ...
3. Find the 17th and 54th terms in the series 10, $11\frac{1}{2}$, 13, ...
4. Find the 20th and 13th terms in the series -3, -2, -1, ...
5. Find the 90th and 16th terms in the series -4, 2·5, 9, ...
6. Find the 37th and 89th terms in the series $-2\cdot8$, 0, $2\cdot8$, ...

Find the last term in the following series:

7. 5, 7, 9, ... to 20 terms.
8. 7, 3, -1, ... to 15 terms.
9. $13\frac{1}{2}$, 9, $4\frac{1}{2}$, ... to 13 terms.
10. ·6, 1·2, 1·8, ... to 12 terms.
11. 2·7, 3·4, 4·1, ... to 11 terms.
12. x, $2x$, $3x$, ... to 25 terms.
13. $a-d$, $a+d$, $a+3d$, ... to 30 terms.
14. $2a-b$, $4a-3b$, $6a-5b$, ... to 40 terms.

Find the last term and sum of the following series:

15. 14, 64, 114, ... to 20 terms.
16. 1, 1·2, 1·4, ... to 12 terms.
17. 9, 5, 1, ... to 100 terms.
18. $\frac{1}{4}$, $-\frac{1}{4}$, $-\frac{3}{4}$, ... to 21 terms.
19. $3\frac{1}{2}$, 1, $-1\frac{1}{2}$, ... to 19 terms.
20. 64, 96, 128, ... to 16 terms.

Find the sum of the following series:

21. 5, 9, 13, ... to 19 terms.
22. 12, 9, 6, ... to 23 terms.
23. 4, $5\frac{1}{4}$, $6\frac{1}{2}$, ... to 37 terms.
24. $10\frac{1}{2}$, 9, $7\frac{1}{2}$, ... to 94 terms.
25. -3, 1, 5, ... to 17 terms.
26. 10, $9\frac{2}{3}$, $9\frac{1}{3}$, ... to 21 terms.
27. p, $3p$, $5p$, ... to p terms.
28. $3a$, a, $-a$, ... to a terms.
29. a, 0, $-a$, ... to a terms.
30. $-3q$, $-q$, q, ... to p terms.

Find the number of terms and the common difference when

31. The first term is 3, the last term 90, and the sum 1395.
32. The first term is 79, the last term 7, and the sum 1075.
33. The sum is 24, the first term 9, the last term -6.
34. The sum is 714, the first term 1, the last term $58\tfrac{1}{2}$.
35. The last term is -16, the sum -133, the first term -3.
36. The first term is -75, the sum -740, the last term 1.
37. The first term is a, the last $13a$, and the sum $49a$.
38. The sum is $-320x$, the first term $3x$, the last term $-35x$.

316. If *any two* terms of an Arithmetical Progression be given, the series can be completely determined; for the data furnish *two* simultaneous equations, the solution of which will give the first term, and the common difference.

Example. Find the series whose 7th and 51st terms are -3 and -355 respectively.

If a be the first term, and d the common difference,

$$-3 = \text{the } 7^{\text{th}} \text{ term}$$
$$= a + 6d;$$

and
$$-355 = \text{the } 51^{\text{st}} \text{ term}$$
$$= a + 50d;$$

whence, by subtraction, $-352 = 44d$;

$\therefore d = -8$; and consequently $a = 45$.

Hence the series is 45, 37, 29, ...

317. DEFINITION. When three quantities are in Arithmetical Progression the middle one is said to be the **arithmetic mean** of the other two.

Thus a is the arithmetic mean between $a - d$ and $a + d$.

318. *To find the arithmetic mean between two given quantities.*

Let a and b be the two quantities; A the arithmetic mean. Then since a, A, b are in A.P. we must have

$$b - A = A - a,$$

each being equal to the common difference;

whence
$$A = \frac{a+b}{2}.$$

319. Between two given quantities it is always possible to insert any number of terms such that the whole series thus formed shall be in A.P.; and by an extension of the definition in Art. 317. the terms thus inserted are called the *arithmetic means*.

Example. Insert 20 arithmetic means between 4 and 67.

Including the extremes the number of terms will be 22; so that we have to find a series of 22 terms in A.P., of which 4 is the first and 67 the last.

Let d be the common difference;

then $\qquad 67 = $ the 22nd term
$$= 4 + 21d;$$

whence $d = 3$, and the series is 4, 7, 10, ... 61, 64, 67;
and the required means are 7, 10, 13, ... 58, 61, 64.

320. *To insert a given number of arithmetic means between two given quantities.*

Let a and b be the given quantities, n the number of means.

Including the extremes the number of terms will be $n+2$; so that we have to find a series of $n+2$ terms in A.P., of which a is the first, and b is the last.

Let d be the common difference;

then $\qquad b = $ the $(n+2)^{\text{th}}$ term
$$= a + (n+1)d;$$

whence $\qquad d = \dfrac{b-a}{n+1};$

and the required means are
$$a + \frac{b-a}{n+1},\ \ a + \frac{2(b-a)}{n+1},\ \ \ldots\ldots\ a + \frac{n(b-a)}{n+1}.$$

Example 1. Find the 30th term of an A.P. of which the first term is 17, and the 100th term -16.

Let d be the common difference;

then $\qquad -16 = $ the 100th term
$$= 17 + 99d;$$
$$\therefore\ d = -\frac{1}{3}.$$

The 30th term $\qquad = 17 + 29\left(-\dfrac{1}{3}\right)$
$$= 7\tfrac{1}{3}.$$

Example 2. The sum of three numbers in A.P. is 33, and their product is 792; find them.

Let a be the *middle* number, d the common difference; then the three numbers are $a-d$, a, $a+d$.

Hence $$a-d+a+a+d=33;$$
whence $a=11$, and the three numbers are $11-d$, 11, $11+d$.
$$\therefore\ 11(11+d)(11-d)=792,$$
$$121-d^2=72,$$
$$d=\pm 7;$$
and the numbers are 4, 11, 18.

Example 3. How many terms of the series 24, 20, 16, ... must be taken that the sum may be 72?

Let the number of terms be n; then, since the common difference is $20-24$, or -4, we have from (3), Art. 314,
$$72=\frac{n}{2}\{2\times 24+(n-1)(-4)\}$$
$$=24n-2n(n-1);$$
whence $\qquad n^2-13n+36=0,$
or $\qquad (n-4)(n-9)=0;$
$$\therefore\ n=4 \text{ or } 9.$$

Both these values satisfy the conditions of the question; for if we write down the first 9 terms, we get 24, 20, 16, 12, 8, 4, 0, -4, -8; and, as the last five terms destroy each other, the sum of 9 terms is the same as that of 4 terms.

Example 4. An A.P. consists of 21 terms; the sum of the three terms in the middle is 129, and of the last three is 237; find the series.

Let a be the first term, and d the common difference. Then
$$237=\text{the sum of the last three terms}$$
$$=a+20d+a+19d+a+18d$$
$$=3a+57d;$$
whence $\qquad a+19d=79$ ....(1).

Again, the three middle terms are the 10th, 11th, 12th; hence $\qquad 129=\text{the sum of the three middle terms}$
$$=a+9d+a+10d+a+11d$$
$$=3a+30d;$$
whence $\qquad a+10d=43$ ....(2).

From (1) and (2), we obtain $d=4$, $a=3$.

Hence the series is 3, 7, 11, ... 83.

EXAMPLES XXXIII. b.

Find the series in which

1. The 27th term is 186, and the 45th term 312.
2. The 5th term is 1, and the 31st term -77.
3. The 15th term is -25, and the 23rd term -41.
4. The 9th term is -11, and the 102nd term $-150\frac{1}{2}$.
5. The 15th term is 25, and the 29th term 46.
6. The 16th term is 214, and the 51st term 739.
7. The 3rd and 7th terms of an A.P. are 7 and 19; find the 15th term.
8. The 54th and 4th terms are -125 and 0; find the 42nd term.
9. The 31st and 2nd terms are $\frac{1}{2}$ and $7\frac{3}{4}$; find the 59th term.
10. Insert 15 arithmetic means between 71 and 23.
11. Insert 17 arithmetic means between 93 and 69.
12. Insert 14 arithmetic means between $-7\frac{1}{5}$ and $-2\frac{1}{5}$.
13. Insert 16 arithmetic means between 7·2 and $-6·4$.
14. Insert 36 arithmetic means between $8\frac{1}{2}$ and $2\frac{1}{6}$.

How many terms must be taken of

15. The series 42, 39, 36, ... to make 315?
16. The series $-16, -15, -14, \ldots\ldots$ to make -100?
17. The series $15\frac{2}{3}, 15\frac{1}{3}, 15, \ldots\ldots$ to make 129?
18. The series $20, 18\frac{3}{4}, 17\frac{1}{2}, \ldots\ldots$ to make $162\frac{1}{2}$?
19. The series $-10\frac{1}{2}, -9, -7\frac{1}{2}, \ldots\ldots$ to make -42?
20. The series $-6\frac{4}{5}, -6\frac{2}{5}, -6, \ldots\ldots$ to make $-52\frac{4}{5}$.
21. The sum of three numbers in A.P. is 39, and their product is 2184; find them.
22. The sum of three numbers in A.P. is 12, and the sum of their squares is 66; find them.
23. The sum of five numbers in A.P. is 75, and the product of the greatest and least is 161; find them.
24. The sum of five numbers in A.P. is 40, and the sum of their squares is 410; find them.
25. The 12th, 85th and last terms of an A.P. are 38, 257, 395 respectively; find the number of terms.

[*Examples for revision will be found in Miscellaneous Examples V., page* 307.]

CHAPTER XXXIV.

Geometrical Progression.

321. Definition. Quantities are said to be in **Geometrical Progression** when they increase or decrease by a *constant factor*.

Thus each of the following series forms a Geometrical Progression:

$$3,\ 6,\ 12,\ 24,\ \ldots\ldots\ldots$$

$$1,\ -\frac{1}{3},\ \frac{1}{9},\ -\frac{1}{27},\ \ldots\ldots\ldots$$

$$a,\ ar,\ ar^2,\ ar^3,\ \ldots\ldots\ldots$$

The constant factor is also called the *common ratio*, and it is found by dividing *any* term by that which immediately *precedes* it. In the first of the above examples the common ratio is 2; in the second it is $-\frac{1}{3}$; in the third it is r.

322. If we examine the series

$$a,\ ar,\ ar^2,\ ar^3,\ ar^4,\ \ldots\ldots$$

we notice that *in any term the index of* r *is always less by one than the number of the term in the series.*

Thus   the 3rd term is ar^2;

         the 6th term is ar^5;

         the 20th term is ar^{19};

and, generally,   the pth term is ar^{p-1}.

If n be the number of terms, and if l denote the last, or n^{th} term, we have $l = ar^{n-1}$.

Example. Find the 8th term of the series $-\frac{1}{3},\ \frac{1}{2},\ -\frac{3}{4},\ \ldots\ldots$

The common ratio is $\frac{1}{2} \div \left(-\frac{1}{3}\right)$, or $-\frac{3}{2}$;

$$\therefore \text{ the 8}^{th}\text{ term} = -\frac{1}{3} \times \left(-\frac{3}{2}\right)^7$$

$$= -\frac{1}{3} \times -\frac{2187}{128}$$

$$= \frac{729}{128}.$$

CHAP. XXXIV.] GEOMETRICAL PROGRESSION.

323. Definition. When three quantities are in Geometrical Progression the middle one is called the **geometric mean** between the other two.

To find the geometric mean between two given quantities.

Let a and b be the two quantities; G the geometric mean. Then since a, G, b are in G.P.,

$$\frac{b}{G} = \frac{G}{a},$$

each being equal to the common ratio;

$$\therefore G^2 = ab;$$

whence $$G = \sqrt{ab}.$$

324. *To insert a given number of geometric means between two given quantities.*

Let a and b be the given quantities, n the number of means.

In all there will be $n+2$ terms; so that we have to find a series of $n+2$ terms in G.P., of which a is the first and b the last.

Let r be the common ratio;

then $$b = \text{the } (n+2)^{\text{th}} \text{ term}$$
$$= ar^{n+1};$$

$$\therefore r^{n+1} = \frac{b}{a};$$

$$\therefore r = \left(\frac{b}{a}\right)^{\frac{1}{n+1}} \quad\quad\quad\quad\quad\quad\quad (1).$$

Hence the required means are ar, ar^2, ... ar^n, where r has the value found in (1).

Example. Insert 4 geometric means between 160 and 5

We have to find 6 terms in G.P. of which 160 is the first, and 5 the sixth.

Let r be the common ratio;

then $5 = $ the sixth term
$$= 160r^5;$$

$$\therefore r^5 = \frac{1}{32};$$

whence, *by trial*, $$r = \frac{1}{2};$$

and the means are 80, 40, 20, 10.

325. *To find the sum of a number of terms in Geometrical Progression.*

Let a be the first term, r the common ratio, n the number of terms, and s the sum required. Then

$$s = a + ar + ar^2 + \ldots + ar^{n-2} + ar^{n-1};$$

multiplying every term by r, we have

$$rs = ar + ar^2 + \ldots + ar^{n-2} + ar^{n-1} + ar^n.$$

Hence by subtraction,

$$rs - s = ar^n - a;$$
$$\therefore (r-1)s = a(r^n - 1);$$
$$\therefore s = \frac{a(r^n - 1)}{r - 1} \quad \ldots\ldots\ldots(1).$$

Changing the signs in numerator and denominator [Art. 170.]

$$s = \frac{a(1 - r^n)}{1 - r} \quad \ldots\ldots\ldots(2).$$

Note. It will be found convenient to remember both forms given above for s, using (2) in all cases except when r is *positive and greater than* 1.

Since $ar^{n-1} = l$, the formula (1) may be written

$$s = \frac{rl - a}{r - 1};$$

a form which is sometimes useful.

Example 1. Sum the series 81, 54, 36, ... to 9 terms.

The common ratio $= \dfrac{54}{81} = \dfrac{2}{3}$, which is less than 1;

hence the sum
$$= \frac{81\left\{1 - \left(\dfrac{2}{3}\right)^9\right\}}{1 - \dfrac{2}{3}}$$

$$= 243\left\{1 - \left(\dfrac{2}{3}\right)^9\right\}$$

$$= 243 - \frac{512}{81}$$

$$= 236\frac{55}{81}.$$

Example 2. Sum the series $\dfrac{2}{3}$, -1, $\dfrac{3}{2}$, ... to 7 terms.

The common ratio $= -\dfrac{3}{2}$; hence by formula (2)

the sum
$$= \dfrac{\dfrac{2}{3}\left\{1 - \left(-\dfrac{3}{2}\right)^7\right\}}{1 + \dfrac{3}{2}}$$

$$= \dfrac{\dfrac{2}{3}\left\{1 + \dfrac{2187}{128}\right\}}{\dfrac{5}{2}}$$

$$= \dfrac{2}{3} \times \dfrac{2315}{128} \times \dfrac{2}{5}$$

$$= \dfrac{463}{96}.$$

EXAMPLES XXXIV. a.

1. Find the 5th and 8th terms of the series 3, 6, 12, ...
2. Find the 10th and 16th terms of the series 256, 128, 64, ...
3. Find the 7th and 11th terms of the series 64, -32, 16, ...
4. Find the 8th and 12th terms of the series 81, -27, 9, ...
5. Find the 14th and 7th terms of the series $\dfrac{1}{64}$, $\dfrac{1}{32}$, $\dfrac{1}{16}$, ...
6. Find the 4th and 8th terms of the series ·008, ·04, ·2, ...

Find the last term in the following series:

7. 2, 4, 8, ... to 9 terms.
8. 2, -6, 18, ... to 8 terms.
9. 2, 3, $4\frac{1}{2}$, ... to 6 terms.
10. 3, -3^2, 3^3, ... to $2n$ terms.
11. x, x^3, x^5, ... to p terms.
12. x, 1, $\dfrac{1}{x}$, ... to 30 terms.
13. Insert 3 geometric means between 486 and 6.
14. Insert 4 geometric means between $\dfrac{1}{8}$ and 128.
15. Insert 6 geometric means between 56 and $-\dfrac{7}{16}$.
16. Insert 5 geometric means between $\dfrac{32}{81}$ and $4\frac{1}{2}$.

Find the last term and the sum of the following series:

17. $3, 6, 12, \ldots$ to 8 terms. 18. $6, -18, 54, \ldots$ to 6 terms.

19. $64, 32, 16, \ldots$ to 10 terms. 20. $8{\cdot}1, 2{\cdot}7, {\cdot}9, \ldots$ to 7 terms.

21. $\dfrac{1}{72}, \dfrac{1}{24}, \dfrac{1}{8}, \ldots$ to 8 terms. 22. $4\tfrac{1}{2}, 1\tfrac{1}{2}, \dfrac{1}{2}, \ldots$ to 9 terms.

Find the sum of the series

23. $3, -1, \dfrac{1}{3}, \ldots$ to 6 terms. 24. $\dfrac{1}{2}, \dfrac{1}{3}, \dfrac{2}{9}, \ldots$ to 7 terms.

25. $-\dfrac{2}{5}, \dfrac{1}{2}, -\dfrac{5}{8}, \ldots$ to 6 terms. 26. $1, -\dfrac{1}{2}, \dfrac{1}{4}, \ldots$ to 12 terms.

27. $9, -6, 4, \ldots$ to 7 terms. 28. $\dfrac{2}{3}, -\dfrac{1}{6}, \dfrac{1}{24}, \ldots$ to 8 terms.

29. $1, 3, 3^2, \ldots$ to p terms. 30. $2, -4, 8, \ldots$ to $2p$ terms.

31. $\dfrac{1}{\sqrt{3}}, 1, \dfrac{3}{\sqrt{3}}, \ldots$ to 8 terms. 32. $\sqrt{a}, \sqrt{a^3}, \sqrt{a^5}, \ldots$ to a terms.

33. $\dfrac{1}{\sqrt{2}}, -2, \dfrac{8}{\sqrt{2}}, \ldots$ to 7 terms. 34. $\sqrt{2}, \sqrt{6}, 3\sqrt{2}, \ldots$ to 12 terms.

326. Consider the series $\quad 1, \dfrac{1}{2}, \dfrac{1}{2^2}, \dfrac{1}{2^3} \ldots\ldots$.

The sum to n terms
$$= \frac{1 - \dfrac{1}{2^n}}{1 - \dfrac{1}{2}}$$
$$= 2\left(1 - \dfrac{1}{2^n}\right)$$
$$= 2 - \dfrac{1}{2^{n-1}}.$$

From this result it appears that however many terms be taken the sum of the above series is always less than 2. Also we see that, by making n sufficiently large, we can make the fraction $\dfrac{1}{2^{n-1}}$ as small as we please. Thus by taking a sufficient number of terms the sum can be made to differ by as little as we please from 2.

In the next article a more general case is discussed.

327. From Art. 325 we have $s = \dfrac{a(1-r^n)}{1-r}$

$$= \dfrac{a}{1-r} - \dfrac{ar^n}{1-r},$$

Suppose r is a proper fraction; then the greater the value of n the smaller is the value of r^n, and consequently of $\dfrac{ar^n}{1-r}$; and therefore by making n sufficiently large, we can make the sum of n terms of the series differ from $\dfrac{a}{1-r}$ by as small a quantity as we please.

This result is usually stated thus: *the sum of an infinite number of terms of a decreasing Geometrical Progression is* $\dfrac{a}{1-r}$; or more briefly, *the sum to infinity is* $\dfrac{a}{1-r}$;

328. Recurring decimals furnish a good illustration of infinite Geometrical Progressions.

Example. Find the value of $.4\dot{2}\dot{3}$.

$$.4\dot{2}\dot{3} = .4232323\ldots\ldots$$

$$= \dfrac{4}{10} + \dfrac{23}{1000} + \dfrac{23}{100000} + \ldots\ldots$$

$$= \dfrac{4}{10} + \dfrac{23}{10^3} + \dfrac{23}{10^5} + \ldots\ldots$$

$$= \dfrac{4}{10} + \dfrac{23}{10^3}\left(1 + \dfrac{1}{10^2} + \dfrac{1}{10^4} + \ldots\ldots\right)$$

$$= \dfrac{4}{10} + \dfrac{23}{10^3} \cdot \dfrac{1}{1 - \dfrac{1}{10^2}}$$

$$= \dfrac{4}{10} + \dfrac{23}{10^3} \cdot \dfrac{100}{99}$$

$$= \dfrac{4}{10} + \dfrac{23}{990}$$

$$= \dfrac{419}{990},$$

which agrees with the value found by the usual arithmetical rule.

EXAMPLES XXXIV. b.

Sum to infinity the following series:

1. $9, 6, 4, \ldots.$
2. $12, 6, 3, \ldots.$
3. $\dfrac{1}{2}, \dfrac{1}{4}, \dfrac{1}{8}, \ldots.$
4. $\dfrac{1}{2}, -\dfrac{1}{4}, \dfrac{1}{8}, \ldots.$
5. $\dfrac{1}{3}, \dfrac{2}{9}, \dfrac{4}{27}, \ldots.$
6. $\dfrac{8}{5}, -1, \dfrac{5}{8}, \ldots.$
7. $\cdot 9, \cdot 03, \cdot 001, \ldots.$
8. $\cdot 8, -\cdot 4, \cdot 2, \ldots.$

Find by the method of Art. 328, the value of

9. $\cdot \dot{3}$ 10. $\cdot 1\dot{6}.$ 11. $\cdot \dot{2}\dot{4}.$ 12. $\cdot \dot{3}7\dot{8}.$ 13. $\cdot 0\dot{3}\dot{7}.$

Find the series in which

14. The 10th term is 320 and the 6th term 20.
15. The 5th term is $\dfrac{27}{16}$ and the 9th term is $\dfrac{1}{3}$.
16. The 7th term is 625 and the 4th term -5.
17. The 3rd term is $\dfrac{9}{16}$ and the 6th term $-4\dfrac{1}{2}$.
18. Divide 183 into three parts in G.P. such that the sum of the first and third is $2\dfrac{1}{20}$ times the second.
19. Shew that the product of any odd number of consecutive terms of a G.P. will be equal to the nth power of the middle term, n being the number of terms.
20. The first two terms of an infinite G.P. are together equal to 1, and every term is twice the sum of all the terms which follow. Find the series.

Sum the following series:

21. $y^2 + 2b, \ y^4 + 4b, \ y^6 + 6b, \ \ldots$ to n terms.
22. $\dfrac{3 + 2\sqrt{2}}{3 - 2\sqrt{2}}, \ 1, \ \dfrac{3 - 2\sqrt{2}}{3 + 2\sqrt{2}}, \ \ldots$ to infinity.
23. $\sqrt{\dfrac{3}{2}}, \ \dfrac{1}{3}\sqrt{2}, \ \dfrac{2}{9}\sqrt{\dfrac{2}{3}}, \ \ldots$ to infinity.
24. $2n - \dfrac{1}{2}, \ 4n + \dfrac{1}{6}, \ 6n - \dfrac{1}{18}, \ \ldots$ to $2n$ terms.

[*Examples for revision will be found in Miscellaneous Examples V., p. 307.*]

CHAPTER XXXV.

Harmonical Progression.

329. Definition. Three quantities a, b, c are said to be in **Harmonical Progression** when $\dfrac{a}{c} = \dfrac{a-b}{b-c}$.

Any number of quantities are said to be in Harmonical Progression when every three consecutive terms are in Harmonical Progression.

330. *The reciprocals of quantities in Harmonical Progression are in Arithmetical Progression.*

By definition, if a, b, c are in Harmonical Progression,

$$\frac{a}{c} = \frac{a-b}{b-c};$$

$$\therefore a(b-c) = c(a-b),$$

dividing every term by abc,

$$\frac{1}{c} - \frac{1}{b} = \frac{1}{b} - \frac{1}{a},$$

which proves the proposition.

331. Harmonical properties are chiefly interesting because of their importance in Geometry and in the Theory of Sound: in Algebra the proposition just proved is the only one of any importance. There is no general formula for the sum of any number of quantities in Harmonical Progression. Questions in H.P. are generally solved by inverting the terms, and making use of the properties of the corresponding A.P.

Example. The 12th term of a H.P. is $\frac{1}{5}$, and the 19th term is $\frac{3}{22}$: find the series.

Let a be the first term, d the common difference of the corresponding A.P.; then
$$5 = \text{the 12}^{\text{th}}\text{ term}$$
$$= a + 11d\,;$$
and
$$\frac{22}{3} = \text{the 19}^{\text{th}}\text{ term}$$
$$= a + 18d\,;$$
whence
$$d = \frac{1}{3},\ a = \frac{4}{3}.$$

Hence the Arithmetical Progression is $\frac{4}{3},\ \frac{5}{3},\ 2,\ \frac{7}{3},\ \ldots\ldots\,;$

and the Harmonical Progression is $\frac{3}{4},\ \frac{3}{5},\ \frac{1}{2},\ \frac{3}{7},\ \ldots\ldots\,.$

332. *To find the harmonic mean between two given quantities.*

Let a, b be the two quantities, H their harmonic mean; then $\frac{1}{a}$, $\frac{1}{H}$, $\frac{1}{b}$ are in A.P.

$$\therefore\ \frac{1}{H} - \frac{1}{a} = \frac{1}{b} - \frac{1}{H},$$
$$\frac{2}{H} = \frac{1}{a} + \frac{1}{b},$$
$$H = \frac{2ab}{a+b}.$$

333. If A, G, H be the arithmetic, geometric, and harmonic means between a and b, we have proved

$$A = \frac{a+b}{2} \quad\ldots\ldots\ldots\ldots\ldots\ldots\ldots(1).$$
$$G = \sqrt{ab} \quad\ldots\ldots\ldots\ldots\ldots\ldots\ldots(2).$$
$$H = \frac{2ab}{a+b} \quad\ldots\ldots\ldots\ldots\ldots\ldots\ldots(3).$$

Therefore
$$AH = \frac{a+b}{2} \cdot \frac{2ab}{a+b}$$
$$= ab$$
$$= G^2\,;$$

that is, G is the geometric mean between A and H.

334. Miscellaneous questions in the Progressions afford scope for much skill and ingenuity, the solution being often very neatly effected by some special artifice. The student will find the following hints useful:

1. If the same quantity be added to, or subtracted from, all the terms of an A.P., the resulting terms will form an A.P. with the same common difference as before. [Art. 312.]

2. If all the terms of an A.P. be multiplied or divided by the same quantity, the resulting terms form an A.P., but with a new common difference. [Art. 312.]

3. If all the terms of a G.P. be multiplied or divided by the same quantity, the resulting terms form a G.P. with the same common ratio as before. [Art. 322.]

4. If $a, b, c, d, \ldots$ be in G.P., they are also in *continued proportion*, since, by definition,

$$\frac{a}{b} = \frac{b}{c} = \frac{c}{d} = \ldots = \frac{1}{r}.$$

Conversely, a series of quantities in continued proportion may be represented by $x, xr, xr^2, \ldots$.

Example 1. Find three quantities in G.P. such that their product is 343, and their sum $30\frac{1}{3}$.

Let $\dfrac{a}{r}, a, ar$ be the three quantities;

then we have $\qquad \dfrac{a}{r} \times a \times ar = 343 \ \ldots\ldots\ldots\ldots\ldots\ldots(1),$

and $\qquad a\left(\dfrac{1}{r} + 1 + r\right) = \dfrac{91}{3} \ \ldots\ldots\ldots\ldots\ldots\ldots(2).$

From (1) $\qquad a^3 = 343,$

$\qquad\qquad\qquad a = 7;$

$\therefore$ from (2) $\qquad 7(1 + r + r^2) = \dfrac{91}{3} r.$

Whence we obtain $\qquad r = 3, \text{ or } \dfrac{1}{3},$

and the numbers are $\dfrac{7}{3}, 7, 21.$

Example 2. If a, b, c be in H.P., prove that $\dfrac{a}{b+c}, \dfrac{b}{c+a}, \dfrac{c}{a+b}$ are also in H.P.

Since $\dfrac{1}{a}, \dfrac{1}{b}, \dfrac{1}{c}$ are in A.P.,

$$\dfrac{a+b+c}{a}, \dfrac{a+b+c}{b}, \dfrac{a+b+c}{c} \text{ are in A.P.;}$$

$$\therefore \quad 1+\dfrac{b+c}{a}, \ 1+\dfrac{a+c}{b}, \ 1+\dfrac{a+b}{c} \text{ are in A.P.;}$$

$$\therefore \quad \dfrac{b+c}{a}, \ \dfrac{a+c}{b}, \ \dfrac{a+b}{c} \text{ are in A.P.;}$$

$$\therefore \quad \dfrac{a}{b+c}, \ \dfrac{b}{c+a}, \ \dfrac{c}{a+b} \text{ are in H.P.}$$

Example 3. The nth term of an A.P. is $\dfrac{n}{5}+2$, find the sum of 49 terms.

Let a be the first term, and l the last; then by putting $n=1$, and $n=49$ respectively, we obtain

$$a = \dfrac{1}{5}+2, \qquad l = \dfrac{49}{5}+2;$$

$$\therefore \quad s = \dfrac{n}{2}(a+l) = \dfrac{49}{2}\left(\dfrac{50}{5}+4\right)$$

$$= \dfrac{49}{2} \times 14 = 343.$$

Example 4. If a, b, c, d, e be in G.P., prove that $b+d$ is the geometric mean between $a+c$ and $c+e$.

Since a, b, c, d, e are in continued proportion,

$$\dfrac{a}{b} = \dfrac{b}{c} = \dfrac{c}{d} = \dfrac{d}{e};$$

$\therefore$ each ratio $\qquad = \dfrac{a+c}{b+d} = \dfrac{b+d}{c+e}.$ [Art. 294.]

Whence $\qquad (b+d)^2 = (a+c)(c+e).$

EXAMPLES XXXV.

1. Find the 6th term of the series $4, \ 2, \ 1\tfrac{1}{3}, \ \ldots$.
2. Find the 21st term of the series $2\tfrac{1}{2}, \ 1\tfrac{12}{13}, \ 1\tfrac{9}{16}, \ \ldots$.
3. Find the 8th term of the series $1\tfrac{1}{3}, \ 1\tfrac{11}{17}, \ 2\tfrac{2}{13}, \ \ldots$.
4. Find the nth term of the series $3, \ 1\tfrac{1}{2}, \ 1, \ \ldots$.

Find the series in which

5. The 15th term is $\frac{1}{25}$, and the 23rd term is $\frac{1}{41}$.

6. The 2nd term is 2, and the 31st term is $\frac{4}{31}$.

7. The 39th term is $\frac{1}{11}$, and the 54th term is $\frac{1}{26}$.

Find the harmonic mean between

8. 2 and 4. 9. 1 and 13. 10. $\frac{1}{4}$ and $\frac{1}{10}$.

11. $\frac{1}{a}$ and $\frac{1}{b}$. 12. $\frac{1}{x+y}$ and $\frac{1}{x-y}$. 13. $x+y$ and $x-y$.

14. Insert two harmonic means between 4 and 12.

15. Insert three harmonic means between $2\frac{2}{5}$ and 12.

16. Insert four harmonic means between 1 and 6.

17. If G be the geometric mean between two quantities A and B, shew that the ratio of the arithmetic and harmonic means of A and G is equal to the ratio of the arithmetic and harmonic means of G and B.

18. To each of three consecutive terms of a G.P. the second of the three is added. Shew that the three resulting quantities are in H.P.

Sum the following series:

19. $1 + 1\frac{3}{4} + 3\frac{1}{16} + \ldots$ to 6 terms.

20. $1 + 1\frac{3}{4} + 2\frac{1}{2} + \ldots$ to 6 terms.

21. $(2a+x) + 3a + (4a-x) + \ldots$ to p terms.

22. $1\frac{4}{5} - 1\frac{1}{5} + \frac{4}{5} - \ldots$ to 8 terms.

23. $1\frac{4}{5} + 1\frac{1}{5} + \frac{3}{5} + \ldots$ to 12 terms.

24. If $x-a$, $y-a$, and $z-a$ be in G.P., prove that $2(y-a)$ is the harmonic mean between $y-x$ and $y-z$.

25. If a, b, c, d be in A.P., a, e, f, d in G.P., a, g, h, d in H.P. respectively; prove that $ad = ef = bh = cg$.

26. If a^2, b^2, c^2 be in A.P., prove that $b+c, c+a, a+b$ are in H.P.

27. If a, b, c be in A.P., and a, β, γ in H.P., shew that
$$\frac{a+c}{b\beta}=\frac{a+\gamma}{a\gamma}.$$

28. If a be the arithmetic mean between b and c, and b the geometric mean between a and c, prove that c will be the harmonic mean between a and b.

29. If $\frac{a+b}{2}, b, \frac{b+c}{2}$ be in H.P., then a, b, c are in G.P.

30. If a, b, c, d, e be in G.P., prove that $c(a+2c+e)=(b+d)^2$.

31. If $a, b, c, d \ldots$ be a series of quantities in G.P., shew that the reciprocals of $a^2-b^2, b^2-c^2, c^2-d^2, \ldots$ are also in G.P.; and find the sum of n terms of this latter series in terms of a and b.

32. If a, b, c be in A.P., and b, c, d in H.P., then $a, \frac{c^2}{d}, c$ are in H.P., and $b, \frac{ad}{b}, d$ are also in H.P.

33. If g be the geometric and a the arithmetic mean between m and n, and if k^2 be the arithmetic mean between m^2 and n^2, prove that a^2 is the arithmetic mean between g^2 and k^2.

34. If a, b, c, d be in G.P., prove that $(b-c)^2=ac+bd-2ad$.

35. If a, b, c, d be in G.P., prove that
$$(a+d)(a-b)^2 : a(a-c)(a-d) = a-b+c : a+b+c.$$

36. If a, b, c be in H.P., prove that
$$\frac{1}{a}+\frac{1}{b+c}, \ \frac{1}{b}+\frac{1}{c+a}, \ \frac{1}{c}+\frac{1}{a+b}$$
are also in H.P.

37. In an infinite G.P., find r when each term is equal to half the sum of the following terms.

38. Find the sum of n terms of a series in which the first term $= x+\frac{1}{2x}$, and the $n^{\text{th}} = nx+\frac{1}{2^n x}$.

39. Find the sum of the first $2n$ terms of the series
$$2+3+5+6+8+9+\ldots.$$

MISCELLANEOUS EXAMPLES V.

(Chiefly on Chapters XXXII.—XXXV.)

1. Simplify $\dfrac{(a^2b^3)^{\frac{1}{6}}b^{-2}c^{\frac{1}{3}}}{a^{\frac{2}{5}}b^{-\frac{5}{4}}c^{\frac{1}{4}}}$, and find its value when $a=2$, $b=3$, and $c=432$.

2. Shew that the ratio $x+y : x-y$ is increased by subtracting y from each term.

3. If $\dfrac{a}{b}=\dfrac{c}{d}$, shew that

(1) $\dfrac{2a+3b}{3a-7b}=\dfrac{2c+3d}{3c-7d}$; (2) $\dfrac{a^2-c^2}{b^2-d^2}=\dfrac{(a+2c)(a+3c)}{(b+2d)(b+3d)}$.

4. If $\dfrac{x}{b-c}=\dfrac{y}{c-a}=\dfrac{z}{a-b}$, prove that

(1) $x+y+z=0$; (2) $(b+c)x+(c+a)y+(a+b)z=0$.

5. If $\dfrac{x}{1}=\dfrac{y}{2}=\dfrac{z}{3}$, prove that $\sqrt{5x^2+8y^2+7z^2}=5y$.

6. If y is the sum of two numbers, of which the first varies directly and the second inversely as x, and if $y=7$, when $x=2$, and $y=-1$, when $x=1$, shew that
$$y=5x-\dfrac{6}{x}.$$

---

7. Simplify $\sqrt{45}+\sqrt{8}-\sqrt{80}+\sqrt{18}+\sqrt{7}-\sqrt{40}$.

8. If $3x+10$ has to $9x+4$ the duplicate ratio of 5 to 7, find x.

9. If $\dfrac{a}{b}=\dfrac{c}{d}=\dfrac{e}{f}$, prove that each ratio is equal to

(1) $\sqrt[3]{\dfrac{4ac^2-3ce^3+2ace}{4bd^2-3cf^3+2bdf}}$; (2) $\sqrt[5]{\dfrac{6a^2c^2e-c^4ef+7ac^5}{6b^2d^2f-d^4f^2+7ad^5}}$.

10. The sides of a triangle are as $1 : 1\frac{1}{2} : 1\frac{3}{4}$, and the perimeter is 221 yards; find the sides.

11. If $3a+5b : 3a-5b = 3c+5d : 3c-5d$, prove that $a:b=c:d$.

12. Reduce to their simplest forms:

(1) $\dfrac{x^{a+b}}{x^{a-b}} + \dfrac{x^{a-b}}{x^{a+b}}$; (2) $\dfrac{(a+b)^{\frac{3}{2}}}{(a-b)^{\frac{1}{2}}} \times \sqrt{a^2-b^2}$.

---

13. When $x = -\dfrac{3a}{4}$, find the value of
$$\dfrac{x^2+ax+a^2}{x^3-a^3} - \dfrac{x^2-ax+a^2}{x^3+a^3}$$

14. Simplify

(1) $\left\{\dfrac{a^{-1}b^3}{\sqrt[3]{a}}\right\}^{\frac{5}{4}} \div \sqrt[6]{\dfrac{a^2\sqrt{b}}{b^2}}$; (2) $\dfrac{2^{n+4} - 2 \times 2^n}{2^{n+2} \times 4}$.

15. Find the ratio compounded of the ratios
$$\dfrac{x-y}{a+b} : \dfrac{x^3-y^3}{a^2-b^2} \text{ and } \dfrac{x^2+xy+y^2}{a^2-b^2} : \dfrac{x^2-y^2}{(a+b)^2}.$$

16. If a, b, c be three proportionals, prove that

(1) $a(a+b) : b(b-a) = b(b+c) : c(c-b)$;

(2) $(a+b+c)(b^2-bc+c^2) = c(a^2+b^2+c^2)$.

17. If $a:b:c = xy:x^2:yz$, prove that $x:y:z = ab:a^2:bc$.

18. If $p:q$ be the duplicate ratio of $p-r : q-r$, prove that r is a mean proportional between p and q.

---

19. If $a:b=c:d$, prove that

(1) $a+c : a+b+c+d = a : a+b$;

(2) $(a-b)-(c-d) = \dfrac{(a-b)(b-d)}{b}$.

20. Shew that any ratio is made more nearly equal to unity by adding the same quantity to each of its terms.

21. If x varies as $y+z$, and z varies as x; and if $x=2$ when $y=4$, find the value of y when $x=1$.

22. If $2x+3y : 2x-3y = 2a^2+3b^2 : 2a^2-3b^2$, then x has to y the duplicate ratio of a to b.

23. Find an A.P. of seven terms whose sum is 28 and common difference 3.

24. The sum of 10 terms of an A.P. is 145, and the sum of its fourth and ninth terms is five times the third term; determine the series.

---

25. Find the value of
$$\sqrt{19+4\sqrt{21}} + \sqrt{7-\sqrt{12}} - \sqrt{29-2\sqrt{28}}.$$

26. Sum to 10 terms each of the series
  (1) $5+10+15+20+\ldots\ldots$;
  (2) $5-10+20-40+\ldots\ldots$.

27. If $\dfrac{p}{bz-cy} = \dfrac{-q}{cx+az} = \dfrac{-r}{ay+bx}$, shew that
$$ap+bq-cr=0, \text{ and } xp-yq+zr=0.$$

28. The sum of five numbers in arithmetical progression is 10, and the sum of their squares is 60; find the numbers.

29. Find the sum of n terms of the progression
$$3 + 2\tfrac{1}{2} + 2\tfrac{1}{12} + \ldots\ldots$$

30. Find the ninth term of the harmonic series whose first and third terms are 3 and 2 respectively.

---

31. Simplify $\dfrac{(a-b)^{\frac{1}{3}} \cdot \sqrt[3]{a^2+2ab+b^2}}{\sqrt[3]{a^2-b^2} \times (a+b)^{-\frac{2}{3}}}.$

32. Sum to n terms
$$\frac{3}{2} + \frac{9}{2} + \frac{15}{2} + \frac{21}{2} + \ldots\ldots,$$
and find five consecutive terms of this progression whose sum is $187\tfrac{1}{2}$.

33. The 8th term of an Arithmetical Progression is double the 13th term; shew that the 2nd term is double the 10th term.

34. Sum the following series:

(1) $(a-2x)+2(a+x)+3(a+2x)+\ldots\ldots$ to 18 terms.

(2) $3\frac{7}{81}-5\frac{25}{27}+8\frac{8}{9}-\ldots\ldots$ to 7 terms.

35. Shew that the sum of $2n$ terms of the series
$$1-\frac{1}{3}-\frac{1}{9}+\frac{1}{27}+\frac{1}{81}-\frac{1}{243}-\frac{1}{729}+\frac{1}{2187}+\ldots\ldots$$
is
$$\frac{3}{5}\{1-(-1)^n 3^{-2n}\}.$$

36. If $\dfrac{1}{b-a}$, $\dfrac{1}{2b}$, $\dfrac{1}{b-c}$ are in A.P., then a, b, c are in G.P.

---

37. The last term of an A.P. is ten times the first, and the last but one is equal to the sum of the 4th and 5th. Find the number of the terms, and shew that the common difference is equal to the first term.

38. Sum to $2n$ terms each of the series

(1) $1-3+9-27+\ldots\ldots$;

(2) $1-3+5-7+\ldots\ldots$,

and write down the last term of each series.

39. Find two numbers whose arithmetic mean exceeds their geometric mean by 2, and whose harmonic mean is one-fifth of the larger number.

40. Find an infinite geometrical progression, whose first term is 1, and in which each term is twice the sum of all the terms that follow it.

41. The arithmetic mean between two numbers is to the geometric mean as 5 to 4, and the difference of their geometric and harmonic means is $\frac{4}{5}$: find the numbers.

42. If x, y, z be in G.P., prove that
$$x^2y^2z^2(x^{-3}+y^{-3}+z^{-3})=x^3+y^3+z^3.$$

CHAPTER XXXVI.

THE THEORY OF QUADRATIC EQUATIONS.

335. In Chapter xxv. it was shewn that after suitable reduction every quadratic equation may be written in the form
$$ax^2+bx+c=0\quad\ldots\ldots\ldots\ldots\ldots\ldots(1),$$
and that the solution of the equation is
$$x=\frac{-b\pm\sqrt{b^2-4ac}}{2a}\ldots\ldots\ldots\ldots\ldots\ldots(2).$$

We shall now prove some important propositions connected with the roots and coefficients of all equations of which (1) is the type.

336. *A quadratic equation cannot have more than two roots.*

For, if possible, let the equation $ax^2+bx+c=0$ have three *different* roots α, β, γ. Then since each of these values must satisfy the equation, we have
$$a\alpha^2+b\alpha+c=0\ldots\ldots\ldots\ldots\ldots\ldots(1),$$
$$a\beta^2+b\beta+c=0\ldots\ldots\ldots\ldots\ldots\ldots(2),$$
$$a\gamma^2+b\gamma+c=0\ldots\ldots\ldots\ldots\ldots\ldots(3),$$

From (1) and (2), by subtraction,
$$a(\alpha^2-\beta^2)+b(\alpha-\beta)=0;$$
divide out by $\alpha-\beta$ which, by hypothesis, is not zero; then
$$a(\alpha+\beta)+b=0.$$

Similarly from (2) and (3)
$$a(\beta+\gamma)+b=0;$$
$\therefore$ by subtraction $\qquad a(\alpha-\gamma)=0;$

which is impossible, since, by hypothesis, a is not zero, and α is not equal to γ. Hence there cannot be three different roots.

337. The terms 'unreal', 'imaginary', and 'impossible' are all used in the same sense: namely, to denote expressions which involve the square root of a negative quantity. It is important that the student should clearly distinguish between the terms *real* and *rational, imaginary* and *irrational*. Thus $\sqrt{25}$ or 5, $3\frac{1}{2}$, $-\frac{3}{8}$ are rational and real; $\sqrt{7}$ is irrational but real; while $\sqrt{-7}$ is irrational and also imaginary.

338. In Art. 335 if the two roots in (2) are denoted by α and β, we have
$$\alpha = \frac{-b+\sqrt{b^2-4ac}}{2a}, \qquad \beta = \frac{-b-\sqrt{b^2-4ac}}{2a}.$$

(1) If b^2-4ac, the quantity under the radical, is positive, α and β are real and unequal.

(2) If b^2-4ac is zero, α and β are real and equal, each reducing in this case to $-\dfrac{b}{2a}$.

(3) If b^2-4ac is negative, α and β are imaginary and unequal.

(4) If b^2-4ac is a perfect square, α and β are rational and unequal.

By applying these tests the nature of the roots of any quadratic may be determined without solving the equation.

Example 1. Shew that the equation $2x^2-6x+7=0$ cannot be satisfied by any real values of x.

Here $a=2$, $b=-6$, $c=7$; so that
$$b^2-4ac = (-6)^2 - 4.2.7 = -20.$$
Therefore the roots are imaginary.

Note. If the equation is solved graphically as in Art. 427, it will be found that the graph does not cut the axis of x. Thus there are no real values of x which make $2x^2-6x+7$ equal to zero.

Example 2. For what value of k will the equation $3x^2-6x+k=0$ have equal roots?

The condition for equal roots gives
$$(-6)^2 - 4.3.k = 0,$$
whence $k=3$.

Example 3. Shew that the roots of the equation
$$x^2 - 2px + p^2 - q^2 + 2qr - r^2 = 0$$
are rational.

The roots will be rational provided $(-2p)^2 - 4(p^2-q^2+2qr-r^2)$ is a perfect square. But this expression reduces to $4(q^2-2qr+r^2)$, or $4(q-r)^2$. Hence the roots are rational.

339. Since $\alpha = \dfrac{-b+\sqrt{b^2-4ac}}{2a}$, $\beta = \dfrac{-b-\sqrt{b^2-4ac}}{2a}$,

we have by addition
$$\alpha+\beta = \dfrac{-b+\sqrt{b^2-4ac}-b-\sqrt{b^2-4ac}}{2a},$$
$$= -\dfrac{2b}{2a} = -\dfrac{b}{a} \quad\dotfill(1);$$

and by multiplication we have
$$\alpha\beta = \dfrac{(-b+\sqrt{b^2-4ac})(-b-\sqrt{b^2-4ac})}{4a^2}$$
$$= \dfrac{(-b)^2-(b^2-4ac)}{4a^2}$$
$$= \dfrac{4ac}{4a^2} = \dfrac{c}{a} \quad\dotfill(2).$$

By writing the equation in the form
$$x^2 + \dfrac{b}{a}x + \dfrac{c}{a} = 0,$$
these results may also be expressed as follows:

In a quadratic equation *where the coefficient of the first term is unity*,

(i) the sum of the roots is equal to the coefficient of x with its sign changed;

(ii) the product of the roots is equal to the third term.

Note. In any equation the term which does not contain the unknown quantity is frequently called *the absolute term*.

340. Since $-\dfrac{b}{a} = \alpha+\beta$, and $\dfrac{c}{a} = \alpha\beta$,

the equation $x^2 + \dfrac{b}{a}x + \dfrac{c}{a}$ may be written
$$x^2 - (\alpha+\beta)x + \alpha\beta = 0 \quad\dotfill(1).$$

Hence any quadratic may also be expressed in the form
$$x^2 - (\text{sum of roots})\,x + \text{product of roots} = 0 \quad\dotfill(2).$$

Again, from (1) we have
$$(x-\alpha)(x-\beta) = 0 \quad\dotfill(3).$$

We may now easily form an equation with given roots.

Example 1. Form the equation whose roots are 3 and -2.
The equation is $(x-3)(x+2)=0$,
or $x^2-x-6=0$.

Example 2. Form the equation whose roots are $\dfrac{3}{7}$ and $-\dfrac{4}{5}$.

The equation is $\left(x-\dfrac{3}{7}\right)\left(x+\dfrac{4}{5}\right)=0$;

that is, $(7x-3)(5x+4)=0$,
or $35x^2+13x-12=0$.

When the roots are irrational it is easier to use the following method:

Example 3. Form the equation whose roots are $2+\sqrt{3}$ and $2-\sqrt{3}$.
We have sum of roots $=4$,
product of roots $=1$;
∴ the equation is $x^2-4x+1=0$,
by using formula (2) of the present article.

341. The results of Art. 339 are most important, and they are generally sufficient to solve problems connected with the roots of quadratics. In such questions *the roots should never be considered singly*, but use should be made of the relations obtained by writing down the sum of the roots, and their product, in terms of the coefficients of the equation.

Example 1. If a and β are the roots of $x^2-px+q=0$, find the value of (1) $a^2+\beta^2$, (2) $a^3+\beta^3$.

We have $a+\beta=p$,
$a\beta=q$.
∴ $a^2+\beta^2=(a+\beta)^2-2a\beta$
$=p^2-2q$.

Again, $a^3+\beta^3=(a+\beta)(a^2+\beta^2-a\beta)$
$=p\{(a+\beta)^2-3a\beta\}$
$=p(p^2-3q)$.

Example 2. If a, β are the roots of the equation $lx^2+mx+n=0$, find the equation whose roots are $\dfrac{a}{\beta}$, $\dfrac{\beta}{a}$.

We have sum of roots $=\dfrac{a}{\beta}+\dfrac{\beta}{a}=\dfrac{a^2+\beta^2}{a\beta}$,

product of roots $=\dfrac{a}{\beta}\cdot\dfrac{\beta}{a}=1$:

XXXVI.] THE THEORY OF QUADRATIC EQUATIONS. 315

∴ by Art. 340 the required equation is

$$x^2 - \left(\frac{\alpha^2+\beta^2}{\alpha\beta}\right)x + 1 = 0,$$

or $\qquad \alpha\beta x^2 - (\alpha^2+\beta^2)x + \alpha\beta = 0.$

As in the last example $\alpha^2+\beta^2 = \dfrac{m^2-2nl}{l^2}$, and $\alpha\beta = \dfrac{n}{l}.$

∴ the equation is $\qquad \dfrac{n}{l}x^2 - \dfrac{m^2-2nl}{l^2}x + \dfrac{n}{l} = 0,$

or $\qquad nlx^2 - (m^2-2nl)x + nl = 0.$

Example 3. Find the condition that the roots of the equation $ax^2+bx+c=0$ should be (1) equal in magnitude and opposite in sign, (2) reciprocals.

The roots will be equal in magnitude and opposite in sign if their sum is zero; therefore $-\dfrac{b}{a} = 0$, or $b=0$.

Again, the roots will be reciprocals when their product is unity; therefore $\dfrac{c}{a} = 1$, or $c=a$.

Example 4. Find the relation which must subsist between the coefficients of the equation $px^2+qx+r=0$, when one root is three times the other.

We have $\qquad \alpha+\beta = -\dfrac{q}{p},\quad \alpha\beta = \dfrac{r}{p};$

but since $\alpha = 3\beta$, we obtain by substitution

$$4\beta = -\frac{q}{p},\quad 3\beta^2 = \frac{r}{p}.$$

From the first of these equations $\beta^2 = \dfrac{q^2}{16p^2}$, and from the second $\beta^2 = \dfrac{r}{3p}.$

$$\therefore \frac{q^2}{16p^2} = \frac{r}{3p},$$

or $\qquad 3q^2 = 16pr,$

which is the required condition.

342. The following example illustrates a useful application of the results proved in Art. 338.

Example. If x is a real quantity, prove that the expression $\dfrac{x^2+2x-11}{2(x-3)}$ can have all numerical values except such as lie between 2 and 6.

Let the given expression be represented by y, so that
$$\frac{x^2+2x-11}{2(x-3)}=y;$$
then multiplying up and transposing, we have
$$x^2+2x(1-y)+6y-11=0.$$
This is a quadratic equation, and if x is to have real values $4(1-y)^2-4(6y-11)$ must be positive; or simplifying and dividing by 4, $y^2-8y+12$ must be positive; that is, $(y-6)(y-2)$ must be positive. Hence the factors of this product must be both positive, or both negative. In the former case y is greater than 6; in the latter y is less than 2. Therefore y cannot lie between 2 and 6, but may have any other value.

In this example it will be noticed that the *expression* $y^2-8y+12$ is positive so long as y does not lie between the roots of the corresponding quadratic *equation* $y^2-8y+12=0$.

This is a particular case of the general proposition investigated in the next article.

343. *For all real values of* x *the expression* ax^2+bx+c *has the same sign as* a, *except when the roots of the equation* $ax^2+bx+c=0$ *are real and unequal, and* x *lies between them.*

CASE I. Suppose that the roots of the equation
$$ax^2+bx+c=0$$
are real; denote them by α and β, and let α be the greater.

Then
$$ax^2+bx+c = a\left(x^2+\frac{b}{a}x+\frac{c}{a}\right)$$
$$= a\{x^2-(\alpha+\beta)x+\alpha\beta\} \qquad \text{[Art. 339.]}$$
$$= a(x-\alpha)(x-\beta).$$

Now if x is greater than α or less than β, the factors $x-\alpha$, $x-\beta$ are either both positive or both negative; therefore the expression $(x-\alpha)(x-\beta)$ is positive, and ax^2+bx+c has the same sign as a. But if x lies between α and β, the expression
$$(x-\alpha)(x-\beta)$$
is negative, and the sign of ax^2+bx+c is opposite to that of a.

CASE II. If α and β are equal, then
$$ax^2+bx+c=a(x-\alpha)^2,$$
and $(x-\alpha)^2$ is positive for all real values of x; hence ax^2+bx+c has the same sign as a.

CASE III. Suppose that the equation $ax^2+bx+c=0$ has imaginary roots; then
$$ax^2+bx+c=a\left\{x^2+\frac{b}{a}x+\frac{c}{a}\right\}$$
$$=a\left\{\left(x+\frac{b}{2a}\right)^2+\frac{4ac-b^2}{4a^2}\right\};$$
but since b^2-4ac is negative [Art. 338], the expression
$$\left(x+\frac{b}{2a}\right)^2+\frac{4ac-b^2}{4a^2}$$
is positive for all real values of x; therefore ax^2+bx+c has the same sign as a.

[*Arts. 426, 427 and 439, Ex. 2, may be read here.*]

EXAMPLES XXXVI.

Find (without actual solution) the nature of the roots of the following equations:

1. $x^2+x-870=0$.
2. $8+6x=5x^2$.
3. $\frac{1}{2}x^2=14-3x^2$.
4. $x^2+7=4x$.
5. $2x=x^2+5$.
6. $(x+2)^2=4x+15$.

Form the equations whose roots are

7. $5, -3$.
8. $-9, -11$.
9. $a+b, a-b$.
10. $\frac{3}{2}, \frac{5}{6}$.
11. $\frac{2}{3}a, -\frac{4}{5}a$.
12. $0, \frac{7}{8}$.

13. If the equation $x^2+2(1+k)x+k^2=0$ has equal roots, what is the value of k?

14. Prove that the equation
$$3mx^2-(2m+3n)x+2n=0$$
has rational roots.

15. Without solving the equation $3x^2-4x-1=0$, find the sum, the difference, and the sum of the squares of the roots.

16. Shew that the roots of $a(x^2-1)=(b-c)x$ are always real.

Form the equations whose roots are

17. $3+\sqrt{5},\ 3-\sqrt{5}$.   18. $-2+\sqrt{3},\ -2-\sqrt{3}$.   19. $-\dfrac{a}{5},\ \dfrac{b}{6}$.

20. $\dfrac{1}{2}(4\pm\sqrt{7})$.   21. $\dfrac{a+b}{a-b},\ \dfrac{a-b}{a+b}$.   22. $\dfrac{a}{2b},\ \dfrac{b}{2a}$.

If $a,\ \beta$ are the roots of the equation $px^2+qx+r=0$, find the values of

23. $a^2+\beta^2$.   24. $(a-\beta)^2$.   25. $a^2\beta+a\beta^2$.

26. $a^4+\beta^4$.   27. $a^5\beta^2+a^2\beta^5$.   28. $\dfrac{a^2}{\beta}+\dfrac{\beta^2}{a}$.

29. If $a,\ \beta$ are the roots of $x^2-px+q=0$, and $a^3,\ \beta^3$ the roots of $x^2-Px+Q=0$, find P and Q in terms of p and q.

30. If $a,\ \beta$ are the roots of $x^2-ax+b=0$, find the equation whose roots are $\dfrac{a}{\beta^2},\ \dfrac{\beta}{a^2}$.

31. Find the condition that one root of the equation
$$ax^2+bx+c=0$$
may be double the other.

32. Form an equation whose roots shall be the cubes of the roots of the equation $2x(x-a)=a^2$.

33. Prove that the roots of the equation
$$(a+b)x^2-(a+b+c)x+\dfrac{c}{2}=0$$
are always real.

34. Shew that $(a+b+c)x^2-2(a+b)x+(a+b-c)=0$
has rational roots.

35. Form an equation whose roots shall be the arithmetic and harmonic means between the roots of $x^2-px+q=0$.

36. In the equation $px^2+qx+r=0$ the roots are in the ratio of l to m, prove that
$$(l^2+m^2)pr+lm(2pr-q^2)=0.$$

37. Shew that if x is real the expression $\dfrac{x^2-15}{2x-8}$ cannot lie between 3 and 5.

38. If x is real, prove that $\dfrac{3x^2+2}{x^2-2x-1}$ can have all values except such as lie between 2 and $-\dfrac{3}{2}$.

CHAPTER XXXVII.

PERMUTATIONS AND COMBINATIONS.

344. Each of the *arrangements* which can be made by taking some or all of a number of things is called a **permutation**.

Each of the *groups* or *selections* which can be made by taking some or all of a number of things is called a **combination**.

Thus the *permutations* which can be made by taking the letters a, b, c, d two at a time are twelve in number: namely,

$$ab, \quad ac, \quad ad, \quad bc, \quad bd, \quad cd,$$
$$ba, \quad ca, \quad da, \quad cb, \quad db, \quad dc;$$

each of these presenting a different *arrangement* of two letters.

The *combinations* which can be made by taking the letters a, b, c, d two at a time are six in number: namely,

$$ab, \quad ac, \quad ad, \quad bc, \quad bd, \quad cd;$$

each of these presenting a different *selection* of two letters.

From this it appears that in forming *combinations* we are only concerned with the number of things each selection contains; whereas in forming *permutations* we have also to consider the order of the things which make up each arrangement; for instance, if from four letters a, b, c, d we make a selection of three, such as abc, this single combination admits of being arranged in the following ways:

$$abc, \quad acb, \quad bca, \quad bac, \quad cab, \quad cba,$$

and so gives rise to six different permutations.

345. Before discussing the general propositions of this chapter the following important principle should be carefully noticed.

If one operation can be performed in m *ways, and (when it has been performed in any one of these ways) a second operation can then be performed in* n *ways; the number of ways of performing the two operations will be* m × n.

If the first operation be performed in *any one* way, we can associate with this any of the n ways of performing the second operation: and thus we shall have n ways of performing the two operations without considering more than *one* way of performing the first; and so, corresponding to *each* of the m ways of performing the first operation, we shall have n ways of performing the two; hence altogether the number of ways in which the two operations can be performed is represented by the product $m \times n$.

Example. There are 10 steamers plying between Liverpool and Dublin; in how many ways can a man go from Liverpool to Dublin and return by a different steamer?

There are *ten* ways of making the first passage; and with each of these there is a choice of *nine* ways of returning (since the man is not to come back by the same steamer); hence the number of ways of making the two journeys is 10×9, or 90.

This principle may easily be extended to the case in which there are more than two operations each of which can be performed in a given number of ways.

346. *To find the number of permutations of* n *dissimilar things taken* r *at a time.*

This is the same thing as finding the number of ways in which we can fill up r places when we have n different things at our disposal.

The first place may be filled up in n ways, for any one of the n things may be taken; when it has been filled up in any one of these ways, the second place can then be filled up in $n-1$ ways; and since each way of filling up the first place can be associated with each way of filling up the second, the number of ways in which the first two places can be filled up is given by the product $n(n-1)$. And when the first two places have been filled up in any way, the third place can be filled up in $n-2$ ways. And reasoning as before, the number of ways in which three places can be filled up is $n(n-1)(n-2)$.

Proceeding thus, and noticing that a new factor is introduced with each new place filled up, and that at any stage the number of factors is the same as the number of places filled up, we shall have the number of ways in which r places can be filled up equal to

$$n(n-1)(n-2) \ldots \ldots \text{ to } r \text{ factors};$$

and the rth factor is $n-(r-1)$, or $n-r+1$.

Therefore the number of permutations of n things taken r at a time is
$$n(n-1)(n-2)\ldots\ldots(n-r+1).$$

Cor. The number of permutations of n things taken all at a time is
$$n(n-1)(n-2)\ldots\ldots \text{ to } n \text{ factors,}$$
or $\qquad n(n-1)(n-2)\ldots\ldots 3.2.1.$

It is usual to denote this product by the symbol $\lfloor n$, which is read "factorial n." Also the symbol $n!$ is sometimes used for $\lfloor n$.

347. We shall in future denote the number of permutations of n things taken r at a time by the symbol ^nP_r, so that
$$^nP_r = n(n-1)(n-2)\ldots\ldots(n-r+1);$$
also $\qquad ^nP_n = \lfloor n.$

In working numerical examples it is useful to notice that the suffix in the symbol ^nP_r always denotes the number of factors in the formula we are using.

Example 1. Four persons enter a railway carriage in which there are six seats; in how many ways can they take their places?

The first person may seat himself in 6 ways; and then the second person in 5; the third in 4; and the fourth in 3; and since each of these ways may be associated with each of the others, the required answer is $6 \times 5 \times 4 \times 3$, or 360.

Example 2. How many different numbers can be formed by using six out of the nine digits 1, 2, 3, ... 9?

Here we have 9 different things and we have to find the number of permutations of them taken 6 at a time;

$$\therefore \text{ the required result} = {}^9P_6$$
$$= 9 \times 8 \times 7 \times 6 \times 5 \times 4$$
$$= 60480.$$

348. *To find the number of combinations of* n *dissimilar things taken* r *at a time.*

Let ^nC_r denote the required number of combinations.

Then each of these combinations consists of a group of r dissimilar things which can be arranged among themselves in $\lfloor r$ ways. [Art. 346. Cor.]

Hence $^nC_r \times \lfloor r$ is equal to the number of *arrangements* of n things taken r at a time; that is,
$$^nC_r \times \lfloor r = {}^nP_r = n(n-1)(n-2)\ldots(n-r+1);$$
$$\therefore \ ^nC_r = \frac{n(n-1)(n-2)\ldots(n-r+1)}{\lfloor r}.$$

COR. This formula for ^nC_r may also be written in a different form; for if we multiply the numerator and the denominator by $\lfloor n-r$ we obtain

$$\frac{n(n-1)(n-2)\ldots(n-r+1)\times\lfloor n-r}{\lfloor r\ \lfloor n-r},\quad \text{or}\quad \frac{\lfloor n}{\lfloor r\ \lfloor n-r};$$

since $n(n-1)(n-2)\ldots(n-r+1)\times\lfloor n-r = \lfloor n$.

Example. From 12 books in how many ways can a selection of 5 be made, (1) when one specified book is always included, (2) when one specified book is always excluded?

(1) Since the specified book is to be included in every selection, we have only to choose 4 out of the remaining 11.

Hence the number of ways $=^{11}C_4 = \dfrac{11\times 10\times 9\times 8}{1\times 2\times 3\times 4} = 330$.

(2) Since the specified book is always to be excluded, we have to select the 5 books out of the remaining 11.

Hence the number of ways $=^{11}C_5 = \dfrac{11\times 10\times 9\times 8\times 7}{1\times 2\times 3\times 4\times 5} = 462$.

349. *The number of combinations of* n *things* r *at a time is equal to the number of combinations of* n *things* n − r *at a time.*

In making all the possible combinations of n things, to each group of r things we select, there is left a corresponding group of $n-r$ things; that is, the number of combinations of n things r at a time is the same as the number of combinations of n things $n-r$ at a time;

$$\therefore\quad ^nC_r = {^nC_{n-r}}.$$

This result is frequently useful in enabling us to abridge arithmetical work.

Example. Out of 14 men in how many ways can an eleven be chosen?

The required number $=^{14}C_{11}$

$$=^{14}C_3 = \frac{14\times 13\times 12}{1\times 2\times 3} = 364.$$

If we had made use of the formula $^{14}C_{11}$, we should have had to reduce an expression whose numerator and denominator each contained 11 factors.

350. In the examples which follow it is important to notice that the formula for *permutations* should not be used until the suitable *selections* required by the question have been made.

Example 1. From 7 Englishmen and 4 Americans a committee of 6 is to be formed: in how many ways can this be done, (1) when the committee contains exactly 2 Americans, (2) at least 2 Americans?

(1) The number of ways in which the Americans can be chosen is $^4C_2$; and the number of ways in which the Englishmen can be chosen is $^7C_4$. Each of the first groups can be associated with each of the second; hence

the required number of ways $= {^4C_2} \times {^7C_4}$

$$= \frac{\lfloor 4}{\lfloor 2 \lfloor 2} \times \frac{\lfloor 7}{\lfloor 4 \lfloor 3} = \frac{\lfloor 7}{\lfloor 2 \lfloor 2 \lfloor 3} = 210.$$

(2) We shall exhaust all the suitable combinations by forming all the groups containing 2 Americans and 4 Englishmen; then 3 Americans and 3 Englishmen; and lastly 4 Americans and 2 Englishmen.

The *sum* of the three results will give the answer. Hence the required number of ways $= {^4C_2} \times {^7C_4} + {^4C_3} \times {^7C_3} + {^4C_4} \times {^7C_2}$

$$= \frac{\lfloor 4}{\lfloor 2 \lfloor 2} \times \frac{\lfloor 7}{\lfloor 4 \lfloor 3} + \frac{\lfloor 4}{\lfloor 3} \times \frac{\lfloor 7}{\lfloor 3 \lfloor 4} + 1 \times \frac{\lfloor 7}{\lfloor 2 \lfloor 5}$$
$$= 210 + 140 + 21 = 371.$$

In this example we have only to make use of the suitable formulæ for *combinations*, for we are not concerned with the possible arrangements of the members of the committee among themselves.

Example 2. Out of 7 consonants and 4 vowels, how many words can be made each containing 3 consonants and 2 vowels?

The number of ways of choosing the three consonants is $^7C_3$, and the number of ways of choosing the 2 vowels is $^4C_2$; and since each of the first groups can be associated with each of the second, the number of combined groups, each containing 3 consonants and 2 vowels, is $^7C_3 \times {^4C_2}$.

Further, each of these groups contains 5 letters, which may be arranged among themselves in $\lfloor 5$ ways. Hence

the required number of words $= \dfrac{\lfloor 7}{\lfloor 3 \lfloor 4} \times \dfrac{\lfloor 4}{\lfloor 2 \lfloor 2} \times \lfloor 5$

$$= 5 \times \lfloor 7 = 25200.$$

EXAMPLES XXXVII. a.

1. Find the value of $^5P_4$, $^7P_6$, $^8C_5$, $^{25}C_{23}$.

2. How many different arrangements can be made by taking (1) five, (2) all of the letters of the word *soldier*?

3. If $^nC_3 : {^{n-1}C_4} = 8 : 5$, find n.

4. How many different selections of four coins can be made from a bag containing a sovereign, a half-sovereign, a half-crown, a florin, a shilling, a franc, a sixpence, a penny, and a farthing?

5. How many numbers between 3000 and 4000 can be made with the digits 9, 3, 4, 6?

6. In how many ways can the letters of the word *volume* be arranged if the vowels can only occupy the even places?

7. If the number of permutations of n things four at a time is fourteen times the number of permutations of $n-2$ things three at a time, find n.

8. From 5 masters and 10 boys how many committees can be selected containing 3 masters and 6 boys?

9. If $^{20}C_r = {}^{20}C_{r-10}$, find $^rC_{12}$, $^{18}C_r$.

10. Out of the twenty-six letters of the alphabet in how many ways can a word be made consisting of five different letters, two of which must be a and e?

11. How many words can be formed by taking 3 consonants and 2 vowels from an alphabet containing 21 consonants and 5 vowels?

12. A railway carriage will accommodate 5 passengers on each side: in how many ways can 10 persons take their seats when two of them decline to face the engine, and a third cannot travel backwards?

351. Hitherto, in the formulæ we have proved, the things have been regarded as *unlike*. Before considering cases in which some one or more sets of things may be *like*, it is necessary to point out exactly in what sense the words *like* and *unlike* are used. When we speak of things being *dissimilar, different, unlike*, we imply that the things are *visibly unlike*, so as to be easily distinguishable from each other. On the other hand we shall always use the term *like* things to denote such as are alike to the eye and cannot be distinguished from each other. For instance, in Ex. 2, Art. 350, the consonants and the vowels may be said each to consist of a group of things united by a common characteristic, and thus in a certain sense to be of the same kind; but they cannot be regarded as like things, because there is an individuality existing among the things of each group which makes them easily distinguishable from each other. Hence, in the final stage of the example we considered each group to consist of five *dissimilar* things and therefore capable of $\lfloor 5$ arrangements among themselves. [Art. 346. Cor.]

352. *To find the number of ways in which* n *things may be arranged among themselves, taking them all at a time, when* p *of the things are exactly alike of one kind,* q *of them exactly alike of another kind,* r *of them exactly alike of a third kind, and the rest all different.*

Let there be n letters; suppose p of them to be a, q of them to be b, r of them to be c, and the rest to be unlike.

Let x be the required number of permutations; then if the p letters a were replaced by p unlike letters different from any of the rest, from *any one* of the x permutations, without altering the position of any of the remaining letters, we could form $\lfloor p$ new permutations. Hence if this change were made in each of the x permutations, we should obtain $x \times \lfloor p$ permutations.

Similarly, if the q letters b were replaced by q unlike letters, the number of permutations would be $x \times \lfloor p \times \lfloor q$.

In like manner, by replacing the r letters c by r unlike letters, we should finally obtain $x \times \lfloor p \times \lfloor q \times \lfloor r$ permutations.

But the things are now all different, and therefore admit of $\lfloor n$ permutations among themselves. Hence

$$x \times \lfloor p \times \lfloor q \times \lfloor r = \lfloor n;$$

that is,
$$x = \frac{\lfloor n}{\lfloor p \lfloor q \lfloor r};$$

which is the required number of permutations.

Any case in which the things are not all different may be treated similarly.

Example 1. How many different permutations can be made out of the letters of the word *assassination* taken all together?

We have here 13 letters of which 4 are s, 3 are a, 2 are i, and 2 are n. Hence the number of permutations

$$= \frac{\lfloor 13}{\lfloor 4 \lfloor 3 \lfloor 2 \lfloor 2}$$
$$= 13 . 11 . 10 . 9 . 8 . 7 . 3 . 5$$
$$= 1001 \times 10800 = 10810800.$$

Example 2. How many numbers can be formed with the digits 1, 2, 3, 4, 3, 2, 1, so that the odd digits always occupy the odd places?

The odd digits 1, 3, 3, 1 can be arranged in their four places in

$$\frac{\lfloor 4}{\lfloor 2 \lfloor 2} \text{ ways} \quad\quad\quad\quad\quad\quad\quad\quad (1).$$

The even digits 2, 4, 2 can be arranged in their three places in

$$\frac{\lfloor 3}{\lfloor 2} \text{ ways} \quad\quad\quad\quad\quad\quad\quad\quad (2).$$

Each of the ways in (1) can be associated with each of the ways in (2).

Hence the required number $= \dfrac{\lfloor 4}{\lfloor 2 \lfloor 2} \times \dfrac{\lfloor 3}{\lfloor 2} = 6 \times 3 = 18.$

353. *To find the number of permutations of n things r at a time, when each thing may be repeated once, twice, ... up to r times in any arrangement.*

Here we have to consider the number of ways in which r places can be filled up when we have n different things at our disposal, each of the n things being used as often as we please in any arrangement.

The first place may be filled up in n ways, and, when it has been filled up in any one way, the second place may also be filled up in n ways, since we are not precluded from using the same thing again. Therefore the number of ways in which the first two places can be filled up is $n \times n$ or n^2.

The third place can also be filled up in n ways, and therefore the first three places in n^3 ways.

Proceeding in this manner, and noticing that at any stage the index of n is always the same as the number of places filled up, we shall have the number of ways in which the r places can be filled up equal to n^r.

Example. In how many ways can 5 prizes be given away to 4 boys, when each boy is eligible for all the prizes?

Any one of the prizes can be given in 4 ways; and then any one of the remaining prizes can also be given in 4 ways, since it may be obtained by the boy who has already received a prize. Thus two prizes can be given away in 4^2 ways, three prizes in 4^3 ways, and so on. Hence the 5 prizes can be given away in 4^5, or 1024 ways.

354. *To find for what value of* r *the number of combinations of* n *things* r *at a time is greatest.*

Since $^nC_r = \dfrac{n(n-1)(n-2)\ldots\ldots(n-r+2)(n-r+1)}{1.2.3\ldots\ldots(r-1)r}$

and $\quad ^nC_{r-1} = \dfrac{n(n-1)(n-2)\ldots\ldots(n-r+2)}{1.2.3\ldots\ldots(r-1)};$

$$\therefore\ ^nC_r = {^nC_{r-1}} \times \dfrac{n-r+1}{r}.$$

The multiplying factor $\dfrac{n-r+1}{r}$ may be written $\dfrac{n+1}{r}-1$, which shews that it decreases as r increases. Hence as r receives the values $1, 2, 3, \ldots\ldots$ in succession, ^nC_r is continually increased, until $\dfrac{n+1}{r}-1$ becomes equal to 1 or less than 1.

Now $\dfrac{n+1}{r}-1 > 1$, so long as $\dfrac{n+1}{r} > 2$; that is, $\dfrac{n+1}{2} > r$.

We have to choose the greatest value of r consistent with this inequality.

(1) Let n be even, and equal to $2m$; then

$$\dfrac{n+1}{2} = \dfrac{2m+1}{2} = m + \dfrac{1}{2};$$

and for all values of r up to m inclusive this is greater than r. Hence by putting $r = m = \dfrac{n}{2}$, we find that the greatest number of combinations is $^nC_{\frac{n}{2}}$.

(2) Let n be odd, and equal to $2m+1$; then

$$\dfrac{n+1}{2} = \dfrac{2m+2}{2} = m+1;$$

and for all values of r up to m inclusive this is greater than r; but when $r = m+1$ the multiplying factor becomes equal to 1, and

$$^nC_{m+1} = {^nC_m}; \text{ that is, } ^nC_{\frac{n+1}{2}} = {^nC_{\frac{n-1}{2}}};$$

and therefore the number of combinations is greatest when the things are taken $\dfrac{n+1}{2}$, or $\dfrac{n-1}{2}$ at a time; the result being the same in the two cases.

EXAMPLES XXXVII. b.

1. Find the number of permutations which can be made from all the letters of the words,

(1) *irresistible*, (2) *phenomenon*, (3) *tittle-tattle*.

2. How many different numbers can be formed by using the seven digits 2, 3, 4, 3, 3, 1, 2? How many with the digits 2, 3, 4, 3, 3, 0, 2?

3. How many words can be formed from the letters of the word *Simoom*, so that vowels and consonants occur alternately in each word?

4. A telegraph has 5 arms and each arm has 4 distinct positions, including the position of rest: find the total number of signals that can be made.

5. In how many ways can n things be given to m persons, when there is no restriction as to the number of things each may receive?

6. How many different arrangements can be made out of the letters of the expression $a^5 b^3 c^6$ when written at full length?

7. There are four copies each of 3 different volumes; find the number of ways in which they can be arranged on one shelf.

8. In how many ways can 6 persons form a ring? Find the number of ways in which 4 gentlemen and 4 ladies can sit at a round table so that no two gentlemen sit together.

9. In how many ways can a word of 4 letters be made out of the letters a, b, e, c, d, o, when there is no restriction as to the number of times a letter is repeated in each word?

10. How many arrangements can be made out of the letters of the word *Toulouse*, so that the consonants occupy the first, fourth, and seventh places?

11. A boat's crew consists of eight men, of whom one can only row on bow side and one only on stroke side: in how many ways can the crew be arranged?

12. Shew that $^{n+1}C_r = {}^nC_r + {}^nC_{r-1}$.

13. A cricket eleven has to be chosen from 13 men of whom only 4 can bowl: in how many ways can the team be made up so as to include *at least* 2 bowlers?

14. In how many ways can n men be arranged in a row if two specified men are neither of them to be at either extremity of the row?

CHAPTER XXXVIII.

Binomial Theorem.

355. It may be shewn by actual multiplication that
$$(x+a)(x+b)(x+c)(x+d)$$
$$= x^4 + (a+b+c+d)x^3 + (ab+ac+ad+bc+bd+cd)x^2$$
$$+ (abc+abd+acd+bcd)x + abcd \quad \dots\dots\dots\dots\dots\dots (1).$$

We may, however, write down this result by inspection; for the complete product consists of the sum of a number of partial products each of which is formed by multiplying together four letters, *one* being taken from *each* of the four factors. If we examine the way in which the various partial products are formed, we see that

(1) the term x^4 is formed by taking the letter x out of *each* of the factors.

(2) the terms involving x^3 are formed by taking the letter x out of *any three* factors, in every way possible, and *one* of the letters a, b, c, d out of the remaining factor.

(3) the terms involving x^2 are formed by taking the letter x out of *any two* factors, in every way possible, and *two* of the letters a, b, c, d out of the remaining factors.

(4) the terms involving x are formed by taking the letter x out of *any one* factor, and *three* of the letters a, b, c, d out of the remaining factors.

(5) the term independent of x is the product of all the letters a, b, c, d.

Example. Find the value of $(x-2)(x+3)(x-5)(x+9)$.

The product
$$= x^4 + (-2+3-5+9)x^3 + (-6+10-18-15+27-45)x^2$$
$$+ (30-54+90-135)x + 270$$
$$= x^4 + 5x^3 - 47x^2 - 69x + 270.$$

356. If in equation (1) of the preceding article we suppose $b=c=d=a$, we obtain
$$(x+a)^4 = x^4 + 4ax^3 + 6a^2x^2 + 4a^3x + a^4.$$

We shall now employ the same method to prove a formula known as the **Binomial Theorem**, by which any binomial of the form $x+a$ can be raised to any assigned positive integral power.

357. *To find the expansion of* $(x+a)^n$ *when* n *is a positive integer.*

Consider the expression
$$(x+a)(x+b)(x+c)\ldots\ldots(x+k),$$
the number of factors being n.

The expansion of this expression is the continued product of the n factors, $x+a, x+b, x+c, \ldots\ldots x+k$, and every term in the expansion is of n dimensions, being a product formed by multiplying together n letters, *one* taken from each of these n factors.

The highest power of x is x^n, and is formed by taking the letter x from *each* of the n factors.

The terms involving x^{n-1} are formed by taking the letter x from *any* $n-1$ of the factors, and *one* of the letters $a, b, c, \ldots k$ from the remaining factor; thus the coefficient of x^{n-1} in the final product is the sum of the letters $a, b, c, \ldots\ldots k$; denote it by S_1.

The terms involving x^{n-2} are formed by taking the letter x from *any* $n-2$ of the factors, and *two* of the letters $a, b, c, \ldots k$ from the two remaining factors; thus the coefficient of x^{n-2} in the final product is the sum of the products of the letters $a, b, c, \ldots k$ taken two at a time; denote it by S_2.

And, generally, the terms involving x^{n-r} are formed by taking the letter x from *any* $n-r$ of the factors, and r of the letters $a, b, c, \ldots k$ from the r remaining factors; thus the coefficient of x^{n-r} in the final product is the sum of the products of the letters $a, b, c, \ldots k$ taken r at a time; denote it by S_r.

The last term in the product is $abc\ldots k$; denote it by S_n.

Hence $(x+a)(x+b)(x+c)\ldots\ldots(x+k)$
$$= x^n + S_1 x^{n-1} + S_2 x^{n-2} + \ldots + S_r x^{n-r} + \ldots + S_{n-1}x + S_n.$$

In S_1 the *number of terms* is n; in S_2 *the number of terms* is the same as the number of combinations of n things 2 at a time; that is, $^nC_2$; in S_3 *the number of terms* is $^nC_3$; and so on.

Now suppose $b, c, \ldots k$, each equal to a; then S_1 becomes $^nC_1 a$; S_2 becomes $^nC_2 a^2$; S_3 becomes $^nC_3 a^3$; and so on; thus
$$(x+a)^n = x^n + {}^nC_1 a x^{n-1} + {}^nC_2 a^2 x^{n-2} + {}^nC_3 a^3 x^{n-3} + \ldots + {}^nC_n a^n;$$

substituting for $^nC_1$, $^nC_2$, ... we obtain

$$(x+a)^n = x^n + nax^{n-1} + \frac{n(n-1)}{1\cdot 2}a^2x^{n-2}$$
$$+ \frac{n(n-1)(n-2)}{1\cdot 2\cdot 3}a^3x^{n-3} + \ldots + a^n,$$

the series containing $n+1$ terms.

This is the *Binomial Theorem*, and the expression on the right is said to be **the expansion** of $(x+a)^n$.

358. The coefficients in the expansion of $(x+a)^n$ are very conveniently expressed by the symbols $^nC_1, ^nC_2, ^nC_3, \ldots ^nC_n$. We shall, however, sometimes further abbreviate them by omitting n, and writing $C_1, C_2, C_3, \ldots C_n$. With this notation we have

$$(x+a)^n = x^n + C_1ax^{n-1} + C_2a^2x^{n-2} + C_3a^3x^{n-3} + \ldots + C_na^n.$$

If we write $-a$ in the place of a, we obtain
$$(x-a)^n = x^n + C_1(-a)x^{n-1} + C_2(-a)^2x^{n-2}$$
$$+ C_3(-a)^3x^{n-3} + \ldots + C_n(-a)^n$$
$$= x^n - C_1ax^{n-1} + C_2a^2x^{n-2} - C_3a^3x^{n-3} + \ldots + (-1)^nC_na^n.$$

Thus the terms in the expansion of $(x+a)^n$ and $(x-a)^n$ are *numerically* the same, but in $(x-a)^n$ they are alternately positive and negative, and the last term is positive or negative according as n is even or odd.

Example 1. Find the expansion of $(x+y)^6$.

By the formula, the expansion
$$= x^6 + {^6C_1}x^5y + {^6C_2}x^4y^2 + {^6C_3}x^3y^3 + {^6C_4}x^2y^4 + {^6C_5}xy^5 + {^6C_6}y^6$$
$$= x^6 + 6x^5y + 15x^4y^2 + 20x^3y^3 + 15x^2y^4 + 6xy^5 + y^6,$$
on calculating the values of $^6C_1, ^6C_2, ^6C_3, \ldots$

Example 2. Find the expansion of $(a-2x)^7$.

$(a-2x)^7 = a^7 - {^7C_1}a^6(2x) + {^7C_2}a^5(2x)^2 - {^7C_3}a^4(2x)^3 + \ldots$ to 8 terms.

Now remembering that $^nC_r = {^nC_{n-r}}$, after calculating the coefficients up to $^7C_3$, the rest may be written down at once; for $^7C_4 = {^7C_3}$; $^7C_5 = {^7C_2}$; and so on. Hence

$$(a-2x)^7 = a^7 - 7a^6(2x) + \frac{7\cdot 6}{1\cdot 2}a^5(2x)^2 - \frac{7\cdot 6\cdot 5}{1\cdot 2\cdot 3}a^4(2x)^3 + \ldots$$
$$= a^7 - 7a^6(2x) + 21a^5(2x)^2 - 35a^4(2x)^3 + 35a^3(2x)^4$$
$$- 21a^2(2x)^5 + 7a(2x)^6 - (2x)^7$$
$$= a^7 - 14a^6x + 84a^5x^2 - 280a^4x^3 + 560a^3x^4$$
$$- 672a^2x^5 + 448ax^6 - 128x^7.$$

359. In the expansion of $(x+a)^n$, the coefficient of the second term is $^nC_1$; of the third term is $^nC_2$; of the fourth term is $^nC_3$; and so on; *the suffix in each term being one less than the number of the term to which it applies;* hence ^nC_r is the coefficient of the $(r+1)^{\text{th}}$ term. This is called the **general term**, because by giving to r different numerical values any of the coefficients may be found from ^nC_r; and by giving to x and a their appropriate indices any assigned term may be obtained. Thus the $(r+1)^{\text{th}}$ term may be written

$$^nC_r x^{n-r} a^r, \text{ or } \frac{n(n-1)(n-2)\ldots\ldots(n-r+1)}{\underline{|r}} x^{n-r} a^r.$$

In applying this formula to any particular case, it should be observed that *the index of* a *is the same as the suffix of* C, *and that the sum of the indices of* x *and* a *is* n.

Example 1. Find the fifth term of $(a+2x^3)^{17}$.

The required term $= {}^{17}C_4 a^{13}(2x^3)^4$

$$= \frac{17 \cdot 16 \cdot 15 \cdot 14}{1 \cdot 2 \cdot 3 \cdot 4} \times 16 a^{13} x^{12}$$

$$= 38080 a^{13} x^{12}.$$

Example 2. Find the fourteenth term of $(3-a)^{15}$.

The required term $= {}^{15}C_{13}(3)^2(-a)^{13}$

$= {}^{15}C_2 \times (-9a^{13})$ [Art. 349.]

$= -945 a^{13}$.

360. The simplest form of the binomial theorem is the expansion of $(1+x)^n$. This is obtained from the general formula of Art. 357, by writing 1 in the place of x, and x in the place of a. Thus

$$(1+x)^n = 1 + {}^nC_1 x + {}^nC_2 x^2 + \ldots + {}^nC_r x^r + \ldots + {}^nC_n x^n$$

$$= 1 + nx + \frac{n(n-1)}{1 \cdot 2} x^2 + \ldots\ldots + x^n;$$

the general term being

$$\frac{n(n-1)(n-2)\ldots\ldots(n-r+1)}{\underline{|r}} x^r.$$

361. The expansion of a binomial may always be made to depend upon the case in which the first term is unity; thus

$$(x+y)^n = \left\{ x\left(1+\frac{y}{x}\right)\right\}^n$$

$$= x^n(1+z)^n, \text{ where } z = \frac{y}{x}.$$

Example. Find the coefficient of x^{16} in the expansion of $(x^2-2x)^{10}$.

We have $(x^2-2x)^{10} = x^{20}\left(1-\dfrac{2}{x}\right)^{10}$;

and, since x^{20} multiplies every term in the expansion of $\left(1-\dfrac{2}{x}\right)^{10}$, we have in this expansion to seek the coefficient of the term which contains $\dfrac{1}{x^4}$.

Hence the required coefficient $= {}^{10}C_4(-2)^4$

$$= \dfrac{10.9.8.7}{1.2.3.4} \times 16$$

$$= 3360.$$

EXAMPLES XXXVIII. a.

Expand the following binomials:

1. $(x+2)^4$.
2. $(x+3)^5$.
3. $(a+x)^7$.
4. $(a-x)^5$.
5. $(1-2y)^5$.
6. $\left(2x+\dfrac{y}{2}\right)^4$.
7. $\left(2-\dfrac{x}{2}\right)^6$.
8. $\left(a-\dfrac{3}{b}\right)^7$.
9. $\left(ax+\dfrac{y}{a}\right)^9$.

Write down and simplify:

10. The 4th term of $(1+x)^{12}$.
11. The 6th term of $(2-y)^8$.
12. The 5th term of $(a-5b)^7$.
13. The 15th term of $(2x-1)^{71}$.
14. The 7th term of $\left(1-\dfrac{1}{x}\right)^{10}$.
15. The 6th term of $\left(3x+\dfrac{a}{2}\right)^9$.
16. The middle term of $\left(\dfrac{2}{3}a-\dfrac{3}{2a}\right)^6$.
17. The 23rd term of $\left(x^2+\dfrac{b}{x}\right)^{25}$.
18. The 10th term of $(x^2-x)^{17}$.

19. Find the value of $(x-\sqrt{3})^4+(x+\sqrt{3})^4$.

20. Expand $(\sqrt{1-x^2}+1)^5-(\sqrt{1-x^2}-1)^5$.

21. Find the coefficient of x^{12} in $(x^2+2x)^{10}$.

22. Find the coefficient of x in $\left(x^2-\dfrac{a}{2x}\right)^{14}$.

23. Find the term independent of x in $\left(2x^2-\dfrac{1}{x}\right)^{12}$.

24. Find the coefficient of x^{-20} in $\left(\dfrac{x^2}{3}-\dfrac{2}{x^3}\right)^{15}$.

362. *In the expansion of $(1+x)^n$ the coefficients of terms equidistant from the beginning and end are equal.*

The coefficient of the $(r+1)^{\text{th}}$ term from the beginning is ^nC_r.

The $(r+1)^{\text{th}}$ term from the end has $n+1-(r+1)$, or $n-r$ terms before it; therefore counting from the beginning it is the $(n-r+1)^{\text{th}}$ term, and its coefficient is $^nC_{n-r}$, which has been shewn to be equal to ^nC_r. [Art. 349.] Hence the proposition follows.

363. *To find the greatest coefficient in the expansion of $(1+x)^n$.*

The coefficient of the general term of $(1+x)^n$ is ^nC_r; and we have only to find for what value of r this is greatest.

By Art. 354, when n is even, the greatest coefficient is $^nC_{\frac{n}{2}}$; and when n is odd, it is $^nC_{\frac{n-1}{2}}$ or $^nC_{\frac{n+1}{2}}$; these two coefficients being equal.

364. *To find the greatest term in the expansion of $(x+a)^n$.*

We have $(x+a)^n = x^n\left(1+\dfrac{a}{x}\right)^n$;

therefore, since x^n multiplies every term in $\left(1+\dfrac{a}{x}\right)^n$, it will be sufficient to find the greatest term in this latter expansion.

Let the rth and $(r+1)^{\text{th}}$ be any two consecutive terms. The $(r+1)^{\text{th}}$ term is obtained by multiplying the rth term by $\dfrac{n-r+1}{r}\cdot\dfrac{a}{x}$; that is, by $\left(\dfrac{n+1}{r}-1\right)\dfrac{a}{x}$. [Art. 359.]

XXXVIII.] BINOMIAL THEOREM. 335

The factor $\dfrac{n+1}{r}-1$ decreases as r increases; hence the $(r+1)^{\text{th}}$ term is not always greater than the rth term, but only until $\left(\dfrac{n+1}{r}-1\right)\dfrac{a}{x}$ becomes equal to 1, or less than 1.

Now $\left(\dfrac{n+1}{r}-1\right)\dfrac{a}{x}>1$, so long as $\dfrac{n+1}{r}-1>\dfrac{x}{a}$;

that is $\dfrac{n+1}{r}>\dfrac{x}{a}+1$, or $\dfrac{(n+1)a}{x+a}>r$ ....(1).

If $\dfrac{(n+1)a}{x+a}$ be an integer, denote it by p; then if $r=p$ the multiplying factor becomes 1, and the $(p+1)^{\text{th}}$ term is equal to the pth; and these are greater than any other term.

If $\dfrac{(n+1)a}{x+a}$ be not an integer, denote its integral part by q; then the greatest value of r consistent with (1) is q; hence the $(q+1)^{\text{th}}$ term is the greatest.

Since we are only concerned with the *numerically greatest term*, the investigation will be the same for $(x-a)^n$; therefore in any numerical example it is unnecessary to consider the sign of the second term of the binomial. Also it will be found best to work each example independently of the general formula.

Example. Find the greatest term in the expansion of $(1+4x)^8$, when x has the value $\dfrac{1}{3}$.

Denote the rth and $(r+1)^{\text{th}}$ terms by T_r and T_{r+1} respectively; then

$$T_{r+1}=\dfrac{8-r+1}{r}\cdot 4x \times T_r = \dfrac{9-r}{r}\times\dfrac{4}{3}\times T_r;$$

hence $T_{r+1}>T_r$, so long as $\dfrac{9-r}{r}\times\dfrac{4}{3}>1$;

that is, $36-4r>3r$, or $36>7r$.

The greatest value of r consistent with this is 5; hence the greatest term is the sixth, and its value

$$={}^8C_5\times\left(\dfrac{4}{3}\right)^5={}^8C_3\times\left(\dfrac{4}{3}\right)^5=\dfrac{57344}{243}.$$

365. *To find the sum of the coefficients in the expansion of* $(1+x)^n$.

In the identity $(1+x)^n = 1 + C_1 x + C_2 x^2 + C_3 x^3 + \ldots + C_n x^n$, put $x = 1$; thus

$$2^n = 1 + C_1 + C_2 + C_3 + \ldots + C_n$$
$$= \text{sum of the coefficients.}$$

COR. $\quad C_1 + C_2 + C_3 + \ldots + C_n = 2^n - 1$;

that is, the total number of combinations of n things *taking some or all of them at a time* is $2^n - 1$.

366. *To prove that in the expansion of* $(1+x)^n$, *the sum of the coefficients of the odd terms is equal to the sum of the coefficients of the even terms.*

In the identity $(1+x)^n = 1 + C_1 x + C_2 x^2 + C_3 x^3 + \ldots + C_n x^n$, put $x = -1$; thus

$$0 = 1 - C_1 + C_2 - C_3 + C_4 - C_5 + \ldots\ldots;$$
$$\therefore\ 1 + C_2 + C_4 + \ldots\ldots = C_1 + C_3 + C_5 + \ldots\ldots$$

367. The Binomial Theorem may also be applied to expand expressions which contain more than two terms.

Example. Find the expansion of $(x^2 + 2x - 1)^3$.

Regarding $2x - 1$ as a single term, the expansion

$$= (x^2)^3 + 3(x^2)^2(2x - 1) + 3x^2(2x - 1)^2 + (2x - 1)^3$$
$$= x^6 + 6x^5 + 9x^4 - 4x^3 - 9x^2 + 6x - 1, \text{ on reduction.}$$

368. For a full discussion of the Binomial Theorem when the index is not restricted to positive integral values the student is referred to the *Higher Algebra*, Chap. XIV. It is there shewn that when x is less than unity, the formula

$$(1+x)^n = 1 + nx + \frac{n(n-1)}{1 \cdot 2} x^2 + \frac{n(n-1)(n-2)}{1 \cdot 2 \cdot 3} x^3 + \ldots$$

is true for any value of n.

When n is negative or fractional the number of terms in the expansion is unlimited, but in any particular case we may write down as many terms as we please, or we may find the coefficient of any assigned term.

Example 1. Expand $(1+x)^{-3}$ to four terms.

$$(1+x)^{-3} = 1 + (-3)x + \frac{(-3)(-3-1)}{1.2}x^2 + \frac{(-3)(-3-1)(-3-2)}{1.2.3}x^3 + \ldots$$

$$= 1 - 3x + \frac{3.4}{1.2}x^2 - \frac{3.4.5}{1.2.3}x^3 + \ldots$$

$$= 1 - 3x + 6x^2 - 10x^3 + \ldots.$$

Example 2. Expand $(4+3x)^{\frac{3}{2}}$ to four terms.

$$(4+3x)^{\frac{3}{2}} = 4^{\frac{3}{2}}\left(1+\frac{3x}{4}\right)^{\frac{3}{2}} = 8\left(1+\frac{3x}{4}\right)^{\frac{3}{2}}$$

$$= 8\left[1 + \frac{3}{2}\cdot\frac{3x}{4} + \frac{\frac{3}{2}\left(\frac{3}{2}-1\right)}{1.2}\left(\frac{3x}{4}\right)^2 + \frac{\frac{3}{2}\left(\frac{3}{2}-1\right)\left(\frac{3}{2}-2\right)}{1.2.3}\left(\frac{3x}{4}\right)^3 + \ldots\right]$$

$$= 8\left[1 + \frac{3}{2}\cdot\frac{3x}{4} + \frac{3}{8}\cdot\frac{9x^2}{16} - \frac{1}{16}\cdot\frac{27x^3}{64} + \ldots\right]$$

$$= 8 + 9x + \frac{27}{16}x^2 - \frac{27}{128}x^3 + \ldots.$$

369. In finding the general term we must now use the formula

$$\frac{n(n-1)(n-2)\ldots\ldots(n-r+1)}{\lfloor r}x^r$$

written in full; for the symbol ^nC_r cannot be employed when n is fractional or negative.

Example 1. Find the general term in the expansion of $(1+x)^{\frac{1}{2}}$.

The $(r+1)^{\text{th}}$ term $= \dfrac{\frac{1}{2}\left(\frac{1}{2}-1\right)\left(\frac{1}{2}-2\right)\ldots\ldots\left(\frac{1}{2}-r+1\right)}{\lfloor r}x^r$

$$= \frac{1(-1)(-3)(-5)\ldots\ldots(-2r+3)}{2^r\lfloor r}x^r.$$

The number of factors in the numerator is r, and $r-1$ of these are negative; therefore, by taking -1 out of each of these negative factors, we may write the above expression

$$(-1)^{r-1}\frac{1.3.5\ldots\ldots(2r-3)}{2^r\lfloor r}x^r.$$

Example 2. Find the general term in the expansion of $(1-x)^{-3}$.

$$\text{The } (r+1)^{\text{th}} \text{ term} = \frac{(-3)(-4)(-5)\ldots\ldots(-3-r+1)}{\lfloor r}(-x)^r$$

$$=(-1)^r\frac{3\cdot 4\cdot 5\ldots\ldots(r+2)}{\lfloor r}(-1)^r x^r$$

$$=(-1)^{2r}\frac{3\cdot 4\cdot 5\ldots\ldots(r+2)}{1\cdot 2\cdot 3\ldots\ldots r}x^r$$

$$=\frac{(r+1)(r+2)}{1\cdot 2}x^r,$$

by removing like factors from the numerator and denominator.

370. The following example illustrates a useful application of the Binomial Theorem.

Example. Find the cube root of 126 to five places of decimals.

$$(126)^{\frac{1}{3}}=(5^3+1)^{\frac{1}{3}}=5\left(1+\frac{1}{5^3}\right)^{\frac{1}{3}}$$

$$=5\left(1+\frac{1}{3}\cdot\frac{1}{5^3}-\frac{1}{9}\cdot\frac{1}{5^6}+\frac{5}{81}\cdot\frac{1}{5^9}-\cdots\right)$$

$$=5+\frac{1}{3}\cdot\frac{1}{5^2}-\frac{1}{9}\cdot\frac{1}{5^5}+\frac{1}{81}\cdot\frac{1}{5^7}-\cdots$$

$$=5+\frac{1}{3}\cdot\frac{2^2}{10^2}-\frac{1}{9}\cdot\frac{2^5}{10^5}+\frac{1}{81}\cdot\frac{2^7}{10^7}-\cdots$$

$$=5+\frac{\cdot 04}{3}-\frac{\cdot 00032}{9}+\frac{\cdot 0000128}{81}-\cdots$$

$$=5+\cdot 013333\ldots-\cdot 000035\ldots+\ldots$$

$$=5\cdot 01329, \text{ to five places of decimals.}$$

EXAMPLES XXXVIII. b.

In the following expansions find which is the greatest term:

1. $(x+y)^{17}$ when $x=4$, $y=3$.
2. $(x-y)^{28}$ when $x=9$, $y=4$.
3. $(1+x)^4$ when $x=\frac{2}{3}$.

BINOMIAL THEOREM.

4. $(a-4b)^{15}$ when $a=12$, $b=2$.

5. $(7x+2y)^{30}$ when $x=8$, $y=14$.

6. $(2x+3)^n$ when $x=\dfrac{5}{2}$, $n=15$.

7. In the expansion of $(1+x)^{25}$ the coefficients of the $(2r+1)^{\text{th}}$ and $(r+5)^{\text{th}}$ terms are equal; find r.

8. Find n when the coefficients of the 16th and 26th terms of $(1+x)^n$ are equal.

9. Find the relation between r and n in order that the coefficients of the $(r+3)^{\text{th}}$ and $(2r-3)^{\text{th}}$ terms of $(1+x)^{3n}$ may be equal.

10. Find the coefficient of x^m in the expansion of $\left(x^2+\dfrac{1}{x}\right)^{2m}$.

11. Find the middle term of $(1+x)^{2n}$ in its simplest form.

12. Find the sum of the coefficients of $(x+y)^{16}$.

13. Find the sum of the coefficients of $(3x+y)^9$.

14. Find the rth term from the beginning and the rth term from the end of $(a+2x)^n$.

15. Expand $(a^2+2a+1)^3$ and $(x^2-4x+2)^3$.

Expand to 4 terms the following expressions:

16. $(1+x)^{\frac{1}{3}}$.　　17. $(1+x)^{\frac{3}{4}}$.　　18. $(1+x)^{\frac{2}{5}}$.

19. $(1+3x)^{-2}$.　　20. $(1-x^2)^{-3}$.　　21. $(1+3x)^{-4}$.

22. $(2+x)^{-3}$.　　23. $(1+2x)^{-\frac{1}{2}}$.　　24. $(a-2x)^{-\frac{3}{2}}$.

Write down and simplify:

25. The 5th term and the 10th term of $(1+x)^{-\frac{3}{2}}$.

26. The 3rd term and the 11th term of $(1+2x)^{\frac{11}{2}}$.

27. The 4th term and the $(r+1)^{\text{th}}$ term of $(1+x)^{-2}$.

28. The 7th term and the $(r+1)^{\text{th}}$ term of $(1-x)^{\frac{1}{2}}$.

29. The $(r+1)^{\text{th}}$ term of $(a-bx)^{-1}$, and of $(1-nx)^{\frac{1}{n}}$.

Find to four places of decimals the value of

30. $\sqrt[3]{122}$.　　31. $\sqrt[4]{620}$.　　32. $\sqrt[5]{31}$.　　33. $1 \div \sqrt{99}$.

CHAPTER XXXIX.

LOGARITHMS.

371. Definition. The **logarithm** of any number to a given **base** is the index of the power to which the base must be raised in order to equal the given number. Thus if $a^x = N$, x is called the logarithm of N to the base a.

Examples. (1) Since $3^4 = 81$, the logarithm of 81 to base 3 is 4.

(2) Since $10^1 = 10$, $10^2 = 100$, $10^3 = 1000$, ...
the natural numbers 1, 2, 3, ... are respectively the logarithms of 10, 100, 1000, ... to base 10.

372. The logarithm of N to base a is usually written $\log_a N$, so that the same meaning is expressed by the two equations
$$a^x = N \ ; \ x = \log_a N.$$

Example. Find the logarithm of $32 \sqrt[5]{4}$ to base $2\sqrt{2}$.

Let x be the required logarithm; then

by definition, $\qquad (2\sqrt{2})^x = 32 \sqrt[5]{4}\ ;$

$\therefore \ (2 \cdot 2^{\frac{1}{2}})^x = 2^5 \cdot 2^{\frac{2}{5}}\ ;$

$\therefore \qquad 2^{\frac{3}{2}x} = 2^{5 + \frac{2}{5}}\ ;$

hence, by equating the indices, $\dfrac{3}{2}x = \dfrac{27}{5}\ ;$

$\therefore \qquad x = \dfrac{18}{5} = 3\cdot 6.$

373. When it is understood that a particular system of logarithms is in use, the suffix denoting the base is omitted. Thus in arithmetical calculations in which 10 is the base, we usually write log 2, log 3, ... instead of $\log_{10} 2$, $\log_{10} 3$, ...

Logarithms to the base 10 are known as **Common Logarithms**; this system was first introduced in 1615 by Briggs, a contemporary of Napier the inventor of Logarithms. Before discussing the properties of common logarithms we shall prove some general propositions which are true for all logarithms independently of any particular base.

374. *The logarithm of 1 is 0.*

For $a^0 = 1$ for all values of a; therefore $\log 1 = 0$, whatever the base may be.

375. *The logarithm of the base itself is 1.*

For $a^1 = a$; therefore $\log_a a = 1$.

376. *To find the logarithm of a product.*

Let MN be the product; let a be the base of the system, and suppose
$$x = \log_a M, \quad y = \log_a N;$$
so that
$$a^x = M, \quad a^y = N.$$

Thus the product
$$MN = a^x \times a^y = a^{x+y};$$
whence, by definition, $\log_a MN = x + y$
$$= \log_a M + \log_a N.$$

Similarly, $\log_a MNP = \log_a M + \log_a N + \log_a P$;
and so on for any number of factors.

Example. $\log 42 = \log(2 \times 3 \times 7) = \log 2 + \log 3 + \log 7$.

377. *To find the logarithm of a fraction.*

Let $\dfrac{M}{N}$ be the fraction, and suppose
$$x = \log_a M, \quad y = \log_a N;$$
so that
$$a^x = M, \quad a^y = N.$$

Thus the fraction
$$\frac{M}{N} = \frac{a^x}{a^y} = a^{x-y};$$

whence, by definition, $\log_a \dfrac{M}{N} = x - y$
$$= \log_a M - \log_a N.$$

E.A.

Example. $\log(2\tfrac{1}{7}) = \log\dfrac{15}{7} = \log 15 - \log 7$

$\qquad\qquad\quad = \log(3 \times 5) - \log 7 = \log 3 + \log 5 - \log 7.$

378. *To find the logarithm of a number raised to any power, integral or fractional.*

Let $\log_a(M^p)$ be required, and suppose

$$x = \log_a M, \text{ so that } a^x = M;$$

then $\qquad\qquad M^p = (a^x)^p = a^{px};$

whence, by definition, $\quad \log_a(M^p) = px;$

that is, $\qquad\qquad\quad \log_a(M^p) = p\log_a M.$

Similarly, $\qquad\qquad \log_a(M^{\tfrac{1}{r}}) = \dfrac{1}{r}\log_a M.$

Example. Express the logarithm of $\dfrac{\sqrt{a^3}}{c^5 b^2}$ in terms of $\log a$, $\log b$, and $\log c$.

$\log\dfrac{\sqrt{a^3}}{c^5 b^2} = \log\dfrac{a^{\tfrac{3}{2}}}{c^5 b^2} = \log a^{\tfrac{3}{2}} - \log(c^5 b^2)$

$\qquad = \dfrac{3}{2}\log a - (\log c^5 + \log b^2) = \dfrac{3}{2}\log a - 5\log c - 2\log b.$

Common Logarithms.

379. From the equation $10^x = N$, it is evident that common logarithms will not in general be integral, and that they will not always be positive.

For instance, $\qquad 3154 > 10^3$ and $< 10^4$;

$\qquad\qquad \therefore\ \log 3154 = 3 + \text{a fraction.}$

Again, $\qquad\qquad \cdot 06 > 10^{-2}$ and $< 10^{-1}$;

$\qquad\qquad \therefore\ \log \cdot 06 = -2 + \text{a fraction.}$

380. DEFINITION. The integral part of a logarithm is called the **characteristic**, and the fractional part when expressed as a decimal is called the **mantissa**.

381. The characteristic of the logarithm of any number to base 10 can be written down by inspection, as we shall now shew.

(i) *To determine the characteristic of the logarithm of any number greater than unity.*

It is clear that a number with two digits in its integral part lies between 10^1 and 10^2; a number with three digits in its integral part lies between 10^2 and 10^3; and so on. Hence a number with n digits in its integral part lies between 10^{n-1} and 10^n.

Let N be a number whose integral part contains n digits; then
$$N = 10^{(n-1)+\text{a fraction}};$$
$$\therefore \log N = (n-1) + \text{a fraction}.$$

Hence the characteristic is $n-1$; that is, *the characteristic of the logarithm of a number greater than unity is less by one than the number of digits in its integral part, and is positive.*

Example. The characteristics of
$$\log 314, \quad \log 87\cdot 263, \quad \log 2\cdot 78, \quad \log 3500$$
are respectively 2, 1, 0, 3.

(ii) *To determine the characteristic of the logarithm of a number less than unity.*

A decimal with one cipher immediately after the decimal point, such as ·0324, being greater than ·01 and less than ·1, lies between 10^{-2} and 10^{-1}; a number with two ciphers after the decimal point lies between 10^{-3} and 10^{-2}; and so on. Hence a decimal fraction with n ciphers immediately after the decimal point lies between $10^{-(n+1)}$ and 10^{-n}.

Let D be a decimal beginning with n ciphers; then
$$D = 10^{-(n+1)+\text{a fraction}};$$
$$\therefore \log D = -(n+1) + \text{a fraction}.$$

Hence the characteristic is $-(n+1)$; this is, *the characteristic of the logarithm of a number less than one is negative and one more than the number of ciphers immediately after the decimal point.*

Example. The characteristics of
$$\log \cdot 4, \quad \log \cdot 3748, \quad \log \cdot 000135, \quad \log \cdot 08$$
are respectively $-1, -1, -4, -2$.

382. *The mantissæ are the same for the logarithms of all numbers which have the same significant digits.*

For if any two numbers have the same sequence of digits, differing only in the position of the decimal point, one must be equal to the other multiplied or divided by some integral power of 10. Hence their logarithms must *differ by an integer*. In other words, their decimal parts or mantissæ are the same.

Examples. (i) $\log 32700 = \log(3\cdot27 \times 10^4) = \log 3\cdot27 + \log 10^4$
$= \log 3\cdot27 + 4.$

(ii) $\log \cdot0327 = \log(3\cdot27 \times 10^{-2}) = \log 3\cdot27 + \log 10^{-2}$
$= \log 3\cdot27 - 2.$

(iii) $\log \cdot000327 = \log(3\cdot27 \times 10^{-4}) = \log 3\cdot27 + \log 10^{-4}$
$= \log 3\cdot27 - 4.$

Thus, $\log 32700$, $\log \cdot0327$, $\log \cdot000327$ differ from $\log 3\cdot27$ only in the *integral* part; that is the mantissa is the same in each case.

Note. The characteristics of the logarithms are 4, -2, -4 respectively. The foregoing examples shew that by introducing a suitable integral power of 10, all numbers can be expressed in one standard form in which *the decimal point always stands after the first significant digit*, and the characteristics are given by the powers of 10, without using the rules of Art. 381.

383. The logarithms of all integers from 1 to 20000 have been found and tabulated. In Chambers' Mathematical Tables they are given to seven places of decimals, but for many practical purposes sufficient accuracy is secured by using four-figure logarithms (available for all numbers from 1 to 9999), such as are contained in the Tables given on pages 348_D to 348_G.

384. Advantages of Common Logarithms. It will now be seen that it is unnecessary to tabulate the characteristics, since they can always be written down by inspection [Art. 381]. Also the Tables need only contain the mantissa of the logarithms of integers [Art. 382].

In order to secure these advantages it is convenient *always to keep the mantissa positive*, and it is usual to write the minus sign over a negative characteristic and not before it, so as to indicate that the characteristic alone is negative. Thus $\bar{4}\cdot30103$, which is the logarithm of $\cdot0002$, is equivalent to $-4 + \cdot30103$, and must be distinguished from $-4\cdot30103$, in which both the integer and the decimal are negative.

LOGARITHMS.

385. In the course of work we sometimes have to deal with a logarithm which is wholly negative. In such a case an arithmetical artifice is necessary in order to write the logarithm with mantissa positive. Thus a result such as $-3\cdot 69897$ may be transformed by subtracting 1 from the integral part and adding 1 to the decimal part. Thus

$$-3\cdot 69897 = -3 - 1 + (1 - \cdot 69897)$$
$$= -4 + \cdot 30103 = \bar{4}\cdot 30103.$$

Example 1. Required the logarithm of $\cdot 0002432$.

In the Tables we find that 3859636 is the mantissa of log 2432 (the decimal point as well as the characteristic being omitted); and, by Art. 382, the characteristic of the logarithm of the given number is -4;

$$\therefore \log \cdot 0002432 = \bar{4}\cdot 3859636.$$

Example 2. Find the value of $\sqrt[5]{\cdot 00000165}$, given
$$\log 165 = 2\cdot 2175, \quad \log 6974 = 5\cdot 8435.$$

Let x denote the value required; then

$$\log x = \log (\cdot 00000165)^{\frac{1}{5}} = \frac{1}{5} \log (\cdot 00000165) = \frac{1}{5}(\bar{6}\cdot 2175);$$

the *mantissa* of log $\cdot 00000165$ being the same as that of log 165, and the *characteristic* being prefixed by the rule.

Now $$\frac{1}{5}(\bar{6}\cdot 2175) = \frac{1}{5}(\overline{10} + 4\cdot 2175) = \bar{2}\cdot 8435$$

and $\cdot 8435$ is the mantissa of log 6974; hence x is a number consisting of these same digits but with one cipher after the decimal point. [Art. 382.]

Thus $$x = \cdot 06974.$$

*386. It is sometimes necessary to transform logarithms from one base to another.

Suppose for example that the logarithms of all numbers to base a are known and tabulated, it is required to find the logarithms to base b.

Let N be any number whose logarithm to base b is required.

Let $y = \log_b N$, so that $b^y = N$;

$$\therefore \log_a(b^y) = \log_a N;$$

that is, $$y \log_a b = \log_a N;$$

$$\therefore y = \frac{1}{\log_a b} \times \log_a N,$$

or $$\log_b N = \frac{1}{\log_a b} \times \log_a N \quad \ldots\ldots\ldots\ldots\ldots\ldots(1).$$

Now since N and b are given, $\log_a N$ and $\log_a b$ are known from the Tables, and thus $\log_b N$ may be found.

Hence it appears that to transform logarithms from base a to base b we have only to multiply them all by $\dfrac{1}{\log_a b}$; this is a constant quantity and is given by the Tables; it is known as the *modulus*.

COR. If in equation (1) we put a for N, we obtain
$$\log_b a = \frac{1}{\log_a b} \times \log_a a = \frac{1}{\log_a b};$$
$$\therefore \log_b a \times \log_a b = 1.$$

387. In the following examples all necessary logarithms will be given. The use of four-figure Tables will be explained in a future section.

Example 1. Given $\log 3 = \cdot 4771213$, find $\log\{(2\cdot7)^3 \times (\cdot 81)^{\frac{4}{5}} \div (90)^{\frac{5}{4}}\}$.

The required value $= 3 \log \dfrac{27}{10} + \dfrac{4}{5} \log \dfrac{81}{100} - \dfrac{5}{4} \log 90$

$$= 3(\log 3^3 - 1) + \frac{4}{5}(\log 3^4 - 2) - \frac{5}{4}(\log 3^2 + 1)$$

$$= \left(9 + \frac{16}{5} - \frac{5}{2}\right) \log 3 - \left(3 + \frac{8}{5} + \frac{5}{4}\right)$$

$$= \frac{97}{10} \log 3 - 5\tfrac{17}{20} = 4\cdot 6280766 - 5\cdot 85$$

$$= \bar{2}\cdot 7780766.$$

Obs. The student should notice that the logarithm of 5 and its powers can always be obtained from $\log 2$; thus
$$\log 5 = \log \frac{10}{2} = \log 10 - \log 2 = 1 - \log 2.$$

Example 2. Find the number of digits in 875^{16}, given
$$\log 2 = \cdot 301, \quad \log 7 = \cdot 845.$$
$$\log(875^{16}) = 16 \log(7 \times 125) = 16(\log 7 + 3 \log 5)$$
$$= 16(\log 7 + 3 - 3 \log 2) = 16 \times 2\cdot 942$$
$$= 47\cdot 072;$$
hence the number of digits is 48. [Art. 381.]

Example 3. Given $\log 2 = \cdot 301$ and $\log 3 = \cdot 477$, find to two places of decimals the value of x from the equation
$$6^{3-4x} \cdot 4^{x+5} = 8.$$

Taking logarithms of both sides, we have
$$(3-4x)\log 6 + (x+5)\log 4 = \log 8;$$
$$\therefore (3-4x)(\log 2 + \log 3) + (x+5)2\log 2 = 3\log 2;$$
$$\therefore x(-4\log 2 - 4\log 3 + 2\log 2) = 3\log 2 - 3\log 2 - 3\log 3 - 10\log 2;$$
$$\therefore x = \frac{10\log 2 + 3\log 3}{2\log 2 + 4\log 3}$$
$$= \frac{4\cdot 44}{2\cdot 51} = 1\cdot 77\ldots.$$

EXAMPLES XXXIX. a.

1. Find the logarithms of $\sqrt{32}$ and $\cdot 03125$ to base $\sqrt[3]{2}$, and 100 and $\cdot 00001$ to base $\cdot 01$.

2. Find the value of $\log_4 512$, $\log_5 \cdot 0016$, $\log_{81}\dfrac{1}{27}$, $\log_{49} 343$.

3. Write down the numbers whose logarithms

to base $25, \quad 3, \quad \cdot 02, \quad 1, \quad -4, \quad 1\cdot 7, \quad 1000$

are $\quad \dfrac{1}{2}, \quad -2, \quad -3, \quad 5, \quad -1, \quad 2, \quad -\dfrac{2}{3}$ respectively.

Simplify the expressions:

4. $\log \dfrac{(ab^2c^4)^{\frac{1}{6}}}{\sqrt[9]{a^{-3}b^3c^6}}.$

5. $\log\left\{\left(\dfrac{x^{\frac{1}{2}}y^{-3}}{x^{-1}y^2}\right)^{-3} \div \left(\dfrac{x^{-2}y^3}{xy^{-1}}\right)^5\right\}.$

6. Find by inspection the characteristics of the logarithms of 3174, $625\cdot 7$, $3\cdot 502$, $\cdot 4$, $\cdot 374$, $\cdot 000135$, $23\cdot 22065$.

7. The mantissa of $\log 37203$ is $\cdot 5705780$: write down the logarithms of $37\cdot 203$, $\cdot 000037203$, 372030000.

8. The logarithm of 7623 is $3\cdot 8821259$: write down the numbers whose logarithms are $\cdot 8821259$, $\bar{6}\cdot 8821259$, $7\cdot 8821259$.

Given $\log 2 = \cdot 3010300$, $\log 3 = \cdot 4771213$, $\log 7 = \cdot 8450980$, find the value of

9. $\log 729$.   **10.** $\log 8400$.   **11.** $\log \cdot 256$.

12. $\log 5\cdot 832$.   **13.** $\log \sqrt[3]{392}$.   **14.** $\log \cdot 304\dot{8}$.

15. Shew that $\log \dfrac{11}{15} + \log \dfrac{490}{297} - 2 \log \dfrac{7}{9} = \log 2$.

16. Find to six decimal places the value of
$$\log \dfrac{225}{224} - 2 \log \dfrac{20}{189} + \log \dfrac{512}{81}.$$

17. Simplify $\log \{(10 \cdot 8)^{\frac{1}{2}} \times (\cdot 24)^{\frac{5}{3}} \div (90)^{-2}\}$, and find its numerical value.

18. Find the value of
$$\log (\sqrt[3]{126} \cdot \sqrt{108} \div \sqrt[6]{1008} \cdot \sqrt[3]{162}).$$

19. Find the value of $\log \sqrt[5]{\dfrac{588 \times 768}{686 \times 972}}$.

20. Find the number of digits in 42^{42}.

21. Shew that $\left(\dfrac{81}{80}\right)^{1000}$ is greater than 100000.

22. How many ciphers are there between the decimal point and the first significant digit in $\left(\dfrac{2}{3}\right)^{1000}$?

23. Find the value of $\sqrt[5]{\cdot 01008}$, having given
$$\log 398742 = 5 \cdot 6006921.$$

24. Find the seventh root of $\cdot 00792$, having given
$$\log 11 = 1 \cdot 0413927 \text{ and } \log 500 \cdot 977 = 2 \cdot 6998179$$

25. Find the value of $2 \log \dfrac{75}{49} + \log \dfrac{135}{32} - 3 \log \dfrac{45}{28}$.

Find the numerical value of x in the following equations, using the values of $\log 2$ and $\log 3$ given in Ex. 3 of Art. 387.

26. $3^{x+2} = 405$. 27. $10^{5-3x} = 27^{-2x}$.

28. $5^{x-3} = 8$. 29. $12^{3x-4} \cdot 18^{7-2x} = 1458$.

Use of Four-Figure Tables.

387$_A$. *To find the logarithm of a given number from the Tables.*

Example 1. Find $\log 38$, $\log 380$, $\log \cdot 0038$.

We first find the number 38 in the left hand column on page 348$_D$. Opposite to this we find the digits 5798. This, with the decimal point prefixed, is the mantissa for the logarithms of all numbers whose significant digits are 38. Hence, prefixing the characteristics we have

$$\log 38 = 1 \cdot 5798, \quad \log 380 = 2 \cdot 5798, \quad \log \cdot 0038 = \bar{3} \cdot 5798.$$

LOGARITHMS.

Example 2. Find $\log 3\cdot 86$, $\log \cdot 0386$, $\log 386000$.

The same line as before will give the mantissa of the logarithms of all numbers which begin with **38**. From this line we choose the mantissa which stands in the column headed 6. This gives $\cdot 5866$ as the mantissa for all numbers whose significant digits are 386. Hence, prefixing the characteristics, we have

$$\log 3\cdot 86 = \cdot 5866, \quad \log \cdot 0386 = \bar{2}\cdot 5866, \quad \log 386000 = 5\cdot 5866.$$

387$_B$. Similarly the logarithm of any number consisting of not more than 3 significant digits can be obtained directly from the Tables. When the number has 4 significant digits, use is made of the principle that when the difference between two numbers is small compared with either of them, the difference between their logarithms is very nearly proportional to the difference between the numbers. It would be out of place to attempt any demonstration of the principle here. It will be sufficient to point out that differences in the logarithms corresponding to small differences in the numbers have been calculated, and are printed ready for use in the *difference columns* at the right hand of the Tables. The way in which these differences are used is shewn in the following example.

Example. Find (i) $\log 3\cdot 864$; (ii) $\log \cdot 003868$.

Here, as before, we can find the mantissa for the sequence of digits 386. This has to be *corrected* by the addition of the figures which stand underneath 4 and 8 respectively in the difference columns

(i) $\log 3\cdot 86 \ = \cdot 5866$
    diff. for $\quad 4 \quad\ \ \underline{\quad 5\quad}$
$\therefore \log 3\cdot 864 = \cdot 5871$

(ii) $\log \cdot 00386 \ = \bar{3}\cdot 5866$
    diff. for $\quad\ 8 \quad\ \ \underline{\quad 9\quad}$
$\therefore \log \cdot 003868 = \bar{3}\cdot 5875$

Note. After a little practice the necessary 'correction' from the difference columns can be performed mentally.

387$_C$. The number corresponding to a given logarithm is called its **antilogarithm**. Thus in the last example $3\cdot 864$ and $\cdot 003868$ are respectively the numbers whose logarithms are $\cdot 5871$ and $\bar{3}\cdot 5875$.

Hence antilog $\cdot 5871 = 3\cdot 864$; antilog $\bar{3}\cdot 5875 = \cdot 003868$.

387$_D$. *To find the antilogarithm of a given logarithm.*

In using the Tables of antilogarithms on pages 348_F, 348_G, it is important to remember that we are seeking *numbers* corresponding to *given logarithms*. Thus in the left hand column we have the first two digits of the given *mantissæ*, with the decimal point prefixed. The characteristics of the given logarithms will fix the position of the decimal point in the numbers taken from the Tables.

Example. Find the antilogarithm of (i) $1 \cdot 583$; (ii) $\bar{2} \cdot 8249$.

(i) We first find $\cdot 58$ in the left hand column and pass along the horizontal line and take the number in the vertical column headed by 3. Thus $\cdot 583$ is the mantissa of the logarithm of a number whose significant digits are 3828.

Hence antilog $1 \cdot 583 = 38 \cdot 28$.

$$
\begin{aligned}
\text{(ii)} \quad \text{antilog}\,\bar{2} \cdot 824 &= \cdot 06668 \\
\text{diff. for} \quad 9 \quad &\underline{14} \\
\therefore \text{antilog}\,\bar{2} \cdot 8249 &= \cdot 06682
\end{aligned}
$$

Here corresponding to the first 3 digits of the mantissa we find the sequence of digits 6668, and the decimal point is inserted in the position corresponding to the characteristic $\bar{2}$. To the number so found we add 14 from the difference column headed 9, placing it under the fourth digit of the given mantissa.

387$_E$. The following examples illustrate the use of logarithms in abbreviating arithmetical calculations.

Example 1. Find the value of $\dfrac{3 \cdot 274 \times \cdot 0059}{14 \cdot 83 \times \cdot 077}$ to four significant digits.

By Art. 377, log *fraction* = log *numerator* − log *denominator*.

Numerator.

$$
\begin{aligned}
\log 3 \cdot 27 &= \cdot 5145 \\
\text{diff. for} \quad 4 \quad &\quad 5 \\
\log \cdot 0059 &= \bar{3} \cdot 7709 \\
\log \textit{numerator} &= \bar{2} \cdot 2859
\end{aligned}
$$

Denominator.

$$
\begin{aligned}
\log 14 \cdot 8 &= 1 \cdot 1703 \\
\text{diff. for} \quad 3 \quad &\quad 9 \\
\log \cdot 077 &= \bar{2} \cdot 8865 \\
\log \textit{denominator} &= \cdot 0577
\end{aligned}
$$

$$
\begin{aligned}
&\bar{2} \cdot 2859 \\
\text{subtract} \quad &\cdot 0577 \\
\log \textit{fraction} = &\bar{2} \cdot 2282
\end{aligned}
$$

$$
\begin{aligned}
\text{antilog}\,\bar{2} \cdot 228 &= \cdot 01690 \\
\text{diff. for} \quad 2 \quad &\quad 1 \\
\text{antilog}\,\bar{2} \cdot 2282 &= \cdot 01691
\end{aligned}
$$

Thus $\dfrac{3 \cdot 274 \times \cdot 0059}{14 \cdot 83 \times \cdot 077} = \cdot 01691$.

Example 2. Find the value of $(1·05)^{17}$ to four significant digits.

$$\log(1·05)^{17} = 17 \log 1·05 \qquad [\text{Art. 378}]$$
$$= ·0212 \times 17, \text{ from the Tables,}$$
$$= ·3604.$$

And antilog $·3604 = 2·293$;

thus $(1·05)^{17} = 2·293.$

Note. Since ·0212 is only the approximate logarithm of 1·05, the error (which may be in excess or defect) is increased when we multiply by 17. Hence there is a corresponding error in the final result. By using seven-figure logarithms it can be shewn that to four decimal figures the correct result is 2·2922.

Example 3. Find a mean proportional between 27·23 and 3·276.

Let x denote the mean proportional; then

$$x = \sqrt{27·23 \times 3·276}; \qquad [\text{Art. 297}]$$
$$\therefore \log x = \tfrac{1}{2}(\log 27·23 + \log 3·276).$$

From the Tables,

$\log 27·23 = 1·4351$	antilog $·975 = 9·441$
$\log 3·276 = ·5153$	diff. for 2 4
$2 \,\vert\, 1·9504$	antilog $·9752 = 9·445$
$\log x = ·9752$	

$$\therefore \quad x = 9·445.$$

EXAMPLES XXXIX. b.

[*For Logarithms and Antilogarithms see* pages 348_D *to* 348_G.]

Find the values of the following products to four significant figures:

1. $1927 \times ·2501.$ 2. $175·6 \times ·2632.$ 3. $·0035 \times 39·87.$
4. $·231 \times 2·394 \times ·0157.$ 5. $5·2 \times 3·81 \times 17·31.$
6. $7·302 \times ·7302 \times ·007302.$ 7. $23 \times 1·7 \times 3·35 \times ·062.$

Divide

8. 2·803 by ·0634. 9. 16·83 by 24·76. 10. 30·56 by 4·105.
11. ·01254 by ·4105. 12. 2417 by 719. 13. 2391 by 3072.

[Continued on page 348_H.

Logarithms.

No.	0	1	2	3	4	5	6	7	8	9	1	2	3	4	5	6	7	8	9
10	0000	0043	0086	0128	0170	0212	0253	0294	0334	0374	4	8	12	17	21	25	29	33	37
11	0414	0453	0492	0531	0569	0607	0645	0682	0719	0755	4	8	11	15	19	23	26	30	34
12	0792	0828	0864	0899	0934	0969	1004	1038	1072	1106	3	7	10	14	17	21	24	28	31
13	1139	1173	1206	1239	1271	1303	1335	1367	1399	1430	3	6	10	13	16	19	23	26	29
14	1461	1492	1523	1553	1584	1614	1644	1673	1703	1732	3	6	9	12	15	18	21	24	27
15	1761	1790	1818	1847	1875	1903	1931	1959	1987	2014	3	6	8	11	14	17	20	22	25
16	2041	2068	2095	2122	2148	2175	2201	2227	2253	2279	3	5	8	11	13	16	18	21	24
17	2304	2330	2355	2380	2405	2430	2455	2480	2504	2529	2	5	7	10	12	15	17	20	22
18	2553	2577	2601	2625	2648	2672	2695	2718	2742	2765	2	5	7	9	12	14	16	19	21
19	2788	2810	2833	2856	2878	2900	2923	2945	2967	2989	2	4	7	9	11	13	16	18	20
20	3010	3032	3054	3075	3096	3118	3139	3160	3181	3201	2	4	6	8	11	13	15	17	19
21	3222	3243	3263	3284	3304	3324	3345	3365	3385	3404	2	4	6	8	10	12	14	16	18
22	3424	3444	3464	3483	3502	3522	3541	3560	3579	3598	2	4	6	8	10	12	14	15	17
23	3617	3636	3655	3674	3692	3711	3729	3747	3766	3784	2	4	6	7	9	11	13	15	17
24	3802	3820	3838	3856	3874	3892	3909	3927	3945	3962	2	4	5	7	9	11	12	14	16
25	3979	3997	4014	4031	4048	4065	4082	4099	4116	4133	2	3	5	7	9	10	12	14	15
26	4150	4166	4183	4200	4216	4232	4249	4265	4281	4298	2	3	5	7	8	10	11	13	15
27	4314	4330	4346	4362	4378	4393	4409	4425	4440	4456	2	3	5	6	8	9	11	13	14
28	4472	4487	4502	4518	4533	4548	4564	4579	4594	4609	2	3	5	6	8	9	11	12	14
29	4624	4639	4654	4669	4683	4698	4713	4728	4742	4757	1	3	4	6	7	9	10	12	13
30	4771	4786	4800	4814	4829	4843	4857	4871	4886	4900	1	3	4	6	7	9	10	11	13
31	4914	4928	4942	4955	4969	4983	4997	5011	5024	5038	1	3	4	6	7	8	10	11	12
32	5051	5065	5079	5092	5105	5119	5132	5145	5159	5172	1	3	4	5	7	8	9	11	12
33	5185	5198	5211	5224	5237	5250	5263	5276	5289	5302	1	3	4	5	6	8	9	10	12
34	5315	5328	5340	5353	5366	5378	5391	5403	5416	5428	1	3	4	5	6	8	9	10	11
35	5441	5453	5465	5478	5490	5502	5514	5527	5539	5551	1	2	4	5	6	7	9	10	11
36	5563	5575	5587	5599	5611	5623	5635	5647	5658	5670	1	2	4	5	6	7	8	10	11
37	5682	5694	5705	5717	5729	5740	5752	5763	5775	5786	1	2	3	5	6	7	8	9	10
38	5798	5809	5821	5832	5843	5855	5866	5877	5888	5899	1	2	3	5	6	7	8	9	10
39	5911	5922	5933	5944	5955	5966	5977	5988	5999	6010	1	2	3	4	5	7	8	9	10
40	6021	6031	6042	6053	6064	6075	6085	6096	6107	6117	1	2	3	4	5	6	8	9	10
41	6128	6138	6149	6160	6170	6180	6191	6201	6212	6222	1	2	3	4	5	6	7	8	9
42	6232	6243	6253	6263	6274	6284	6294	6304	6314	6325	1	2	3	4	5	6	7	8	9
43	6335	6345	6355	6365	6375	6385	6395	6405	6415	6425	1	2	3	4	5	6	7	8	9
44	6435	6444	6454	6464	6474	6484	6493	6503	6513	6522	1	2	3	4	5	6	7	8	9
45	6532	6542	6551	6561	6571	6580	6590	6599	6609	6618	1	2	3	4	5	6	7	8	9
46	6628	6637	6646	6656	6665	6675	6684	6693	6702	6712	1	2	3	4	5	6	7	7	8
47	6721	6730	6739	6749	6758	6767	6776	6785	6794	6803	1	2	3	4	5	5	6	7	8
48	6812	6821	6830	6839	6848	6857	6866	6875	6884	6893	1	2	3	4	4	5	6	7	8
49	6902	6911	6920	6928	6937	6946	6955	6964	6972	6981	1	2	3	4	4	5	6	7	8
50	6990	6998	7007	7016	7024	7033	7042	7050	7059	7067	1	2	3	3	4	5	6	7	8
51	7076	7084	7093	7101	7110	7118	7126	7135	7143	7152	1	2	3	3	4	5	6	7	8
52	7160	7168	7177	7185	7193	7202	7210	7218	7226	7235	1	2	2	3	4	5	6	7	7
53	7243	7251	7259	7267	7275	7284	7292	7300	7308	7316	1	2	2	3	4	5	6	6	7
54	7324	7332	7340	7348	7356	7364	7372	7380	7388	7396	1	2	2	3	4	5	6	6	7

Logarithms.

No.	0	1	2	3	4	5	6	7	8	9	1	2	3	4	5	6	7	8	9
55	7404	7412	7419	7427	7435	7443	7451	7459	7466	7474	1	2	2	3	4	5	5	6	7
56	7482	7490	7497	7505	7513	7520	7528	7536	7543	7551	1	2	2	3	4	5	5	6	7
57	7559	7566	7574	7582	7589	7597	7604	7612	7619	7627	1	2	2	3	4	5	5	6	7
58	7634	7642	7649	7657	7664	7672	7679	7686	7694	7701	1	1	2	3	4	4	5	6	7
59	7709	7716	7723	7731	7738	7745	7752	7760	7767	7774	1	1	2	3	4	4	5	6	7
60	7782	7789	7796	7803	7810	7818	7825	7832	7839	7846	1	1	2	3	4	4	5	6	6
61	7853	7860	7868	7875	7882	7889	7896	7903	7910	7917	1	1	2	3	4	4	5	6	6
62	7924	7931	7938	7945	7952	7959	7966	7973	7980	7987	1	1	2	3	3	4	5	6	6
63	7993	8000	8007	8014	8021	8028	8035	8041	8048	8055	1	1	2	3	3	4	5	5	6
64	8062	8069	8075	8082	8089	8096	8102	8109	8116	8122	1	1	2	3	3	4	5	5	6
65	8129	8136	8142	8149	8156	8162	8169	8176	8182	8189	1	1	2	3	3	4	5	5	6
66	8195	8202	8209	8215	8222	8228	8235	8241	8248	8254	1	1	2	3	3	4	5	5	6
67	8261	8267	8274	8280	8287	8293	8299	8306	8312	8319	1	1	2	3	3	4	5	5	6
68	8325	8331	8338	8344	8351	8357	8363	8370	8376	8382	1	1	2	3	3	4	4	5	6
69	8388	8395	8401	8407	8414	8420	8426	8432	8439	8445	1	1	2	2	3	4	4	5	6
70	8451	8457	8463	8470	8476	8482	8488	8494	8500	8506	1	1	2	2	3	4	4	5	6
71	8513	8519	8525	8531	8537	8543	8549	8555	8561	8567	1	1	2	2	3	4	4	5	5
72	8573	8579	8585	8591	8597	8603	8609	8615	8621	8627	1	1	2	2	3	4	4	5	5
73	8633	8639	8645	8651	8657	8663	8669	8675	8681	8686	1	1	2	2	3	4	4	5	5
74	8692	8698	8704	8710	8716	8722	8727	8733	8739	8745	1	1	2	2	3	4	4	5	5
75	8751	8756	8762	8768	8774	8779	8785	8791	8797	8802	1	1	2	2	3	3	4	5	5
76	8808	8814	8820	8825	8831	8837	8842	8848	8854	8859	1	1	2	2	3	3	4	5	5
77	8865	8871	8876	8882	8887	8893	8899	8904	8910	8915	1	1	2	2	3	3	4	4	5
78	8921	8927	8932	8938	8943	8949	8954	8960	8965	8971	1	1	2	2	3	3	4	4	5
79	8976	8982	8987	8993	8998	9004	9009	9015	9020	9025	1	1	2	2	3	3	4	4	5
80	9031	9036	9042	9047	9053	9058	9063	9069	9074	9079	1	1	2	2	3	3	4	4	5
81	9085	9090	9096	9101	9106	9112	9117	9122	9128	9133	1	1	2	2	3	3	4	4	5
82	9138	9143	9149	9154	9159	9165	9170	9175	9180	9186	1	1	2	2	3	3	4	4	5
83	9191	9196	9201	9206	9212	9217	9222	9227	9232	9238	1	1	2	2	3	3	4	4	5
84	9243	9248	9253	9258	9263	9269	9274	9279	9284	9289	1	1	2	2	3	3	4	4	5
85	9294	9299	9304	9309	9315	9320	9325	9330	9335	9340	1	1	2	2	3	3	4	4	5
86	9345	9350	9355	9360	9365	9370	9375	9380	9385	9390	1	1	2	2	3	3	4	4	5
87	9395	9400	9405	9410	9415	9420	9425	9430	9435	9440	0	1	1	2	2	3	3	4	4
88	9445	9450	9455	9460	9465	9469	9474	9479	9484	9489	0	1	1	2	2	3	3	4	4
89	9494	9499	9504	9509	9513	9518	9523	9528	9533	9538	0	1	1	2	2	3	3	4	4
90	9542	9547	9552	9557	9562	9566	9571	9576	9581	9586	0	1	1	2	2	3	3	4	4
91	9590	9595	9600	9605	9609	9614	9619	9624	9628	9633	0	1	1	2	2	3	3	4	4
92	9638	9643	9647	9652	9657	9661	9666	9671	9675	9680	0	1	1	2	2	3	3	4	4
93	9685	9689	9694	9699	9703	9708	9713	9717	9722	9727	0	1	1	2	2	3	3	4	4
94	9731	9736	9741	9745	9750	9754	9759	9763	9768	9773	0	1	1	2	2	3	3	4	4
95	9777	9782	9786	9791	9795	9800	9805	9809	9814	9818	0	1	1	2	2	3	3	4	4
96	9823	9827	9832	9836	9841	9845	9850	9854	9859	9863	0	1	1	2	2	3	3	4	4
97	9868	9872	9877	9881	9886	9890	9894	9899	9903	9908	0	1	1	2	2	3	3	4	4
98	9912	9917	9921	9926	9930	9934	9939	9943	9948	9952	0	1	1	2	2	3	3	4	4
99	9956	9961	9965	9969	9974	9978	9983	9987	9991	9996	0	1	1	2	2	3	3	3	4

Antilogarithms.

Log.	0	1	2	3	4	5	6	7	8	9	1	2	3	4	5	6	7	8	9
·00	1000	1002	1005	1007	1009	1012	1014	1016	1019	1021	0	0	1	1	1	1	2	2	2
·01	1023	1026	1028	1030	1033	1035	1038	1040	1042	1045	0	0	1	1	1	1	2	2	2
·02	1047	1050	1052	1054	1057	1059	1062	1064	1067	1069	0	0	1	1	1	1	2	2	2
·03	1072	1074	1076	1079	1081	1084	1086	1089	1091	1094	0	0	1	1	1	1	2	2	2
·04	1096	1099	1102	1104	1107	1109	1112	1114	1117	1119	0	1	1	1	1	2	2	2	2
·05	1122	1125	1127	1130	1132	1135	1138	1140	1143	1146	0	1	1	1	1	2	2	2	2
·06	1148	1151	1153	1156	1159	1161	1164	1167	1169	1172	0	1	1	1	1	2	2	2	2
·07	1175	1178	1180	1183	1186	1189	1191	1194	1197	1199	0	1	1	1	1	2	2	2	2
·08	1202	1205	1208	1211	1213	1216	1219	1222	1225	1227	0	1	1	1	1	2	2	2	3
·09	1230	1233	1236	1239	1242	1245	1247	1250	1253	1256	0	1	1	1	1	2	2	2	3
·10	1259	1262	1265	1268	1271	1274	1276	1279	1282	1285	0	1	1	1	1	2	2	2	3
·11	1288	1291	1294	1297	1300	1303	1306	1309	1312	1315	0	1	1	1	2	2	2	2	3
·12	1318	1321	1324	1327	1330	1334	1337	1340	1343	1346	0	1	1	1	2	2	2	2	3
·13	1349	1352	1355	1358	1361	1365	1368	1371	1374	1377	0	1	1	1	2	2	2	3	3
·14	1380	1384	1387	1390	1393	1396	1400	1403	1406	1409	0	1	1	1	2	2	2	3	3
·15	1413	1416	1419	1422	1426	1429	1432	1435	1439	1442	0	1	1	1	2	2	2	3	3
·16	1445	1449	1452	1455	1459	1462	1466	1469	1472	1476	0	1	1	1	2	2	2	3	3
·17	1479	1483	1486	1489	1493	1496	1500	1503	1507	1510	0	1	1	1	2	2	2	3	3
·18	1514	1517	1521	1524	1528	1531	1535	1538	1542	1545	0	1	1	1	2	2	2	3	3
·19	1549	1552	1556	1560	1563	1567	1570	1574	1578	1581	0	1	1	1	2	2	3	3	3
·20	1585	1589	1592	1596	1600	1603	1607	1611	1614	1618	0	1	1	1	2	2	3	3	3
·21	1622	1626	1629	1633	1637	1641	1644	1648	1652	1656	0	1	1	2	2	2	3	3	3
·22	1660	1663	1667	1671	1675	1679	1683	1687	1690	1694	0	1	1	2	2	2	3	3	3
·23	1698	1702	1706	1710	1714	1718	1722	1726	1730	1734	0	1	1	2	2	2	3	3	4
·24	1738	1742	1746	1750	1754	1758	1762	1766	1770	1774	0	1	1	2	2	2	3	3	4
·25	1778	1782	1786	1791	1795	1799	1803	1807	1811	1816	0	1	1	2	2	2	3	3	4
·26	1820	1824	1828	1832	1837	1841	1845	1849	1854	1858	0	1	1	2	2	3	3	3	4
·27	1862	1866	1871	1875	1879	1884	1888	1892	1897	1901	0	1	1	2	2	3	3	3	4
·28	1905	1910	1914	1919	1923	1928	1932	1936	1941	1945	0	1	1	2	2	3	3	4	4
·29	1950	1954	1959	1963	1968	1972	1977	1982	1986	1991	0	1	1	2	2	3	3	4	4
·30	1995	2000	2004	2009	2014	2018	2023	2028	2032	2037	0	1	1	2	2	3	3	4	4
·31	2042	2046	2051	2056	2061	2065	2070	2075	2080	2084	0	1	1	2	2	3	3	4	4
·32	2089	2094	2099	2104	2109	2113	2118	2123	2128	2133	0	1	1	2	2	3	3	4	4
·33	2138	2143	2148	2153	2158	2163	2168	2173	2178	2183	0	1	1	2	2	3	3	4	4
·34	2188	2193	2198	2203	2208	2213	2218	2223	2228	2234	1	1	2	2	3	3	4	4	5
·35	2239	2244	2249	2254	2259	2265	2270	2275	2280	2286	1	1	2	2	3	3	4	4	5
·36	2291	2296	2301	2307	2312	2317	2323	2328	2333	2339	1	1	2	2	3	3	4	4	5
·37	2344	2350	2355	2360	2366	2371	2377	2382	2388	2393	1	1	2	2	3	3	4	4	5
·38	2399	2404	2410	2415	2421	2427	2432	2438	2443	2449	1	1	2	2	3	3	4	4	5
·39	2455	2460	2466	2472	2477	2483	2489	2495	2500	2506	1	1	2	2	3	3	4	5	5
·40	2512	2518	2523	2529	2535	2541	2547	2553	2559	2564	1	1	2	2	3	4	4	5	5
·41	2570	2576	2582	2588	2594	2600	2606	2612	2618	2624	1	1	2	2	3	4	4	5	5
·42	2630	2636	2642	2649	2655	2661	2667	2673	2679	2685	1	1	2	2	3	4	4	5	6
·43	2692	2698	2704	2710	2716	2723	2729	2735	2742	2748	1	1	2	3	3	4	4	5	6
·44	2754	2761	2767	2773	2780	2786	2793	2799	2805	2812	1	1	2	3	3	4	4	5	6
·45	2818	2825	2831	2838	2844	2851	2858	2864	2871	2877	1	1	2	3	3	4	5	5	6
·46	2884	2891	2897	2904	2911	2917	2924	2931	2938	2944	1	1	2	3	3	4	5	5	6
·47	2951	2958	2965	2972	2979	2985	2992	2999	3006	3013	1	1	2	3	3	4	5	5	6
·48	3020	3027	3034	3041	3048	3055	3062	3069	3076	3083	1	1	2	3	4	4	5	6	6
·49	3090	3097	3105	3112	3119	3126	3133	3141	3148	3155	1	1	2	3	4	4	5	6	6

Antilogarithms.

Log.	0	1	2	3	4	5	6	7	8	9	1	2	3	4	5	6	7	8	9
·50	3162	3170	3177	3184	3192	3199	3206	3214	3221	3228	1	1	2	3	4	4	5	6	7
·51	3236	3243	3251	3258	3266	3273	3281	3289	3296	3304	1	2	2	3	4	5	5	6	7
·52	3311	3319	3327	3334	3342	3350	3357	3365	3373	3381	1	2	2	3	4	5	5	6	7
·53	3388	3396	3404	3412	3420	3428	3436	3443	3451	3459	1	2	2	3	4	5	6	6	7
·54	3467	3475	3483	3491	3499	3508	3516	3524	3532	3540	1	2	2	3	4	5	6	6	7
·55	3548	3556	3565	3573	3581	3589	3597	3606	3614	3622	1	2	2	3	4	5	6	7	7
·56	3631	3639	3648	3656	3664	3673	3681	3690	3698	3707	1	2	3	3	4	5	6	7	8
·57	3715	3724	3733	3741	3750	3758	3767	3776	3784	3793	1	2	3	3	4	5	6	7	8
·58	3802	3811	3819	3828	3837	3846	3855	3864	3873	3882	1	2	3	4	4	5	6	7	8
·59	3890	3899	3908	3917	3926	3936	3945	3954	3963	3972	1	2	3	4	5	5	6	7	8
·60	3981	3990	3999	4009	4018	4027	4036	4046	4055	4064	1	2	3	4	5	6	6	7	8
·61	4074	4083	4093	4102	4111	4121	4130	4140	4150	4159	1	2	3	4	5	6	7	8	9
·62	4169	4178	4188	4198	4207	4217	4227	4236	4246	4256	1	2	3	4	5	6	7	8	9
·63	4266	4276	4285	4295	4305	4315	4325	4335	4345	4355	1	2	3	4	5	6	7	8	9
·64	4365	4375	4385	4395	4406	4416	4426	4436	4446	4457	1	2	3	4	5	6	7	8	9
·65	4467	4477	4487	4498	4508	4519	4529	4539	4550	4560	1	2	3	4	5	6	7	8	9
·66	4571	4581	4592	4603	4613	4624	4634	4645	4656	4667	1	2	3	4	5	6	7	9	10
·67	4677	4688	4699	4710	4721	4732	4742	4753	4764	4775	1	2	3	4	5	7	8	9	10
·68	4786	4797	4808	4819	4831	4842	4853	4864	4875	4887	1	2	3	4	6	7	8	9	10
·69	4898	4909	4920	4932	4943	4955	4966	4977	4989	5000	1	2	3	5	6	7	8	9	10
·70	5012	5023	5035	5047	5058	5070	5082	5093	5105	5117	1	2	4	5	6	7	8	9	11
·71	5129	5140	5152	5164	5176	5188	5200	5212	5224	5236	1	2	4	5	6	7	8	10	11
·72	5248	5260	5272	5284	5297	5309	5321	5333	5346	5358	1	2	4	5	6	7	9	10	11
·73	5370	5383	5395	5408	5420	5433	5445	5458	5470	5483	1	3	4	5	6	8	9	10	11
·74	5495	5508	5521	5534	5546	5559	5572	5585	5598	5610	1	3	4	5	6	8	9	10	12
·75	5623	5636	5649	5662	5675	5689	5702	5715	5728	5741	1	3	4	5	7	8	9	10	12
·76	5754	5768	5781	5794	5808	5821	5834	5848	5861	5875	1	3	4	5	7	8	9	11	12
·77	5888	5902	5916	5929	5943	5957	5970	5984	5998	6012	1	3	4	5	7	8	10	11	12
·78	6026	6039	6053	6067	6081	6095	6109	6124	6138	6152	1	3	4	6	7	8	10	11	13
·79	6166	6180	6194	6209	6223	6237	6252	6266	6281	6295	1	3	4	6	7	9	10	11	13
·80	6310	6324	6339	6353	6368	6383	6397	6412	6427	6442	1	3	4	6	7	9	10	12	13
·81	6457	6471	6486	6501	6516	6531	6546	6561	6577	6592	2	3	5	6	8	9	11	12	14
·82	6607	6622	6637	6653	6668	6683	6699	6714	6730	6745	2	3	5	6	8	9	11	12	14
·83	6761	6776	6792	6808	6823	6839	6855	6871	6887	6902	2	3	5	6	8	9	11	13	14
·84	6918	6934	6950	6966	6982	6998	7015	7031	7047	7063	2	3	5	6	8	10	11	13	15
·85	7079	7096	7112	7129	7145	7161	7178	7194	7211	7228	2	3	5	7	8	10	12	13	15
·86	7244	7261	7278	7295	7311	7328	7345	7362	7379	7396	2	3	5	7	8	10	12	13	15
·87	7413	7430	7447	7464	7482	7499	7516	7534	7551	7568	2	3	5	7	9	10	12	14	16
·88	7586	7603	7621	7638	7656	7674	7691	7709	7727	7745	2	4	5	7	9	11	12	14	16
·89	7762	7780	7798	7816	7834	7852	7870	7889	7907	7925	2	4	5	7	9	11	13	14	16
·90	7943	7962	7980	7998	8017	8035	8054	8072	8091	8110	2	4	6	7	9	11	13	15	17
·91	8128	8147	8166	8185	8204	8222	8241	8260	8279	8299	2	4	6	8	9	11	13	15	17
·92	8318	8337	8356	8375	8395	8414	8433	8453	8472	8492	2	4	6	8	10	12	14	15	17
·93	8511	8531	8551	8570	8590	8610	8630	8650	8670	8690	2	4	6	8	10	12	14	16	18
·94	8710	8730	8750	8770	8790	8810	8831	8851	8872	8892	2	4	6	8	10	12	14	16	18
·95	8913	8933	8954	8974	8995	9016	9036	9057	9078	9099	2	4	6	8	10	12	15	17	19
·96	9120	9141	9162	9183	9204	9226	9247	9268	9290	9311	2	4	6	8	11	13	15	17	19
·97	9333	9354	9376	9397	9419	9441	9462	9484	9506	9528	2	4	7	9	11	13	15	17	20
·98	9550	9572	9594	9616	9638	9661	9683	9705	9727	9750	2	4	7	9	11	13	16	18	20
·99	9772	9795	9817	9840	9863	9886	9908	9931	9954	9977	2	5	7	9	11	14	16	18	20

Evaluate the following expressions to four significant figures:

14. $\dfrac{2\cdot 38 \times 3\cdot 901}{4\cdot 83}$.

15. $\dfrac{14\cdot 72 \times 38\cdot 05}{387\cdot 9}$.

16. $\dfrac{925\cdot 9 \times 1\cdot 597}{74\cdot 03}$.

17. $\dfrac{15\cdot 38 \times \cdot 0137}{276 \times \cdot 0038}$.

18. $(\cdot 097)^4$. 19. $(1\cdot 73)^{11}$. 20. $\sqrt{\cdot 51}$. 21. $\sqrt{8\tfrac{1}{3}}$.

22. $\sqrt[3]{127}$. 23. $\sqrt[5]{27\cdot 2}$. 24. $\sqrt[11]{1772}$. 25. $\sqrt[13]{27\cdot 82}$.

26. Find a mean proportional between $2\cdot 87$ and $30\cdot 08$; and a third proportional to $\cdot 0238$ and $7\cdot 805$.

Evaluate

27. $\sqrt[3]{\left(\dfrac{294 \times 125}{42 \times 32}\right)^2}$.

28. $\dfrac{(330 \times \tfrac{1}{49})^4}{\sqrt[3]{22} \times 70}$

29. Find the value of $\sqrt{\dfrac{\cdot 678 \times 9\cdot 01}{\cdot 0234}}$ to the nearest integer.

30. Find a mean proportional between
$$\sqrt[3]{347\cdot 3} \quad \text{and} \quad \sqrt[5]{256\cdot 4}.$$

31. Find a fourth proportional to
$$\sqrt[3]{32\cdot 78}, \quad \sqrt[5]{357\cdot 8}, \quad \sqrt[4]{7836}.$$

[*Before attempting the following Examples the student should read* Arts. 403–405.]

32. Find to the nearest pound the amount of £35 in 25 years at 3 p.c. Compound Interest.

33. Find to the nearest pound the Present Value of £1000 due 17 years hence at 4 p.c. Compound Interest.

34. Find in how many years £1130 will amount to £3000 at 5 p.c. Compound Interest.

35. If a farthing is put out at Compound Interest for 1000 years at 5 p.c., how many digits will be required to express the amount in pounds?

36. A train starts with velocity ·001 ft. per second, and at the end of each second its velocity is greater by one-third than at the end of the preceding second; find to two places of decimals the rate of the train in miles per hour at the end of 25 seconds.

CHAPTER XL.

SCALES OF NOTATION.

388. The ordinary numbers with which we are acquainted in Arithmetic are expressed by means of multiples of powers of 10; for instance

$$25 = 2 \times 10 + 5;$$
$$4705 = 4 \times 10^3 + 7 \times 10^2 + 0 \times 10 + 5.$$

This method of representing numbers is called the **common** or **denary scale of notation,** and ten is said to be the **radix** of the scale. The symbols employed in this system of notation are the nine digits and zero.

In like manner any number other than ten may be taken as the radix of a scale of notation; thus if 7 is the radix, a number expressed by 2453 represents $2 \times 7^3 + 4 \times 7^2 + 5 \times 7 + 3$; and in this scale no digit higher than 6 can occur.

389. The names Binary, Ternary, Quaternary, Quinary, Senary, Septenary, Octenary, Nonary, Denary, Undenary, and Duodenary are used to denote the scales corresponding to the values *two, three, ... twelve* of the radix.

In the undenary, duodenary, ... scales we shall require symbols to represent the digits which are greater than nine. It is unusual to consider any scale higher than that with radix twelve; when necessary we shall employ the symbols t, e, T as digits to denote 'ten,' 'eleven' and 'twelve.'

It is especially worthy of notice that in every scale 10 is the symbol not for 'ten', but for the radix itself.

390. The ordinary operations of Arithmetic may be performed in any scale; but, bearing in mind that the successive powers of the radix are no longer powers of ten, in determining the *carrying figures* we must not divide by ten, but by the radix of the scale in question.

Example 1. In the scale of eight subtract 371532 from 530225, and multiply the difference by 7.

$$\begin{array}{rr} 530225 & 136473 \\ 371532 & 7 \\ \hline 136473 & 1226235 \end{array}$$

Explanation. After the first figure of the subtraction, since we cannot take 3 from 2 we add 8; thus we have to take 3 from ten, which leaves 7; then 6 from ten, which leaves 4; then 2 from eight which leaves 6; and so on.

Again, in multiplying by 7, we have
$$3 \times 7 = \text{twenty-one} = 2 \times 8 + 5;$$
we therefore put down 5 and carry 2.

Next $\qquad 7 \times 7 + 2 = \text{fifty-one} = 6 \times 8 + 3;$
put down 3 and carry 6; and so on.

Example 2. Divide 15*et*20 by 9 in the scale of twelve.

$$9\,)\,15et20$$
$$\overline{1ee96\ \ldots\ 6.}$$

Explanation. Since $15 = 1 \times T + 5 = \text{seventeen} = 1 \times 9 + 8$, we put down 1 and carry 8.

Also $8 \times T + e = \text{one hundred and seven} = e \times 9 + 8;$
we therefore put down e and carry 8; and so on.

391. *To express a given integral number in any proposed scale.*

Let N be the given number, and r the radix of the proposed scale.

Let $a_0, a_1, a_2, \ldots a_n$ be the required digits by which N is to be expressed, beginning with that in the unit's place; then
$$N = a_n r^n + a_{n-1} r^{n-1} + \ldots + a_2 r^2 + a_1 r + a_0.$$

We have now to find the values of $a_0, a_1, a_2, \ldots a_n$.

Divide N by r, then the remainder is a_0, and the quotient is
$$a_n r^{n-1} + a_{n-1} r^{n-2} + \ldots + a_2 r + a_1.$$

If this quotient is divided by r, the remainder is a_1;
if the next quotient .... a_2;
and so on, until there is no further quotient.

Thus all the required digits $a_0, a_1, a_2, \ldots a_n$ are determined by successive divisions by the radix of the proposed scale.

Example 1. Express the denary number 5213 in the scale of seven.

$$\begin{array}{r}7\,)\,5213\\ 7\,)\,\overline{744}\,\ldots\,5\\ 7\,)\,\overline{106}\,\ldots\,2\\ 7\,)\,\overline{15}\,\ldots\,1\\ \overline{2}\,\ldots\,1\end{array}$$

Thus $\quad 5213 = 2\times 7^4 + 1\times 7^3 + 1\times 7^2 + 2\times 7 + 5$;
and the number required is 21125.

Example 2. Transform 21125 from scale seven to scale eleven.

$$\begin{array}{r}e\,)\,21125\\ e\,)\,\overline{1244}\,\ldots\,t\\ e\,)\,\overline{61}\,\ldots\,0\\ \overline{3}\,\ldots\,t\end{array}$$

∴ the required number is $30t$.

Explanation. In the first line of work
$$21 = 2\times 7 + 1 = \text{fifteen} = 1\times e + 4\,;$$
therefore on dividing by e we put down 1 and carry 4.

Next $4\times 7 + 1 = \text{twenty-nine} = 2\times e + 7$;
therefore we put down 2 and carry 7; and so on.

392. Hitherto we have only discussed whole numbers; but fractions may also be expressed in any scale of notation; thus

$$\cdot25 \text{ in scale ten denotes } \frac{2}{10} + \frac{5}{10^2};$$

$$\cdot25 \text{ in scale six denotes } \frac{2}{6} + \frac{5}{6^2};$$

$$\cdot25 \text{ in scale } r \text{ denotes } \frac{2}{r} + \frac{5}{r^2}.$$

Fractions thus expressed in a form analogous to that of ordinary decimal fractions are called **radix-fractions**, and the point is called the **radix-point**. The general type of such fractions in scale r is

$$\frac{b_1}{r} + \frac{b_2}{r^2} + \frac{b_3}{r^3} + \ldots\ldots\,;$$

where $b_1, b_2, b_3, \ldots$ are integers, all less than r, of which any one or more may be zero.

393. *To express a given radix-fraction in any proposed scale.*

Let F be the given fraction, and r the radix of the scale.

Let $b_1, b_2, b_3, \ldots$ be the required digits beginning from the left; then

$$F = \frac{b_1}{r} + \frac{b_2}{r^2} + \frac{b_3}{r^3} + \ldots\ldots$$

We have now to find the values of $b_1, b_2, b_3, \ldots\ldots$.

Multiply both sides of the equation by r; then

$$rF = b_1 + \frac{b_2}{r} + \frac{b_3}{r^2} + \ldots\ldots$$

Hence b_1 is equal to the integral part of rF; and, if we denote the fractional part by F_1, we have

$$F_1 = \frac{b_2}{r} + \frac{b_3}{r^2} + \ldots\ldots$$

Multiply again by r; then b_2 is the integral part of rF_1. Similarly by successive multiplications by r, each of the digits may be found, and the fraction expressed in the proposed scale.

Example 1. Express $\frac{7}{8}$ as a radix-fraction in scale six.

$$\frac{7}{8} \times 6 = \frac{7 \times 3}{4} = 5 + \frac{1}{4};$$

$$\frac{1}{4} \times 6 = \frac{1 \times 3}{2} = 1 + \frac{1}{2};$$

$$\frac{1}{2} \times 6 = 3.$$

$\therefore$ the required fraction $= \frac{5}{6} + \frac{1}{6^2} + \frac{3}{6^3} = \cdot 513$.

Example 2. Transform $1606 \cdot 7$ from scale eight to scale five.

Treating the integral and the fractional parts separately, we have

```
5 ) 1606              ·7
   ───────            5
   5 ) 264 ... 2      ───
      ─────           4·3
      5 ) 44 ... 0    5
         ────         ───
         5 ) 7 ... 1  1·7
            ───
              1 ... 2
```

After this the digits in the fractional part recur; hence the required number is $12102 \cdot \dot{4}\dot{1}$.

XL.] SCALES OF NOTATION. 353

Example 3. In what scale is the septenary number 2403 represented by 735?

Let r be the radix of the scale required; then
$$7r^2 + 3r + 5 = 2 \times 7^3 + 4 \times 7^2 + 3 = 885;$$
that is, $\quad 7r^2 + 3r - 880 = 0;$ whence $r = 11$ or $-\dfrac{80}{7}$.

Thus the scale is the undenary.

394. *In any scale of notation of which the radix is* r, *the sum of the digits of any whole number divided by* r − 1 *will leave the same remainder as the whole number divided by* r − 1.

Let N denote the number, $a_0, a_1, a_2, \ldots\ldots a_n$ the digits beginning with that in the unit's place, and S the sum of the digits; then
$$N = a_0 + a_1 r + a_2 r^2 + \ldots\ldots + a_{n-1} r^{n-1} + a_n r^n;$$
$$S = a_0 + a_1 + a_2 + \ldots\ldots + a_{n-1} + a_n.$$
$$\therefore\ N - S = a_1(r-1) + a_2(r^2-1) + \ldots\ldots + a_{n-1}(r^{n-1}-1) + a_n(r^n-1).$$

Now every term on the right-hand side is divisible by $r-1$;
$$\therefore\ \frac{N-S}{r-1} = \text{an integer} = I \text{ suppose};$$
that is, $\quad \dfrac{N}{r-1} = I + \dfrac{S}{r-1};$ which proves the proposition.

Hence a number in scale r will be divisible by $r-1$ when the sum of its digits is divisible by $r-1$. For example, in the ordinary scale a number is divisible by 9 when the sum of its digits is divisible by 9.

EXAMPLES XL.

1. Add together 352, 21435, 3505, 35 in the scale of six.
2. From 35260013 take 7471235 in the scale of eight.
3. Multiply 31044 by 4302 in the quinary scale.
4. Find the product of the undenary numbers $9t83$ and $3t7$.

5. Divide 31664435 by 6541 in the scale of seven.

6. Find the square of 3024 in the quinary scale.

7. Express 75013 in the nonary, and 5210 in the quaternary scale.

8. Transform 987504 to the scale of twelve.

9. Express the octenary number 76543 in the denary scale.

10. Transform 54321 from scale six to scale seven.

11. Express the duodenary number te in the binary scale.

12. Express a thousand and one in powers of two, and one hundred thousand in powers of eleven.

13. Express the sum of the septenary numbers 532, 2106, 3261, 53 in the undenary scale; also express the difference of the ternary numbers 2021121 and 1221212 in the same scale, and find the product of the two results.

14. Find the difference between 53774 in the scale of 8 and 32875 in the scale of 9, expressing the result in the denary scale.

15. Express 131·890625 in scale eight.

16. Transform 1001·12211 from the ternary to the nonary scale.

17. Express the octenary fraction ·2037 in the scale of 4.

18. Express $\dfrac{27}{32}$ and $\dfrac{500}{729}$ as radix fractions in scale 6.

19. Reduce the undenary fraction $\dfrac{587}{749}$ to its lowest terms.

20. In what scale is a hundred denoted by 400?

21. In what scale is 647 the square of 25?

22. In what scale are the numbers denoted by 432, 565, 708 in arithmetical progression?

23. In what scale are the numbers denoted by 22, 2·6, 34 in geometrical progression?

24. Find the square root of 443001 in the scale of 5; 2434524 in the scale of 7; and t985679 in the scale of eleven.

CHAPTER XLI.

EXPONENTIAL AND LOGARITHMIC SERIES.

395. The advantages of common logarithms have been explained in Art. 383, and in practice no other system is used. But in the first place these logarithms are calculated to another base and then transformed to the base 10.

In the present chapter we shall prove certain formulæ known as the **Exponential and Logarithmic Series**, and give a brief explanation of the way in which they are used in constructing a table of logarithms.

396. *To expand a^x in ascending powers of x.*

By the Binomial Theorem, if $n > 1$,

$$\left(1 + \frac{1}{n}\right)^{nx}$$

$$= 1 + nx \cdot \frac{1}{n} + \frac{nx(nx-1)}{\underline{2}} \cdot \frac{1}{n^2} + \frac{nx(nx-1)(nx-2)}{\underline{3}} \cdot \frac{1}{n^3} + \ldots$$

$$= 1 + x + \frac{x\left(x - \frac{1}{n}\right)}{\underline{2}} + \frac{x\left(x - \frac{1}{n}\right)\left(x - \frac{2}{n}\right)}{\underline{3}} + \ldots \ldots \ldots (1)$$

By putting $x = 1$, we obtain

$$\left(1 + \frac{1}{n}\right)^n = 1 + 1 + \frac{1 - \frac{1}{n}}{\underline{2}} + \frac{\left(1 - \frac{1}{n}\right)\left(1 - \frac{2}{n}\right)}{\underline{3}} + \ldots \ldots (2).$$

But $$\left(1 + \frac{1}{n}\right)^{nx} = \left\{\left(1 + \frac{1}{n}\right)^n\right\}^x;$$

hence the series (1) is the xth power of the series (2); that is,

$$1 + x + \frac{x\left(x - \frac{1}{n}\right)}{\underline{2}} + \frac{x\left(x - \frac{1}{n}\right)\left(x - \frac{2}{n}\right)}{\underline{3}} + \ldots$$

$$= \left\{1 + 1 + \frac{1 - \frac{1}{n}}{\underline{2}} + \frac{\left(1 - \frac{1}{n}\right)\left(1 - \frac{2}{n}\right)}{\underline{3}} + \ldots\right\}^x;$$

and this is true however great n may be. If therefore n be indefinitely increased we have

$$1+x+\frac{x^2}{\lfloor 2}+\frac{x^3}{\lfloor 3}+\ldots\ldots = \left(1+1+\frac{1}{\lfloor 2}+\frac{1}{\lfloor 3}+\ldots\ldots\right)^x.$$

The series $\qquad 1+1+\dfrac{1}{\lfloor 2}+\dfrac{1}{\lfloor 3}+\dfrac{1}{\lfloor 4}+\ldots\ldots$

is usually denoted by e; hence

$$e^x = 1 + x + \frac{x^2}{\lfloor 2} + \frac{x^3}{\lfloor 3} + \frac{x^4}{\lfloor 4} + \ldots\ldots$$

Write cx for x, then

$$e^{cx} = 1 + cx + \frac{c^2 x^2}{\lfloor 2} + \frac{c^3 x^3}{\lfloor 3} + \ldots\ldots$$

Now let $e^c = a$, so that $c = \log_e a$; by substituting for c we obtain

$$a^x = 1 + x \log_e a + \frac{x^2 (\log_e a)^2}{\lfloor 2} + \frac{x^3 (\log_e a)^3}{\lfloor 3} + \ldots\ldots$$

This is the *Exponential Theorem*.

397. The series

$$1+1+\frac{1}{\lfloor 2}+\frac{1}{\lfloor 3}+\frac{1}{\lfloor 4}+\ldots\ldots,$$

which we have denoted by e, is very important as it is the base to which logarithms are first calculated. Logarithms to this base are known as the Napierian system, so named after Napier their inventor. They are also called *natural* logarithms from the fact that they are the first logarithms which naturally come into consideration in algebraical investigations.

When logarithms are used in theoretical work it is to be remembered that the base e is always understood, just as in arithmetical work the base 10 is invariably employed.

From the series the approximate value of e can be determined to any required degree of accuracy; to 10 places of decimals it is found to be 2·7182818284.

EXPONENTIAL AND LOGARITHMIC SERIES.

Example 1. Find the sum of the infinite series
$$1+\frac{1}{\lfloor 2}+\frac{1}{\lfloor 4}+\frac{1}{\lfloor 6}+\ldots\ldots.$$

We have $\quad e=1+1+\dfrac{1}{\lfloor 2}+\dfrac{1}{\lfloor 3}+\dfrac{1}{\lfloor 4}+\ldots\ldots;$

and by putting $x=-1$ in the series for e^x, we obtain
$$e^{-1}=1-1+\frac{1}{\lfloor 2}-\frac{1}{\lfloor 3}+\frac{1}{\lfloor 4}-\ldots\ldots.$$

$$\therefore\ e+e^{-1}=2\left(1+\frac{1}{\lfloor 2}+\frac{1}{\lfloor 4}+\frac{1}{\lfloor 6}+\ldots\ldots\right);$$

hence the sum of the series is $\dfrac{1}{2}(e+e^{-1})$.

Example 2. Find the coefficient of x^r in the expansion of $\dfrac{a-bx}{e^x}$.

$$\frac{a-bx}{e^x}=(a-bx)e^{-x}$$
$$=(a-bx)\left\{1-x+\frac{x^2}{\lfloor 2}-\frac{x^3}{\lfloor 3}+\ldots+\frac{(-1)^r x^r}{\lfloor r}+\ldots\right\}.$$

The coefficient required $=\dfrac{(-1)^r}{\lfloor r}\cdot a-\dfrac{(-1)^{r-1}}{\lfloor r-1}\cdot b$

$$=\frac{(-1)^r}{\lfloor r}(a+rb).$$

398. *To expand* $\log_e(1+x)$ *in ascending powers of* x.

From Art. 396,
$$a^y=1+y\log_e a+\frac{y^2(\log_e a)^2}{\lfloor 2}+\frac{y^3(\log_e a)^3}{\lfloor 3}+\ldots\ldots,$$

In this series write $1+x$ for a; thus $(1+x)^y$
$$=1+y\log_e(1+x)+\frac{y^2}{\lfloor 2}\{\log_e(1+x)\}^2+\frac{y^3}{\lfloor 3}\{\log_e(1+x)\}^3+\ldots\ldots\ (1).$$

Also by the Binomial Theorem, when $x<1$ we have
$$(1+x)^y=1+yx+\frac{y(y-1)}{\lfloor 2}x^2+\frac{y(y-1)(y-2)}{\lfloor 3}x^3+\ldots\ldots\ (2).$$

Now in (2) the coefficient of y is
$$x+\frac{(-1)}{1.2}x^2+\frac{(-1)(-2)}{1.2.3}x^3+\frac{(-1)(-2)(-3)}{1.2.3.4}x^4+\ldots\ldots;$$
that is, $\qquad x-\dfrac{x^2}{2}+\dfrac{x^3}{3}-\dfrac{x^4}{4}+\ldots\ldots.$

Equate this to the coefficient of y in (1); thus we have
$$\log_e(1+x)=x-\frac{x^2}{2}+\frac{x^3}{3}-\frac{x^4}{4}+\ldots\ldots$$
This is known as the *Logarithmic Series*.

399. Except when x is very small the series for $\log_e(1+x)$ is of little use for numerical calculations. We can, however, deduce from it other series by the aid of which Tables of Logarithms may be constructed.

400. In Art. 398 we have proved that
$$\log_e(1+x)=x-\frac{x^2}{2}+\frac{x^3}{3}-\ldots;$$
replacing x by $-x$, we have
$$\log_e(1-x)=-x-\frac{x^2}{2}-\frac{x^3}{3}-\ldots.$$
By subtraction,
$$\log_e\frac{1+x}{1-x}=2\left(x+\frac{x^3}{3}+\frac{x^5}{5}+\ldots\right).$$
Put $\dfrac{1+x}{1-x}=\dfrac{n+1}{n}$, so that $x=\dfrac{1}{2n+1}$; we thus obtain
$$\log_e(n+1)-\log_e n=2\left\{\frac{1}{2n+1}+\frac{1}{3(2n+1)^3}+\frac{1}{5(2n+1)^5}+\ldots\right\}.$$

From this formula by putting $n=1$ we can obtain $\log_e 2$. Again, by putting $n=2$ we obtain $\log_e 3-\log_e 2$; whence $\log_e 3$ is found, and therefore also $\log_e 9$ is known.

Now by putting $n=9$ we obtain $\log_e 10-\log_e 9$; thus the value of $\log_e 10$ is found to be $2\cdot 30258509\ldots$.

To convert Napierian logarithms into logarithms to base 10 we multiply by $\dfrac{1}{\log_e 10}$, which is the *modulus* [Art. 386] of the common system, and its value is $\dfrac{1}{2\cdot 30258509\ldots}$, or $\cdot 43429448\ldots$; we shall denote this modulus by μ.

By multiplying the last series throughout by μ we obtain a formula adapted to the calculation of common logarithms. Thus

$$\mu\log_e(n+1)-\mu\log_e n = 2\mu\left\{\frac{1}{2n+1}+\frac{1}{3(2n+1)^3}+\frac{1}{5(2n+1)^5}+\ldots\right\};$$

that is,

$$\log_{10}(n+1)-\log_{10}n = 2\left\{\frac{\mu}{2n+1}+\frac{\mu}{3(2n+1)^3}+\frac{\mu}{5(2n+1)^5}+\ldots\right\}.$$

From this result we see that if the logarithm of one of two consecutive numbers be known, the logarithm of the other may be found, and thus a table of logarithms can be constructed.

EXAMPLES XLI.

1. Shew that

(1) $\quad e^{-2} = 1 - \dfrac{2^3}{\underline{|3}} + \dfrac{2^4}{\underline{|4}} - \dfrac{2^5}{\underline{|5}} + \ldots\ldots$.

(2) $\quad \dfrac{e^2-1}{2e} = 1 + \dfrac{1}{\underline{|3}} + \dfrac{1}{\underline{|5}} + \dfrac{1}{\underline{|7}} + \ldots\ldots$.

2. Expand $\log\sqrt{1+x}$ in ascending powers of x.

3. Prove that $\log_e 2 = \dfrac{1}{2} + \dfrac{1}{12} + \dfrac{1}{30} + \dfrac{1}{56} + \ldots\ldots$.

4. Shew that

$$\log_{10}\left(\frac{1}{1-x}\right) = \frac{1}{\log_e 10}\left(x + \frac{x^2}{2} + \frac{x^3}{3} + \ldots\right).$$

5. Prove that

$$\log\frac{1+x}{1-3x} = 4x + 4x^2 + \frac{28}{3}x^3 + 20x^4 + \ldots\ldots.$$

6. Shew that if $x > 1$,

$$\log\sqrt{x^2-1} = \log x - \frac{1}{2x^2} - \frac{1}{4x^4} - \frac{1}{6x^6} - \ldots\ldots.$$

7. Shew that

$$\log\sqrt{\frac{1+x}{1-x}} - \log\sqrt{\frac{1-x}{1+x}} = 2\left(x + \frac{x^3}{3} + \frac{x^5}{5} + \ldots\right).$$

8. If $$a = b - \frac{b^2}{2} + \frac{b^3}{3} - \frac{b^4}{4} + \ldots,$$
express b in ascending powers of a.

9. Calculate the value of $\sqrt{e}$ to 4 places of decimals.

10. Prove that
$$\log_e(1 + x - 2x^2) = x - \frac{5x^2}{2} + \frac{7x^3}{3} - \frac{17x^4}{4} + \ldots,$$
and find the general term of the series.

11. Prove that $e^{-1} = 2\left(\dfrac{1}{\lfloor 3} + \dfrac{2}{\lfloor 5} + \dfrac{3}{\lfloor 7} + \ldots\right).$

12. Prove that $\log_e 3 = 1 + \dfrac{1}{3 \cdot 2^2} + \dfrac{1}{5 \cdot 2^4} + \dfrac{1}{7 \cdot 2^6} + \ldots.$

13. Shew that
$$\log_e(1 - 3x + 2x^2)^{-1} = 3x + \frac{5x^2}{2} + 3x^3 + \frac{17x^4}{4} + \ldots,$$
and find the general term of the series.

14. Prove that the expansion of $\log_e(1 - x + x^2)$ is
$$-x + \frac{x^2}{2} + \frac{2x^3}{3} + \frac{x^4}{4} - \frac{x^5}{5} - \frac{x^6}{3} - \ldots.$$

15. If $x > 1$, prove that
$$\frac{1}{x} + \frac{1}{2x^2} + \frac{1}{3x^3} + \ldots = \frac{1}{x-1} - \frac{1}{2(x-1)^2} + \frac{1}{3(x-1)^3} - \ldots.$$

CHAPTER XLII.

Miscellaneous Equations.

401. Many kinds of miscellaneous equations may be solved by the ordinary rules for quadratic equations as explained in Art. 202; but others require some special artifice for their solution. These will be illustrated in the present chapter.

Example 1. Solve $\dfrac{x^2-6}{x}+\dfrac{5x}{x^2-6}=6$.

Write y for $\dfrac{x^2-6}{x}$; thus

$$y+\frac{5}{y}=6, \text{ or } y^2-6y+5=0;$$

whence $\qquad y=5,\text{ or } 1,$

$$\therefore \frac{x^2-6}{x}=5, \text{ or } \frac{x^2-6}{x}=1\,;$$

that is, $\qquad x^2-5x-6=0,\text{ or } x^2-x-6=0.$

Thus $\qquad x=6,\ -1\,;\text{ or } x=3,\ -2.$

Example 2. Solve $3^{2x+3}-55=28(3^x-2)$.

This equation may be written $3^3 \cdot 3^{2x} - 28 \cdot 3^x + 1 = 0$.

By writing y for 3^x, we obtain

$$27y^2 - 28y + 1 = 0\,;\text{ that is, } (27y-1)(y-1)=0\,;$$

whence $\qquad y=\dfrac{1}{27},\text{ or } 1.$

Thus $\qquad 3^x=\dfrac{1}{27}=3^{-3},\text{ or } 3^x=1=3^0.$

and therefore $\qquad x=-3,\text{ or } 0.$

Example 3. Solve $\quad 2x^2 - 3\sqrt{2x^2 - 7x + 7} = 7x - 3$.

On transposition, $(2x^2 - 7x) - 3\sqrt{2x^2 - 7x + 7} = -3$.

By putting $\sqrt{2x^2 - 7x + 7} = y$, so that $2x^2 - 7x + 7 = y^2$, we obtain
$$(y^2 - 7) - 3y = -3, \text{ or } y^2 - 3y - 4 = 0;$$
whence $\quad\quad\quad\quad\quad y = 4, \text{ or } -1$.

Thus $\quad\quad\quad \sqrt{2x^2 - 7x + 7} = 4, \text{ or } \sqrt{2x^2 - 7x + 7} = -1$;

that is, $\quad\quad 2x^2 - 7x - 9 = 0, \text{ or } 2x^2 - 7x + 6 = 0$.

From the first of these quadratics we obtain $x = \dfrac{9}{2}$, or -1, and from the second $x = 2$, or $\dfrac{3}{2}$.

It should be noticed that in this solution we have tacitly assumed y to be the *positive* value of the expression $\sqrt{2x^2 - 7x + 7}$, so that the roots obtained from the solution of $\sqrt{2x^2 - 7x + 7} = -1$ will only satisfy the original equation in the modified form obtained by changing the sign of the radical.

Thus $x = \dfrac{9}{2}$, or -1 satisfies $2x^2 - 3\sqrt{2x^2 - 7x + 7} = 7x - 3$,

and $\quad x = 2$, or $\dfrac{3}{2}$ satisfies $2x^2 + 3\sqrt{2x^2 - 7x + 7} = 7x - 3$.

EXAMPLES XLII. a.

Solve the following equations :

1. $x^2 + x + 1 = \dfrac{42}{x^2 + x}$.

2. $\dfrac{x}{x^2 - 1} + \dfrac{x^2 - 1}{x} = 2\dfrac{1}{6}$.

3. $\left(x + \dfrac{1}{x}\right)^2 - 4\left(x + \dfrac{1}{x}\right) = 5$.

4. $8x^6 + 65x^3 + 8 = 0$.

5. $\dfrac{x + 8}{x + 12} + \dfrac{5}{x + 4} = \dfrac{3x + 14}{3x + 8}$.

6. $4^x + 8 = 9 \cdot 2^x$.

7. $\dfrac{3x - 6}{5 - x} + \dfrac{11 - 2x}{10 - 4x} = 3\dfrac{1}{2}$.

8. $3\sqrt{x} - 3x^{-\frac{1}{2}} = 8$.

9. $\left(x - \dfrac{6}{x}\right)^2 + 4x - \dfrac{24}{x} = 5$.

10. $27x^{\frac{3}{2}} - 1 = 26x^{\frac{3}{4}}$.

11. $7\sqrt{x - 8} - \sqrt{21x + 12} = 2\sqrt{3}$.

12. $4^{2x+1} + 16 = 65 \cdot 4^x$.

13. $x + 2 = \sqrt{4 + x\sqrt{8 - x}}$.

14. $3^{\frac{x}{2}} + 3^{-\frac{x}{2}} = 2$.

15. $2x^2 - 2x + 2\sqrt{2x^2 - 7x + 6} = 5x - 6$.
16. $x^2 + 6\sqrt{x^2 - 2x + 5} = 11 + 2x$.
17. $2\sqrt{x^2 - 6x + 2} + 4x + 1 = x^2 - 2x$.
18. $\sqrt{4x^2 + 2x + 7} = 12x^2 + 6x - 119$.
19. $3x(3-x) = 11 - 4\sqrt{x^2 - 3x + 5}$.
20. $x^2 - x + 3\sqrt{2x^2 - 3x + 2} = \dfrac{x}{2} + 7$.
21. $\sqrt{\dfrac{2-x}{3x}} - \sqrt{\dfrac{3x}{2-x}} = \dfrac{3}{2}$.
22. $\sqrt{\dfrac{a}{x}} - \sqrt{\dfrac{x}{a}} = \dfrac{a^2 - 1}{a}$.
23. $(a-b)x^2 + (b-c)x + c - a = 0$.
24. $a(b-c)x^2 + b(c-a)x + c(a-b) = 0$.
25. $\sqrt{a-x} + \sqrt{b-x} = \sqrt{a+b-2x}$.
26. $\dfrac{1}{a-x} + \dfrac{1}{b-x} = \dfrac{1}{a-c} + \dfrac{1}{b-c}$.
27. $\sqrt{x-p} + \sqrt{x-q} = \dfrac{p}{\sqrt{x-q}} + \dfrac{q}{\sqrt{x-p}}$.
28. $\sqrt{(x-2)(x-3)} + 5\sqrt{\dfrac{x-2}{x-3}} = \sqrt{x^2 + 6x + 8}$.
29. $\sqrt{x^2 + 4x - 4} + \sqrt{x^2 + 4x - 10} = 6$.
30. $\sqrt[3]{x-a} - \sqrt[3]{x-b} = \sqrt[3]{b-a}$.

402. No general methods can be given for the solution of simultaneous equations containing two or more unknowns. The simpler cases have been considered in Chapter XXVI.; the following examples illustrate useful artifices to be employed in special cases.

Example 1. Solve $\quad x + y = 4 \quad \ldots\ldots\ldots\ldots\ldots\ldots\ldots(1)$,
$$(x^2 + y^2)(x^3 + y^3) = 280 \quad \ldots\ldots\ldots\ldots\ldots\ldots(2).$$

We have $\quad x^2 + y^2 = (x+y)^2 - 2xy = 16 - 2xy$;
and $\quad x^3 + y^3 = (x+y)^3 - 3xy(x+y) = 64 - 12xy$.

By substituting in (2), we obtain
$$(16 - 2xy)(64 - 12xy) = 280 \text{ ; that is, } 3x^2y^2 - 40xy + 93 = 0;$$

whence $\quad xy = 3, \text{ or } \dfrac{31}{3}$.

Thus $\quad \begin{aligned} x+y &= 4, \\ xy &= 3; \end{aligned}\Big\}$ whence we obtain $\begin{aligned} x &= 3, \text{ or } 1; \\ y &= 1, \text{ or } 3; \end{aligned}\Big\}$

or $\quad \begin{aligned} x+y &= 4, \\ xy &= \tfrac{31}{3}; \end{aligned}\Big\}$ whence $\begin{aligned} x &= 2 \pm \sqrt{-\tfrac{19}{3}}; \\ y &= 2 \mp \sqrt{-\tfrac{19}{3}}. \end{aligned}\Big\}$

Example 2. Solve $x^2y^2z = 225$, $xy^2z^2 = 75$, $x^2yz^2 = 45$.

By multiplying the three equations together, we have
$$x^5 y^5 z^5 = 225 \times 75 \times 45 = 5^5 \times 3^5;$$
whence $\qquad xyz = 5 \times 3 = 15.$

By squaring this equation and dividing by each of the given equations in succession, we obtain
$$z = 1, \quad x = 3, \quad y = 5.$$

Example 3. Solve the equations
$$x^2 + xy + xz = 48, \quad xy + y^2 + yz = 12, \quad xz + yz + z^2 = 84.$$

These equations may be written
$$x(x+y+z) = 48, \quad y(x+y+z) = 12, \quad z(x+y+z) = 84.$$
By addition, $\qquad (x+y+z)(x+y+z) = 144;$
whence $\qquad x+y+z = \pm 12.$

On dividing each of the given equations in turn by this last equation, we obtain
$$x = \pm 4, \quad y = \pm 1, \quad z = \pm 7.$$

It is clear that the roots must be taken either all positively or all negatively.

Example 4. Solve $\quad x + y - z = 14 \dotfill (1),$
$\qquad\qquad\qquad\qquad y^2 + z^2 - x^2 = 46 \dotfill (2),$
$\qquad\qquad\qquad\qquad yz = 9 \dotfill (3).$

From (2) and (3), $\quad (y-z)^2 - x^2 = 28.$

Put u for $y - z$; then this equation becomes
$$u^2 - x^2 = 28.$$
Also from (1), $\qquad u + x = 14;$
by division, $\qquad u - x = 2;$
whence $\qquad x = 6, \text{ and } u = 8.$

Thus $y - z = 8$, and $yz = 9$; whence $y = 9$, or -1; $z = 1$, or -9; and the solution is $x = 6$, $y = 9$, $z = 1$; or $x = 6$, $y = -1$, $z = -9$.

EXAMPLES XLII. b.

1. $3x - 2y = 11,$
   $9x^2 - 4y^2 = 209.$

2. $x^3 + y^3 = 91,$
   $x^2y + xy^2 = 84.$

3. $x^3 - y^3 = 335,$
   $x^2y - xy^2 = 70.$

4. $x^2 + xy + y^2 = 84,$
   $x + \sqrt{xy} + y = 14.$

5. $x^2 + xy + y^2 = 189,$
   $x - \sqrt{xy} + y = 9.$

6. $\dfrac{3}{x^2} - \dfrac{1}{xy} - \dfrac{2}{y^2} = \dfrac{2}{9},$
   $\dfrac{3}{x} + \dfrac{2}{y} = \dfrac{4}{3}.$

7. $\dfrac{2}{x^2} - \dfrac{3}{xy} - \dfrac{2}{y^2} = 17,$
   $\dfrac{1}{x} - \dfrac{2}{y} = 1.$

8. $x^2y + y^2x = 20,$
   $\dfrac{1}{x} + \dfrac{1}{y} = \dfrac{5}{4}.$

9. $x^2 - 7xy + 4y^2 = 34,$
   $\dfrac{2x+y}{x-3y} - \dfrac{x-3y}{2x+y} = 2\dfrac{2}{3}.$

10. $x^2 - xy + x = 35,$
    $xy - y^2 + y = 15.$

11. $(x+y)^2 + 3(x-y) = 30,$
    $xy + 3(x-y) = 11.$

12. $(x-y)^2 = 3 - 2x - 2y,$
    $y(x-y+1) = x(y-x+1).$

13. $x^3 + 1 = 81(y^2 + y)$
    $x^2 + x = 9(y^3 + 1).$

14. Find the rational roots of
    (1) $x + y = 5$
        $(x^2+y^2)(x^3+y^3) = 455$,
    (2) $x - y = 2$
        $(x^2+y^2)(x^3-y^3) = 260.$

15. $\sqrt{\dfrac{x}{y}} + \sqrt{\dfrac{y}{x}} = 4\dfrac{1}{4},$
    $\dfrac{x}{\sqrt{y}} + \dfrac{y}{\sqrt{x}} = 16\dfrac{1}{4}.$

16. $\sqrt{\dfrac{x+y}{x-y}} + \sqrt{\dfrac{x-y}{x+y}} = \dfrac{34}{15},$
    $\dfrac{x+y}{\sqrt{x-y}} + \dfrac{x-y}{\sqrt{x+y}} = \dfrac{152}{15}.$

17. $x^2yz = 72, \quad xy^2z = 48, \quad xyz^2 = 96.$

18. $xyz = 30, \quad xyu = 120, \quad xzu = 20, \quad yzu = 24.$

19. $yz + zx = 13, \quad zx + xy = 25, \quad xy + yz = 20.$

20. $y(x+z)=112$, $z(x+y)=132$, $x(y+z)=90$.

21. $(x+a)(y-b)=2$, $(y-b)(z+c)=3$, $(z+c)(x+a)=6$.

22. $(x+y)(x+z)=63$, $(y+z)(y+x)=42$, $(z+x)(z+y)=54$.

23. $x^2-xy-xz=14$, $xy-y^2-yz=6$, $xz-yz-z^2=4$.

24. $x(3z-2y)=42$, $y(x-2z)=4$, $z(x+5y)+34=0$.

25. $xy+x+y=11$, $yz+y+z=3$, $zx+z+x=2$.

26. $(x+y)^2-z^2=65$, $x^2-(y+z)^2=13$, $x+y-z=5$.

27. $z+x=9xyz$, $x+y=5xyz$, $y+z=8xyz$.

28. $x^2+y^2+z^2=84$, $x+y+z=14$, $xy=z^2$.

29. $x+y-z=1$, $x^2-y^2+z^2=15$, $xz=12$.

30. $y+z-x=9$, $x^2-y^2-z^2=15$, $yz=3$.

31. $x^2+y^2+z^2=133$, $y+z-x=7$, $yz=x^2$.

32. $3^x=9^{y-1}$, $16^{3-x}=8^{y-2}$.

33. $2^{y-1}=16^{x-1}$, $3^{\frac{1}{x}}=9^{\frac{1}{z}}$, $\sqrt[x]{2^{y-3}}=\sqrt[2x]{8^{z-2}}$.

34. $x^2-(y-z)^2=a^2$, $y^2-(z-x)^2=b^2$, $z^2-(x-y)^2=c^2$.

CHAPTER XLIII.

INTEREST AND ANNUITIES.

403. QUESTIONS involving Simple Interest are easily solved by the ordinary rules of Arithmetic; but in Compound Interest the calculations are often extremely laborious. We shall now shew how these arithmetical calculations may be simplified by the aid of logarithms. Instead of taking as the rate of interest the interest on £100 for one year, it will be found more convenient to take the interest on £1 for one year. If this be denoted by £r, and the amount of £1 for 1 year by £R, we have $R = 1 + r$.

404. *To find the interest and amount of a given sum in a given time at compound interest.*

Let P denote the principal, R the amount of £1 in one year, n the number of years, I the interest, and M the amount.

The amount of P at the end of the first year is PR; and, since this is the principal for the second year, the amount at the end of the second year is $PR \times R$ or PR^2. Similarly the amount at the end of the third year is PR^3, and so on; hence the amount in n years is PR^n; that is,

$$M = PR^n;$$

and therefore
$$I = P(R^n - 1).$$

Example. Find the amount of £100 in a hundred years, allowing compound interest at the rate of 5 per cent., payable quarterly; having given

$$\log 2 = \cdot 3010300, \quad \log 3 = \cdot 4771213, \quad \log 14\cdot 3906 = 1\cdot 15808.$$

The amount of £1 in a quarter of a year is £$\left(1 + \frac{1}{4} \cdot \frac{5}{100}\right)$ or £$\frac{81}{80}$.

The number of payments is 400. If M be the amount, we have

$$M = 100 \left(\frac{81}{80}\right)^{400};$$

$$\therefore \log M = \log 100 + 400(\log 81 - \log 80)$$
$$= 2 + 400(4\log 3 - 1 - 3\log 2)$$
$$= 2 + 400(\cdot 0053952) = 4\cdot 15808 \text{ ;}$$

whence $M = 14390\cdot\cdot$.

Thus the amount is £14390. 12s.

Note. At simple interest the amount is £600

405. *To find the present value and discount of a given sum due in a given time, allowing compound interest.*

Let P be the given sum, V the present value, D the discount, R the amount of £1 for one year, n the number of years.

Since V is the sum which, put out to interest at the present time, will in n years amount to P, we have
$$P = VR^n;$$
$$\therefore V = PR^{-n},$$
and
$$D = P - V = P(1 - R^{-n}).$$

Annuities.

406. An **annuity** is a fixed sum paid periodically under certain stated conditions; the payment may be made either once a year or at more frequent intervals. Unless it is otherwise stated we shall suppose the payments annual.

407. *To find the amount of an annuity left unpaid for a given number of years allowing compound interest.*

Let A be the annuity, R the amount of £1 for one year, n the number of years, M the amount.

At the end of the first year A is due, and the amount of this sum in the remaining $n-1$ years is AR^{n-1}; at the end of the second year another A is due, and the amount of this sum in the remaining $n-2$ years is AR^{n-2}; and so on.

$$\therefore M = AR^{n-1} + AR^{n-2} + \ldots\ldots + AR^2 + AR + A$$
$$= A(1 + R + R^2 + \ldots\ldots \text{ to } n \text{ terms})$$
$$= A\frac{R^n - 1}{R - 1}.$$

408. *To find the present value of an annuity to continue for a given number of years allowing compound interest.*

Let A be the annuity, R the amount of £1 in one year, n the number of years, V the required present value.

The present value of A due in 1 year is AR^{-1};

the present value of A due in 2 years is AR^{-2};

the present value of A due in 3 years is AR^{-3}; and so on.

[Art. 405.]

Now V is the sum of the present values of the different payments;

$$V = AR^{-1} + AR^{-2} + AR^{-3} + \ldots \text{ to } n \text{ terms}$$
$$= AR^{-1}\frac{1 - R^{-n}}{1 - R^{-1}}$$
$$= A\frac{1 - R^{-n}}{R - 1}.$$

Note. This result may also be obtained by dividing the value of M, given in Art. 407, by R^n. [Art. 404.]

Cor. If we make n infinite we obtain for the present value of a perpetual annuity

$$V = \frac{A}{R - 1} = \frac{A}{r}.$$

409. If mA is the present value of an annuity A, the annuity is said to be worth m **years' purchase.**

In the case of a perpetual annuity $mA = \frac{A}{r}$; hence

$$m = \frac{1}{r} = \frac{100}{\text{rate per cent.}};$$

that is, the number of years' purchase of a perpetual annuity is obtained by dividing 100 by the rate per cent.

A good test of the credit of a Government is furnished by the number of years' purchase of its Stocks; thus the $2\frac{1}{2}$ p.c. Consols at $92\frac{1}{2}$ are worth 37 years' purchase; Russian 4 p.c. Stock at 96 is worth 24 years' purchase; while Austrian 5 p.c. Stock at 80 is only worth 16 years' purchase.

410. A freehold estate is an estate which yields a perpetual annuity called the *rent*; and thus the value of the estate is equal to the present value of a perpetual annuity equal to the rent.

It follows from Art. 409 that if we know the number of years' purchase that a tenant pays in order to buy his farm, we obtain the rate per cent. at which interest is reckoned by dividing 100 by the number of years' purchase.

EXAMPLES XLIII.

[*The Examples marked* * *may be solved directly by use of the Tables.*]

1. If in the year 1600 a sum of £1000 had been left to accumulate for 300 years, find its amount in the year 1900, reckoning compound interest at 4 per cent. per annum. Given

$$\log 104 = 2\cdot0170333 \text{ and } \log 12885\cdot5 = 4\cdot10999.$$

*2. Find in how many years a sum of money will amount to one hundred times its value at $5\frac{1}{2}$ per cent. per annum compound interest. Given $\log 1055 = 3\cdot023$.

3. Find the present value of £6000 due in 20 years, allowing compound interest at 8 per cent. per annum. Given

$$\log 2 = \cdot30103, \ \log 3 = \cdot47712, \text{ and } \log 12875 = 4\cdot10975.$$

4. Find at what rate per cent. per annum £1200 will amount to £20000 in 15 years at compound interest. Given

$$\log 2 = \cdot30103, \ \log 3 = \cdot47712, \text{ and } \log 12063 = 4\cdot08145.$$

5. Find the amount of an annuity of £100 in 15 years, allowing compound interest at 4 per cent. per annum. Given

$$\log 1\cdot04 = \cdot01703, \text{ and } \log 180075 = 5\cdot25545.$$

6. A freehold estate worth £280 a year is sold for £7000; find the rate of interest.

*7. If a perpetual annuity is worth 40 years' purchase, find what an annuity of £300 will amount to in 10 years at the same rate of interest. Given $\log 10\cdot25 = 1\cdot01072$, and $\log 1280 = 3\cdot1072$.

8. Find the present value of an annuity of £900 to continue for 20 years at $4\frac{1}{2}$ per cent. compound interest. Given

$$\log 1\cdot045 = \cdot01912, \text{ and } \log 41458 = 4\cdot6176.$$

*9. A man borrows £20000 at 5 per cent. compound interest. If the principal and interest are to be paid by 20 equal annual instalments, find the amount of each of these; having given

$$\log 105 = 2\cdot0212, \text{ and } \log 3767 = 3\cdot576.$$

*10. A man has a capital of £100000, for which he receives interest at $3\frac{1}{2}$ per cent.; if he spends £7000 a year, find in what time he will be ruined. Given

$$\log 2 = \cdot301, \ \log 3 = \cdot477, \text{ and } \log 23 = 1\cdot362.$$

CHAPTER XLIV.

GRAPHICAL REPRESENTATION OF FUNCTIONS.

[*A considerable portion of this chapter may be taken at an early stage. For example, Arts. 411–416 may be read as soon as the student has had sufficient practice in substitutions involving negative quantities. Arts. 417–424 may be read in connection with Easy Simultaneous Equations. With the exception of a few articles the rest of the chapter should be postponed until the student is acquainted with quadratic equations.*]

411. DEFINITION. Any expression which involves a variable quantity x, and whose value is dependent on that of x, is called a **function of x**.

Thus $3x+8$, $2x^2+6x-7$, $x^4-3x^3+x^2-9$ are functions of x of the first, second, and fourth degree respectively. [Art. 24.]

412. The symbol $f(x)$ is often used to briefly denote a function of x. If $y=f(x)$, by substituting a succession of numerical values for x we can obtain a corresponding succession of values for y which stands for the value of the function. Hence in this connection it is sometimes convenient to call x the **independent variable**, and y the **dependent variable**.

413. Consider the function $x(9-x^2)$, and let its value be represented by y.

Then, when $\quad x=0, \quad y=0\times 9= 0,$
,, $\quad\quad\quad\quad\quad x=1, \quad y=1\times 8= 8,$
,, $\quad\quad\quad\quad\quad x=2, \quad y=2\times 5=10,$
,, $\quad\quad\quad\quad\quad x=3, \quad y=3\times 0= 0,$
,, $\quad\quad\quad\quad\quad x=4, \quad y=4\times(-7)=-28,$

and so on.

By proceeding in this way we can find as many values of the function as we please. But we are often not so much concerned with the actual values which a function assumes for different values of the variable as with *the way in which the value of the function changes.* These variations can be very conveniently represented by a **graphical** method which we shall now explain.

414. Two straight lines XOX', YOY' are taken intersecting at right angles in O, thus dividing the plane of the paper into four spaces XOY, YOX', $X'OY'$, $Y'OX$, which are known as the first, second, third, and fourth quadrants respectively.

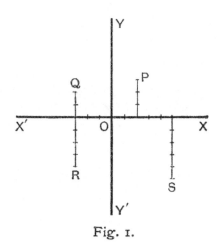

Fig. 1.

The lines $X'OX$, YOY' are usually drawn horizontally and vertically; they are taken as lines of reference and are known as the **axis of x and y** respectively. The point O is called the **origin.** Values of x are measured from O along the axis of x, according to some convenient scale of measurement, and are called **abscissæ,** *positive* values being drawn to the *right* of O along OX, and *negative* values to the *left* of O along OX'.

Values of y are drawn (on the same scale) parallel to the axis of y, from the ends of the corresponding abscissæ, and are called **ordinates.** These are *positive* when drawn *above* $X'X$, *negative* when drawn *below* $X'X$.

415. The abscissa and ordinate of a point taken together are known as its **coordinates.** A point whose coordinates are x and y is briefly spoken of as "the point (x, y)."

The coordinates of a point completely determine its position in the plane. Thus if we wish to mark the point $(2, 3)$, we take $x=2$ units measured to the right of O, $y=3$ units measured perpendicular to the x-axis and above it. The resulting point P is in the first quadrant. The point $(-3, 2)$ is found by taking $x=3$ units to the left of O, and $y=2$ units above the x-axis. The resulting point Q is in the second quadrant. Similarly the points $(-3, -4)$, $(5, -5)$ are represented by R and S in Fig. 1, in the third and fourth quadrants respectively.

This process of marking the position of a point in reference to the coordinate axes is known as **plotting the point.**

416. In practice it is convenient to use **squared paper**; that is, paper ruled into small squares by two sets of equi-distant parallel straight lines, the one set being horizontal and the other vertical. After selecting two of the intersecting lines as axes (and slightly thickening them to aid the eye) one or more of the divisions may be chosen as our unit, and points may be readily plotted when their coordinates are known. Conversely, if the position of a point in any of the quadrants is marked, its coordinates can be measured by the divisions on the paper.

In the following pages we have used paper ruled to tenths of an inch, but a larger scale will sometimes be more convenient. See Art. 436.

Example. Plot the points $(5, 2)$, $(-3, 2)$, $(-3, -4)$, $(5, -4)$ on squared paper. Find the area of the figure determined by these points, assuming the divisions on the paper to be tenths of an inch.

Taking the points in the order given, it is easily seen that they are represented by P, Q, R, S in Fig. 2, and that they form a rectangle which contains 48 squares. Each of these is *one-hundredth* part of a *square* inch. Thus the area of the rectangle is ·48 of a square inch.

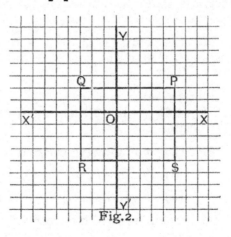

EXAMPLES XLIV. a.

[*The following examples are intended to be done mainly by actual measurement on squared paper; where possible, they should also be verified by calculation.*]

Plot the following pairs of points and draw the line which joins them:

1. $(3, 0), (0, 6)$.
2. $(-2, 0), (0, -8)$.
3. $(3, -8), (-2, 6)$.
4. $(5, 5), (-2, -2)$.
5. $(-2, 6), (1, -3)$.
6. $(4, 5), (-1, 5)$.

7. Plot the points $(3, 3), (-3, 3), (-3, -3), (3, -3)$, and find the number of squares contained by the figure determined by these points.

8. Plot the points $(4, 0), (0, 4), (-4, 0), (0, -4)$, and find the number of units of area in the resulting figure.

9. Plot the points $(0, 0), (0, 10), (5, 5)$, and find the number of units of area in the triangle.

10. Shew that the triangle whose vertices are $(0, 0), (0, 6), (4, 3)$ contains 12 units of area. Shew also that the points $(0, 0), (0, 6), (4, 8)$ determine a triangle of the same area.

11. Plot the points $(5, 6), (-5, 6), (5, -6), (-5, -6)$. If one millimetre is taken as unit, find the area of the figure in square centimetres.

12. Plot the points $(1, 3), (-3, -9)$, and shew that they lie on a line passing through the origin. Name the coordinates of other points on this line.

13. Plot the eight points $(0, 5), (3, 4), (5, 0), (4, -3), (-5, 0), (0, -5), (-4, 3), (-4, -3)$, and shew that they are all equidistant from the origin.

14. Plot the two following series of points:

(i) $(5, 0), (5, 2), (5, 5), (5, -1), (5, -4)$;

(ii) $(-4, 8), (-1, 8), (0, 8), (3, 8), (6, 8)$.

Shew that they lie on two lines respectively parallel to the axis of y, and the axis of x. Find the coordinates of the point in which they intersect.

15. Plot the points (13, 0), (0, −13), (12, 5), (−12, 5), (−13, 0), (−5, −12), (5, −12). Find their locus, (i) by measurement, (ii) by calculation.

16. Plot the points (2, 2), (−3, −3), (4, 4), (−5, −5), shewing that they all lie on a certain line through the origin. Conversely, shew that for *every* point on this line the abscissa and ordinate are equal.

Graph of a Function.

417. Let $f(x)$ represent a function of x, and let its value be denoted by y. If we give to x a series of numerical values we get a corresponding series of values for y. If these are set off as abscissæ and ordinates respectively, we plot a succession of points. If *all* such points were plotted we should arrive at a line, straight or curved, which is known as the **graph** of the *function* $f(x)$, or the **graph** of the *equation* $y = f(x)$. The variation of the function for different values of the variable x is exhibited by the variation of the ordinates as we pass from point to point.

In practice a few points carefully plotted will usually enable us to draw the graph with sufficient accuracy.

418. The student who has worked intelligently through the preceding examples will have acquired for himself some useful preliminary notions which will be of service in the examples on simple graphs which we are about to give. In particular, before proceeding further he should satisfy himself with regard to the following statements:

(i) The coordinates of the origin are (0, 0).

(ii) The abscissa of every point on the axis of y is 0.

(iii) The ordinate of every point on the axis of x is 0.

(iv) The graph of all points which have the same abscissa is a line parallel to the axis of y. (*e.g.* $x = 2$.)

(v) The graph of all points which have the same ordinate is a line parallel to the axis of x. (*e.g.* $y = 5$.)

(vi) The distance of any point $P(x, y)$ from the origin is given by $OP^2 = x^2 + y^2$.

Example 1. Plot the graph of $y=x$.

When $x=0$, $y=0$; thus the origin is one point on the graph.

Also, when $\quad x=1, 2, 3, \ldots -1, -2, -3, \ldots,$

$\qquad\qquad y=1, 2, 3, \ldots -1, -2, -3, \ldots.$

Thus the graph passes through O, and represents a series of points each of which has its ordinate equal to its abscissa, and is clearly represented by POP' in Fig. 3.

Example 2. Plot the graph of $y=x+3$.

Arrange the values of x and y as follows:

x	3	2	1	0	-1	-2	-3	...
y	6	5	4	3	2	1	0	...

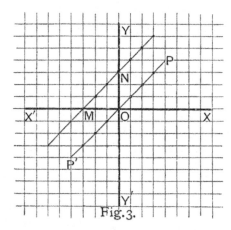
Fig. 3.

By joining these points we obtain a line MN parallel to that in Example 1.

The results printed in larger and deeper type should be specially noted and compared with the graph. They shew that the distances ON, OM (usually called the *intercepts on the axes*) are obtained by separately putting $x=0$, $y=0$ in the equation of the graph.

Note. By observing that in Example 2 each ordinate is 3 units greater than the corresponding ordinate in Example 1, the graph of $y=x+3$ may be obtained from that of $y=x$ by simply producing each ordinate 3 units in the positive direction.

In like manner the equations

$$y=x+5, \quad y=x-5$$

represent two parallel lines on opposite sides of $y=x$ and equidistant from it, as the student may easily verify for himself.

Example 3. Plot the graphs represented by the following equations:

(i) $y=2x$; (ii) $y=2x+4$; (iii) $y=2x-5$.

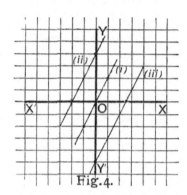

Fig. 4.

Here we only give the diagram which the student should verify in detail for himself, following the method explained in the two preceding examples.

EXAMPLES XLIV. b.

[*In the following examples Nos. 1-18 are arranged in groups of three; each group should be represented on the same diagram so as to exhibit clearly the position of the three graphs relatively to each other.*]

Plot the graphs represented by the following equations:

1. $y=5x$.
2. $y=5x-4$.
3. $y=5x+6$.
4. $y=-3x$.
5. $y=-3x+3$.
6. $y=-3x-2$.
7. $y+x=0$.
8. $y+x=8$.
9. $y+4=x$.
10. $4x=3y$.
11. $3y=4x+6$.
12. $4y+3x=8$.
13. $x-5=0$.
14. $y-6=0$.
15. $5y=6x$.
16. $3x+4y=10$.
17. $4x+y=9$.
18. $5x-2y=8$.

19. Shew by careful drawing that the three last graphs have a common point whose coordinates are 2, 1.

20. Shew by careful drawing that the equations
$$x+y=10, \quad y=x-4$$
represent two straight lines at right angles.

21. Draw on the same axes the graphs of $x=5$, $x=9$, $y=3$, $y=11$. Find the number of units of area enclosed by these lines.

22. Taking one-tenth of an inch as the unit of length, find the area included between the graphs of $x=7$, $x=-3$, $y=-2$, $y=8$.

23. Find the area included by the graphs of
$$y=x+6, \quad y=x-6, \quad y=-x+6, \quad y=-x-6.$$

24. With one millimetre as linear unit, find in square centimetres the area of the figure enclosed by the graphs of
$$y=2x+8, \quad y=2x-8, \quad y=-2x+8, \quad y=-2x-8.$$

419. The student should now be prepared for the following statements:

(i) For all numerical values of a the equation $y=ax$ represents a straight line through the origin.

(ii) For all numerical values of a and b the equation $y=ax+b$ represents a line parallel to $y=ax$, and cutting off an intercept b from the axis of y.

420. Conversely, since every equation involving x and y only in the first degree can be reduced to one of the forms $y=ax$, $y=ax+b$, it follows that *every simple equation connecting two variables represents a straight line.* For this reason an expression of the form $ax+b$ is said to be a **linear function** of x, and an equation such as $y=ax+b$, or $ax+by+c=0$, is said to be a **linear equation**.

Example. Shew that the points $(3, -4)$, $(9, 4)$, $(12, 8)$ lie on a straight line, and find its equation.

Assume $y=ax+b$ as the equation of the line. If it passes through the first two points given, their coordinates must satisfy the above equation. Hence
$$-4=3a+b, \quad 4=9a+b.$$

These equations give $\quad a=\dfrac{4}{3}, \quad b=-8$.

Hence $\quad y=\dfrac{4}{3}x-8, \quad$ or $\quad 4x-3y=24,$

is the equation of the line passing through the first two points. Since $x=12$, $y=8$ satisfies this equation, the line also passes through $(12, 8)$. This example may be verified graphically by plotting the line which joins *any two* of the points and shewing that it passes through the third.

Application to Simultaneous Equations.

421. It was shewn in Art. 100 that in the case of a simple equation between x and y, it is possible to find as many pairs of values of x and y as we please which satisfy the given equation. We now see that this is equivalent to saying that we may find as many points as we please on any given straight line. If, however, we have two simultaneous equations between x and y, there can only be one pair of values which will satisfy both equations. This is equivalent to saying that two straight lines can have only one common point.

Example. Solve graphically the equations:

$$3x + 7y = 27, \quad 5x + 2y = 16.$$

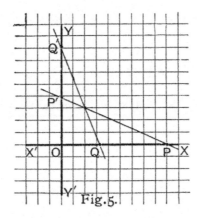
Fig. 5.

If carefully plotted it will be found that these two equations represent the lines in the annexed diagram. On measuring the coordinates of the point at which they intersect it will be found that $x=2$, $y=3$, thus verifying the solution given in Art. 103, Ex. 1.

422. It will now be seen that the process of solving two simultaneous equations is equivalent to finding the coordinates of the point (or points) at which their graphs meet.

423. Since a straight line can always be drawn by joining *any* two points on it, in solving *linear* simultaneous equations graphically, it is only necessary to plot two points on each line. The points where the lines meet the axes will usually be the most convenient to select.

424. Two simultaneous equations lead to no finite solution if they are inconsistent with each other. For example, the equations
$$x+3y=2, \quad 3x+9y=8$$
are inconsistent, for the second equation can be written $x+3y=2\tfrac{2}{3}$, which is clearly inconsistent with $x+3y=2$. The graphs of these two equations will be found to be two parallel straight lines which have no finite point of intersection.

Again, two simultaneous equations must be independent. The equations
$$4x+3y=1, \quad 16x+12y=4$$
are not independent, for the second can be deduced from the first by dividing throughout by 4. Thus *any pair of values* which will satisfy one equation will satisfy the other. Graphically these two equations represent two coincident straight lines which of course have an unlimited number of common points.

EXAMPLES XLIV. c.

Solve the following equations, in each case verifying the solution graphically :

1. $y=2x+3,$
   $y+x=6.$

2. $y=3x+4,$
   $y=x+8.$

3. $y=4x,$
   $2x+y=18.$

4. $2x-y=8,$
   $4x+3y=6.$

5. $3x+2y=16,$
   $5x-3y=14.$

6. $6y-5x=18,$
   $4x=3y.$

7. $2x+y=0,$
   $y=\dfrac{4}{3}(x+5).$

8. $2x-y=3,$
   $3x-5y=15.$

9. $2y=5x+15,$
   $3y-4x=12.$

10. Prove by graphical representation that the three points $(3, 0)$, $(2, 7)$, $(4, -7)$ lie on a straight line. Where does this line cut the axis of y?

11. Prove that the three points $(1, 1)$, $(-3, 4)$, $(5, -2)$ lie on a straight line. Find its equation. Draw the graph of this equation, shewing that it passes through the given points.

12. Shew that the three points $(3, 2)$, $(8, 8)$, $(-2, -4)$ lie on a straight line. Prove algebraically and graphically that it cuts the axis of x at a distance $1\tfrac{1}{3}$ from the origin.

425. We shall now give some graphs of functions of higher degree than the first.

Example 1. Plot the graph of $2y = x^2$.

Corresponding values of x and y may be tabulated as follows:

x	...	3	2·5	2	1·5	1	0	-1	-2	-3	...
y	...	4·5	3·125	2	1·125	·5	0	·5	2	4·5	...

Here, in order to obtain a figure on a sufficiently large scale, it will be found convenient to take two divisions on the paper for our unit.

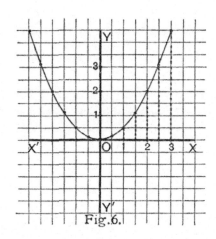
Fig. 6.

If the above points are plotted and connected by a line drawn freehand, we shall obtain the curve shewn in Fig. 6. This curve is called a **parabola**.

There are two facts to be specially noted in this example.

(i) Since from the equation we have $x = \pm\sqrt{2y}$, it follows that for every value of the ordinate we have two values of the abscissa, *equal in magnitude and opposite in sign*. Hence the graph is symmetrical with respect to the axis of y; so that after plotting with care enough points to determine the form of the graph in the first quadrant, its form in the second quadrant can be inferred without actually plotting any points in this quadrant. At the same time, in this and similar cases beginners are recommended to plot a few points in each quadrant through which the graph passes.

E.A. 2 c

(ii) We observe that all the plotted points lie above the axis of x. This is evident from the equation; for since x^2 must be positive for all values of x, every ordinate obtained from the equation $y=\dfrac{x^2}{2}$ must be positive.

In like manner the student may shew that the graph of $2y= -x^2$ is a curve similar in every respect to that in Fig. 6, but lying entirely below the axis of x.

Note. Some further remarks on the graph of this and the next example will be found in Art. 431.

Example 2. Find the graph of $y=2x+\dfrac{x^2}{4}$.

Here the following arrangement will be found convenient:

x	3	2	1	0	-1	-2	-3	-4	-5	-6	-7	-8
$2x$	6	4	2	0	-2	-4	-6	-8	-10	-12	-14	-16
$\dfrac{x^2}{4}$	2·25	1	·25	0	·25	1	2·25	4	6·25	9	12·25	16
y	8·25	5	2·25	0	$-1·75$	-3	$-3·75$	-4	$-3·75$	-3	$-1·75$	0

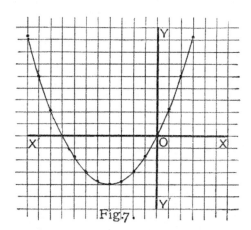
Fig. 7.

From the form of the equation it is evident that every positive value of x will yield a positive value of y, and that as x increases y also increases. Hence the portion of the curve in the first quadrant lies as in Fig. 7, and can be extended indefinitely in this quadrant. In the present case only two or three positive values of x and y need be plotted, but more attention must be paid to the results arising out of negative values of x.

When $y=0$, we have $\frac{x^2}{4}+2x=0$; thus the two values of x in the graph which correspond to $y=0$ furnish the roots of the equation $\frac{x^2}{4}+2x=0$.

426. If $f(x)$ represent a function of x, an approximate solution of the equation $f(x)=0$ may be obtained by plotting the graph of $y=f(x)$, and then measuring the intercepts made on the axis of x. These intercepts are values of x which make y equal to zero, and are therefore roots of $f(x)=0$.

427. If $f(x)$ gradually increases till it reaches a value a, which is algebraically greater than neighbouring values on either side, a is said to be a **maximum value** of $f(x)$.

If $f(x)$ gradually decreases till it reaches a value b, which is algebraically less than neighbouring values on either side, b is said to be a **minimum** value of $f(x)$.

When $y=f(x)$ is treated graphically, it is now evident that maximum and minimum values of $f(x)$ occur at points where the ordinates are algebraically greatest and least in the immediate vicinity of such points.

Example. Solve the equation $x^2-7x+11=0$ graphically, and find the minimum value of the function $x^2-7x+11$.

Put $y=x^2-7x+11$, and find the graph of this equation.

x	0	1	2	3	3·5	4	5	6	7
y	11	5	1	−1	−1·25	−1	1	5	11

The values of x which make the function $x^2-7x+11$ vanish are those which correspond to $y=0$. By careful measurement it will be found that the intercepts OM and ON are approximately equal to 2·38 and 4·62.

The algebraical solution of
$$x^2-7x+11=0$$
gives $x=\frac{1}{2}(7\pm\sqrt{5})$.

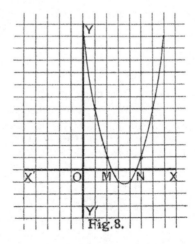

Fig. 8.

If we take 2·236 as the approximate value of $\sqrt{5}$, the values of x will be found to agree with those obtained from the graph.

384　　　　　　　　ALGEBRA.　　　　　　　[CHAP.

Again, $x^2-7x+11 = \left(x-\dfrac{7}{2}\right)^2 - \dfrac{5}{4}$. Now $\left(x-\dfrac{7}{2}\right)^2$ must be positive for all real values of x except $x=\dfrac{7}{2}$, in which case it vanishes, and the value of the function reduces to $-\dfrac{5}{4}$, which is the least value it can have.

The graph shews that when $x=3{\cdot}5$, $y=-1{\cdot}25$, and that this is the algebraically least ordinate in the plotted curve.

428. The following example shews that points selected for graphical representation must sometimes be restricted within certain limits.

Example. Find the graph of $x^2+y^2=36$.

The equation may be written in either of the following forms:

(i) $y=\pm\sqrt{36-x^2}$;　　(ii) $x=\pm\sqrt{36-y^2}$.

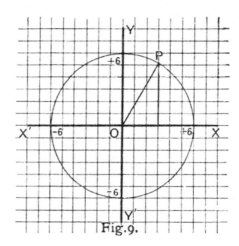

Fig. 9.

In order that y may be a real quantity we see from (i) that $36-x^2$ must be positive. Thus x can only have values between -6 and $+6$. Similarly from (ii) it is evident that y must also lie between -6 and $+6$. Between these limits it will be found that all plotted points will lie at a distance 6 from the origin. Hence the graph is a circle whose centre is O and whose radius is 6.

This is otherwise evident, for the distance of any point $P(x, y)$ from the origin is given by $OP=\sqrt{x^2+y^2}$. [Art. 418.] Hence the equation $x^2+y^2=36$ asserts that the graph consists of a series of points all of which are at a distance 6 from the origin.

Note. To plot the curve from equation (ii), we should select a succession of values for y and then find corresponding values of x. In other words we make y the *independent* and x the *dependent* variable. The student should be prepared to do this in some of the examples which follow.

EXAMPLES XLIV. d.

1. Draw the graphs of $y=x^2$, and $x=y^2$, and shew that they have only one common chord. Find its equation.

2. From the graphs, and also by calculation, shew that $y=\dfrac{x^2}{8}$ cuts $x=-y^2$ in only two points, and find their coordinates.

3. Draw the graphs of
   (i) $y^2=-4x$;   (ii) $y=2x-\dfrac{x^2}{4}$;   (iii) $y=\dfrac{x^2}{4}+x-2$.

4. Draw the graph of $y=x+x^2$. Shew also that it may be deduced from that of $y=x^2$, obtained in example 1.

5. Shew (i) graphically, (ii) algebraically, that the line $y=2x-3$ meets the curve $y=\dfrac{x^2}{4}+x-2$ in one point only. Find its coordinates.

6. Find graphically the roots of the following equations to 2 places of decimals:
   (i) $\dfrac{x^2}{4}+x-2=0$;   (ii) $x^2-2x=4$;   (iii) $4x^2-16x+9=0$;

and verify the solutions algebraically.

7. Find the minimum value of x^2-2x-4, and the maximum value of $5+4x-2x^2$.

8. Draw the graph of $y=(x-1)(x-2)$ and find the minimum value of $(x-1)(x-2)$. Measure, as accurately as you can, the values of x for which $(x-1)(x-2)$ is equal to 5 and 9 respectively. Verify algebraically.

9. Solve the simultaneous equations
$$x^2+y^2=100, \quad x+y=14;$$
and verify the solution by plotting the graphs of the equations and measuring the coordinates of their common points.

10. Plot the graphs of $x^2+y^2=25$, $3x+4y=25$, and examine their relation to each other where they intersect. Verify the result algebraically.

429. Infinite and zero values. Consider the fraction $\dfrac{a}{x}$ in which the numerator a has a *certain fixed value*, and the denominator is a *quantity subject to change*; then it is clear that the smaller x becomes the larger does the value of the fraction $\dfrac{a}{x}$ become. For instance

$$\frac{a}{\frac{1}{10}} = 10a, \quad \frac{a}{\frac{1}{1000}} = 1000a, \quad \frac{a}{\frac{1}{1000000}} = 1000000a.$$

By making the denominator x sufficiently small the value of the fraction $\dfrac{a}{x}$ can be made as large as we please; that is, if x is made *less than any quantity that can be named*, the value of $\dfrac{a}{x}$ will become *greater than any quantity that can be named*.

A quantity less than any assignable quantity is called **zero** and is denoted by the symbol 0.

A quantity greater than any assignable quantity is called **infinity** and is denoted by the symbol ∞.

We may now say briefly

when $x = 0$, *the value of* $\dfrac{a}{x}$ *is* ∞.

Again if x is a quantity which gradually increases and finally becomes *greater than any assignable quantity* the fraction becomes *smaller than any assignable quantity*. Or more briefly

when $x = \infty$, *the value of* $\dfrac{a}{x}$ *is* 0.

430. It should be observed that when the symbols for zero and infinity are used in the sense above explained, they are subject to the rules of signs which affect other algebraical symbols. Thus we shall find it convenient to use a concise statement such as "when $x = +0$, $y = +\infty$" to indicate that when a *very small and positive* value is given to x, the corresponding value of y is *very large and positive*.

431. If we now return to the examples worked out in Art. 425, in Example 1, we see that when $x = \pm\infty$, $y = +\infty$; hence the curve extends upwards to infinity in both the first and second quadrants. In Example 2, when $x = +\infty$, $y = +\infty$. Again y is negative between the values 0 and -8 of x. For all

XLIV.] INFINITE AND ZERO VALUES. 387

negative values of x numerically greater than 8, y is positive, and when $x = -\infty$, $y = +\infty$. Hence the curve extends to infinity in both the first and second quadrants.

The student should now examine the nature of the graphs in Examples XLIV. d. when x and y are infinite.

Example. Find the graph of $xy = 4$.

The equation may be written in the form
$$y = \frac{4}{x},$$
from which it appears that when $x = 0$, $y = \infty$ and when $x = \infty$, $y = 0$. Also y is positive when x is positive, and negative when x is negative. Hence the graph must lie entirely in the first and third quadrants.

It will be convenient in this case to take the positive and negative values of the variables separately.

(1) *Positive values:*

x	0	1	2	3	4	5	6	...	∞
y	∞	4	2	$1\frac{1}{3}$	1	·8	$\frac{2}{3}$	...	0

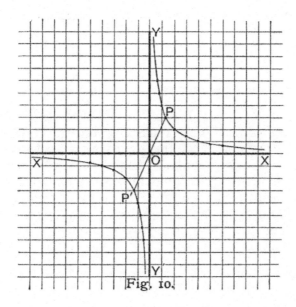

Fig. 10.

Graphically these values shew that as we recede further and further from the origin on the x-axis in the positive direction, the values of y are positive and become smaller and smaller. That is the graph is continually approaching the x-axis in such a way that by taking a sufficiently great positive value of x we obtain a point on the graph as near as we please to the x-axis but never actually reaching it until $x = \infty$. Similarly, as x becomes smaller and smaller the graph approaches more and more nearly to the positive end of the y-axis, never actually reaching it as long as x has any finite positive value, however small.

(2) *Negative values:*

x	-0	-1	-2	-3	-4	-5	...	$-\infty$
y	$-\infty$	-4	-2	$-1\frac{1}{3}$	-1	$-\cdot 8$	...	-0

The portion of the graph obtained from these values is in the third quadrant as shewn in Fig. 10, and exactly similar to the portion already traced in the first quadrant. It should be noticed that as x passes from $+0$ to -0 the value of y changes from $+\infty$ to $-\infty$. Thus the graph, which in the first quadrant has run away to an infinite distance on the positive side of the y-axis, reappears in the third quadrant coming from an infinite distance on the negative side of that axis. Similar remarks apply to the graph in its relation to the x-axis.

432. When a curve continually approaches more and more nearly to a line without actually meeting it until an infinite distance is reached, such a line is said to be an **asymptote** to the curve. In the above case each of the axes is an asymptote.

433. Every equation of the form $y = \dfrac{c}{x}$, or $xy = c$, where c is constant, will give a graph similar to that exhibited in the example of Art. 431. The resulting curve is known as a **rectangular hyperbola,** and has many interesting properties. In particular we may mention that from the form of the equation it is evident that for every point (x, y) on the curve there is a corresponding point $(-x, -y)$ which satisfies the equation. Graphically this amounts to saying that any line through the origin meeting the two branches of the curve in P and P' is bisected at O.

434. In the simpler cases of graphs, sufficient accuracy can usually be obtained by plotting a few points, and there is little difficulty in selecting points with suitable coordinates. But in other cases, and especially when the graph has infinite branches, more care is needed. The most important things to observe are (1) the values for which the function $f(x)$ becomes zero or infinite; and (2) the values which the function assumes for zero and infinite values of x. In other words, we determine the *general character* of the curve in the neighbourhood of the origin, the axes, and infinity. Greater accuracy of detail can then be secured by plotting points at discretion. The selection of such points will usually be suggested by the earlier stages of our work.

The existence of symmetry about either of the axes should also be noted. When an equation contains no *odd* powers of x, the graph is symmetrical with regard to the axis of y. Similarly the absence of odd powers of y indicates symmetry about the axis of x. Compare Art. 425, Ex. 1.

Example. Draw the graph of $y = \dfrac{2x+7}{x-4}$. [See fig. on next page.]

We have $y = \dfrac{2x+7}{x-4} = \dfrac{2+\dfrac{7}{x}}{1-\dfrac{4}{x}}$, the latter form being convenient for infinite values of x.

(i) When $\quad\quad\quad y=0, \quad x=-\dfrac{7}{2},$
,, $\quad\quad\quad\quad\quad y=\infty, \quad x=4;$

∴ the curve cuts the axis of x at a distance $-3 \cdot 5$ from the origin, and meets the line $x=4$ at an infinite distance.

If x is positive and very little greater than 4, y is very great and positive. If x is positive and very little less than 4, y is very great and negative. Thus the infinite points on the graph near to the line $x=4$ have positive ordinates to the right, and negative ordinates to the left of this line.

(ii) When $\quad\quad\quad x=0, \quad y=-1\cdot 75,$
,, $\quad\quad\quad\quad\quad x=\infty, \quad y=2;$

∴ the curve cuts the axis of y at a distance $-1 \cdot 75$ from the origin, and meets the line $y=2$ at an infinite distance.

By taking positive values of y very little greater and very little less than 2, it appears that the curve lies above the line $y=2$ when $x=+\infty$, and below this line when $x=-\infty$.

The general character of the curve is now determined: the lines $PO'P'$ ($x=4$) and $QO'Q'$ ($y=2$) are asymptotes; the two branches of the curve lie in the compartments $PO'Q$, $P'O'Q'$, and the lower branch cuts the axes at distances $-3\cdot 5$ and $-1\cdot 75$ from the origin.

390   ALGEBRA.   [CHAP.

To examine the lower branch in detail values of x may be selected between $-\infty$ and $-3·5$ and between $-3·5$ and 4.

x	$-\infty$	...	-16	-8	-6	$-3·5$	-1	0	2	3	...	4
y	2	...	$1·25$	$·75$	$·5$	0	-1	$-1·75$	$-5·5$	-13	...	$-\infty$

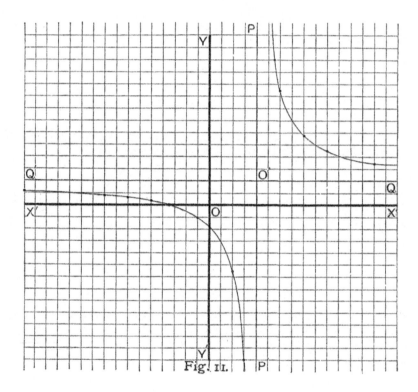

Fig. 11.

The upper branch may now be dealt with in the same way, selecting values of x between 4 and ∞. The graph will be found to be as represented in Fig. 11.

435. When the equation of a curve contains the square or higher power of y, the calculation of the values of y corresponding to selected values of x will have to be obtained by evolution, or else by the aid of logarithms. We give one example to illustrate the way in which a table of four-figure logarithms may be employed in such cases.

USE OF LOGARITHMS.

Example. Draw the graph of $y^3 = x(9 - x^2)$.

For the sake of brevity we shall confine our attention to that part of the curve which lies to the right of the axis of y, leaving the other half to be traced in like manner by the student.

When $x=0$, $y=0$: therefore the curve passes through the origin. Again, y is positive for all values of x between 0 and 3, and vanishes when $x=3$; for values of x greater than 3, y is negative and continually increases numerically.

x	0	1	2	3	4	5	6	...
x^2	0	1	4	9	16	25	36	...
$9 - x^2$	9	8	5	0	-7	-16	-27	...
y^3	0	8	10	0	-28	-80	-162	...
$\log y^3$			1		1·4472*	1·9031*	2·2095*	...
$\log y$			·3333		·4824	·6344	·7365	...
y	0	2	2·15	0	$-3·04$	$-4·31$	$-5·45$	...

These points will be sufficient to give a rough approximation to the curve. For greater accuracy a few intermediate values such as $x = 1·5$, $2·5$, $3·5$ ... should be taken, and the resulting curve will be as in Fig. 12, in which we have taken *two-tenths of an inch as our linear unit*.

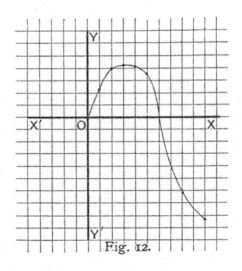

Fig. 12.

* In taking logarithms of the successive values of y^3, the negative sign is disregarded, but care must be taken to insert the proper signs in the last line which gives the successive values of y.

Measurement on Different Scales.

436 For convenience on the printed page we have supposed the paper to be ruled to tenths of an inch, generally using one of the divisions as our linear unit. In practice, however, it will often be advisable to choose a unit much larger than this in order to get a satisfactory graph. For the sake of simplicity we have hitherto measured abscissæ and ordinates on the same scale, but there is no necessity for so doing, and it will often be found convenient to measure the variables on different scales suggested by the particular conditions of the question.

As an illustration let us take the graph of $y=\dfrac{x^2}{2}$, given in Art. 425. If with the same unit as before we plot the graph of $y=x^2$, it will be found to be a curve similar to that drawn on page 11, but elongated in the direction of the axis of y. In fact, it will be the same as if the former graph were stretched to twice its length in the direction of the y-axis.

437. Any equation of the form $y=ax^2$, where a is constant, will represent a parabola elongated more or less according to the value of a; and the larger the value of a the more rapidly will y increase in comparison with x. We might have very large ordinates corresponding to very small abscissæ, and the graph might prove quite unsuitable for practical applications. In such a case the inconvenience is obviated by measuring the values of y on a considerably smaller scale than those of x.

Speaking generally, whenever one variable increases much more rapidly than the other, a small unit should be chosen for the rapidly increasing variable and a large one for the other. Further modifications will be suggested in the examples which follow.

438. On the opposite page we give for comparison the graphs of
$$y=x^2 \text{ (Fig. 13)}, \text{ and } y=8x^2 \text{ (Fig. 14)}.$$

In Fig. 13 the unit for x is twice as great as that for y.

In Fig. 14 the x-unit is ten times the y-unit.

It will be useful practice for the student to plot other similar graphs on the same or a larger scale. For example, in Fig. 14 the graphs of $y=16x^2$ and $y=2x^2$ may be drawn and compared with that of $y=8x^2$.

XLIV.] GRAPHS OF $y = x^2$ AND $y = 8x^2$. 393

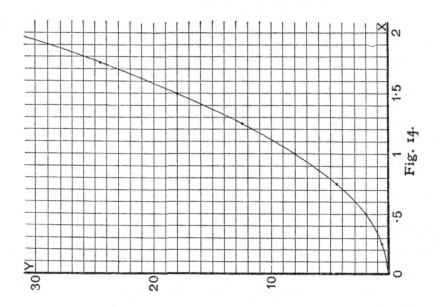

Fig. 14.

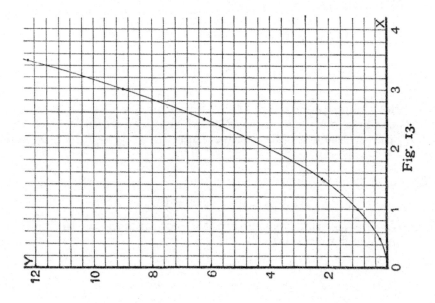

Fig. 13.

EXAMPLES XLIV. e.

1. Plot the graph of $y=x^3$. Shew that it consists of a continuous curve lying in the first and third quadrants, crossing the axis of x at the origin. Deduce the graphs of

(i) $y=-x^3$;   (ii) $y=\frac{1}{2}x^3$.

2. Plot the graph of $y=x-x^3$. Verify it from the graphs of $y=x$, and $y=x^3$.

3. Plot the graph of $y=\frac{1}{x^2}$, shewing that it consists of two branches lying entirely in the first and second quadrants. Examine and compare the nature and position of the graph as it approaches the axes.

4. Discuss the general character of the graph of $y=\frac{a}{x^2}$ where a has some constant integral value. Distinguish between two cases in which a has numerical values, equal in magnitude but opposite in sign.

5. Plot the graphs of

(i) $y=1+\frac{1}{x}$,   (ii) $y=2+\frac{10}{x^2}$.

Verify by deducing them from the graphs of $y=\frac{1}{x}$, and $y=\frac{10}{x^2}$.

6. Plot the graph of $y=x^3-3x$. Examine the character of the curve at the points $(1, -2)$, $(-1, 2)$, and shew graphically that the roots of the equation $x^3-3x=0$ are approximately $-1{\cdot}732$, 0, and $1{\cdot}732$.

7. Solve the equations:
$$3x+2y=16, \qquad xy=10,$$
and verify the solution by finding the coordinates of the points where their graphs intersect.

8. Plot the graphs of

(i) $y=\dfrac{15-x^2}{x}$,   (ii) $x=\dfrac{10-y^2}{y}$,

and thus verify the algebraical solution of the equations $x^2+xy=15$, $y^2+xy=10$.

9. Trace the curve whose equation is $y = \dfrac{x}{2-x}$, shewing that it has two branches, one lying in the first and third quadrants, and the other entirely in the fourth. Find the equations of its asymptotes.

Plot the graphs of

10. $y = \dfrac{1+x}{1-x}$.

11. $y = \dfrac{1+x^2}{1-x}$.

12. $y = \dfrac{x^2 - 15}{x - 4}$.

13. $y = \dfrac{(x-1)(x-2)}{x-3}$.

14. $y = \dfrac{x^2 + x + 1}{x^2 - x + 1}$.

15. $y = \dfrac{x^2 + 5x + 6}{x^2 + 1}$.

16. $y = x^3 - 6x^2 + 11x - 6$.

17. $10y = x^3 - 5x^2 + x - 5$.

18. $y = \dfrac{20}{x^2 + 2}$.

19. $y = \dfrac{40x}{x^2 + 10}$.

20. $y = \dfrac{x(8-x)}{x+5}$.

21. $y = \dfrac{(x-2)(x-3)}{x-5}$.

22. $y = \dfrac{(x-1)(x-2)(x+1)}{4}$.

23. $y^2 = x^2 - 5x + 4$.

24. $4y^2 = x^2(5-x)$.

25. $y^2 = \dfrac{x(3-x)(x-8)}{x^2 + 5}$.

26. $y^2 = \dfrac{(x+7)(x-4)(x-10)}{x^2 + 5}$.

27. $y^2 = \dfrac{x^2(49-x^2)}{50}$.

28. $y^2 = \dfrac{(81-x^2)(x^2-4)}{100}$.

29. $5y^3 = x(x^2 - 64)$.

30. $5y^3 = x^2(36 - x^2)$.

31. Plot the graphs of $y = x^3$, and of $y = 2x^2 + x - 2$. Hence find the roots of the equation $x^3 - 2x^2 - x + 2 = 0$.

32. Find graphically the roots of the equation
$$x^3 - 4x^2 - 5x + 14 = 0$$
to three significant figures.

439. Besides the instances already given there are several of the ordinary processes of Arithmetic and Algebra which lend themselves readily to graphical illustration.

For example, the graph of $y=x^2$ may be used to furnish numerical square roots. For since $x=\sqrt{y}$, each ordinate and corresponding abscissa give a number and its square root. Similarly cube roots may be found from the graph of $y=x^3$.

Example 1. Find graphically the cube root of 10 to 3 places of decimals.

The required root is clearly a little greater than 2. Hence it will be enough to plot the graph of $y=x^3$ taking $x=2\cdot1$, $2\cdot2$, ... The corresponding ordinates are $9\cdot26$, $10\cdot65$, ...

When $x=2$, $y=8$. Take the axes through this point and let the units for x and y be 10 inches and $\cdot5$ inch respectively. On this scale the portion of the graph differs but little from a straight line, and yields results to a high degree of accuracy.

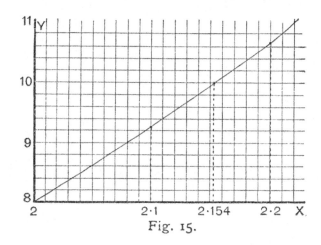

Fig. 15.

When $y=10$, the measured value of x will be found to be $2\cdot154$.

Example 2. Shew graphically that the expression $4x^2+4x-3$ is negative for all real values of x between $\cdot5$ and $-1\cdot5$, and positive for all real values of x outside these limits. [Fig. 16.]

Put $y=4x^2+4x-3$, and proceed as in the example given in Art. 16, taking the unit for x four times as great as that for y. It will be found that the graph cuts the axis of x at points whose abscissæ are $\cdot5$ and $-1\cdot5$; and that it lies below the axis of x between these points. That is, the value of y is negative so long as x lies between $\cdot5$ and $-1\cdot5$, and positive for all other values of x.

XLIV.] ILLUSTRATIVE EXAMPLES. 397

Or we may proceed as follows:

Put $y_1 = 4x^2$, and $y_2 = -4x+3$, and plot the graphs of these two equations. At their points of intersection $y_1 = y_2$, and the values of x at these points are found to be $\cdot 5$ and $-1\cdot 5$. Hence for these values of x we have

$$4x^2 = -4x+3, \quad \text{or} \quad 4x^2+4x-3=0.$$

Thus the roots of the equation $4x^2+4x-3=0$ are furnished by the abscissæ of the common points of the graphs of $4x^2$ and $-4x+3$.

Again, between the values $\cdot 5$ and $-1\cdot 5$ for x it will be found graphically that y_1 is less than y_2, hence $y_1 - y_2$, or $4x^2+4x-3$ is negative.

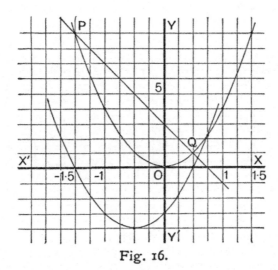

Fig. 16.

Both solutions are here exhibited.

The upper curve is the graph of $y = 4x^2$; PQ is the graph of $y = -4x+3$; and the lower curve is the graph of $y = 4x^2+4x-3$.

440. Of the two methods in the last Example the first is the more direct and instructive; but the second has this advantage:

If a number of equations of the form $x^2 = px+q$ have to be solved graphically, $y = x^2$ can be plotted once for all on a convenient scale, and $y = px+q$ can then be readily drawn for different values of p and q.

Equations of higher degree may be treated similarly.

E.A. 2 D

For example, the solution of such equations as
$$x^3 = px + q, \quad \text{or} \quad x^3 = ax^2 + bx + c$$
can be made to depend on the intersection of $y = x^3$ with other graphs.

Example. Find the real roots of the equations

(i) $x^3 - 2\cdot 5x - 3 = 0$;   (ii) $x^3 - 3x + 2 = 0$.

Here we have to find the points of intersection of

(i) $y = x^3$,   (ii) $y = x^3$,
$y = 2\cdot 5x + 3$; $y = 3x - 2$.

Plot the graphs of these equations, choosing the unit for x five times as great as that for y.

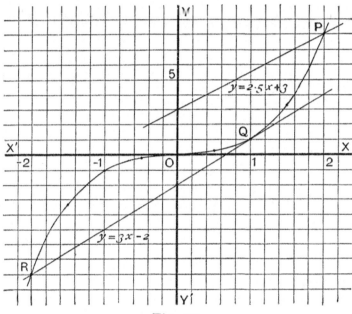

Fig. 17.

It will be seen that $y = 2\cdot 5x + 3$ meets $y = x^3$ only at the point for which $x = 2$. Thus 2 is the only real root of equation (i).

Again $y = 3x - 2$ *touches* $y = x^3$ at the point for which $x = 1$, and cuts it where $x = -2$.

Corresponding to the former point the equation $x^3 - 3x + 2 = 0$ has two equal roots. Thus the roots of (ii) are 1, 1, -2.

XLIV.] SIMULTANEOUS QUADRATICS. 399

441. In Art. 421 we have given the graphical solution of two *linear* simultaneous equations. As the principle is the same for equations of any degree, the few examples of this kind on pages 385, 392 have been given without special explanation. It may, however, be instructive here to shew the graphical solution of some of the equations discussed in Chap. XXVI.

Example. Solve the following equations graphically:

(i) $\left.\begin{array}{r}x-y=2\\xy=35\end{array}\right\}.$ (ii) $\left.\begin{array}{r}x^2+y^2=74\\xy=35\end{array}\right\}.$

(Compare Art. 203, Ex. 2.)   (Compare Art. 204, Ex. 1.)

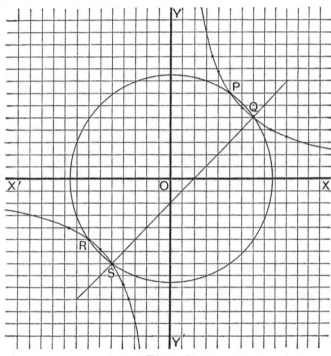

Fig. 18.

Here $xy=35$ is represented by a rectangular hyperbola [Art. 431]; $x-y=2$ is the line QS, and $x^2+y^2=74$ is represented by the circle.

The roots of (i) are the coordinates of Q and S; that is,
$$x=7,\ y=5;\ \text{ or }\ x=-5,\ y=-7.$$
The roots of (ii) are the coordinates of P, Q, R, and S; that is,
$$x=5,\ y=7;\ \ x=7,\ y=5;\ \ x=-7,\ y=-5;\ \ x=-5,\ y=-7.$$

EXAMPLES XLIV. f.

1. Draw the graph of $y=x^2$ on a scale twice as large as that in Fig. 13, and employ it to find the squares of $\cdot72$, $1\cdot7$, $3\cdot4$; and the square roots of $7\cdot56$, $5\cdot29$, $9\cdot61$.

2. Draw the graph of $y=\sqrt{x}$ taking the unit for y five times as great as that for x.

By means of this curve check the values of the square roots found in Example 1.

3. From the graph of $y=x^3$ (on the scale of the diagram of Art. 29) find the values of $\sqrt[3]{9}$ and $\sqrt[3]{9\cdot8}$ to 4 significant figures.

4. A boy who was ignorant of the rule for cube root required the value of $\sqrt[3]{14\cdot71}$. He plotted the graph of $y=x^3$, using for x the values $2\cdot2$, $2\cdot3$, $2\cdot4$, $2\cdot5$, and found $2\cdot45$ as the value of the cube root. Verify this process in detail. From the same graph find the value of $\sqrt[3]{13\cdot8}$.

5. Find graphically the values of x for which the expression x^2-2x-8 vanishes. Shew that for values of x between these limits the expression is negative and for all other values positive. Find the least value of the expression.

6. From the graph in the preceding example shew that for any value of a greater than 1 the equation $x^2-2x+a=0$ cannot have real roots.

7. Shew graphically that the expression x^2-4x+7 is positive for all real values of x.

8. On the same axes draw the graphs of
$$y=x^2, \quad y=x+6, \quad y=x-6, \quad y=-x+6, \quad y=-x-6.$$
Hence discuss the roots of the four equations
$$x^2-x-6=0, \quad x^2-x+6=0, \quad x^2+x-6=0, \quad x^2+x+6=0.$$

9. If x is real, prove graphically that $5-4x-x^2$ is not greater than 9; and that $4x^2-4x+3$ is not less than 2. Between what values of x is the first expression positive?

10. Solve the equation $x^3=3x^2+6x-8$ graphically, and shew that the function x^3-3x^2-6x+8 is positive for all values of x between -2 and 1, and negative for all values of x between 1 and 4.

11. Shew graphically that the equation $x^3+px+q=0$ has only one real root when p is positive.

12. Trace the curve whose equation is $y=2^x$. Find the approximate values of $2^{4\cdot75}$ and $2^{5\cdot25}$. Express 12 as a power of 2 approximately.

Prove also that $\log_2 26\cdot9 + \log_2 38 = 10$.

13. By repeated evolution find the values of $10^{\frac{1}{2}}, 10^{\frac{1}{4}}, 10^{\frac{1}{8}}, 10^{\frac{1}{16}}$. By multiplication find the values of $10^{\frac{3}{16}}, 10^{\frac{5}{16}}, 10^{\frac{6}{16}}, 10^{\frac{7}{16}}, 10^{\frac{9}{16}}$. Use these values to plot a portion of the curve $y=10^x$ on a large scale. Find correct to three places of decimals the values of $\log 3$, $\log 1\cdot68$, $\log 2\cdot24$, $\log 34\cdot3$. Also by choosing numerical values for a and b, verify the laws

$$\log ab = \log a + \log b; \quad \log \frac{a}{b} = \log a - \log b.$$

[*By using paper ruled to tenths of an inch, if 10 in. and 1 in. be taken as units for x and y respectively, a diagonal scale will give values of x correct to three decimal places and values of y correct to two.*]

14. Calculate the values of $x(9-x)^2$ for the values $0, 1, 2, 3, \ldots 9$ of x. Draw the graph of $x(9-x)^2$ from $x=0$ to $x=9$.

If a very thin elastic rod, 9 inches in length, fixed at one end, swings like a pendulum, the expression $x(9-x)^2$ measures the tendency of the rod to break at a place x inches from the point of suspension. From the graph find where the rod is most likely to break.

15. If a man spends $22s.$ a year on tea whatever the price of tea is, what amounts will he receive when the price is 12, 16, 18, 20, 24, 28, 33, and 36 pence respectively? Give your results to the nearest quarter of a pound. Draw a curve to the scale of 4 lbs. to the inch and 10 pence to the inch, to shew the number of pounds that he would receive at intermediate prices.

16. The reciprocal of a number is multiplied by $2\cdot25$ and the product is added to the number. Find graphically what the number must be if the resulting expression has the least possible value.

17. Shew graphically that the expression $4x^2 + 2x - 8\cdot75$ is positive for all real values of x except such as lie between $1\cdot25$ and $-1\cdot75$. For what value of x is the expression a minimum?

18. Find graphically the real roots of the equations:

(i) $x^3 + x - 2 = 0$. (ii) $x^3 - 7x + 6 = 0$.

19. Draw the graphs of

$$x+y=9\tfrac{1}{2}, \quad xy=12, \quad x^2-y^2=32,$$

on the same axes. Hence find the solutions of the following pairs of simultaneous equations:

(i) $\left.\begin{array}{l} x+y=9\tfrac{1}{2} \\ xy=12 \end{array}\right\}.$ (ii) $\left.\begin{array}{l} x^2-y^2=32 \\ x+y=9\tfrac{1}{2} \end{array}\right\}.$ (iii) $\left.\begin{array}{l} x^2-y^2=32 \\ xy=12 \end{array}\right\}.$

402   ALGEBRA.   [CHAP.

20. Draw the graphs of $y=x^3$ and $y=3x^2-4$ on the same axes, and find the roots of the equation $x^3-3x^2+4=0$.

Shew that the expression x^3-3x^2+4 is negative for values of x less than -1, and positive for all other values of x.

21. From a graphical consideration of the following pairs of simultaneous equations:

(i) $x^2+y^2=a$,   (ii) $x+y=a$,
    $xy=b$, $xy=b$,

explain why (i) has either *four* solutions or none, while (ii) has *two* solutions or none.

22. Draw the graphs of $y=x^3$ and $y=x^2+3x-3$ on the same axes.

Hence find the roots of the equation $x^3-x^2-3x+3=0$ to three places of decimals, and discuss the sign of the expression x^3-x^2-3x+3 for different values of x.

Practical Applications.

442. In all the cases hitherto considered the equation of the curve has been given, and its graph has been drawn by first selecting values of x and y which satisfy the equation, and then drawing a line so as to pass through the plotted points. We thus determine accurately the position of as many points as we please, and the process employed assures us that they all lie on the graph we are seeking. We could obtain the same result without knowing the equation of the curve provided that we were furnished with a sufficient number of corresponding values of the variables *accurately calculated*.

Sometimes from the nature of the case the form of the equation which connects two variables is known. For example, if a quantity y is directly proportional to another quantity x it is evident that we may put $y=ax$, where a is some constant quantity. Hence in all cases of direct proportionality between two quantities the graph which exhibits their variations is a straight line through the origin. Also since two points are sufficient to determine a straight line, it follows that in the cases under consideration we only require to know the position of one point besides the origin, and this will be furnished by any pair of simultaneous values of the variables.

Example 1. Given that 5·5 kilograms are roughly equal to 12·125 pounds, shew graphically how to express any number of pounds in kilograms. Express $7\frac{1}{2}$ lbs. in kilograms, and $4\frac{1}{4}$ kilograms in pounds.

Here measuring pounds horizontally and kilograms vertically, the required graph is obtained at once by joining the origin to the point whose coordinates are 12·125 and 5·5.

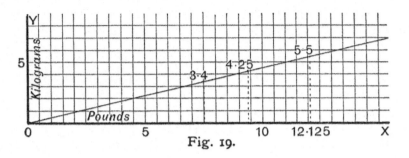

Fig. 19.

By measurement it will be found that 7½ lbs. = 3·4 kilograms, and 4¼ kilograms = 9·37 lbs.

Example 2. The expenses of a school are partly constant and partly proportional to the number of boys. The expenses were £650 for 105 boys, and £742 for 128. Draw a graph to represent the expenses for any number of boys; find the expenses for 115 boys, and the number of boys that can be maintained at a cost of £710.

If the expenses for x boys are represented by £y, it is evident that x and y satisfy a linear equation $y = ax + b$, where a and b are constants. Hence the graph is a straight line.

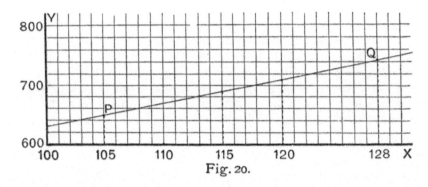

Fig. 20.

As the numbers are large, it will be convenient if we begin measuring ordinates at 600, and abscissæ at 100. This enables us to bring the requisite portion of the graph into a smaller compass. The points P and Q are determined by the data of the question, and the line PQ is the graph required.

By measurement we find that when $x = 115$, $y = 690$; and that when $y = 710$, $x = 120$. Thus the required answers are £690, and 120 boys.

443. Sometimes corresponding values of two variables are obtained by observation or experiment. In such cases the data cannot be regarded as free from error; the position of the plotted points cannot be absolutely relied on; and we cannot correct irregularities in the graph by plotting other points selected at discretion. All we can do is to draw a curve to lie as evenly as possible among the plotted points, passing through some perhaps, and with the rest fairly distributed on either side of the curve. As an aid to drawing an even continuous curve a thin piece of wood or other flexible material may be bent into the requisite curve, and held in position while the line is drawn.* When the plotted points lie approximately on a straight line, the simplest plan is to use a piece of tracing paper or celluloid on which a straight line has been drawn. When this has been placed in the right position the extremities can be marked on the squared paper, and by joining these points the approximate graph is obtained.

Example 1. The following table gives statistics of the population of a certain country, where P is the number of millions at the beginning of each of the years specified.

Year	1830	1835	1840	1850	1860	1865	1870	1880
P	20	22·1	23·5	29·0	34·2	38·2	41·0	49·4

Let t be the time in years from 1830. Plot the values of P vertically and those of t horizontally and exhibit the relation between P and t by a simple curve passing fairly evenly among the plotted points. Find what the population was at the beginning of the years 1848 and 1875.

The graph is given in Fig. 21 on the opposite page. The populations in 1848 and 1875, at the points A and B respectively, will be found to be 27·8 millions and 45·3 millions.

Example 2. Corresponding values of x and y are given in the following table:

x	1	4	6·8	8	9·5	12	14·4
y	4	8	12·2	13	14·8	20	24·8

Supposing these values to involve errors of observation, draw the graph approximately and determine the most probable equation between x and y. [See Fig. 22 on p. 406.]

* One of "Brooks' Flexible Curves" will be found very useful.

PRACTICAL APPLICATIONS.

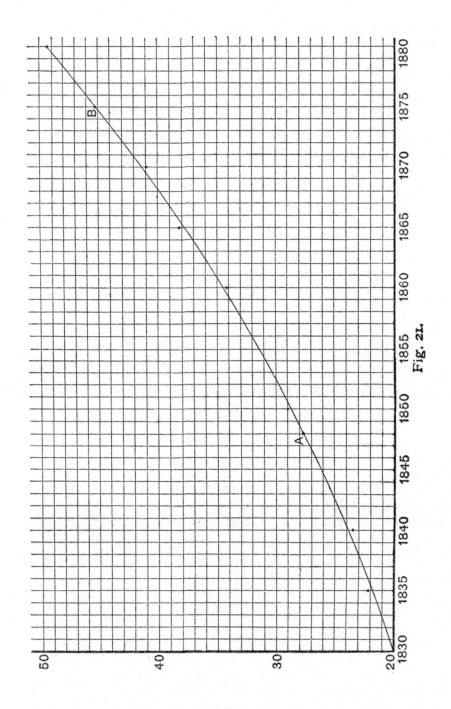

Fig. 21.

After carefully plotting the given points we see that a straight line can be drawn passing through three of them and lying evenly among the others. This is the required graph.

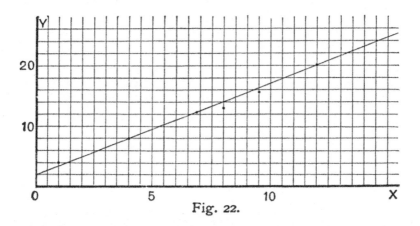

Fig. 22.

Assuming $y = ax + b$ for its equation, we find the values of a and b by selecting two pairs of simultaneous values of x and y.

Thus substituting $x = 4$, $y = 8$, and $x = 12$, $y = 20$ in the equation, we obtain $a = 1\cdot5$, $b = 2$. Thus the equation of the graph is $y = 1\cdot5x + 2$.

444. In the last Example as the graph is linear it can be produced to any extent within the limits of the paper, and so any value of one of the variables being determined, the corresponding value of the other can be read off. When large values are in question this method is not only inconvenient but unsafe, owing to the fact that any divergence from accuracy in the portion of the graph drawn is increased when the curve is produced beyond the limits of the plotted points. The following Example illustrates the method of procedure in such cases.

Example. In a certain machine P is the force in pounds required to raise a weight of W pounds. The following corresponding values of P and W were obtained experimentally:

P	2·48*	3·9	6·8	8·8	9·2	11*	13·3
W	21	36·25	66·2	87·5	103·75	120	152·5

By plotting these values on squared paper draw the graph connecting P and W, and read off the value of P when $W = 70$. Also determine a linear law connecting P and W; find the force necessary to raise a weight of 310 lbs., and also the weight which could be raised by a force of 180·6 lbs.

As the page is too small to exhibit the graphical work on a convenient scale we shall merely indicate the steps of the solution, which is similar in detail to that of the last example.

Plot the values of P vertically and the values of W horizontally. It will be found that a straight line can be drawn through the points corresponding to the results marked with an asterisk, and lying evenly among the other points. From this graph we find that when $W=70$, $P=7$.

Assume $P=aW+b$, and substitute for P and W from the values corresponding to the two points through which the line passes. By solving the resulting equations we obtain $a=·08$, $b=1·4$. Thus the linear equation connecting P and W is $P=·08W+1·4$.

This is called the **Law of the Machine.**

From this equation, when $W=310$, $P=26·2$, and when $P=180·6$, $W=2240$.

Thus a force of 26·2 lbs. will raise a weight of 310 lbs.; and when a force of 180·6 lbs. is applied the weight raised is 2240 lbs. or 1 ton.

Note. The equation of the graph is not only useful for determining results difficult to obtain graphically, but it can always be used to check results found by measurement.

445. The example in the last article is a simple illustration of a method of procedure which is common in the laboratory or workshop, the object being to determine the law connecting two variables when a certain number of simultaneous values have been determined by experiment or observation.

Though we can always draw a graph to lie fairly among the plotted points corresponding to the observed values, unless the graph is a straight line it may be difficult to find its equation except by some indirect method.

For example, suppose x and y are quantities which satisfy an equation of the form $xy=ax+by$, and that this law has to be discovered.

By writing the equation in the form

$$\frac{a}{y}+\frac{b}{x}=1, \text{ or } au+bv=1;$$

where $u=\dfrac{1}{y}$, $v=\dfrac{1}{x}$, it is clear that u, v satisfy the equation of a straight line. In other words, if we were to plot the points corresponding to the reciprocals of the given values, their linear connection would be at once apparent. Hence the values of a and b could be found as in previous examples, and the required law in the form $xy=ax+by$ could be determined.

Again, suppose x and y satisfy an equation of the form $x^n y = c$, where n and c are constants.

By taking logarithms, we have
$$n \log x + \log y = \log c.$$

The form of this equation shews that $\log x$ and $\log y$ satisfy the equation to a straight line. If, therefore, the values of $\log x$ and $\log y$ are plotted, a linear graph can be drawn, and the constants n and c can be found as before.

Example. The weight, y grammes, necessary to produce a given deflection in the middle of a beam supported at two points, x centimetres apart, is determined experimentally for a number of values of x with results given in the following table:

x	50	60	70	80	90	100
y	270	150	100	60	47	32

Assuming that x and y are connected by the equation $x^n y = c$, find n and c.

From pages 348_D, 348_E we obtain the annexed values of $\log x$ and $\log y$ corresponding to the observed values of x and y. By plotting these we obtain the graph given in Fig. 23, and its equation is of the form
$$n \log x + \log y = \log c.$$

$\log x$	$\log y$
1·699	2·431
1·778	2·176
1·845	2·000
1·903	1·778
1·954	1·672
2·000	1·519

To obtain n and c, choose *two extreme points through which the line passes.* It will be found that when
$$\log x = 1·642, \quad \log y = 2·6$$
and when
$$\log x = 2·1, \quad \log y = 1·21.$$

Substituting these values, we have
$$2·6 + n \times 1·642 = \log c \quad \text{...............................(i)},$$
$$1·21 + n \times 2·1 = \log c \quad \text{...............................(ii)};$$
$$\therefore \ 1·39 - 0·458 n = 0;$$
whence
$$n = 3·04.$$
$\therefore$ from (ii) $\log c = 6·38 + 1·21$
$$= 7·59;$$
$\therefore c = 39 \times 10^6$, from the tables.

Thus the required equation is $x^3 y = 39 \times 10^6$.

The student should work through this example in detail on a larger scale. The adjoining figure was drawn on paper ruled to tenths of an inch and then reduced to half the original scale.

PRACTICAL APPLICATIONS.

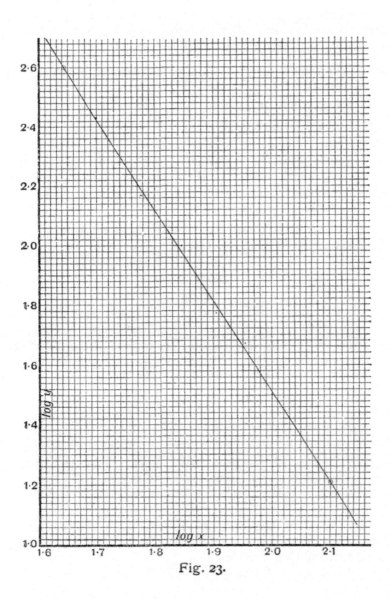

Fig. 23.

EXAMPLES XLIV. g.

1. Given that $6{\cdot}01$ yards $= 5{\cdot}5$ metres, draw the graph shewing the equivalent of any number of yards when expressed in metres.

Shew that $22{\cdot}2$ yards $= 20{\cdot}3$ metres approximately.

2. Draw a graph shewing the relation between equal weights in grains and grammes, having given that $10{\cdot}8$ grains $= 1{\cdot}17$ grammes.

Express (i) $3{\cdot}5$ grammes in grains.

(ii) $3{\cdot}09$ grains as a decimal of a gramme.

3. If $3{\cdot}26$ inches are equivalent to $8{\cdot}28$ centimetres, shew how to determine graphically the number of inches corresponding to a given number of centimetres. Obtain the number of inches in a metre, and the number of centimetres in a yard. What is the equation of the graph?

4. The following table gives approximately the circumferences of circles corresponding to different radii:

C	15·7	20·1	31·4	44	52·2
r	2·5	3·2	5	7	8·3

Plot the values on squared paper, and from the graph determine the diameter of a circle whose circumference is $12{\cdot}1$ inches and the circumference of a circle whose radius is $2{\cdot}8$ inches.

5. For a given temperature, C degrees on a Centigrade are equal to F degrees on a Fahrenheit thermometer. The following table gives a series of corresponding values of F and C:

C	−10	−5	0	5	10	15	25	40
F	14	23	32	41	50	59	77	104

Draw a graph to shew the Fahrenheit reading corresponding to a given Centigrade temperature, and find the Fahrenheit readings corresponding to $12{\cdot}5°$ C. and $31°$ C.

By observing the form of the graph find the algebraical relation between F and C.

6. For a certain book it costs a publisher £100 to prepare the type and 2s. to print each copy. Find an expression for the total cost in pounds of x copies. Make a diagram on a scale of 1 inch to 1000 copies, and 1 inch to £100 to shew the total cost of any number of copies up to 5000. Read off the cost of 2500 copies, and the number of copies costing £525.

7. At different ages the mean after-lifetime ("expectation of life") of males, calculated on the death rates of 1871-1880, was given by the following table:

Age	6	10	14	18	22	26	27
Expectation	50·38	47·60	44·26	40·96	37·89	34·96	34·24

Draw a graph to shew the expectation of any male between the ages of 6 and 27, and from it determine the expectation of persons aged 12 and 20.

8. In the Clergy Mutual Assurance Society the premium ($£P$) to insure £100 at different ages is given approximately by the following table:

Age	20	22	25	30	35	40	45	50	55
P	1·8	1·9	2·0	2·3	2·7	3·1	3·6	4·4	5·5

Illustrate the same statistics graphically, and estimate to the nearest shilling the premiums for persons aged 34 and 43.

9. If W is the weight in ounces required to stretch an elastic string till its length is l inches, plot the following values of W and l:

W	2·5	3·75	6·25	7·5	10	11·25
l	8·5	8·7	9·1	9·3	9·7	9·9

From the graph determine the unstretched length of the string, and the weight the string will support when its length is 1 foot.

10. In the following table P and A (expressed in hundreds of pounds) represent the Principal and corresponding Amount for 1 year at 3 per cent. simple interest.

P	2·3	2·7	3·0	3·5	3·9	5·2	7·6
A	2·369	2·781	3·090	3·605	4·017	5·356	7·828

Plot the values of P and A on a large scale, and from the graph determine the Principal which will amount to (i) £329. 12s.; (ii) £597. 8s.

11. The highest and lowest marks gained in an examination are 297 and 132 respectively. These have to be reduced in such a way that the maximum for the paper (200) shall be given to the first candidate, and that there shall be a range of 150 marks between the first and last. Find the equation between x, the actual marks gained, and y, the corresponding marks when reduced.

Draw the graph of this equation, and read off the marks which should be given to candidates who gained 200, 262, 163 marks in the examination.

12. A body starting with an initial velocity, and subject to an acceleration in the direction of motion, has a velocity of v feet per second after t seconds. If corresponding values of v and t are given by the annexed table,

v	9	13	17	21	25	29	33	37	41	45
t	1	2	3	4	5	6	7	8	9	10

plot the graph exhibiting the velocity at any given time. Find from it (i) the initial velocity, (ii) the time which has elapsed when the velocity is 28 feet per second. Also find the equation between v and t.

13. The connection between the areas of equilateral triangles and their bases (in corresponding units) is given by the following table:

Area	·43	1·73	3·90	6·93	10·82	15·59
Base	1	2	3	4	5	6

Illustrate these results graphically, and determine the area of an equilateral triangle on a base of 2·4 ft.

14. A body falling freely under gravity drops s feet in t seconds from the time of starting. If corresponding values of s and t at intervals of half a second are as follows:

t	·5	1	1·5	2	2·5	3	3·5	4
s	4	16	36	64	100	144	196	256

draw the curve connecting s and t, and find from it

(i) the distance through which the body has fallen after 1·8″.

(ii) the depth of a well if a stone takes 3·16″ to reach the bottom.

15. A body is projected with a given velocity at a given angle to the horizon, and the height in feet reached after t seconds is given by the equation $h = 64t - 16t^2$. Find the values of h at intervals of $\frac{1}{4}$th of a second and draw the path described by the body. Find the maximum value of h, and the time after projection before the body reaches the ground.

16. The keeper of a hotel finds that when he has G guests a day his total daily profit is P pounds. If the following numbers are averages obtained by comparison of many days' accounts determine a simple relation between P and G.

G	21	27	29	32	35
P	−1·8	2	3·2	4·5	6·6

For what number of guests would he just have no profit?

17. A man wishes to place in his catalogue a list of a certain class of fishing rods varying from 9 ft. to 16 ft. in length. Four sizes have been made at prices given in the following table:

9 ft.	11 ft. 9 in.	14 ft. 4 in.	16 ft.
15s.	22s.	31s.	38s.

Draw a graph to exhibit prices for rods of intermediate lengths, and from it determine the probable prices for rods of 13 ft. and 15 ft. 8 in.

18. The following table gives the sun's position at 7 A.M. on different dates:

Mar. 23	Ap. 3	Ap. 20	May 8	May 27	June 22	July 18	Aug. 5	Aug. 25
80° E.	82° E.	85° E.	89° E.	92° E.	95° E.	94° E.	91° E.	85° E.

Shew these results graphically, and estimate approximately the sun's position at the same hour on June 8th.

19. At a given temperature p lbs. per square inch represents the pressure of a gas which occupies a volume of v cubic inches. Draw a curve connecting p and v from the following table of corresponding values:

p	36	30	25·7	22·5	20	18	16·4	15
v	5	6	7	8	9	10	11	12

20. Plot on squared paper the following measured values of x and y, and determine the most probable equation between x and y:

x	3	5	8·3	11	13	15·5	18·6	23	28
y	2	2·2	3·4	3·8	4	4·6	5·4	6·2	7·25

21. Corresponding values of x and y are given in the following table:

x	1	3·1	6	9·5	12·5	16	19	23
y	2	2·8	4·2	5·3	6·6	8·3	9	10·8

Supposing these values to involve errors of observation, draw the graph approximately, and determine the most probable equation between x and y. Find the correct value of y when $x=19$, and the correct value of x when $y=2\cdot 8$.

22. The following corresponding values of x and y were obtained experimentally:

x	0·5	1·7	3·0	4·7	5·7	7·1	8·7	9·9	10·6	11·8
y	148	186	265	326	388	436	529	562	611	652

It is known that they are connected by an equation of the form $y=ax+b$, but the values of x and y involve errors of measurement. Find the most probable values of a and b, and estimate the error in the measured value of y when $x=9\cdot 9$.

23. In a certain machine P is the force in pounds required to raise a weight of W pounds. The following corresponding values of P and W were obtained experimentally:

P	2·8	3·7	4·8	5·5	6·5	7·3	8	9·5	10·4	11·75
W	20	25	31·7	35·6	45	52·4	57·5	65	71	82·5

Draw the graph connecting P and W, and read off the value of P when $W=60$. Also determine the law of the machine, and find from it the weight which could be raised by a force of 31·7 lbs.

24. The following values of x and y, some of which are slightly inaccurate, are connected by an equation of the form $y = ax^2 + b$.

x	1	1·6	3	3·7	4	5	5·7	6	6·3	7
y	3·25	4	5	6·5	7·4	9·25	10·5	11·6	14	15·25

By plotting these values draw the graph, and find the most probable values of a and b.

Find the true value of x when $y = 4$, and the true value of y when $x = 6$.

25. The following table gives corresponding values of two variables x and y:

x	2·75	3	3·2	3·5	4·3	4·5	5·3	6	7	8	10
y	11	9·8	8	6·5	6·1	5·4	5	4·3	4·1	4	3·9

These values involve errors of observation, but the true values are known to satisfy an equation of the form $xy = ax + by$. Draw the graph by plotting the points determined by the above table, and find the most probable values of a and b. Find the correct values of y corresponding to $x = 3·5$, and $x = 7$.

26. Observed values of x and y are given as follows:

x	100	90	70	60	50	40
y	30	31·08	33·5	35·56	37·8	40·7

Assuming that x and y are connected by an equation of the form $xy^n = c$, find n and c.

27. The following values of x and y involve errors of observation:

x	66·83	63·10	58·88	51·52	48·53	44·16	40·36
y	144·5	158·5	177·8	208·9	236·0	264·9	309·0

If x and y satisfy an equation of the form $x^n y = c$, find n and c.

MISCELLANEOUS EXAMPLES VI.

1. Simplify $b - \{b - (a+b) - [b - (b - \overline{a-b})] + 2a\}$.

2. Find the sum of
$$a + b - 2(c+d), \quad b + c - 3(d+a), \quad \text{and} \quad c + d - 4(a+b).$$

3. Multiply $\frac{1}{2}x + \frac{2}{3}y$ by $x - \frac{1}{3}y$.

4. If $x = 6$, $y = 4$, $z = 3$, find the value of $\sqrt[3]{2x + 3y + z}$.

5. Find the square of $2 - 3x + x^2$.

6. Solve $\dfrac{x+3}{x-1} + \dfrac{x-4}{x-6} = 2$.

7. Find the H.C.F. of $a^3 - 2a - 4$ and $a^3 - a^2 - 4$.

8. Simplify $\dfrac{2a}{a+b} + \dfrac{2b}{a-b} - \dfrac{a^2 + b^2}{a^2 - b^2}$.

9. Solve
$$\left. \begin{array}{l} \dfrac{3}{5}x + \dfrac{y}{4} = 13 \\[2pt] \dfrac{1}{3}x - \dfrac{y}{8} = 3 \end{array} \right\}.$$

10. Two digits, which form a number, change places when 18 is added to the number, and the sum of the two numbers thus formed is 44 : find the digits.

11. If $a = 1$, $b = -2$, $c = 3$, $d = -4$, find the value of
$$\dfrac{a^2 b^2 + b^2 c + d(a-b)}{10a - (c+b)^2}.$$

12. Subtract $-x^2 + y^2 - z^2$ from the sum of
$$\tfrac{1}{3}x^2 + \tfrac{1}{4}y^2, \quad \tfrac{1}{5}y^2 + \tfrac{1}{3}z^2, \quad \text{and} \quad \tfrac{1}{3}z^2 - \tfrac{1}{4}x^2.$$

13. Write down the cube of $x + 8y$.

14. Simplify $\dfrac{x^2 + xy}{x^2 + y^2} \times \dfrac{x^4 - y^4}{xy + y^2} \times \dfrac{y}{x}$.

15. Solve $\dfrac{3}{5}(2x - 7) - \dfrac{2}{3}(x - 8) = \dfrac{4x + 1}{15} + 4$.

16. Find the H.C.F. and L.C.M. of
$$x^4 + x^3 + 2x - 4 \quad \text{and} \quad x^3 + 3x^2 - 4.$$

MISCELLANEOUS EXAMPLES. VI.

17. Find the square root of $4a^4 + 9(1-2a) + 3a^2(7-4a)$.

18. Solve
$$\left. \begin{array}{c} y = \dfrac{x+a}{2} + \dfrac{b}{3} \\ x = \dfrac{y+b}{2} + \dfrac{a}{3} \end{array} \right\}.$$

19. Simplify $\left(\dfrac{a}{x+a} - \dfrac{x}{x-a} \right) \div \dfrac{x^2 + a^2}{x^2 + ax}$.

20. When 1 is added to the numerator and denominator of a certain fraction the result is equal to $\dfrac{3}{2}$; and when 1 is subtracted from its numerator and denominator the result is equal to 2: find the fraction.

21. Shew that the sum of $12a + 6b - c$, $-7a - b + c$, and $a + b + 6c$, is six times the sum of $25a + 13b - 8c$, $-13a - 13b - c$, and $-11a + b + 10c$.

22. Divide $x^2 - xy + \dfrac{3}{16}y^2$ by $x - \dfrac{1}{4}y$.

23. Add together $18 \left\{ \dfrac{2x}{9} - \dfrac{1}{6} \left(\dfrac{2y}{3} + z \right) \right\}$,
$24 \left(\dfrac{3x}{8} - \dfrac{2y - 3z}{12} \right)$, and $30 \left\{ \dfrac{7z}{15} - \dfrac{4}{5}(2x - y) \right\}$.

24. Find the factors of
   (1) $10x^2 + 79x - 8$.   (2) $729x^6 - y^6$.

25. Solve $\dfrac{2x-1}{5} + \dfrac{5x+3}{17} = 3 - \dfrac{4x - 118}{11}$.

26. Find the value of
$(5a - 3b)(a - b) - b\{3a - c(4a - b) - b^2(a + c)\}$,
when $a = 0$, $b = -1$, $c = \dfrac{1}{2}$.

27. Find the H.C.F. of
$7x^3 - 10x^2 - 7x + 10$ and $2x^3 - x^2 - 2x + 1$.

28. Simplify $\dfrac{x^2 - 7xy + 12y^2}{x^2 + 5xy + 6y^2} \div \dfrac{x^2 - 5xy + 4y^2}{x^2 + xy - 2y^2}$.

29. Solve
$$\left. \begin{array}{c} 3abx + y = 9b \\ 4abx + 3y = 17b \end{array} \right\}.$$

30. Find the two times between 7 and 8 o'clock when the hands of a watch are separated by 15 minutes.

31. If $a=1$, $b=-2$, $c=3$, $d=-4$, find the value of
$$\sqrt{d^2-4b+a^2}-\sqrt{c^3+b^3+a+d}.$$

32. Multiply the product of $\frac{1}{4}x^2-\frac{1}{2}xy+y^2$ and $\frac{1}{2}x+y$ by x^3-8y^3.

33. Simplify by removing brackets
$$a^4-\{4a^3-(6a^2-4a+1)\}$$
$$-[-2-\{a^4-(-4a^3-\overline{6a^2-4a})\}-(8a-1)].$$

34. Find the remainder when $5x^4-7x^3+3x^2-x+8$ is divided by $x-4$.

35. Simplify $\dfrac{x^2+y^2}{x^2-xy} \times \dfrac{xy-y^2}{x^4-y^4} \times \dfrac{x}{y}$.

36. Solve
$$\left.\begin{array}{r}\dfrac{x-11}{3}+y=18\\2x+\dfrac{y-13}{4}=29\end{array}\right\}.$$

37. Find the square root of $4x^6-12x^4+28x^3+9x^2-42x+49$.

38. Solve $\cdot006x - \cdot491 + \cdot723x = -\cdot005$.

39. Find the L.C.M. of x^3+y^3, $3x^2+2xy-y^2$, and $x^3-x^2y+xy^2$.

40. A bill of 25 guineas is paid with crowns and half-guineas, and twice the number of half-guineas exceeds three times that of the crowns by 17: how many of each are used?

41. Simplify
$$(a+b+c)^2-(a-b+c)^2+(a+b-c)^2-(-a+b+c)^2.$$

42. Find the remainder when $a^4-3a^3b+2a^2b^2-b^4$ is divided by $a^2-ab+2b^2$.

43. If $a=0$, $b=1$, $c=-2$, $d=3$, find the value of
$$(3abc-2bcd)\sqrt[3]{a^3bc-c^3bd+3}.$$

44. Find an expression which will divide both $4x^2+3x-10$ and $4x^3+7x^2-3x-15$ without remainder.

45. Simplify $\dfrac{a+\dfrac{ab}{a-b}}{a^2-\dfrac{2a^2b^2}{a^2+b^2}} \times \dfrac{\dfrac{1}{a^2}-\dfrac{1}{b^2}}{\dfrac{1}{a}-\dfrac{1}{b}}$.

46. Find the cube root of $8x^3-2x^2y+\dfrac{xy^2}{6}-\dfrac{y^3}{216}$.

47. Solve
$$\left.\begin{array}{r}9x+8y=43xy\\8x+9y=42xy\end{array}\right\}.$$

MISCELLANEOUS EXAMPLES. VI.

48. Simplify $\dfrac{3}{x-4} - \dfrac{2}{x-5} - \dfrac{x-7}{(x-2)(x-3)}$.

49. Find the L.C.M. of $8x^3 + 38x^2 + 59x + 30$
    and $6x^3 - 13x^2 - 13x + 30$.

50. A boy spent half of his money in one shop, one-third of the remainder in a second, and one-fifth of what he had left in a third. He had one shilling at last: how much had he at first?

51. Find the remainder when $x^7 - 10x^6 + 8x^5 - 7x^3 + 3x - 11$ is divided by $x^2 - 5x + 4$.

52. Simplify $4\left\{a - \dfrac{3}{2}\left(b - \dfrac{4c}{3}\right)\right\}\left\{\dfrac{1}{2}(2a-b) + 2(b-c)\right\}$.

53. If $a = \dfrac{25}{16}$, $b = 1$, $c = \dfrac{3}{4}$, prove that
$$(a - \sqrt{b})(\sqrt{a} + b)\sqrt{a-b} = \dfrac{3c^4}{\sqrt{a-c^2}}.$$

54. Find the L.C.M. of $x^2 - 7x + 12$, $3x^2 - 6x - 9$, and $2x^2 - 6x - 8$.

55. Find the sum of the squares of $ax + by$, $bx - ay$, $ay + bx$, $by - ax$; and express the result in factors.

56. Solve $\dfrac{x}{6} + \dfrac{y}{4} = \dfrac{3x-5z}{4} = \dfrac{z}{8} + \dfrac{7y}{16} = 1$.

57. Simplify $\dfrac{a^3 + b^3}{a^4 - b^4} - \dfrac{a+b}{a^2 - b^2} - \dfrac{1}{2}\left\{\dfrac{a-b}{a^2+b^2} - \dfrac{1}{a-b}\right\}$.

58. Solve $x - \left(3x - \dfrac{2x+5}{10}\right) = \dfrac{1}{6}(2x + 67) + \dfrac{5}{3}\left(1 + \dfrac{x}{5}\right)$.

59. Add together the following fractions:
$$\dfrac{2}{x^2 + xy + y^2}, \quad \dfrac{-4x}{x^3 - y^3}, \quad \dfrac{x^2}{y^2(x-y)^2}, \quad \dfrac{-x^2}{x^3y - y^4}.$$

60. A man agreed to work for 30 days, on condition that for every day's work he should receive 3s. 4d., and that for every day's absence from work he should forfeit 1s. 6d.; at the end of the time he received £3. 11s.: how many days did he work?

61. Divide $\dfrac{3x^5}{4} + 27 - \dfrac{43x^2}{4} - 4x^4 + \dfrac{77x^3}{8} - \dfrac{33x}{4}$ by $\dfrac{x^2}{2} + 3 - x$.

62. Find the value of
$$\dfrac{4y}{5}(y - x) - 35\left[\dfrac{3x - 4y}{5} - \dfrac{1}{10}\left\{3x - \dfrac{5}{7}(7x - 4y)\right\}\right]$$
when $x = -\dfrac{1}{2}$ and $y = 2$.

63. Simplify $\dfrac{10x-11}{3(x^2-1)} - \dfrac{10x-1}{3(x^2+x+1)} + \dfrac{x^2-2x+5}{(x^3-1)(x+1)}$.

64. Find the cube root of $\dfrac{a^3c^3}{b^3}x^6 - \dfrac{3a^2c}{b}x^5 + \dfrac{3ab}{c}x^4 - \dfrac{b^3}{c^3}x^3$.

65. Solve $\dfrac{4x-17}{x-4} + \dfrac{10x-13}{2x-3} = \dfrac{8x-30}{2x-7} + \dfrac{5x-4}{x-1}$.

66. Find the factors of
 (1) $x^3 + 5x^2 + x + 5$. (2) $x^2 - 2xy - 323y^2$.

67. Solve
$$\left.\begin{array}{l} \dfrac{1}{3}(x+y) + 2z = 21 \\[4pt] 3x - \dfrac{1}{2}(y+z) = 65 \\[4pt] x + \dfrac{1}{2}(x+y-z) = 38 \end{array}\right\}.$$

68. Simplify $\dfrac{x+2y}{\frac{2}{7}x - y} - \dfrac{3x^2 + 63xy + 70y^2}{2x^2 + 3xy - 35y^2}$.

69. Find the square root of $-(3b-2c-2a)^3\{2(a+c)-3b\}$.

70. The united ages of a man and his wife are six times the united ages of their children. Two years ago their united ages were ten times the united ages of their children, and six years hence their united ages will be three times the united ages of the children. How many children have they?

71. Find the sum of
$x^2 - 3xy - \dfrac{2}{3}y^2$, $2y^2 - \dfrac{2}{3}y^3 + z^2$, $xy - \dfrac{1}{3}y^2 + y^3$, and $2xy - \dfrac{1}{3}y^3$.

72. From $\{(a+b)(a-x) - (a-b)(b-x)\}$ subtract $(a+b)^2 - 2bx$.

73. If $a=5$, $b=4$, $c=3$, find the value of
$\sqrt[3]{6abc + (b+c)^3 + (c+a)^3 + (a+b)^3 - (a+b+c)^3}$.

74. Find the factors of
 (1) $3x^3 + 6x^2 - 189x$. (2) $a^2 + 2ab + b^2 + a + b$.

75. Solve
$$\left.\begin{array}{l} px = qy \\ (p+q)x - (q-p)y = r \end{array}\right\}.$$

76. Simplify $\dfrac{x + \frac{y}{2}}{2x^2 + xy + \frac{y^2}{2}} - \dfrac{x^2 - \frac{y^2}{2}}{4\left(x^3 - \frac{y^3}{8}\right)}$.

MISCELLANEOUS EXAMPLES. VI.

77. Solve $\dfrac{x-7}{x+7}+\dfrac{1}{2(x+7)}=\dfrac{2x-15}{2x-6}$.

78. Reduce $\dfrac{x^4-x^2-2x+2}{2x^3-x-1}$ to its lowest terms.

79. Add together the fractions:
$$\dfrac{1}{2x^2-4x+2}, \quad \dfrac{1}{2x^2+4x+2}, \quad \text{and} \quad \dfrac{1}{1-x^2}.$$

80. A number consists of three digits, the right-hand one being zero. If the left-hand and middle digits be interchanged the number is diminished by 180; if the left-hand digit be halved, and the middle and right-hand digit be interchanged, the number is diminished by 336: find the number.

81. Divide $1-5x+\dfrac{152}{15}x^3-\dfrac{106}{225}x^4-\dfrac{28}{9}x^5$ by $1-x-\dfrac{14}{15}x^2$.

82. If $p=1$, $q=\dfrac{1}{2}$, find the value of
$$\dfrac{(p^2+q^2)-(p-q)\sqrt{p^2+2pq+q^2}}{2p+q-\{p-(q-p)\}}.$$

83. Multiply $\dfrac{3x^3}{2}-5x^2+\dfrac{x}{4}+9$ by $\dfrac{x^2}{2}-x+3$.

84. Find the L.C.M. of
$(a^2b-2ab^2)^2$, $\quad 2a^2-3ab-2b^2$, $\quad$ and $\quad 2(2a^2+ab)^2$.

85. Solve $\dfrac{2x+3}{x+1}=\dfrac{4x+5}{4x+4}+\dfrac{3x+3}{3x+1}$.

86. Reduce $\dfrac{5x^3-14x^2+16}{3x^3-2x^2+16x-48}$ to its lowest terms.

87. Find the square root of
$$4a^4+9\left(a^2+\dfrac{1}{a^2}\right)+12a(a^2+1)+18.$$

88. Solve
$$\left.\begin{array}{l}\dfrac{x}{2a}+\dfrac{y}{3b}=a+b \\ \dfrac{3x}{a}-\dfrac{2y}{b}=6(b-a)\end{array}\right\}.$$

89. Multiply $3x+4y+\dfrac{11xy}{x-\dfrac{3}{2}y}$ by $10x-3y-\dfrac{11xy}{\dfrac{x}{4}+y}$.

90. A bag contained five pounds in shillings and half-crowns; after 17 shillings and 6 half-crowns were taken out, thrice as many half-crowns as shillings were left: find the number of each coin.

91. Find the value of
$$5(a-b)-2\{3a-(a+b)\}+7\{(a-2b)-(5a-2b)\},$$
when $a = -\dfrac{1}{9}b$.

92. Divide $3x^4 - 5x^3 + 7x^2 - 11x - 13$ by $3x - 2$.

93. Find the L.C.M. of
$$15(p^3+q^3),\quad 5(p^2-pq+q^2),\quad 4(p^2+pq+q^2),\quad \text{and}\quad 6(p^2-q^2).$$

94. Resolve into factors:
    (1) $a^3 - 8b^{15}$. (2) $-x^2 + 2x - 1 + x^4$.

95. Solve $\dfrac{x+a}{x+b} = \dfrac{x+3a}{x+a+b}$.

96. Simplify
    (1) $\dfrac{35a^2b^2c^2 - 49b^3c^3}{65a^5bc - 91a^3b^2c^2}$. (2) $\dfrac{y^4 - 7y^3 + 8y^2 - 12y}{2y^2 - 2y - 60}$.

97. Solve
$$\left.\begin{array}{l}7x - 9y + 4z = 16 \\ \dfrac{x+y}{3} = \dfrac{x+y+z}{2} \\ 2x - 3y + 4z - 5 = 0\end{array}\right\}.$$

98. Simplify $\dfrac{y^2 - \dfrac{2y}{y-1}}{y^2 - \dfrac{2y}{y+1}} \div \left(\dfrac{y^2 - 5y - 6}{y^2 - 6y + 5} \times \dfrac{y-2}{y+2}\right)$.

99. Find the square root of
$$\dfrac{4a^2 - 12ab - 6bc + 4ac + 9b^2 + c^2}{4a^2 + 9c^2 - 12ac}.$$

100. The express leaves Bristol at 3 p.m. and reaches London at 6; the ordinary train leaves London at 1.30 p.m. and arrives at Bristol at 6. If both trains travel uniformly, find the time when they will meet.

101. Solve (1) $\cdot\dot{6}x + \cdot 75x - \cdot 1\dot{6} = x - \cdot 58\dot{3}x + 5$.
    (2) $\dfrac{37}{x^2 - 5x + 6} + \dfrac{4}{x-2} = \dfrac{7}{3-x}$.

MISCELLANEOUS EXAMPLES. VI.

102. Simplify (1) $\dfrac{a+x}{a^2+ax+x^2}+\dfrac{a-x}{a^2-ax+x^2}+\dfrac{2x^3}{a^4+a^2x^2+x^4}$.

(2) $(1+x)^2 \div \left\{1+\dfrac{x}{1-x+\dfrac{x}{1+x+x^2}}\right\}$.

103. Find the square root of
$$a^6+\dfrac{1}{a^6}-6\left(a^4+\dfrac{1}{a^4}\right)+15\left(a^2+\dfrac{1}{a^2}\right)-20\ ;$$
also the cube root of the result.

104. Divide $1-2x$ by $1+3x$ to 4 terms.

105. I bought a horse and carriage for £75; I sold the horse at a gain of 5 per cent., and the carriage at a gain of 20 per cent., making on the whole a gain of 16 per cent. Find the original cost of the horse.

106. Find the divisor when $(4a^2+7ab+5b^2)^2$ is the dividend, $8(a+2b)^2$ the quotient, and $b^2(9a+11b)^2$ the remainder.

107. Solve (1) $5x(x-3)=2(x-7)$.

(2) $\dfrac{1}{(x-1)(x-2)}+6=\dfrac{3}{x-2}+\dfrac{2}{x-1}$.

108. If $x=a+b+\dfrac{(a-b)^2}{4(a+b)}$, and $y=\dfrac{a+b}{4}+\dfrac{ab}{a+b}$.
prove that $(x-a)^2-(y-b)^2=b^2$.

109. Find the square root of
$$49x^4+\dfrac{1051x^2}{25}-\dfrac{14x^3}{5}-\dfrac{6x}{5}+9.$$

110. Solve $\dfrac{a+x}{a^2+ax+x^2}+\dfrac{a-x}{a^2-ax+x^2}=\dfrac{3a}{x(a^4+a^2x^2+x^4)}$.

111. Subtract $\dfrac{x+3}{x^2+x-12}$ from $\dfrac{x+4}{x^2-x-12}$,
and divide the difference by $1+\dfrac{2(x^2-12)}{x^2+7x+12}$.

112. Find the H.C.F. and L.C.M. of
$2x^2+(6a-10b)x-30ab$ and $3x^2-(9a+15b)x+45ab$.

113. Solve (1) $2cx^2-abx+2abd=4cdx$.

(2) $\dfrac{x}{2(x+3)}-2\dfrac{5}{24}=\dfrac{x^2}{x^2-9}-\dfrac{8x-1}{4(x-3)}$.

114. If $a=1$, $b=2$, $c=3$, $d=4$, find the value of
$$\frac{a^b+b^c+c^d}{b^a+c^b+d^c+(a+b)(b+c)}+3(a^a+b^b+c^c)\left(\frac{1}{a}+\frac{1}{b}+\frac{1}{c}\right).$$

115. I rode one-third of a journey at 10 miles an hour, one-third more at 9, and the rest at 8 miles an hour; if I had ridden half the journey at 10, and the other half at 8 miles per hour, I should have been half a minute longer on the way: what distance did I ride?

116. The product of two factors is $(3x+2y)^3-(2x+3y)^3$, and one of the factors is $x-y$; find the other factor.

117. If $a+b=1$, prove that $(a^2-b^2)^2=a^3+b^3-ab$.

118. Resolve into factors:
   (1) $x^3+y^3+3xy(x+y)$. (2) $m^3-n^3-m(m^2-n^2)+n(m-n)^2$.

119. Solve (1) $\left.\begin{array}{l}x^3-y^3=28\\x^2+xy+y^2=7\end{array}\right\}$. (2) $\left.\begin{array}{l}x^2-6xy+11y^2=9\\x-3y=1\end{array}\right\}$.

120. Find the square root of
$$(a-b)^4-2(a^2+b^2)(a-b)^2+2(a^4+b^4).$$

121. Simplify the fractions:
   (1) $\dfrac{1}{a^2-\dfrac{a^3-1}{a+\dfrac{1}{a+1}}}$. (2) $\dfrac{\left(1+\dfrac{1}{x}\right)\times\left(1-\dfrac{1}{x}\right)^2}{x-\dfrac{1}{x}}$.

122. Find the H.C.F. of
$$a^2b+b^2c-abc-ab^2 \text{ and } ax^2+ab-a^2-bx^2.$$

123. A constituency had two-thirds of its number Conservatives: in an election 25 refused to vote, and 60 went over to the Liberals; the voters were now equal. How many voters were there altogether?

124. Solve (1) $\dfrac{x^2}{a+b}+(a-b)=\dfrac{2ax}{a+b}$.

   (2) $\dfrac{3}{x}+\dfrac{2}{y}=6\left(\dfrac{1}{y}-\dfrac{1}{2x}\right)=2$.

125. Simplify (1) $\left(1+\dfrac{y^2+z^2-x^2}{2yz}\right)\div\left(1-\dfrac{x^2+y^2-z^2}{2xy}\right)$.

   (2) $\dfrac{(x+1)^3-(x-1)^3}{(x+1)^4-(x-1)^4}$.

126. Divide
$$x^4+(a-1)x^3-(2a+1)x^2+(a^2+4a-5)x+3a+6$$
by
$$x^2-3x+a+2.$$

MISCELLANEOUS EXAMPLES. VI.

127. Resolve into factors:

(1) $x^2 + 5xy - 24y^2 + x - 3y$. (2) $x^3 - \dfrac{4}{x}$.

128. Find the square root of $p^2 - 3q$ to three terms.

129. Solve (1) $\dfrac{x-5}{x-6} - \dfrac{x-6}{x-7} = \dfrac{x-1}{x-2} - \dfrac{x-2}{x-3}$.

(2) $ax + 1 = by + 1 = ay + bx$.

130. Find the H.C.F. of $3x^2 + (4a - 2b)x - 2ab + a^2$ and
$$x^3 + (2a - b)x^2 - (2ab - a^2)x - a^2 b.$$

131. Simplify

(1) $\dfrac{(x^a)^3}{x^{b+c}} \times \dfrac{(x^b)^3}{x^{c+a}} \times \dfrac{(x^c)^3}{x^{a+b}}$. (2) $x^{\frac{1}{2}} y^{\frac{1}{3}} \left(\dfrac{y^{\frac{1}{4}}}{x^{\frac{1}{6}}} \right)^2 \div \dfrac{y^{-\frac{1}{4}}}{x^{\frac{1}{4}}}$.

132. At a cricket match the contractor provided dinner for 24 persons, and fixed the price so as to gain $12\frac{1}{2}$ per cent. upon his outlay. Three of the cricketers being absent, the remaining 21 paid the fixed price for their dinner, and the contractor lost 1s.: what was the charge for the dinner?

133. Prove that $x(y+2) + \dfrac{x}{y} + \dfrac{y}{x}$ is equal to a, if
$$x = \dfrac{y}{y+1} \text{ and } y = \dfrac{a-2}{2}.$$

134. Find the cube root of
$$x^3 - 12x^2 + 54x - 112 + \dfrac{108}{x} - \dfrac{48}{x^2} + \dfrac{8}{x^3}.$$

135. Find the H.C.F. and L.C.M. of
$$x^3 + 2ax^2 + a^2 x + 2a^3 \text{ and } x^3 - 2ax^2 + a^2 x - 2a^3.$$

136. Simplify

(1) $42 \left\{ \dfrac{4x - 3y}{6} - \dfrac{3x - 4y}{7} \right\} - 56 \left\{ \dfrac{3x - 2y}{7} - \dfrac{2x - 3y}{8} \right\}$,

(2) $\dfrac{4b + a}{3b + a} + \dfrac{a - 4b}{a - 3b} + \dfrac{a^2 - 3b^2}{a^2 - 9b^2}$.

137. Resolve $4a^2(x^3 + 18ab^2) - (32a^5 + 9b^2 x^3)$ into four factors.

138. Solve (1) $5\sqrt{3x} - 1 = \sqrt{75x - 29}$.

(2) $\dfrac{xy}{x+y} = 70, \quad \dfrac{xz}{x+z} = 84, \quad \dfrac{yz}{y+z} = 140$.

139. Shew that the difference between
$$\frac{x}{x-a}+\frac{x}{x-b}+\frac{x}{x-c} \text{ and } \frac{a}{x-a}+\frac{b}{x-b}+\frac{c}{x-c}$$
is the same whatever value x may have.

140. Multiply $x^{\frac{3}{2}}+2y^{\frac{3}{2}}+3z^{\frac{3}{2}}$ by $x^{\frac{3}{2}}-2y^{\frac{3}{2}}-3z^{\frac{3}{2}}$.

141. Walking $4\frac{1}{4}$ miles an hour, I start $1\frac{1}{2}$ hours after a friend whose pace is 3 miles an hour: how long shall I be in overtaking him?

142. Express in the simplest form

(1) $(8^{\frac{2}{3}}+4^{\frac{3}{2}})\times 16^{-\frac{3}{4}}$. (2) $\dfrac{\left\{9^n\cdot 3^2\times\dfrac{1}{3^{-n}}\right\}-27^n}{3^{3n}\times 9}$.

143. Find the square root of
$$\frac{x}{y}+\frac{y}{x}+3-2\sqrt{\frac{x}{y}}-2\sqrt{\frac{y}{x}}.$$

144. Simplify

(1) $\left(\dfrac{x}{x-1}-\dfrac{1}{x+1}\right)\cdot\dfrac{x^3-1}{x^6+1}\cdot\dfrac{(x-1)^2(x+1)^2+x^2}{x^4+x^2+1}$.

(2) $\left\{\dfrac{a^4-y^4}{a^2-2ay+y^2}\div\dfrac{a^2+ay}{a-y}\right\}\times\left\{\dfrac{a^5-a^3y^2}{a^3+y^3}\div\dfrac{a^4-2a^3y+a^2y^2}{a^2-ay+y^2}\right\}$

145. Find the value of

(1) $\sqrt{8}+\sqrt{50}-\sqrt{18}+\sqrt{48}$. (2) $\sqrt{35+14\sqrt{6}}$.

146. Solve (1) $\dfrac{x-b}{x-a}-\dfrac{x-a}{x-b}=\dfrac{2(a-b)}{x-(a+b)}$.

(2) $\left.\begin{array}{r}2x+3y=1\frac{1}{2}\\4x^2+9xy+9y^2=11\end{array}\right\}$.

147. Shew that
$$\frac{(a+b)^3-c^3}{(a+b)-c}+\frac{(b+c)^3-a^3}{b+c-a}+\frac{(c+a)^3-b^3}{c+a-b}$$
is equal to $2(a+b+c)^2+a^2+b^2+c^2$.

148. Divide $a-x+4a^{\frac{1}{4}}x^{\frac{3}{4}}-4a^{\frac{1}{2}}x^{\frac{1}{2}}$

by $a^{\frac{1}{2}}+2a^{\frac{1}{4}}x^{\frac{1}{4}}-x^{\frac{1}{2}}$.

149. Find the square root of
$$(a-1)^4+2(a^4+1)-2(a^2+1)(a-1)^2.$$

MISCELLANEOUS EXAMPLES. VI.

150. How much are pears a gross when 120 more for a sovereign lowers the price 2d. a score?

151. Shew that if a number of two digits is six times the sum of its digits, the number formed by interchanging the digits is five times their sum.

152. Find the value of
$$\frac{1}{(a-b)(b-c)} - \frac{1}{(b-c)(a-c)} - \frac{1}{(c-a)(b-a)}.$$

153. Multiply
$$3+5x-\frac{12+41x+36x^2}{4+7x} \text{ by } 5-2x+\frac{26x-8x^2-14}{3-4x}.$$

154. If $x-\frac{1}{x}=1$, prove that $x^2+\frac{1}{x^2}=3$, and $x^3-\frac{1}{x^3}=4$.

155. Solve (1) $\frac{3x}{11}+\frac{23}{x+4}=\frac{1}{3}(x+5)$.

   (2) $\left. \begin{array}{l} 2x^2-3y^2=23 \\ 2xy-3y^2=3 \end{array} \right\}.$

156. Simplify

   (1) $1\frac{3}{5}\sqrt{20}-3\sqrt{5}-\sqrt{\frac{1}{5}}$.   (2) $\frac{\sqrt{x}}{y^{-\frac{1}{3}}}\left(\frac{\sqrt[4]{y}}{x^{\frac{1}{3}}}\right) \div \frac{y^{-\frac{1}{4}}}{x^{\frac{1}{4}}}$.

157. Find the H.C.F. of $(p^2-1)x^2+(3p-1)x-p(p-1)$ and
$$p(p+1)x^2-(p^2-2p-1)x-(p-1).$$

158. Reduce to its simplest form
$$\frac{ax+\frac{a}{y}}{x-\frac{1}{y}} \times \frac{x^2+\frac{1}{y^2}}{bx^2-\frac{b}{y^2}} \times \frac{\frac{1}{5}(xy-1)^2}{\frac{1}{3}(x^4y^4-1)}.$$

159. Find the square root of
   (1) $1-2^{2n+1}+4^{2n}$.   (2) $9^n-2\cdot 6^n+4^n$.

160. A clock gains 4 minutes a day. What time should it indicate at 6 o'clock in the morning, in order that it may be right at 7.15 p.m. on the same day?

161. If $x=2+\sqrt{2}$, find the value of $x^2+\frac{4}{x^2}$.

162. Solve

(1) $\dfrac{\sqrt{x+a}}{\sqrt{x-b}} = \dfrac{\sqrt{x-a}}{\sqrt{x}}.$

(2) $\dfrac{\sqrt{1+x}+\sqrt{1-x}}{\sqrt{1+x}-\sqrt{1-x}} = 3.$

163. Simplify
$$\dfrac{a^2}{(b-a)(c-a)} + \dfrac{b^2}{(c-b)(a-b)} + \dfrac{c^2}{(a-c)(b-c)}.$$

164. Find the product of $\dfrac{1}{5}\sqrt{5}$, $\dfrac{1}{2}\sqrt[3]{2}$, $\sqrt[6]{80}$, $\sqrt[3]{5}$, and divide
$$\dfrac{8-4\sqrt{5}}{\sqrt{5}+1} \text{ by } \dfrac{3\sqrt{5}-7}{5+\sqrt{7}}.$$

165. Resolve $9x^6y^2 - 576y^2 - 4x^8 + 256x^2$ into six factors.

166. Simplify

(1) $\dfrac{1 - \dfrac{a^2}{(x+a)^2}}{(x+a)(x-a)} \div \dfrac{x(x+2a)}{(x^2-a^2)(x+a)^2}.$

(2) $\dfrac{6x^2y^2}{m+n} \div \left[\dfrac{3(m-n)x}{7(r+s)} \div \left\{ \dfrac{4(r-s)}{21xy^2} \div \dfrac{r^2-s^2}{4(m^2-n^2)} \right\} \right].$

167. Simplify (1) $\left(a^{1+\frac{q}{p}}\right)^{\frac{p}{p+q}} \div \sqrt[p]{\dfrac{a^{2p}}{(a^{-1})^{-p}}}.$

(2) $\sqrt{14 - \sqrt{132}}.$

168. Find the H.C.F. and L.C.M. of
$$20x^4 + x^2 - 1,\ 25x^4 + 5x^3 - x - 1,\ 25x^4 - 10x^2 + 1.$$

169. Solve (1) $a + x + \sqrt{2ax+x^2} = b.$

(2) $x + 9\tfrac{5}{8} + \dfrac{1}{\dfrac{x}{7} + \dfrac{11}{8}} = 8.$

170. The price of photographs is raised 3s. per dozen, and customers consequently receive seven less than before for a guinea: what were the prices charged?

171. If $\left(a + \dfrac{1}{a}\right)^2 = 3$, prove that $a^3 + \dfrac{1}{a^3} = 0.$

172. Find the value of
$$\dfrac{x+2a}{2b-x} + \dfrac{x-2a}{2b+x} + \dfrac{4ab}{x^2-4b^2}, \text{ when } x = \dfrac{ab}{a+b}.$$

173. Reduce to fractions in their lowest terms

(1) $\left(\dfrac{1}{x}+\dfrac{1}{y}+\dfrac{1}{z}\right) \div \left(\dfrac{x+y+z}{x^2+y^2+z^2-xy-yz-zx} - \dfrac{1}{x+y+z}\right) + 1.$

(2) $\left(1 - \dfrac{56}{x+4} + \dfrac{42}{x+3}\right)\left(1 + \dfrac{56}{x-4} - \dfrac{42}{x-3}\right).$

174. Express as a whole number
$$(27)^{\frac{2}{3}} + (16)^{\frac{3}{4}} - \dfrac{2}{(8)^{-\frac{2}{3}}} + \dfrac{\sqrt[5]{2}}{(4)^{-\frac{2}{5}}}.$$

175. Simplify

(1) $\dfrac{n}{1-x^n} + \dfrac{n}{1-x^{-n}}$ (2) $\sqrt[4]{97 - 56\sqrt{3}}.$

176. Solve

(1) $\dfrac{x-4a}{x-3a} + \dfrac{x-5a}{x-4a} = \dfrac{x+6a}{x-4a} + \dfrac{x+5a}{x-3a}.$

(2) $\left. \begin{array}{l} 3x^2 + xy + 3y^2 = 8\tfrac{1}{4} \\ 8x^2 - 3xy + 8y^2 = 17\tfrac{3}{4} \end{array} \right\}.$

177. Find the square root of $\dfrac{a^2x^2 + 2ab^2x^3 + b^4x^4}{a^{2m} + 2a^m x^n + x^{2n}}.$

178. Simplify

(1) $\dfrac{b}{\sqrt{a}} \times \sqrt[3]{ac} \times \dfrac{\sqrt[4]{c^3}}{\sqrt{b}} \times \dfrac{\sqrt{b-1}}{a^{-\frac{1}{6}}}.$ (2) $\left\{ \dfrac{(9^{n+\frac{1}{4}}) \times \sqrt{3 \cdot 3^n}}{3\sqrt{3^{-n}}} \right\}^{\frac{1}{n}}.$

179. A boat's crew can row 8 miles an hour in still water; what is the speed of a river's current if it take them 2 hours and 40 minutes to row 8 miles up and 8 miles down?

180. If $a = x^2 - yz$, $b = y^2 - zx$, $c = z^2 - xy$, prove that
$$a^2 - bc = x(ax + by + cz).$$

181. Find a quantity such that when it is subtracted from each of the quantities a, b, c, the remainders are in continued proportion.

182. Simplify

(1) $\left(x + y - \dfrac{1}{x+y - \dfrac{xy}{x+y}} \right) \times \dfrac{x^3 - y^3}{x^2 - y^2}.$

(2) $\dfrac{2(7x-4)}{6x^2 - 7x + 2} + \dfrac{x-10}{6x^2 - x - 2} - \dfrac{2(4x-1)}{4x^2 - 1}.$

183. Find the sixth root of
$$729 - 2916x^2 + 4860x^4 - 4320x^6 + 2160x^8 - 576x^{10} + 64x^{12}.$$

184. Simplify

(1) $\dfrac{1}{x+\sqrt{x^2-1}} + \dfrac{1}{x-\sqrt{x^2-1}}.$

(2) $\sqrt[4]{16} + \sqrt[3]{81} - \sqrt[3]{-512} + \sqrt[3]{192} - 7\sqrt[6]{9}.$

185. Solve (1) $\dfrac{5}{6 - \dfrac{5}{6 - \dfrac{5}{6-x}}} = x.$

(2) $\left.\begin{array}{l} x^2y^2 + 192 = 28xy \\ x + y = 8 \end{array}\right\}.$

186. Simplify
$$\dfrac{b-c}{a^2-(b-c)^2} + \dfrac{c-a}{b^2-(c-a)^2} + \dfrac{a-b}{c^2-(a-b)^2}.$$

187. Solve (1) $x - 15\tfrac{3}{4} + \dfrac{5}{x - 15\tfrac{3}{4}} = 6.$

(2) $2(x + y^{-1}) = 3(x^{-1} - y) = 4.$

188. If $xy = ab(a+b)$ and $x^2 - xy + y^2 = a^3 + b^3$ prove that
$$\left(\dfrac{x}{a} - \dfrac{y}{b}\right)\left(\dfrac{x}{b} - \dfrac{y}{a}\right) = 0.$$

189. Find the H.C.F. of
$$(2a^2 - 3a - 2)x^2 + (a^2 + 7a + 2)x - a^2 - 2a$$
and $(4a^2 + 4a + 1)x^2 - (4a^2 + 2a)x + a^2.$

190. Multiply $\sqrt{2x} + \sqrt{2(2x-1)} - \dfrac{1}{\sqrt{2x}}$

by $\dfrac{1}{\sqrt{2x}} + \sqrt{2(2x-1)} - \sqrt{2x}.$

191. Divide $a^4b^2 + b^4c^2 + c^4a^2 - a^2b^4 - b^2c^4 - c^2a^4$
by $a^2b + b^2c + c^2a - ab^2 - bc^2 - ca^2.$

192. Simplify

(1) $\dfrac{7}{2(x+1)} - \dfrac{1}{6(x-1)} - \dfrac{10x-1}{3(x^2+x+1)}.$

(2) $\left\{\dfrac{\sqrt{x+a}}{\sqrt{x-a}} - \dfrac{\sqrt{x-a}}{\sqrt{x+a}}\right\} \times \dfrac{\sqrt{x^3-a^3}}{\sqrt{(x+a)^2 - ax}}.$

MISCELLANEOUS EXAMPLES. VI.

193. If p be the difference between any quantity and its reciprocal, q the difference between the square of the same quantity and the square of its reciprocal, shew that
$$p^2(p^2+4) = q^2.$$

194. A man started for a walk when the hands of his watch were coincident between three and four o'clock. When he finished, the hands were again coincident between five and six o'clock. What was the time when he started, and how long did he walk?

195. If n be an integer, shew that $7^{2n+1}+1$ is always divisible by 8.

196. Simplify $\dfrac{\left(p+\dfrac{1}{q}\right)^p \left(p-\dfrac{1}{q}\right)^q}{\left(q+\dfrac{1}{p}\right)^p \left(q-\dfrac{1}{p}\right)^q}.$

197. Find the value of

(1) $\dfrac{7+3\sqrt{5}}{7-3\sqrt{5}} + \dfrac{7-3\sqrt{5}}{7+3\sqrt{5}}.$

(2) $\dfrac{\sqrt{1+x}+\sqrt{1-x}}{\sqrt{1+x}-\sqrt{1-x}}$ when $x = \dfrac{2b}{b^2+1}.$

198. If $a+b+c+d = 2s$, prove that
$$4(ab+cd)^2 - (a^2+b^2-c^2-d^2)^2 = 16(s-a)(s-b)(s-c)(s-d).$$

199. A man buys a number of articles for £1, and sells for £1. 1s. all but two at 2d. apiece more than they cost: how many did he buy?

200. Find the square root of
$$2(81x^4+y^4) - 2(9x^2+y^2)(3x-y)^2 + (3x-y)^4.$$

201. If $x:a::y:b::z:c$, prove that
$$(bc+ca+ab)^2(x^2+y^2+z^2) = (bz+cx+ay)^2(a^2+b^2+c^2).$$

202. If a man save £10 more than he did the previous year, and if he saved £20 the first year, in how many years will his savings amount to £1700?

203. Given that 4 is a root of the quadratic $x^2-5x+q=0$, find the value of q and the other root.

ALGEBRA.

204. A person having 7 miles to walk increases his speed one mile an hour after the first mile, and finds that he is half an hour less on the road than he would have been had he not altered his rate. How long did he take?

205. If $(a+b+c)x = (-a+b+c)y = (a-b+c)z = (a+b-c)w$,

shew that $\quad \dfrac{1}{y} + \dfrac{1}{z} + \dfrac{1}{w} = \dfrac{1}{x}$.

206. Find a Geometrical Progression of which the sum of the first two terms is $2\frac{2}{3}$, and the sum to infinity $4\frac{1}{6}$.

207. Simplify $\quad \dfrac{\left(1+\dfrac{x}{y}\right)^m \left(1-\dfrac{y}{x}\right)^n}{\left(1+\dfrac{y}{x}\right)^n \left(1-\dfrac{x}{y}\right)^m}$

208. A man has a stable containing 10 stalls; in how many ways could he stable 5 horses?

209. In boring a well 400 feet deep the cost is 2s. 3d. for the first foot and an additional penny for each subsequent foot: what is the cost of boring the last foot, and also of boring the entire well?

210. If α, β are the roots of $x^2 + px + q = 0$, shew that p, q are the roots of the equation

$$x^2 + (\alpha + \beta - \alpha\beta)x - \alpha\beta(\alpha + \beta) = 0.$$

211. Multiply together the duodenary numbers *tete* and *ete*.

212. If $\dfrac{x+z}{y} = \dfrac{z}{x} = \dfrac{x}{z-y}$; determine the ratios $x : y : z$.

213. If a, b, c are in H.P. shew that

$$\left(\dfrac{3}{a} + \dfrac{3}{b} - \dfrac{2}{c}\right)\left(\dfrac{3}{c} + \dfrac{3}{b} - \dfrac{2}{a}\right) + \dfrac{9}{b^2} = \dfrac{25}{ac}.$$

214. Find the number of permutations which can be made from all the letters of the words

(1) *Consequences*, (2) *Acarnania*.

MISCELLANEOUS EXAMPLES. VI.

215. Expand by the Binomial Theorem $(2a-3x)^5$; and find the numerically greatest term in the expansion of $(1+x)^n$, if $x=\dfrac{3}{5}$, and $n=7$.

216. When $x=\dfrac{\sqrt{3}}{4}$, find the value of
$$\dfrac{1+2x}{1+\sqrt{1+2x}}+\dfrac{1-2x}{1-\sqrt{1-2x}}.$$

217. Simplify $\dfrac{x^2-bc}{(a-b)(a-c)}+\dfrac{x^2-ca}{(b-c)(b-a)}+\dfrac{x^2-ab}{(c-a)(c-b)}.$

218. Solve the equations:
   (1) $(x^2-5x+2)^2=x^2-5x+22.$
   (2) $\left(x^2+\dfrac{1}{x^2}\right)^2+4\left(x^2+\dfrac{1}{x^2}\right)=12.$

219. Prove that
$$(y-z)^3+(x-y)^3+3(x-y)(x-z)(y-z)=(x-z)^3.$$

220. Out of 16 consonants and 5 vowels, how many words can be formed each containing 4 consonants and 2 vowels?

221. If $b-a$ is a harmonic mean between $c-a$ and $d-a$, shew that $d-c$ is a harmonic mean between $a-c$ and $b-c$.

222. In how many ways may 2 red balls, 3 black, 1 white, 2 blue be selected from 4 red, 6 black, 2 white and 5 blue; and in how many ways may they be arranged?

223. The sum of a certain number of terms of an arithmetical series is 36, and the first and last of these terms are 1 and 11 respectively: find the number of terms, and the common difference of the series.

224. Expand by the Binomial Theorem
   (1) $\left(2-\dfrac{3a}{4}\right)^5$; (2) $\left(1-\dfrac{2}{3}x\right)^{\frac{3}{2}}$ to 5 terms.

225. In what scale is the denary number 418 represented by 1534?

226. Simplify $\dfrac{2\sqrt{10}}{3\sqrt{27}} \times \dfrac{15\sqrt{21}}{4\sqrt{15}} \div \dfrac{5\sqrt{14}}{7\sqrt{48}}$; and find the value of $\dfrac{1}{3\sqrt{5}-6}$, given that $\sqrt{5}=2\cdot236$.

227. By the Binomial Theorem find the cube root of 128 to six places of decimals.

228. There are 9 books of which 4 are Greek, 3 are Latin, and 2 are English; in how many ways could a selection be made so as to include at least one of each language?

229. Simplify

(1) $\dfrac{\sqrt{45x^3} - \sqrt{80x^3} + \sqrt{5a^2x}}{a-x}$;

(2) $\left\{ \dfrac{x^{\frac{1}{2}} + x^{-\frac{1}{2}}}{x^2 - x + 1} - \dfrac{x^{\frac{1}{2}} - x^{-\frac{1}{2}}}{x^2 + x + 1} \right\} \div \left\{ \dfrac{x^{\frac{1}{2}} + 2x^{-\frac{1}{2}}}{x^3 - 1} - \dfrac{x^{\frac{1}{2}} - 2x^{-\frac{1}{2}}}{x^3 + 1} \right\}$.

230. Form the quadratic equation whose roots are $5 \pm \sqrt{6}$.

If the roots of $x^2 - px + q = 0$ are two consecutive integers prove that $p^2 - 4q - 1 = 0$.

231. Subtract 4·72473 from 7·641 in the scale of eight, and find the square root of t08404 in the scale of twelve.

232. Find $\log_{16} 128$, $\log_4 \sqrt{128}$, $\log_2 \tfrac{1}{4}$; and having given
$$\log 2 = \cdot3010300 \text{ and } \log 3 = \cdot4771213,$$
find the logarithm of ·00001728.

233. A and B start from the same point, B five days after A; A travels 1 mile the first day, 2 miles the second, 3 miles the third, and so on; B travels 12 miles a day. When will they be together? Explain the double answer.

234. Solve the equations:

(1) $2^x = 8^{y+1}$, $9^y = 3^{x-9}$;

(2) $z^x = y^{2x}$, $2^z = 2 \times 4^x$, $x + y + z = 16$.

235. The sum of the first 10 terms of an arithmetical series is to the sum of the first 5 terms as 13 is to 4. Find the ratio of the first term to the common difference.

MISCELLANEOUS EXAMPLES. VI.

236. Find the greatest term in the expansion of $(1-x)^{-\frac{4}{3}}$ when $x = \frac{12}{13}$.

237. Five gentlemen and one lady wish to enter an omnibus in which there are only three vacant places; in how many ways can these places be occupied (1) when there is no restriction, (2) when one of the places is to be occupied by the lady?

238. Given $\log 2 = \cdot 301030$, $\log 3 = \cdot 477121$, and $\log 7 = \cdot 845098$, find the logarithms of $\cdot 005$, $6 \cdot 3$, and $\left(\frac{49}{216}\right)^{\frac{1}{3}}$.

Find x from the equation $18^{8-4x} = (54\sqrt{2})^{3x-2}$.

239. If P and Q vary respectively as $y^{\frac{1}{2}}$ and $y^{\frac{1}{3}}$ when z is constant, and as $z^{\frac{1}{2}}$ and $z^{\frac{1}{3}}$ when y is constant, and if $x = P + Q$, find the equation between x, y, z; it being known that when $y = z = 64$, $x = 12$; and that when $y = 4z = 16$, $x = 2$.

240. Simplify
$$\log \frac{133}{65} + 2 \log \frac{13}{7} - \log \frac{143}{90} + \log \frac{77}{171}.$$

241. If the number of permutations of n things 4 at a time is to the number of combinations of $2n$ things 3 at a time as 22 to 3, find n.

242. If $\frac{1}{a} + \frac{1}{c} = \frac{1}{2b-a} + \frac{1}{2b-c}$, prove that $2b$ is either the arithmetic mean between $2a$ and $2c$, or the harmonic mean between a and c.

243. If ^nC_r denote the number of combinations of n things taken r together, prove that
$$^{n+2}C_{r+1} = {^nC_{r+1}} + {^nC_{r-1}} + (2 \times {^nC_r}).$$

244. Find (1) the characteristic of $\log 54$ to base 3;

(2) $\log_{10}(\cdot 0125)^{\frac{1}{3}}$; (3) the number of digits in 3^{45}.

Given $\log_{10} 2 = \cdot 30103$, $\log_{10} 3 = \cdot 47712$.

ALGEBRA.

245. Write down the $(r+1)^{\text{th}}$ term of $(2ax^2 - x^3)^{\frac{5}{3}}$, and express it in its simplest form.

246. At a meeting of a Debating Society there were 9 speakers; 5 spoke for the Government, and 4 for the Opposition. In how many ways could the speeches have been made, if a member of the Government always speaks first, and the speeches are alternately for the Government and the Opposition?

247. Form the quadratic equation whose roots are
$$a+b+\sqrt{a^2+b^2} \text{ and } \frac{2ab}{a+b+\sqrt{a^2+b^2}}.$$

248. A point moves with a speed which is different in different miles, but invariable in the same mile, and its speed in any mile varies inversely as the number of miles travelled before it commences this mile. If the second mile be described in 2 hours, find the time taken to describe the nth mile.

249. Solve the equations:
(1) $x^2(b-c) + ax(c-a) + a^2(a-b) = 0$,
(2) $(x^2 - px + p^2)(qx + pq + p^2) = qx^3 + p^2q^2 + p^4$.

250. Prove by the Binomial Theorem that
$$1 + \frac{3}{4} + \frac{3\cdot 5}{4\cdot 8} + \frac{3\cdot 5\cdot 7}{4\cdot 8\cdot 12} + \ldots\ldots \text{ ad inf.} = \sqrt{8}.$$

ANSWERS.

I. a. PAGE 4.

1.	70.	2.	125.	3.	105.	4.	343.	5.	30.
6.	32.	7.	12.	8.	70.	9.	6.	10.	108.
11.	7.	12.	144.	13.	48.	14.	189.	15.	200.
16.	27.	17.	1000.	18.	3.	19.	1.	20.	567.
21.	4.	22.	125.	23.	81.	24.	1.	25.	243.
26.	512.	27.	5.	28.	4096.	29.	64.	30.	90.
31.	24.	32.	2.	33.	81.	34.	64.	35.	8.
36.	2401.	37.	56.	38.	3.	39.	48.	40.	16.

I. b. PAGE 5.

1.	700.	2.	686.	3.	96.	4.	135.	5.	15.
6.	60.	7.	162.	8.	0.	9.	0.	10.	3000.
11.	98.	12.	225.	13.	3.	14.	1.	15.	$5\frac{1}{3}$.
16.	2.	17.	36.	18.	160.	19.	$1\frac{1}{3}$.	20.	40.
21.	0.	22.	72.	23.	2048.	24.	81.	25.	$11\frac{1}{4}$.
26.	$13\frac{1}{2}$.	27.	$\frac{1}{64}$.	28.	$3\frac{3}{8}$.				

I. c. PAGE 6.

1.	4.	2.	6.	3.	8.	4.	36.	5.	3.
6.	8.	7.	8.	8.	0.	9.	32.	10.	60.
11.	0.	12.	16.	13.	$2\frac{2}{3}$.	14.	$3\frac{1}{3}$.	15.	$1\frac{1}{7}$.
16.	0.	17.	$\frac{2}{3}$	18.	$2\frac{2}{3}$.	19.	0.	20.	3.	21.	8.

I. d. PAGE 8.

1.	19.	2.	0.	3.	7.	4.	11.	5.	21.
6.	6.	7.	18.	8.	36.	9.	6.	10.	14.
11.	85.	12.	96.	13.	36.	14.	0.	15.	0.
16.	12.	17.	24.	18.	43.	19.	4.	20.	8.
21.	12.	22.	0.	23.	1.	24.	6000.	25.	17.
26.	16.	27.	18.	28.	9.	29.	$3\frac{1}{2}$.	30.	49.
31.	$\frac{1}{4}$.	32.	$\frac{3}{16}$.	33.	0.	34.	$1\frac{5}{6}$.		

438   ALGEBRA.

I. e. PAGE 8ᴀ.

1. 20, 2, 2, 12, 30.   2. 3, 5·25, 8, 11·25, 15.   3. 6.
4. 4·48, 17·76, 27·52, 40.   6. 504.   8. The first by 14.
10. 21, 0.   12. ·196.

II. a. PAGE 10.

1. $-£12$.   2. 4, -2.   3. 20.   4. $-6°$.
5. -3 feet.   6. 24, -4.   7. A, C, B with $+4, 0, -2$ points.

II. b. PAGE 12.

1. $47a$.   2. $24x$.   3. $39b$.   4. $151c$.   5. $-26x$.
6. $-40b$.   7. $-17y$.   8. $-66c$.   9. $-20b$.   10. $2x$.
11. 0.   12. $-16f$.   13. $-s$.   14. $7y$.   15. 0.
16. $2ab$.   17. x^2.   18. $-14a^2x$.   19. $-21a^5$.   20. $-10x^0$.
21. 0.   22. $-19x^4$.   23. $-43abcd$.   24. $\frac{11}{6}x$.   25. $\frac{8}{5}a$.
26. $-3b$.   27. $-x^2$.   28. $-\frac{5}{6}ab$.   29. $\frac{5}{4}x$.   30. $-5x^2$.

III. a. PAGE 16.

1. 0.   2. $4a+4b+4c$.   3. 0.
4. $4x+4y+4z$.   5. $3a+5b-2c$.   6. $b-c$.
7. $39a-5b+4c$.   8. $5c$.   9. $3ax-3by+3cz$.
10. $22p-18q-20r$.   11. ab.   12. $-20ab+ca$.
13. $5ab+bc$.   14. $pq+qr+rp$.   15. $6x$.
16. $20a$.   17. $2xy+2zx$.   18. $14ab-11bc$.
19. $13z$.   20. $a+b+c$.

III. b. PAGE 18.

1. abc.   2. x^2+xy+y^2.   3. $a^2+3ab-2b^2$.
4. $yz+zx+xy$.   5. $3x^2+2xy-y^2$.   6. $-2x^3+x^2+4x+2$.
7. x^2+7x.   8. $15x^2-32x-18$.   9. $15x^3-4x^2+3x-1$.
10. $a^3+b^3+c^3$.   11. $a^3+b^3+c^3+d^3$.   12. x^3+x^2+x+3.
13. $9a^3-3a^2$.   14. $3x^2-2y^2-2xy-4yz-3xz$.
15. $-x^3+x^2+2y^2+y$.   16. $3x^2y+xy^2$.   17. $2a^2b$.
18. $x^5-x^4y-y^5$.   19. $a^3+b^3+c^3-3abc$.

ANSWERS. 439

20. $x^3 + x^2y + 7xy^2 + 3y^3$.   21. $\dfrac{1}{4}a - \dfrac{2}{3}b$.

22. $-3a - \dfrac{1}{2}b$.   23. $-\dfrac{7}{3}a + \dfrac{2}{3}b - \dfrac{1}{2}c$.   24. $\dfrac{3}{8}a - \dfrac{4}{5}b - \dfrac{15}{4}c$.

25. $\dfrac{1}{3}x^2 - \dfrac{4}{3}xy + \dfrac{1}{2}y^2$.   26. $\dfrac{5}{6}a^2 + \dfrac{3}{5}ab - \dfrac{1}{6}b^2$.

27. $\dfrac{3}{8}x^2 - \dfrac{2}{5}xy - \dfrac{1}{2}y^2$.   28. $-\dfrac{1}{4}x^3 + \dfrac{3}{8}ax^2 + \dfrac{5}{8}a^2x$.

29. $-\dfrac{1}{4}x^2 - xy + \dfrac{3}{5}y^2$.   30. $-a^3 - \dfrac{1}{2}a^2b + \dfrac{1}{4}ab^2 + b^3$.

IV. a. Page 21.

1. $-2a - 2c$.   2. $3a - 5b - 4c$.   3. $13x + 18y - 19z$.
4. $-5a + 30b - 4c$.   5. $11x + 13y - 16z$.   6. $12ab - 10bc - 10cd$.
7. $21a - 13b - 33c$.   8. $11x + 26y + 22z$.   9. $2ac + 2bd$.
10. $2ab - 2cd + 2ac - 2bd$.   11. $-cd - ac - bd$.
12. $2xy$.   13. $-3x^3 - x^2 - 2x + 1$.
14. $-12x^2y + 21xy^2 + 15xyz$.   15. $\dfrac{1}{6}a - \dfrac{3}{2}b + \dfrac{5}{6}c$.
16. $\dfrac{1}{4}x + \dfrac{3}{2}y - \dfrac{2}{3}z$.   17. $-\dfrac{5}{2}a - \dfrac{10}{3}b + \dfrac{1}{2}c$.
18. $x - y + \dfrac{1}{5}z$.   19. $-\dfrac{4}{3}x - \dfrac{4}{3}z$.   20. $-\dfrac{5}{6}x + \dfrac{13}{6}y$.

IV. b. Page 22.

1. $7xy - 7yz + 18xz$.   2. $-12x^2y^2 + 8x^3y + 21xy^3$.
3. $-12 + 9ab + 6a^2b^2$.   4. $-2a^2bc + 6b^2ca + 5c^2ab$.
5. $-12a^2b + 15ab^2 - 5cd$.   6. $-16x^2y + 10xy^2 - 2x^2y^2$.
7. $20a^2b^2 + 16a^2b$.   8. $9x^2 - 9x + 9$.
9. $x^3 + 3x^2 + 5x + 7$.   10. $-17a^2x^2 + 13x^2 + 20$.
11. $2x^2 - 2x$.   12. $6x^2y + 2y^3$.   13. $a^3 - c^3 - abc$.
14. $3x^3 + 10x^2y - 10xy^2$.   15. $4x^4 - 5x^3 - 2x^2 - x + 2$.
16. $-4a^3 + 4b^3 - 2c^3 + 10abc$.   17. $-x^5 + 2x^4 + x^3 - x^2 + 2x - 2$.
18. $4a^5 - 7a^4 - 5a^3 + 9a^2 - a - 7$.   19. $-5a^2b - 14ab^2 + a^3b^3 + b^4$.
20. $-a^3 + 22a^2b - 16ab^2 + 2b^3$.   21. $2x^2 - \dfrac{4}{3}xy - \dfrac{1}{2}y^2$.
22. $\dfrac{4}{3}a^2 - \dfrac{7}{2}a - \dfrac{1}{2}$.   23. $-\dfrac{1}{6}x^2 - \dfrac{5}{6}x + \dfrac{7}{6}$.   24. $\dfrac{5}{8}x^2 + \dfrac{1}{6}ax - \dfrac{1}{3}$.
25. $\dfrac{3}{4}x^3 - \dfrac{1}{2}x^2y - \dfrac{1}{6}y^2$.   26. $-\dfrac{1}{8}a^3 - \dfrac{2}{3}a^2x - \dfrac{1}{2}ax^2$.

440 ALGEBRA

Miscellaneous Examples I. Page 23.

1. (1) $2x+x^2$; (2) $-3a+b$. 2. $2a+2c$.
3. (1) 21; (2) 108. 4. (1) 11; (2) 18. 5. $7x^3-10x^2$.
6. $8a^3-2a$. 9. $2x^3-2x^2$. 11. $2a-(3b+5c)$.
12. 47; 12. 13. $-y^2+y$. 15. 36. 16. 0. 17. a^2b.
18. $x+2z$. 20. $7xy$. 21. 8. 23. $4a$. 24. 118.
25. 30 B.C. 26. $2x^3-2x$. 28. $a+b-(c-d)$.
29. $a+3b$ miles south of O. 30. $2x^2+7x-3$.

V. a. Page 27.

1. $35x^7$. 2. $20a^{11}$. 3. $56a^4b^3$. 4. $30x^4y^2$.
5. $8a^3b^6$. 6. $6a^2bc^4$. 7. $4a^6b^6$. 8. $10a^3b$.
9. $28a^7b^3$. 10. $5a^4b^3x^2y^2$. 11. $6a^2x^7y^3$. 12. $abcxyz$.
13. $15a^7b^8x^4$. 14. $28a^3b^3x^5$. 15. $40a^2cx^2$. 16. $30a^3x^6y^3$.
17. $2x^7y^8$. 18. $3a^5x^9y^{16}$. 19. $a^4b^2+a^3b^2c$.
20. $20a^3b^2x^3-28a^2b^2x^4$. 21. $10x^3+6x^2y$.
22. $a^5b+a^3b^3-a^3bc^2$. 23. $ab^2c^2+a^2bc^2-a^2b^2c$.
24. $20a^4bc^3+12a^2b^3c^3-8a^2bc^5$. 25. $15x^5y+3x^4y^2-21x^5y^2$.
26. $48x^5y^3-40x^4y^4+56x^3y^5$. 27. $6a^5b^3c-7a^3b^4c^2$.

V. b. Page 30.

1. 36. 2. -48. 3. 5. 4. 24. 5. -16.
6. -12. 7. -9. 8. -24. 9. -168. 10. 480.
11. -16. 12. 375. 13. 500. 14. 140. 15. -2000.
16. 500. 17. -180. 18. -56. 19. -1000. 20. -224.
21. 40. 22. -63. 23. 118. 24. -130. 25. -54.
26. 3. 27. 1. 28. 0. 29. 29. 30. -13.

V. c. Page 31.

1. $-3a^2x^2$. 2. $14a^2b^2x^2$. 3. $-a^3b^3$. 4. $-60x^3y^2$.
5. $3a^3b^4c^5d^6$. 6. $-5x^3y^4z^2$. 7. $-36x^2y^2z-48xy^2z^2$.
8. $a^3b^2c^5-a^2b^2c^4$. 9. $3x^2+3xy+3xz$. 10. $a^3bc-ab^3c+abc^3$.
11. $a^2b^2c-ab^2c^2+a^2bc^2$. 12. $14a^4b^3+28a^3b^4$.
13. $15x^3y^2-18x^2y^3+24x^3y^3$. 14. $56x^6y^4+40x^4y^6$.
15. $-5x^2y^3z^2+3x^2y^2z^3-8x^3y^2z^2$. 16. $-48x^5y^3z^5+96x^4y^2z^4$.
17. $91x^4y^5+105x^5y^4$. 18. $-8x^2y^2z^2+10x^4y^2z^4$.
19. $-a^2b^2c^2+a^3b^2c^2+a^2b^3c^2$. 20. $a^3b^2c-a^2b^3c+a^2b^2c^2$.

ANSWERS. 441

21. $-3a^2 + \dfrac{9}{2}ab - 6ac$.   22. $-\dfrac{5}{2}x^2 + \dfrac{5}{3}xy + \dfrac{10}{3}x$.

23. $\dfrac{1}{4}a^2x - \dfrac{1}{16}abx - \dfrac{3}{8}acx$.   24. $-2a^5x^3 + \dfrac{7}{2}a^4x^4$.

25. $\dfrac{5}{2}a^4x^2 - \dfrac{5}{3}a^3x^3 + a^2x^4$.   26. $\dfrac{21}{2}x^3y - x^2y^2$.

27. $\dfrac{1}{2}x^5y^2 - 3x^3y^5$.   28. $-x^8y^3 + \dfrac{16}{49}x^5y^6$.

V. d. PAGE 32.

1. $x^2 + 15x + 50$.   2. $x^2 - 25$.   3. $x^2 - 17x + 70$.
4. $x^2 + 3x - 70$.   5. $x^2 - 3x - 70$.   6. $x^2 + 17x + 70$.
7. $x^2 - 36$.   8. $x^2 + 4x - 32$.   9. $x^2 - 13x + 12$.
10. $x^2 + 11x - 12$.   11. $x^2 - 225$.   12. $-x^2 + 18x - 45$.
13. $x^2 + 5x + 6$.   14. $-x^2 + 14x - 49$.   15. $x^2 - 25$.
16. $x^2 + x - 182$.   17. $x^2 + x - 306$.   18. $x^2 - x - 380$.
19. $x^2 - 256$.   20. $-x^2 + 42x - 441$.   21. $2x^2 + 13x - 24$.
22. $2x^2 - 13x - 24$.   23. $2x^2 - 11x + 5$.   24. $2x^2 - 7x + 5$.
25. $6x^2 + 11x - 35$.   26. $6x^2 - 11x - 35$.   27. $10x^2 + 3x - 18$.
28. $10x^2 - 3x - 18$.   29. $9x^2 - 25y^2$.   30. $9x^2 - 30xy + 25y^2$.
31. $a^2 + ab - 6b^2$.   32. $a^2 + ab - 56b^2$.   33. $3a^2 - 30ab + 48b^2$.
34. $a^2 - 4ab - 45b^2$.   35. $x^2 + ax - bx - ab$.   36. $x^2 - ax + bx - ab$.
37. $x^2 - 2ax + 3bx - 6ab$.   38. $a^2x^2 - b^2y^2$.
39. $x^2y^2 - a^2b^2$.   40. $4p^2q^2 - 9r^2$.

V. e. PAGE 34.

1. $a^2 + 2ab + b^2 - c^2$.   2. $a^2 - 4b^2 + 4bc - c^2$.
3. $a^4 + a^2b^2 + b^4$.   4. $x^3 + 4x^2y + 3xy^2 + 12y^3$.
5. $x^4 - 4x^2 + 8x + 16$.   6. $x^6 + y^6$.   7. $x^3 - y^3$.
8. $a^4 + 4a^2x^2 + 16x^4$.   9. $64a^3 - 27b^3$.   10. $x^4 - a^4$.
11. $x^4 + 2x^3 - 7x^2 - 8x + 12$.   12. $4x^5 - x^3 + 4x$.
13. $a^6 + a^3b^3$.   14. $x^5 - 2x^4 - 4x^3 + 19x^2 - 31x + 15$.
15. $a^5 + 4ab^4$.   16. $8x^3 - 27y^3$.
17. $-x^4 + 4x^3y - x^2y^2 - 4xy^3 - y^4$.   18. $a^6 - a^4b^4 + 2a^3b^3 + b^6$.
19. $x^4 - 2x^2y^2 + y^4$.   20. $a^2b^2 + c^2d^2 - a^2c^2 - b^2d^2$.
21. $75a^5b^3 - 28a^3b^5 + 13a^2b^6 - 12ab^7$.   22. $81x^4 - 256a^4$.
23. $a^4 - 25a^2b^2 - 10ab^3 - b^4$.   24. $x^3 + 3xy + y^3 - 1$.
25. $a^3 + b^3 + c^3 - 3abc$.   26. $x^5 + y^5$.   27. $x^{15} + y^{10}$.

28. $a^6 - 2a^3 + 1$.   29. $a^2x^3 + 27a^2y^6$.   30. $x^6 + 2x^3y^3 + y^6$.

31. $\frac{1}{4}a^3 + \frac{1}{72}a - \frac{1}{12}$.   32. $\frac{1}{4}x^3 - \frac{5}{6}x^2 + \frac{1}{12}x + \frac{1}{2}$.

33. $\frac{2}{9}x^3 - \frac{3}{4}y^3$.   34. $\frac{9}{8}x^4 - \frac{3}{2}ax^3 + \frac{1}{2}a^2x^2 - \frac{2}{9}a^4$.

35. $\frac{1}{4}x^4 - \frac{43}{36}x^2 + \frac{9}{16}$.   36. $\frac{1}{4}a^4 + x^4$.

V. f. Page 36.

1. $x^2 + 3x - 40$.   2. $x^2 + 5x - 6$.   3. $x^2 + 7x - 30$.
4. $x^2 + 4x - 5$.   5. $x^2 - 2x - 63$.   6. $x^2 - 18x + 80$.
7. $x^2 + 7x - 44$.   8. $x^2 + 2x - 8$.   9. $x^2 - 4$.
10. $a^2 - 1$.   11. $a^2 + 4a - 45$.   12. $a^2 + 9a - 36$.
13. $a^2 - 4a - 32$.   14. $a^2 - 64$.   15. $a^2 + 7a - 78$.
16. $a^2 + 6a + 9$.   17. $a^2 - 121$.   18. $a^2 - 16a + 64$.
19. $x^2 - ax - 6a^2$.   20. $x^2 + ax - 30a^2$.   21. $x^2 - 9a^2$.
22. $x^2 + 2xy - 8y^2$.   23. $x^2 - 49y^2$.   24. $x^2 - 6xy + 9y^2$.
25. $a^2 + 6ab + 9b^2$.   26. $a^2 + 5ab - 50b^2$.   27. $a^2 - 17ab + 72b^2$.
28. $2x^2 - x - 10$.   29. $2x^2 - 9x + 10$.   30. $2x^2 - 3x - 9$.
31. $3x^2 + 2x - 1$.   32. $4x^2 + 8x - 5$.   33. $6x^2 + 5x - 21$.
34. $8x^2 + 6x - 9$.   35. $9x^2 - 64$.   36. $4x^2 - 20x + 25$.
37. $9x^2 - 3xy - 2y^2$.   38. $9x^2 + 12xy + 4y^2$.   39. $4x^2 + 4xy - 35y^2$.
40. $25x^2 - 9a^2$.   41. $2x^2 + 5ax - 25a^2$.   42. $4x^2 + 4ax + a^2$.

VI. a. Page 40.

1. $3x$.   2. $-3x$.   3. $-5x^3$.   4. $-bx$.
5. xy^2.   6. $-a^2$.   7. $4ac$.   8. $-4a^2b^4c^5$.
9. a^4c^6.   10. $3x^3y^5z^2$.   11. $4x^2$.   12. $6a^6$.
13. $5a^4$.   14. $7a^2b^3$.   15. -1.   16. $-7ab^2$.
17. $-8b^2x$.   18. $10y^2$.   19. $x - 2y$.   20. $x^2 - 3x + 1$.
21. $x^4 - 7x^3 + 4x^2$.   22. $10x^4 - 8x^3 + 3x$.
23. $-3x^2 + 5x$.   24. $3x - 4$.   25. $3x^3 + 4x$.   26. $2x^2y - 3xy^2$.
27. $-a + b + c$.   28. $a - b - b^2$.   29. $-x^2 + 3xy + 4y^2$.
30. $-2x^3y^3 + 4x^2y - 3y^2$.   31. $2a - 3b + 4c$.
32. $-\frac{1}{3}x^2 + 2y^2$.   33. $3x - 2y - 4$.
34. $-\frac{6}{7}a^2x^2 + \frac{3}{2}ax^3$.   35. $\frac{2}{3}a - \frac{1}{6}b - c$.

ANSWERS. 443

VI. b. Page 42.

1. $x+2$. 2. $x-4$. 3. $a-6$. 4. $a-24$.
5. $3x+1$. 6. $x+5$. 7. $5x+1$. 8. $x+7$.
9. $5x+1$. 10. $x+11$. 11. $x+5$. 12. $3x+1$.
13. $3x+7$. 14. $3x-7$. 15. $3x-5$. 16. $4x-7$.
17. $4a+3x$. 18. $5a-x$. 19. $3a+4c$. 20. $3a-5c$.
21. $6x+5y$. 22. $8x+3y$. 23. x^2+14x. 24. $4x^2-x$.
25. $9x^2+9x+5$. 26. $2a^2-5a+3$.
27. $3+3a+a^2$. 28. $8-36x+54x^2-27x^3$.

VI. c. Page 44.

1. $x-4$. 2. $y+1$. 3. $2m-3$. 4. $2a^2-3a$.
5. x^2-x+1. 6. a^2-3a+1; rem. $a-6$. 7. a^2+3a+2.
8. $2x^2+x-1$; rem. $3x+4$. 9. x^3-2x^2+x+1.
10. x^3-3x^2+2x-1. 11. $10x^2-3x-12$; rem. $7x-45$.
12. $7y^2+5y-3$; rem. $-39y+27$. 13. $2k^2-5k+2$.
14. $5-7m-m^3$. 15. x^2+5x+6.
16. x^2-2x+3; rem. $31x-15$. 17. $12+8x+x^2$.
18. $7x^2+5xy+2y^2$. 19. x^2-xy+y^2; rem. x^2.
20. x^3+x-y. 21. $a^6+a^3b^3+b^6$.
22. $x^7-x^6y+x^4y^3-x^3y^4+xy^6-y^7$.
23. $x^6+2x^5y^2-3x^4y^4-6x^3y^6+2x^2y^8+4xy^{10}+y^{12}$.
24. $a^2+2ab+b^2+a+b+1$. 25. $x^5-x^4y+xy^4-y^5$.
26. $a^{10}+a^8b^2+a^6b^4+a^4b^6+a^2b^8+b^{10}$.
27. $a^8-2a^6b^2+3a^4b^4-2a^2b^6+b^8$. 28. $1+a+a^2+2x-2ax+4x^2$.
29. $\frac{1}{4}a^2-3ax+9x^2$. 30. $\frac{1}{9}a^2-\frac{1}{6}a+\frac{1}{16}$.
31. $\frac{6}{25}a^3-\frac{3}{5}a^2c+\frac{3}{2}ac^2$. 32. $\frac{3}{8}a^2-\frac{1}{4}a-\frac{2}{3}$.
33. $6x-\frac{1}{3}y-\frac{1}{2}$. 34. $\frac{4}{9}a^4+\frac{1}{2}a^3x+\frac{9}{16}a^2x^2+\frac{81}{128}ax^3$.

VII. a. Page 47.

1. $a+b-c$. 2. a. 3. $a+3b-4c$. 4. $3a-b-c$.
5. $-2a-4b-2c$. 6. $-a+b-c$. 7. $b-a$.
8. $x-y$. 9. $2a-2b$. 10. $-2x-5y$. 11. $x-a$.
12. $2a-b-d$. 13. $-3c+4y$. 14. $-x+2y+6z$.
15. $-5x$. 16. $-25x+2y$. 17. $11x-36y$.
18. $2x-2z$. 19. $2a$. 20. a.

VII. b. Page 48.

1. $3a$.
2. a.
3. $6a + 2b - 2c - 2d$.
4. $2x - 3y + 12z$.
5. b.
6. $21a + b$.
7. $2b + 4c$.
8. $-a^2 + 8b^2 - 9c^2$.
9. $-2a + 6b + 2c - 2d$.
10. $4a + b + c$.
11. $-50c$.
12. $-11a - 2b$.
13. $-a + b + 5c$.
14. $-2a + 10b - 11c$.
15. $-227a + 216b + 84$.
16. $2a - 12c + 84d$.
17. $3a + 4x$.
18. $-10a$.
19. $4a$.
20. 0.
21. $\dfrac{11}{5}a - 2b$.
22. $12x - 30y$.
23. $a - \dfrac{13}{3}b + \dfrac{10}{3}c$.
24. 0.

VII. c. Page 50.

1. $(a+2)x^4 + (b-5)x^2 + (2b-3)x + 5$.
2. $(5a-b)x^3 + (3b-4)x^2 + (c-2)x + ab - 7$.
3. $(9a-7)x^3 + (5a-3)x^2 + (7-2c)x + 2$.
4. $(2c-a^2)x^5 + (1-3b)x^4 + (4d-3ab)x$.
5. $-(a^2+b)x^4 - (2b-5)x^3 - (3-a)x^2$.
6. $-(ab-7)x^5 - (abc-7)x^3 - (3c^2-5a)x$.
7. $-(c-a^2)x^3 - (b+5-a)x^2$.
8. $-(a+c+7-3b^2)x^4 - (b+5c^2)x$.
9. $(a-b)x^3 - (b+2c)x^2 - (b+c+d)x$.
10. $(5a+4c)x^3 + (3a-6b+7c)x^2 + (2a-7b)x$.
11. $(3a+2c)x^3 + (a+8b)x^2 - (8a+9b)x$.
12. $(6b+1)x^5 - (a+2b)x^4 - (2a+3c)x$.
13. $(a+b)x^3 - (a+b)x^2 + (a-b)x$.

VII. d. Page 51.

1. $(a-c)x^3 + (b+c)x^2 - (2c+1)x$.
2. $(1-b)x^3 + (a+1)x^2 + (b-1)x - 1$.
3. $(a^2-5a+2)x^3 + (2a-b)x^2 - (a+5)x$.
4. $(a-p+1)x^2 + (b+q+2)x - c - r + 3$.
5. $(p+q-1)x^3 + (p+q)x^2 - (p+q)x + q$.
6. $acx^3 + (2a+bc)x^2 + (2b+c)x + 2$.
7. $acx^3 - (2a+bc)x^2 + (3a+2b)x - 3b$.
8. $apx^3 + (aq-bp)x^2 - (bq+cp)x - cq$.
9. $2bx^3 - (3b-2c)x^2 - (b+3c)x - c$.
10. $ax^3 - (a+2b)x^2 + (2b+3c)x - 3c$.

ANSWERS. 445

11. $apx^3 - (2a+3p)x^2 + (6-aq)x + 3q.$
12. $x^6 - (a^2+2b)x^4 + (2ac+b^2)x^2 - c^2.$
13. $a^2x^6 + (6a-1)x^4 + (9-2b)x^2 - b^2.$
14. $x^8 - (a^2+2b)x^6 + (2ac+b^2+2d)x^4 - (2bd+c^2)x^2 + d^2.$

VIII. (1). Page 54$_A$.

1. 3.	2. 5.	3. 4.	4. 6.	5. 2.	6. 5.
7. 2.	8. 3.	9. 5.	10. 7.	11. $3\frac{1}{2}$.	12. 2.
13. 2.	14. 4.	15. 3.	16. 5.	17. $\frac{1}{2}$	18. $2\frac{1}{2}$.
19. 1.	20. $1\frac{1}{2}$.	21. 2.	22. 1.	23. $2\frac{1}{2}$.	24. $6\frac{2}{3}$.
25. $\frac{5}{8}$.	26. $\frac{7}{12}$.	27. $\frac{8}{21}$.	28. $1\frac{13}{27}$.	29. 12.	30. 15.
31. $4\frac{1}{2}$.	32. -4.				

VIII. a. Page 55.

1. 5.	2. 4.	3. 7.	4. 4.	5. 3.	6. 1.
7. 5.	8. 3.	9. 15.	10. 13.	11. 13.	12. 5.
13. 1.	14. 16.	15. 10.	16. 30.	17. 5.	18. 1.
19. 2.	20. 1.	21. 1.	22. 2.	23. 3.	24. 1.
25. 4.	26. 3.	27. 3.	28. 3.	29. 1.	30. 4.
31. 7.	32. 3.	33. 4.	34. 4.	35. 1.	36. 1.
37. 2.	38. 2.	39. 1.	40. 2.		

VIII. b. Page 58.

1. 20.	2. 15.	3. 8.	4. 16.	5. 25.	6. 17.
7. 13.	8. 10.	9. 7.	10. 4.	11. $-\frac{1}{7}$.	12. $\frac{1}{7}$.
13. 5.	14. 7.	15. 6.	16. 10.	17. 6.	18. 8.
19. 7.	20. 25.	21. $3\frac{1}{7}$.	22. 8.	23. 12.	24. 5.
25. 5.	26. 12.	27. $\frac{4}{7}$	28. $-5\frac{1}{6}$.	29. 8.	30. $66\frac{2}{3}$.
31. 7.	32. 7.	33. 2.	34. 12.	35. 27.	36. 5.

VIII. c. PAGE 60.

1. $2\frac{2}{3}$. 2. 6. 3. 10. 4. -6. 5. $9\frac{2}{3}$. 6. $1\frac{1}{3}$.
7. -12. 8. $\frac{3}{8}$. 9. $1\frac{4}{5}$. 10. $-\frac{3}{4}$. 11. $\frac{4}{5}$. 12. $-\frac{2}{21}$.
13. $1\frac{1}{2}$. 14. $-\frac{2}{3}$. 15. $1\frac{3}{4}$. 16. 12. 17. $3\frac{6}{7}$. 18. $2\frac{1}{4}$.
19. $\frac{5}{7}$. 20. $1\frac{2}{5}$. 21. $\frac{3}{7}$.

IX. a. PAGE 62.

1. $y-x$. 2. $\frac{a}{3}$. 3. $5b$. 4. $3d-2c$.
5. $2k$. 6. $100-x$. 7. $\frac{b}{a}$. 8. $20-c$.
9. $\frac{5}{6}a$. 10. $\frac{600}{x}$. 11. $x+11$. 12. $c-20$.
13. $90-x$. 14. $x-30$. 15. 20. 16. $2x$.
17. $36-x$. 18. $x+a$. 19. $5x$ days. 20. 4.
21. $\frac{x}{2}$. 22. $\frac{x}{4}$. 23. xy miles. 24. $\frac{y}{x}$ miles.
25. $\frac{60x}{a}$. 26. $\frac{120}{x}$ hours. 27. $5p$. 28. $\frac{44}{x}$.
29. $5a+2b$. 30. $400-x$. 31. $240a+12b-c$.
32. $x-6$. 33. b. 34. $40x$. 35. $20a+2b-c$.
36. $\frac{100}{xy}$. 37. $y-\frac{13x}{5}$. 38. $100-x-y-z$.
39. $240x+12y+z-30$. 40. $2y+2z-x$.

IX. b. PAGE 65.

1. x, $x+1$, $x+2$, $x+3$. 2. $y-2$, $y-1$, y.
3. $x-2$, $x-1$, x, $x+1$, $x+2$. 4. $2n+2$. 5. $2x-1$.
6. $6n+3$. 7. $x-a-b$ miles. 8. $n(a+b)$. 9. $x+y+5$.
10. $2x+5$. 11. $mx+y$. 12. $6x$. 13. £$10bc$.
14. £$\frac{ax}{20}$. 15. £$\frac{a^3}{2}$. 16. £$\frac{x^2y^2}{50}$. 17. $3xy$.
18. $\frac{x^2}{9}$. 19. $\frac{px}{2}$. 20. $\frac{abc}{60}$. 21. $\frac{4yx}{x}$.

ANSWERS. 447

22. $ab - \dfrac{c^2}{9}$. 23. $\dfrac{3a}{4x}$. 24. $\dfrac{bc}{20}$ hours. 25. $\dfrac{22a}{15b}$.

26. $\dfrac{15xy}{22}$. 27. $\dfrac{y}{xz}$ days. 28. yz. 29. $\dfrac{y}{10r}$. 30. $\dfrac{100p}{ar}$.

31. $p(p-1)(p-2) = y$. 32. $6n = x$. 33. $pq = 5(a-b)$.

34. $\dfrac{x}{y} = m + n + 10$. 35. $a + x + 5 = 2(a+5)$; 35; 24.

36. $20(p - x) = 3(q + 20x)$. 37. $p - 5 = 7(q - 5)$.

IX. c. PAGE 68$_A$.

1. (i) 272 sq. ft.; (ii) 16 ft.; (iii) 6 ch. 84 lks.
2. (i) 50 cu. ft.; (ii) $4\tfrac{1}{2}$ cu. ft.; (iii) 5 ft.
3. (i) 49; (ii) 1 hr. 20 min.; (iii) $37\tfrac{1}{2}$.
4. (i) 144·9 ft.; (ii) 5 secs.
5. 22 in., 38·5 sq. in.; 11 ft., 9·625 sq. ft.
6. (i) 24·64 sq. in.; (ii) 1 ft. 9 in.
7. (i) $2(x+y)$ ft.; (ii) xy sq. ft.; (iii) $2z(x+y)$ sq. ft.
8. 59 ft. 10 in.; 210 sq. ft.; 718 sq. ft. 9. 10 ft. 6 in.
10. (i) 22 sq. cm.; (ii) 3·6 sq. in. 11. 1·5 in. 12. 27 sq. ft.
13. 328. 14. 15. 15. 55. 16. (i) and (iii).
20. (i) 17; (ii) 24; (iii) 40; (iv) 1·6.
21. (i) ·7854; (ii) 96·6; (iii) 294. 22. 40. 23. 12.
24. (i) 9780; (ii) 1; (iii) 12; (iv) $-40·5$.
25. 4, $5\tfrac{1}{5}$, $6\tfrac{2}{5}$, $7\tfrac{3}{5}$, $8\tfrac{4}{5}$, 10.

X. a. PAGE 71.

1. 17, 12. 2. 13, 5. 3. 75. 4. 20 miles.
5. 15, 43. 6. 162. 7. 1. 8. 50, 55.
9. 27, 28, 29. 10. 3, 5. 11. 15, 5. 12. £20.
13. 5. 14. 60, 61. 15. 6, 3.
16. A £100, B £130, C £150. 17. 53 florins, 71 shillings.
18. Silk 6s., Linen 1s. 19. 48, 12. 20. 65, 40.
21. 60, 10. 22. 20 half-crowns, 5 crowns, 10 shillings.
23. 25, 5. 24. 123 runs, 10 byes, 5 wides.
25. 15 ft., 12 ft. 26. 18 ft., 10 ft.

X. b. Page 73.

1. 54. 2. 24. 3. 60. 4. 35.
5. 75. 6. 24, 25. 7. 224, 252. 8. 49, 50.
9. 50, 51, 52. 10. £33. 11. 27.
12. 90 Port, 150 Claret. 13. A £450, B £180, C £140.
14. A £525, B £600, C £160. 15. £49.
16. 12 ft. 18 ft. 17. £12000. 18. 44.

XI. a. Page 74.

1. $2ab$. 2. x^2y^2. 3. $2xy^2z$. 4. abc.
5. $5ab$. 6. $3xy^2z$. 7. $2a^2b^2c^2$. 8. $7ab^2c^3$.
9. $3x^2yz^2$. 10. $2ax$. 11. $7a$. 12. $17abc$.
13. xy. 14. $8a^2b^2c^2$. 15. $25xy$. 16. bx.
17. $5a^3b^3c^2$. 18. abc.

XI. b. Page 75.

In examples 19–29 the H.C.F. stands first; the L.C.M. second.

1. $2a^2bc$. 2. x^3y^2z. 3. $12x^3y^3z$. 4. $20a^2b^2c^3$.
5. $15a^4b^3c^5$. 6. $24abxy$. 7. abc. 8. $a^2b^2c^2$.
9. $12abc$. 10. $12xyz$. 11. $12x^2y^2z^2$. 12. $42a^2b^3$.
13. $a^2b^2c^3$. 14. $30a^2b^2c^2$. 15. $12x^3y^4$. 16. $56x^4y^5$.
17. $210a^3b^3c^3$. 18. $264a^4b^4c^4$. 19. ac, $12abc$. 20. $2y$, $12xyz$.
21. bc, $9ab^2c$. 22. $13a^2bc$, $39a^3bc^2$. 23. $17xy$, $51x^2yz^2$.
24. $5xy^3z$, $75x^3y^3z^2$. 25. b, $30abc$. 26. $17m^2p^2$, $51m^4n^4p^4$.
27. y^2, $x^3y^5z^4$. 28. $5p^2$, $60m^2p^3q^4$. 29. $36k^2m^2n^4$, $216k^3m^3n^5$.

XII. a. Page 76.

1. $\dfrac{1}{2b}$. 2. $\dfrac{a}{4b}$. 3. $\dfrac{2y}{5x}$. 4. $\dfrac{1}{5ab}$.
5. $\dfrac{z^2}{xy}$. 6. $\dfrac{3a}{5c}$. 7. $\dfrac{3x^2}{4z^3}$. 8. $\dfrac{2a^2}{3bc}$.
9. $\dfrac{4n}{5mp}$. 10. $\dfrac{5m^2p^2}{6n^4}$. 11. $\dfrac{c}{a^2b}$. 12. $\dfrac{3xz}{5y^3}$.
13. $\dfrac{yz^3}{2x}$. 14. $\dfrac{a^2c^3}{3b^2}$. 15. $\dfrac{nq}{mp^3}$. 16. $\dfrac{2np^3}{3m}$.
17. $\dfrac{3x^2}{5ay^4}$. 18. $\dfrac{3}{4abc}$. 19. $\dfrac{2p^2m^2}{3k}$. 20. $\dfrac{2xyz}{3}$.

ANSWERS. 449

XII. b. PAGE 77.

1. $\dfrac{2cd^2}{3b}$. 2. $\dfrac{a^2}{bc}$. 3. $\dfrac{9ax^2z^2}{bc}$. 4. $\dfrac{14b^3}{15c^3y}$.

5. $\dfrac{3mnz^2}{2x}$. 6. $\dfrac{9mnp}{4k}$. 7. $\dfrac{x}{2a^2}$. 8. $\dfrac{400x}{441y^3}$.

9. $\dfrac{y^3z^2}{nx^4}$. 10. 3. 11. $\dfrac{7b}{4a}$. 12. $\dfrac{d^2}{4a^2c}$. 13. $\dfrac{7acy}{8bdx}$.

14. $\dfrac{6x^2yz}{5a}$. 15. $\dfrac{9b^2cz^2}{4x^3y}$. 16. 8. 17. $\dfrac{p^2q^2y}{10x^2}$. 18. y^2.

XII. c. PAGE 78.

1. $\dfrac{4x,\ y}{2a}$. 2. $\dfrac{4x^3,\ 3y^2}{3x^2y}$. 3. $\dfrac{ac,\ 2b^2}{2bc}$.

4. $\dfrac{ad,\ bc,\ 2bd}{bd}$. 5. $\dfrac{6ac,\ b^2}{3bc}$. 6. $\dfrac{5m,\ 4p}{20n}$.

7. $\dfrac{3k,\ 2p}{6x}$. 8. $\dfrac{2m,\ n}{6x}$. 9. $\dfrac{a^2,\ b^2}{abc}$. 10. $\dfrac{ax,\ b}{x^2}$.

11. $\dfrac{2y,\ 3x}{xy}$. 12. $\dfrac{x^2,\ y^2,\ 3x^2y}{xy}$. 13. $\dfrac{4x^2,\ 9y^2}{6xy}$.

14. $\dfrac{8ac,\ 3ab}{10bc}$. 15. $\dfrac{9ac,\ 5b^2}{21bc}$. 16. $\dfrac{18,\ 3ab,\ a^2}{9a}$.

XII. d. PAGE 79.

1. $\dfrac{5x}{6}$. 2. $\dfrac{y}{20}$. 3. $\dfrac{a}{12}$. 4. $\dfrac{2x^2-15}{3x}$.

5. $\dfrac{5x+2y}{10}$. 6. $\dfrac{3a-2b}{12}$. 7. $\dfrac{3m-2n}{24}$. 8. $\dfrac{2m-3n}{15}$.

9. $\dfrac{3x-y}{21}$. 10. $\dfrac{3a+b}{39}$. 11. $\dfrac{3p-q}{48}$. 12. $\dfrac{15m-n}{36}$.

13. $\dfrac{22x}{15}$. 14. $\dfrac{9x}{20}$. 15. $\dfrac{x}{4}$. 16. $\dfrac{6a-4b}{15}$.

17. $\dfrac{11a}{30}$. 18. $\dfrac{5x}{24}$. 19. $\dfrac{7x}{18}$. 20. $\dfrac{5x}{4}$.

21. $\dfrac{31x}{36}$. 22. $\dfrac{17x}{24}$. 23. $\dfrac{bx-ay}{ab}$. 24. $\dfrac{9bx+2ay}{3ab}$.

25. $\dfrac{ac+b}{c}$. 26. $\dfrac{xz-y}{z}$. 27. $\dfrac{a^2-3b^2}{3a}$. 28. $\dfrac{a^3+b^3}{a}$.

29. $\dfrac{x^3-2y^2}{2x^2}$. 30. $\dfrac{p^5-k^5}{p^2}$.

Miscellaneous Examples II. Page 80.

1. $3x^2+7x-8$.
2. $13z$.
3. 20.
4. $a^2+b^2+c^2$.
5. $x^5-11x-10$.
6. (1) $\frac{1}{2}$; (2) -3.
7. x^2+2x-3.
8. $-4a+5b$.
9. $5x$.
10. $4x^2-6x-1$.
11. (1) $x^2+14x-51$; (2) $24x^2-55x-24$.
12. (1) $\frac{1}{4}$; (2) 1.
13. $-ab$.
14. $\frac{8}{5}$.
15. $16a^2+2ab$.
16. $x^5+4x^4+48x-32$.
17. $29a$.
18. (1) -2; (2) 41.
19. $3p^3-5p^2+2p$.
20. $6a+2c-2d$.
21. $2x^3-x^2-x$.
22. 1.
23. 1935.
24. 4.
25. $4m-5n$.
26. $3x-9$.
27. 0.
28. (1) -15; (2) 4.
29. $3y^3-9y^2+2y-1$.
30. A £800, B £320.
31. 14.
32. $6m^4-96$.
33. $x-2$.
34. $ap+bq$ miles; $\frac{ap+bq}{c}$ hours. Numerically, 55 miles; 5 hours.
35. (1) $\frac{1}{7}$; (2) $7\frac{1}{13}$.
36. 4320.

XIII. a. Page 86.

1. $x=2$, $y=1$.
2. $x=3$, $y=5$.
3. $x=2$, $y=3$.
4. $x=4$, $y=-1$.
5. $x=1$, $y=2$.
6. $x=3$, $y=4$.
7. $x=5$, $y=6$.
8. $x=1$, $y=2$.
9. $x=3$, $y=1$.
10. $x=2$, $y=1$.
11. $x=1$, $y=3$.
12. $x=1$, $y=1$.
13. $x=7$, $y=5$.
14. $x=10$, $y=3$.
15. $x=5$, $y=12$.
16. $x=7$, $y=8$.
17. $x=6$, $y=8$.
18. $x=5$, $y=8$.
19. $x=-7$, $y=-3$.
20. $x=17$, $y=-19$.
21. $x=1$, $y=2$.

XIII. b. Page 87.

1. $x=12$, $y=8$.
2. $x=10$, $y=6$.
3. $x=18$, $y=12$.
4. $x=20$, $y=15$.
5. $x=45$, $y=35$.
6. $x=51$, $y=17$.
7. $x=20$, $y=60$.
8. $x=14$, $y=15$.
9. $x=-2$, $y=4$.
10. $x=3$, $y=5$.
11. $x=7$, $y=3$.
12. $x=5$, $y=4$.
13. $x=3$, $y=-4$.
14. $x=19$, $y=3$.
15. $x=12$, $y=-4$.
16. $x=13$, $y=7$.

ANSWERS. 451

XIII. c. Page 90.

1. $x=1, y=2, z=3$.
2. $x=-2, y=4, z=1$.
3. $x=2, y=3, z=1$.
4. $x=1, y=2, z=3$.
5. $x=9, y=2, z=-4$.
6. $x=3, y=2, z=1$.
7. $x=5, y=6, z=7$.
8. $x=1, y=2, z=3$.
9. $x=2, y=-2, z=5$.
10. $x=4, y=-3, z=2$.
11. $x=8, y=10, z=14$.
12. $x=3, y=9, z=15$.
13. $x=6, y=8, z=5$.
14. $x=\frac{3}{2}, y=\frac{2}{3}, z=\frac{5}{6}$.
15. $x=6, y=2, z=1$.
16. $x=35, y=30, z=25$.

XIII. d. Page 92.

1. $x=5, y=3$.
2. $x=2, y=7$.
3. $x=3, y=2$.
4. $x=\frac{1}{3}, y=\frac{1}{4}$.
5. $x=7, y=6$.
6. $x=\frac{1}{3}, y=\frac{1}{5}$.
7. $x=2, y=-3$.
8. $x=-5, y=4$.
9. $x=\frac{2}{3}, y=\frac{3}{4}$.
10. $x=9, y=25$.
11. $x=\frac{1}{4}, y=\frac{1}{3}$.
12. $x=\frac{1}{5}, y=\frac{1}{6}$.
13. $x=\frac{1}{2}, y=\frac{1}{3}, z=\frac{1}{4}$.
14. $x=\frac{1}{8}, y=\frac{1}{12}, z=\frac{1}{16}$.
15. $x=3, y=-2, z=1$.

XIV. Page 95.

1. 22, 12.
2. 55, 18.
3. 25, 17.
4. 53, 23.
5. 23. 17.
6. Tea 3s. 4d., Sugar 4d.
7. Horse £23, Cow £16.
8. A £140, B £60, C £70, D £20.
9. A £99, B 115, C £33, D £23.
10. A 36 years, B 14 years.
11. A 55 years, B 21 years.
12. A 5 miles, B 4 miles.
13. C $3\frac{1}{2}$ miles, D $4\frac{1}{4}$ miles.
14. $\frac{13}{25}$.
15. $\frac{15}{26}$.
16. $\frac{2}{15}$.
17. $\frac{3}{14}$.
18. 28, 82.
19. 85, 58.
20. 27.
21. 72.
22. £5. 11s.
23. 8 white, 12 black.
24. 860.
25. Man 2s. 6d., Boy 1s. 6d.
26. 20 lbs., 40 lbs.
27. 15 miles.
28. 8 hours.
29. 6 miles, 3 miles an hour.
30. 10s., 1s. 6d.
31. £500.
32. 3 miles, $4\frac{2}{7}$ miles an hour.

XV. a. Page 99.

1. $9a^2b^6$.
2. a^6c^2.
3. $49a^2b^4$.
4. $121b^4c^6$.
5. $16a^8b^{10}x^4$.
6. $25x^4y^{10}$.
7. $4a^2b^2c^4$.
8. $9c^2x^6$.
9. $16x^2y^2z^6$.
10. $\dfrac{4}{9}a^4b^6$.
11. $\dfrac{4x^4}{9y^6}$.
12. $\dfrac{16}{9x^4y^2}$.
13. $\dfrac{49a^2b^2}{9}$.
14. $\dfrac{9a^4b^6}{16c^{10}x^8}$.
15. $\dfrac{1}{4x^2y^2}$.
16. $4x^2y^4$.
17. $\dfrac{25a^2b^6}{4x^2y^2}$.
18. $169c^{10}x^6$.
19. $\dfrac{1}{16a^8}$.
20. $\dfrac{9a^{10}}{25x^6}$.
21. $8a^3b^6$.
22. $27x^9$.
23. $64x^{12}$.
24. $-27a^9b^3$.
25. $-125a^3b^6$.
26. $-b^9c^6x^3$.
27. $-216a^{18}$.
28. $-8a^{21}c^6$.
29. $\dfrac{1}{27y^6}$.
30. $-\dfrac{27x^{15}}{125a^9}$.
31. $343x^9y^{12}$.
32. $-\dfrac{8}{27}a^{15}$.
33. $81a^8b^{12}$.
34. $a^{10}x^9$.
35. $32m^{15}y^5$.
36. $\dfrac{1}{128a^{14}}$.
37. $\dfrac{243x^{20}}{32y^{15}}$.
38. $\dfrac{256x^{24}}{6561y^8}$.
39. $-\dfrac{x^{21}}{2187}$.
40. $\dfrac{64x^{30}}{729a^{24}}$.

XV. b. Page 101.

1. $a^2+6ab+9b^2$.
2. $a^2-6ab+9b^2$.
3. $x^2-10xy+25y^2$.
4. $4x^2+12xy+9y^2$.
5. $9x^2-6xy+y^2$.
6. $9x^2+30xy+25y^2$.
7. $81x^2-36xy+4y^2$.
8. $25a^2b^2-10abc+c^2$.
9. $p^2q^2-2pqr+r^2$.
10. $x^2-2abcx+a^2b^2c^2$.
11. $a^2x^2+4abxy+4b^2y^2$.
12. x^4-2x^2+1.
13. $a^2+b^2+c^2-2ab-2ac+2bc$.
14. $a^2+b^2+c^2+2ab-2ac-2bc$.
15. $a^2+4b^2+c^2+4ab+2ac+4bc$.
16. $4a^2+9b^2+16c^2-12ab+16ac-24bc$.
17. $x^4+y^4+z^4-2x^2y^2-2x^2z^2+2y^2z^2$.
18. $x^2y^2+y^2z^2+z^2x^2+2xy^2z+2x^2yz+2xyz^2$.
19. $9p^2+4q^2+16r^2-12pq+24pr-16qr$.
20. $x^4-2x^3+3x^2-2x+1$.
21. $4x^4+12x^3+5x^2-6x+1$.
22. $x^2+y^2+a^2+b^2-2xy+2ax-2bx-2ay+2by-2ab$.
23. $4x^2+9y^2+a^2+4b^2+12xy+4ax-8bx+6ay-12by-4ab$.
24. $m^2+n^2+p^2+q^2-2mn-2mp-2mq+2np+2nq+2pq$.
25. $\dfrac{a^2}{4}+4b^2+\dfrac{c^2}{16}-2ab+\dfrac{ac}{4}-bc$.
26. $\dfrac{a^2}{9}+9b^2+\dfrac{9}{4}-2ab-a+9b$.
27. $\dfrac{4x^4}{9}-\dfrac{4x^3}{3}+3x^2-3x+\dfrac{9}{4}$.

ANSWERS.

XV. c. PAGE 101.

1. $x^3 + 3ax^2 + 3a^2x + a^3$.
2. $x^3 - 3ax^2 + 3a^2x - a^3$.
3. $x^3 - 6x^2y + 12xy^2 - 8y^3$.
4. $8x^3 + 12x^2y + 6xy^2 + y^3$.
5. $27x^3 - 135x^2y + 225xy^2 - 125y^3$.
6. $a^3b^3 + 3a^2b^2c + 3abc^2 + c^3$.
7. $8a^3b^3 - 36a^2b^2c + 54abc^2 - 27c^3$.
8. $125a^3 - 75a^2bc + 15ab^2c^2 - b^3c^3$.
9. $x^6 + 12x^4y^2 + 48x^2y^4 + 64y^6$.
10. $64x^6 - 240x^4y^2 + 300x^2y^4 - 125y^6$.
11. $8a^9 - 36a^6b^2 + 54a^3b^4 - 27b^6$.
12. $125x^{15} - 300x^{10}y^4 + 240x^5y^8 - 64y^{12}$.
13. $a^3 - 2a^2b + \dfrac{4}{3}ab^2 - \dfrac{8}{27}b^3$.
14. $\dfrac{1}{27}a^3 + \dfrac{2}{3}a^2 + 4a + 8$.
15. $\dfrac{1}{27}x^6 - x^5 + 9x^4 - 27x^3$.
16. $\dfrac{1}{216}a^3 + \dfrac{1}{6}a^2x + 2ax^2 + 8x^3$.

XVI. a. PAGE 103.

1. $2ab^2$.
2. $3x^3y$.
3. $5x^2y^3$.
4. $4a^2bc^3$.
5. $9a^3b^4$.
6. $10x^4$.
7. $a^{10}b^3c^2$.
8. a^4bc^6.
9. $8x^3y^9$.
10. $\dfrac{6}{a^{18}}$.
11. $\dfrac{a^8b^4}{4}$.
12. $\dfrac{17y^2}{5}$.
13. $\dfrac{18x^6}{13y^3}$.
14. $\dfrac{9a^9}{6b^6}$.
15. $\dfrac{16xy^2}{17p^7}$.
16. $\dfrac{20a^{20}b^{10}}{9x^5y^9}$.
17. $3a^2bc$.
18. $-2a^4b^3$.
19. $4x^2yz^4$.
20. $-7a^4bc$.
21. $-\dfrac{x^4y^3}{5}$.
22. $\dfrac{2x^3}{9y^5}$.
23. $\dfrac{5ab^2}{6x^2y^3}$.
24. $-\dfrac{3x^9}{4y^{21}}$.
25. a^2x^3.
26. x^2y^3.
27. $2xy^2$.
28. $3a^3b$.
29. $2ax^8$.
30. $-x^2y^3$.
31. $\dfrac{2}{a^9b^8}$.
32. $\dfrac{a^3x^5}{b^{10}}$.
33. $\dfrac{a^2}{b^3c^4}$.

XVI. b. PAGE 105.

1. $x + 2y$.
2. $3a + 2b$.
3. $x - 5y$.
4. $2x - 3y$.
5. $9x + y$.
6. $5x - 3y$.
7. $x^2 - y^2$.
8. $1 - a^3$.
9. $a^2 - a + 1$.
10. $2x^2 - 3x + 5$.
11. $3x^2 - 2x - 1$.
12. $x^2 - 2x + 1$.
13. $2a^2 + a - 2$.
14. $1 - 5x + x^2$.
15. $2x + 3y - 5z$.
16. $4x^3 + 2x^4 - x^5$.
17. $x^3 - 11x + 17$.
18. $5x^2 - 3ax + 4a^2$.
19. $2x^2 + y^2 - 3z^2$.
20. $ab - 2ac + 3bc$.
21. $2a^2 + b^2 - 3c^2$.
22. $2x^2 - xy + 3y^2$.
23. $3x^2 - 5x + 7$.
24. $1 - 2x + 3x^2 - 4x^3$.
25. $ax^5 - 2bx^2 + 3c$.

XVI. c. Page 106.

1. $\dfrac{x}{2}-3$. 2. $2-\dfrac{x}{y}$. 3. $\dfrac{x}{5}+y$. 4. $\dfrac{x}{y}+5$. 5. $\dfrac{x}{2y}-2$.

6. $\dfrac{x}{y}-\dfrac{a}{b}$. 7. $\dfrac{8x}{3y}+2$. 8. $\dfrac{3x}{5}-\dfrac{5}{3x}$. 9. $\dfrac{a^2}{8}+\dfrac{a}{2}-1$.

10. $x^2+x-\dfrac{1}{2}$. 11. $a^2-\dfrac{3}{2}a+\dfrac{5}{3}$. 12. $x^2-3x+\dfrac{1}{3}$.

13. $\dfrac{a^2}{2}+\dfrac{a}{x}-\dfrac{x}{a}$. 14. $x^2-x+\dfrac{1}{4}$. 15. $\dfrac{x^2}{2}-2x+\dfrac{a}{3}$.

16. $\dfrac{3a}{x}-\dfrac{1}{5}+\dfrac{2x}{3a}$. 17. $4m^2+\dfrac{2}{3}n+1$. 18. $2x^2+8+\dfrac{8}{x^2}$.

XVI. d. Page 109.

1. $a+1$. 2. $x+2$. 3. $ax-y^2$.
4. $2m-1$. 5. $4a-3b$. 6. $1+x+x^2$.
7. $1-2x+3x^2$. 8. $a+2b-c$. 9. $2a^2-3a+1$.
10. y^2-y+1. 11. $2x^2+x-3$. 12. $3x^2-2xa+3a^2$.
13. $3x^3-x-1$. 14. $x^2-2xy+4y^2$. 15. $3x^2-x+6$.

XVI. e. Page 110.

1. $\dfrac{x}{2}-1$. 2. $\dfrac{x}{3}+2$. 3. $2x-\dfrac{y^2}{3}$.

4. $\dfrac{3x}{4y}-2$. 5. $x-\dfrac{3}{x}$. 6. $\dfrac{x^2}{y}-2y^2$.

7. $\dfrac{x}{y}+2-\dfrac{y}{x}$. 8. $\dfrac{x}{3}-1+\dfrac{3}{x}$. 9. $\dfrac{x}{a}-4+\dfrac{2a}{x}$.

10. $\dfrac{4a}{x}-4+\dfrac{x}{a}$. 11. $\dfrac{a}{b}-1+\dfrac{b}{a}$. 12. $\dfrac{2x^2}{y^2}+\dfrac{4x}{y}-3$.

XVII. a. Page 112.

1. $a(a^2-x)$. 2. $x^2(x-1)$. 3. $2a(1-a)$. 4. $a(a-b^2)$.
5. $p(7p+1)$. 6. $2x(4-x)$. 7. $5ax(1-a^2x)$.
8. $x^2(3+x^3)$. 9. $x(x+y)$. 10. $x^2(x-y)$.
11. $5x(1-5xy)$. 12. $5(3+5x^2)$. 13. $16x(1+4xy)$.
14. $15a^2(1-15a^2)$. 15. $27(2-3x)$. 16. $5x^3(2-5xy)$.
17. $x(3x^2-x+1)$. 18. $2x^3(3+x+2x^2)$. 19. $x(x^2-xy+y^2)$.
20. $3a^2(a^2-ab+2b^2)$. 21. $2xy^2(xy-3x+y)$.
22. $3x(2x^2-3xy+4y^2)$. 23. $5x^3(x^2-2a^2-3a^3)$.
24. $7a(1-a^2+2a^3)$. 25. $19a^3x^2(2x^3+3a)$.

ANSWERS. 455

XVII. b. PAGE 113.

1. $(a+b)(a+c)$.
2. $(a-c)(a+b)$.
3. $(ac+d)(ac+b)$.
4. $(a+3)(a+c)$.
5. $(2+c)(x+c)$.
6. $(x-a)(x+5)$.
7. $(5+b)(a+b)$.
8. $(a-y)(b-y)$.
9. $(a-b)(x-z)$.
10. $(p+q)(r-s)$.
11. $(x-y)(m-n)$.
12. $(x-a)(m+n)$.
13. $(2x+y)(a+b)$.
14. $(3a-b)(x-y)$.
15. $(2x+y)(3x-a)$.
16. $(x-2y)(m-n)$.
17. $(ax-3by)(x-y)$.
18. $(x+my)(x-4y)$.
19. $(a+b)(x^2+2)$.
20. $(x-3)(x-y)$.
21. $(2x-1)(x^3+2)$.
22. $(3x+5)(x^2+1)$.
23. $(x+1)(x^3+2)$.
24. $(y-1)(y^2+1)$.
25. $(a+bc)(xy-z)$.
26. $(f^2+g^2)(x^2-a)$.
27. $(2x+3y)(ax-by)$.
28. $(ax+by)(mx-ny)$.
29. $(a-b-c)(x-y)$.
30. $(a+b)(ax+by+c)$.

XVII. c. PAGE 115.

1. $(a+1)(a+2)$.
2. $(a+1)(a+1)$.
3. $(a+3)(a+4)$.
4. $(a-4)(a-3)$.
5. $(x-5)(x-6)$.
6. $(x-7)(x-8)$.
7. $(x-9)(x-10)$.
8. $(x+6)(x+7)$.
9. $(x-10)(x-11)$.
10. $(x-9)(x-12)$.
11. $(x-5)(x-16)$.
12. $(x+6)(x+15)$.
13. $(x-7)(x-12)$.
14. $(x-6)(x-13)$.
15. $(x-3)(x-15)$.
16. $(x+8)(x+12)$.
17. $(x-11)(x-15)$.
18. $(x-13)(x-8)$.
19. $(x+17)(x+6)$.
20. $(a-19)(a-5)$.
21. $(a-16)(a-16)$.
22. $(a+15)(a+15)$.
23. $(a+27)(a+27)$.
24. $(a-19)(a-19)$.
25. $(a-7b)(a-7b)$.
26. $(a+2b)(a+3b)$.
27. $(m-5n)(m-8n)$.
28. $(m-7n)(m-15n)$.
29. $(x-11y)(x-12y)$.
30. $(x-13y)(x-13y)$.
31. $(x^2+1)(x^2+7)$.
32. $(x^2+2y^2)(x^2+7y^2)$.
33. $(xy-3)(xy-13)$.
34. $(x+24y)(x+25y)$.
35. $(xy+17)(xy+17)$.
36. $(a^2b^2+25)(a^2b^2+12)$.
37. $(a-5bx)(a-15bx)$.
38. $(x+13y)(x+30y)$.
39. $(a-2b)(a-27b)$.
40. $(x^2+81)(x^2+81)$.
41. $(4-x)(3-x)$.
42. $(5+x)(4+x)$.
43. $(12-x)(11-x)$.
44. $(8+x)(11+x)$.
45. $(26+xy)(5+xy)$.
46. $(13-xa)(11-xa)$.
47. $(17-x^2)(12-x^2)$.
48. $(27+x)(8+x)$.

XVII. d. PAGE 116.

1. $(x+1)(x-2)$.
2. $(x+2)(x-1)$.
3. $(x+2)(x-3)$.
4. $(x+3)(x-2)$.
5. $(x+1)(x-3)$.
6. $(x+3)(x-1)$.
7. $(x+8)(x-7)$.
8. $(x+8)(x-5)$.
9. $(x+2)(x-6)$.
10. $(a+4)(a-5)$.
11. $(a+3)(a-7)$.
12. $(a+5)(a-4)$.
13. $(a+9)(a-13)$.
14. $(x+12)(x-3)$.
15. $(x+13)(x-12)$.
16. $(x+11)(x-10)$.
17. $(x+6)(x-15)$.
18. $(x+15)(x-16)$.

19. $(a+5)(a-17)$.
20. $(a+8)(a-19)$.
21. $(xy+3)(xy-8)$.
22. $(x+12y)(x-5y)$.
23. $(x+7a)(x-6a)$.
24. $(x+3y)(x-35y)$.
25. $(a+14y)(a-15y)$.
26. $(x+23)(x-5)$.
27. $(x+4y)(x-24y)$.
28. $(x+26)(x-10)$.
29. $(a+2)(a-13)$.
30. $(ay+24)(ay-10)$.
31. $(a^2+7b^2)(a^2-8b^2)$.
32. $(x^2+3)(x^2-17)$.
33. $(y^2+9x^2)(y^2-3x^2)$.
34. $(ab+2c)(ab-5c)$.
35. $(a+14bx)(a-2bx)$.
36. $(a+9xy)(a-27xy)$.
37. $(x^2+25a^2)(x^2-12a^2)$.
38. $(x^2+11a^2)(x^2-12a^2)$.
39. $(x^2+21a^2)(x^2-22a^2)$.
40. $(x^3+30)(x^3-29)$.
41. $(1+x)(2-x)$.
42. $(2+x)(3-x)$.
43. $(11+x)(10-x)$.
44. $(20+x)(19-x)$.
45. $(15+ax)(8-ax)$.
46. $(5+xy)(13-xy)$.
47. $(14+x)(7-x)$.
48. $(17+x)(12-x)$.

XVII. e. PAGE 119.

1. $(x+1)(2x+1)$.
2. $(x+1)(3x+2)$.
3. $(x+2)(2x+1)$.
4. $(x+3)(3x+1)$.
5. $(x+4)(2x+1)$.
6. $(x+2)(3x+2)$.
7. $(x+2)(2x+3)$.
8. $(x+5)(2x+1)$.
9. $(x+3)(3x+2)$.
10. $(x+2)(5x+1)$.
11. $(x+2)(2x-1)$.
12. $(x+1)(3x-2)$.
13. $(x+3)(4x-1)$.
14. $(x+5)(3x-1)$.
15. $(x+8)(2x-1)$.
16. $(2x+1)(x-1)$.
17. $(x+3)(3x-2)$.
18. $(x+4)(2x-7)$.
19. $(x+6)(3x-5)$.
20. $(2x+3)(3x-1)$.
21. $(3x+1)(2x-3)$.
22. $(3x+4)(x+1)$.
23. $(x+7)(3x+2)$.
24. $(2x+5)(x-3)$.
25. $(x+7)(3x-2)$.
26. $(x-7)(3x+2)$.
27. $(3x-5)(2x-7)$.
28. $(4x-7)(x+2)$.
29. $(x-2)(3x-7)$.
30. $(x+13)(3x+2)$.
31. $(x+5)(4x+3)$.
32. $(2x+y)(x-3y)$.
33. $(2x-7)(4x-5)$.
34. $(3x-2y)(4x-5y)$.
35. $(15x-1)(x+15)$.
36. $(15x-2)(x-5)$.
37. $(12x+5)(x-3)$.
38. $(12x-7)(2x+3)$.
39. $(8x-9)(9x-8)$.
40. $(8x+y)(3x-4y)$.
41. $(2+x)(1-2x)$.
42. $(3-x)(1+4x)$.
43. $(2+3x)(3-2x)$.
44. $(4+3x)(1-2x)$.
45. $(1+7x)(5-3x)$.
46. $(7+3x)(1+x)$.
47. $(6-x)(3-5x)$.
48. $(4+5x)(2-x)$.
49. $(5+4x)(4-5x)$.
50. $(8-9x)(3+8x)$.

XVII. f. PAGE 120.

1. $(x+2)(x-2)$.
2. $(a+9)(a-9)$.
3. $(y+10)(y-10)$.
4. $(c+12)(c-12)$.
5. $(3+a)(3-a)$.
6. $(7+c)(7-c)$.
7. $(11+x)(11-x)$.
8. $(20+a)(20-a)$.
9. $(x+3a)(x-3a)$.
10. $(y+5x)(y-5x)$.
11. $(6x+5b)(6x-5b)$.
12. $(3x+1)(3x-1)$.
13. $(6p+7q)(6p-7q)$.
14. $(2k+1)(2k-1)$.
15. $(7+10k)(7-10k)$.
16. $(1+5x)(1-5x)$.
17. $(a+2b)(a-2b)$.
18. $(3x+y)(3x-y)$.

ANSWERS.

19. $(pq+6)(pq-6)$. 20. $(ab+2cd)(ab-2cd)$.
21. $(x^2+3)(x^2-3)$. 22. $(3a^2+11)(3a^2-11)$.
23. $(5x+8)(5x-8)$. 24. $(9a^2+7x^2)(9a^2-7x^2)$.
25. $(x^3+5)(x^3-5)$. 26. $(1+6a^3)(1-6a^3)$. 27. $(3x^2+a)(3x^2-a)$.
28. $(9x^3+5a)(9x^3-5a)$. 29. $(x^2a+7)(x^2a-7)$.
30. $(a+8x^3)(a-8x^3)$. 31. $(ab+3x^3)(ab-3x^3)$. 32. $(x^3y^3+2)(x^3y^3-2)$.
33. $(1+ab)(1-ab)$. 34. $(2+x)(2-x)$. 35. $(3+2a)(3-2a)$.
36. $(3a^2+5b^2)(3a^2-5b^2)$. 37. $(x^2+4b)(x^2-4b)$.
38. $(x+5y)(x-5y)$. 39. $(1+10b)(1-10b)$. 40. $(5+8x)(5-8x)$.
41. $(11a+9x)(11a-9x)$. 42. $(pq+8a^2)(pq-8a^2)$.
43. $(8x+5z^3)(8x-5z^3)$. 44. $(7x^2+4y^2)(7x^2-4y^2)$.
45. $(9p^2z^3+5b)(9p^2z^3-5b)$. 46. $(4x^8+3y^3)(4x^8-3y^3)$.
47. $(6x^{18}+7a^7)(6x^{18}-7a^7)$. 48. $(1+10a^3b^2c)(1-10a^3b^2c)$.
49. $(5x^5+4a^4)(5x^5-4a^4)$. 50. $(ab^2c^3+x^8)(ab^2c^3-x^8)$.
51. $1000 \times 150 = 150000$. 52. $241 \times 1 = 241$.
53. $1000 \times 500 \times 500000$. 54. $658 \times 20 = 13160$.
55. $1006 \times 500 = 503000$. 56. $200 \times 2 = 400$.
57. $2000 \times 1446 = 2892000$. 58. $2378 \times 900 = 2140200$.
59. $2500 \times 1122 = 2805000$. 60. $3000 \times 2462 = 7386000$.
61. $16264 \times 2 = 32528$. 62. $10002 \times 10000 = 100020000$.

XVII. g. Page 121.

1. $(a+b+c)(a+b-c)$. 2. $(a-b+c)(a-b-c)$.
3. $(x+y+2z)(x+y-2z)$. 4. $(x+2y+a)(x+2y-a)$.
5. $(a+3b+4x)(a+3b-4x)$. 6. $(x+5a+3y)(x+5a-3y)$.
7. $(x+5c+1)(x+5c-1)$. 8. $(a-2x+b)(a-2x-b)$.
9. $(2x-3a+3c)(2x-3a-3c)$. 10. $(a+b-c)(a-b+c)$.
11. $(x+y+z)(x-y-z)$. 12. $(2a+y-z)(2a-y+z)$.
13. $(3x+2a-3b)(3x-2a+3b)$. 14. $(1+a-b)(1-a+b)$.
15. $(c+5a-3b)(c-5a+3b)$. 16. $(a+b+c+d)(a+b-c-d)$.
17. $(a-b+x+y)(a-b-x-y)$. 18. $(7x+y+1)(7x+y-1)$.
19. $(a+b+m-n)(a+b-m+n)$. 20. $(a-n+b+m)(a-n-b-m)$.
21. $(b-c+a-x)(b-c-a+x)$. 22. $(4a+x+b+y)(4a+x-b-y)$.
23. $(a+2b+3x+4y)(a+2b-3x-4y)$.
24. $(1+7a-3b)(1-7a+3b)$. 25. $(a-b+x-y)(a-b-x+y)$.
26. $(a-3x+4y)(a-3x-4y)$. 27. $(2a-5x+1)(2a-5x-1)$.
28. $(a+b-c+x-y+z)(a+b-c-x+y-z)$.
29. $(3a+2b+c+x-2y)(3a+2b-c-x+2y)$. 30. $y(2x+y)$.

31. $y(2x-y)$. 32. $(x+5y)(x+y)$. 33. $47x(x+2y)$.
34. $(8x+y)(2x+3y)$. 35. $5y(6x-5y)$. 36. $(12x-1)(2x+7)$.
37. $5a(a+2)$. 38. $(7a+1)(a-1)$. 39. $3a(a+2b-2c)$.
40. $x(x-14y+2z$. 41. $y(2x+y-16)$. 42. $a(4x+a-6)$.

XVII. h. Page 122.

1. $(x+y+a)(x+y-a)$. 2. $(a-b+x)(a-b-x)$.
3. $(x-3a+4b)(x-3a-4b)$. 4. $(2a+b+3c)(2a+b-3c)$.
5. $(x+a+y)(x+a-y)$. 6. $(a+y+x)(a+y-x)$.
7. $(x+a+b)(x-a-b)$. 8. $(y+c-x)(y-c+x)$.
9. $(1+x+y)(1-x-y)$. 10. $(c+x-y)(c-x+y)$.
11. $(x+y+2xy)(x+y-2xy)$. 12. $(a-2b+3ac)(a-2b-3ac)$.
13. $(x+y+a+b)(x+y-a-b)$. 14. $(a-b+c+d)(a-b-c-d)$.
15. $(x-2a+b-y)(x-2a-b+y)$. 16. $(y+b+a+3x)(y+b-a-3x)$.
17. $(x-1+a+2b)(x-1-a-2b)$. 18. $(3a-1+x+4d)(3a-1-x-4d)$.
19. $(x-y+a-b)(x-y-a+b)$. 20. $(a-b+c+d)(a-b-c-d)$.
21. $(2x-3a+c+k)(2x-3a-c-k)$.
22. $(a-5b+3bx-1)(a-5b-3bx+1)$.
23. $(a^2+4x^2+5x^3-3)(a^2+4x^2-5x^3+3)$.
24. $(x^2-a^2+x-3)(x^2-a^2-x+3)$.
25. $(a^2+ab+b^2)(a^2-ab+b^2)$. 26. $(x^2+2xy+4y^2)(x^2-2xy+4y^2)$.
27. $(p^2+3pq+9q^2)(p^2-3pq+9q^2)$. 28. $(c^2+cd+2d^2)(c^2-cd+2d^2)$.
29. $(x^2+3xy-y^2)(x^2-3xy-y^2)$.
30. $(2m^2+3mn+n^2)(2m^2-3mn+n^2)$.

XVII. k. Page 123.

1. $(x-y)(x^2+xy+y^2)$. 2. $(x+y)(x^2-xy+y^2)$.
3. $(x-1)(x^2+x+1)$. 4. $(1+a)(1-a+a^2)$.
5. $(2x-y)(4x^2+2xy+y^2)$. 6. $(x+2y)(x^2-2xy+4y^2)$.
7. $(3x+1)(9x^2-3x+1)$. 8. $(1-2y)(1+2y+4y^2)$.
9. $(ab-c)(a^2b^2+abc+c^2)$. 10. $(2x+3y)(4x^2-6xy+9y^2)$.
11. $(1-7x)(1+7x+49x^2)$. 12. $(4+y)(16-4y+y^2)$.
13. $(5+a)(25-5a+a^2)$. 14. $(6-a)(36+6a+a^2)$.
15. $(ab+8)(a^2b^2-8ab+64)$. 16. $(10y-1)(100y^2+10y+1)$.
17. $(x+4y)(x^2-4xy+16y^2)$. 18. $(3-10x)(9+30x+100x^2)$.
19. $(ab+6c)(a^2b^2-6abc+36c^2)$. 20. $(7-2x)(49+14x+4x^2)$.
21. $(a+3b)(a^2-3ab+9b^2)$. 22. $(3x-4y)(9x^2+12xy+16y^2)$.
23. $(5x-1)(25x^2+5x+1)$. 24. $(6p-7)(36p^2+42p+49)$.
25. $(xy+z)(x^2y^2-xyz+z^2)$. 26. $(abc-1)(a^2b^2c^2+abc+1)$.

ANSWERS. 459

27. $(7x+10y)(49x^2-70xy+100y^2)$.
28. $(9a-4b)(81a^2+36ab+16b^2)$.
29. $(2ab+5x)(4a^2b^2-10abx+25x^2)$.
30. $(xy-6z)(x^2y^2+6xyz+36z^2)$.
31. $(x^2-3y)(x^4+3x^2y+9y^2)$.
32. $(4x^2+5y)(16x^4-20x^2y+25y^2)$.
33. $(2x-z^2)(4x^2+2xz^2+z^4)$.
34. $(6x^2-b)(36x^4+6x^2b+b^2)$.
35. $(a+7b)(a^2-7ab+49b^2)$.
36. $(a^2+9b)(a^4-9a^2b+81b^2)$.
37. $(2x-9y^2)(4x^2+18xy^2+81y^4)$.
38. $(pq-3x)(p^2q^2+3pqx+9x^2)$.
39. $(z-4y^2)(z^2+4zy^2+16y^4)$.
40. $(xy-8)(x^2y^2+8xy+64)$.

XVII. 1. Page 124$_A$.

1. $(x-1)(x-2)$.
2. $(a+2)(a+5)$.
3. $(b+4)(b-3)$.
4. $(y-7)(y+3)$.
5. $(c+1)(c+11)$.
6. $(x-5)(x+1)$.
7. $(n+2)(n+10)$.
8. $(y+10)(y-1)$.
9. $(p-6q)(p+4q)$.
10. $(y+11)(y-10)$.
11. $(z-15)(z+6)$.
12. $(k-6)(k-8)$.
13. $(a+9)(a+9)$.
14. $(b-27)(b+3)$.
15. $(c+27)(c+3)$.
16. $(x-7)(x-7)$.
17. $(y+7z)(y+3z)$.
18. $(z+9)(z-7)$.
19. $(n+8)(n+3)$.
20. $(p-8)(p+3)$.
21. $(l+12)(l-3)$.
22. $(ab-2)(ab-2)$.
23. $(ab+8)(ab+2)$.
24. $(b-9)(b+5)$.
25. $(m+11)(m-8)$.
26. $(n-15)(n+3)$.
27. $(p+13)(p-3)$.
28. $(xy-9)(xy+8)$.
29. $(z-5)(z+4)$.
30. $(x+8y)(x-7y)$.
31. $(a-13b)(a+2b)$.
32. $(ab-8)(ab+7)$.
33. $(y^2+13)(y^2-12)$.
34. $(z^2-13)(z^2+6)$.
35. $(y^2+5)(y^2-7)$.
36. $(x+13y)(x-7y)$.
37. $m^2n^2(m-3n)$.
38. $5x^3(2+5xy)$.
39. $(y-5)(y+3)$.
40. $(a+b)(x+y)$.
41. $(x+y)(x-z)$.
42. $(3c-2)(c+1)$.
43. $(2b+1)(b+5)$.
44. $(x-3y)(x-3y)$.
45. $(3x-1)(x-3)$.
46. $(cd+1)(cd-2)$.
47. $(2x+3)(3x-1)$.
48. $(a-b)(4-c)$.
49. $(a^3+2)(a+1)$.
50. $2c^2d(c-3d+d^2)$.
51. $xy(x+9)(x-7)$.
52. $(2y-3)(3y+1)$.
53. $(2x-3)(2x-3)$.
54. $(3+4p)(1-3p)$.
55. $(4+pq)(4+pq)$.
56. $z(4z-3)(z+2)$.
57. $a(a+7)(a-6)$.
58. $(m^3+2)(2m-1)$.
59. $a^2(a-b)(a-3)$.
60. $(7+x)(2-x)$.
61. $(17-z)(1-z)$.
62. $(2m^2+3)(m^2-7)$.
63. $(5x-3y)(x+2y)$.
64. $(3m^3-5)(2m^3+9)$.
65. $(3m-4)(3m-4)$.
66. $(5+9a)(5-9a)$.

67. $(a^2b^2+3)(a^2b^2-3)$.
68. $(3+l)(9-3l+l^2)$.
69. $(1-4m)(1+4m+16m^2)$.
70. $(k^2+5l)(k^2-5l)$.
71. $(pq-1)(p^2q^2+pq+1)$.
72. $(2z+1)(4z^2-2z+1)$.
73. $(1+8x)(1-8x)$.
74. $2(5p+1)(25p^2-5p+1)$.
75. $4(5ab+1)(5ab-1)$.
76. $(9+cd)(81-9cd+c^2d^2)$.
77. $(a+x+1)(a+x-1)$.
78. $(4+b-c)(4-b-c)$.
79. $x(3x+2y)(3x-2y)$.
80. $(p-5q)(p+4q)$.
81. $l(l-7)(l+6)$.
82. $(abc+9d)(abc-9d)$.
83. $(4x^2-3y)(16x^4+12x^2y+9y^2)$.
84. $(x-17)(x+19)$.
85. $(x^2+17)(x^2-17)$.
86. $(l+17)(l-16)$.
87. $(10z-3)(100z^2+30z+9)$.
88. $(a+23)(a-13)$.
89. $(a+b+c)(a-b-c)$.
90. $(1+x-3y)(1-x+3y)$.
91. $(x^2+y^2+3xy)(x^2+y^2-3xy)$.
92. $(a^2+a+2)(a^2-a+2)$.
93. $(b-29)(b+27)$.
94. $(x+2)(x^2-2x+4)(x-2)(x^2+2x+4)$.
95. $(3y+2x)(9y^2-6xy+4x^2)(3y-2x)(9y^2+6xy+4x^2)$.
96. $(x^4+1)(x^2+1)(x+1)(x-1)$.
97. $ab(3a+b)(9a^2-3ab+b^2)(3a-b)(9a^2+3ab+b^2)$.
98. $a^2(ax+2y)(a^2x^2-2axy+4y^2)(ax-2y)(a^2x^2+2axy+4y^2)$.
99. $(a^2+b^2)(a^4-a^2b^2+b^4)(a+b)(a^2-ab+b^2)(a-b)(a^2+ab+b^2)$.
100. $(x^2+2y^2z^2)(x^2+2y^2z^2)$.
101. $(ab+8)(a^2b^2-8ab+64)$.
102. $(2x+7)(x+5)$.
103. $20y(5x+y)(5x-y)$.
104. $\{(a+b)^2+1\}(a+b+1)(a+b-1)$.
105. $(c+d-1)\{(c+d)^2+c+d+1\}$.
106. $(1-x+y)\{1+x-y+(x-y)^2\}$.
107. $(x-19)(x+13)$.
108. $(a+9)(a-31)$.
109. $2\{5(a-b)+1\}\{25(a-b)^2-5(a-b)+1\}$.
110. $2c(c^2+3d^2)$.
111. $9y(4x^2+2xy+y^2)$.
112. $(x-2y)(x+2y+1)$.
113. $(a-b)(a+b+1)$.
114. $(a+b)(a+b+1)$.
115. $(a+b)(a^2-ab+b^2+1)$.
116. $(a+3b)(a-3b+1)$.
117. $(x-y)\{2(x-y)+1\}\{2(x-y)-1\}$.
118. $xy(x+y)(x-y)(x-y)$.

ANSWERS. 461

Miscellaneous Examples III. PAGE 126.

1. $x^3 - 2x$.
2. $42a - 40b + 30c$.
3. $a^6 - c^6$.
4. (1) 12; (2) $x = 5$, $y = 6$.
5. $x^2 + 4x - 1$.
6. 72.
7. $\dfrac{109}{210}$.
8. $x^2 + \dfrac{3}{4}x + \dfrac{5}{4}$.
9. $2x^2 - x$.
10. (1) $(ax - 5)(ax + 3)$; (2) $(2m^2 + 9pq)(2m^2 - 9pq)$.
11. (1) $x = -2$, $y = 4$; (2) $x = 5$, $y = -2$.
12. $\dfrac{am}{pb}$ miles.
13. $84x^4 + 25x^3 + 101x - 30$.
14. (1) 7; (2) $-1\tfrac{1}{2}$.
15. $x^4 + 14x^3 + 27x^2 - 154x + 121$.
16. (1) $(x + 2a)(x - b)$; (2) $(x^2 + 14y)(x^2 - 4y)$.
17. L.C.F. = 7; L.C.M. = $3528 a^3 b^2 c^3$.
18. £14.
19. $3p$.
21. (1) $m - n = a + c$; (2) $3a^2 b^2 + c^3 = p(m + n)$.
22. 8.
23. $6x^2 - xy - y^2$.
24. Apples 4d. a dozen; eggs 1s. 4d. a score.
25. $(x - 5)(2x - 3)(x + 1)$.
26. 33.
27. $2x^2 + 9xy - 7y^2$.
28. $x = 2$, $y = 3$, $z = 0$.
29. (1) $xy(x + 2y)(x - 2y)$; (2) $(m + n)(m - n)(2m^2 + 3n^2)$.
30. $\dfrac{bc}{am}$ days, $2\tfrac{5}{8}$.

XVIII. a. PAGE 129.

1. $a + b$.
2. $x + y$.
3. $x(x - y)$.
4. $2x - 3y$.
5. $x + y$.
6. $ab(a - b)$.
7. $a(a - x)$.
8. $a + 2x$.
9. $b(a + b)$.
10. $x - 3y$.
11. $a - x$.
12. $2x + y$.
13. $2(5x - 1)$.
14. $3x + 2y$.
15. $x + 1$.
16. $y(x - 1)$.
17. $(x - y)^2$.
18. $x^2 + a^2$.
19. $x + 2y$.
20. $x - 3a$.
21. $x + 2$.
22. $x - 5$.
23. $x - 3$.
24. $x - 3$.
25. $3x + 1$.
26. $x - 1$.
27. $cx + d$.
28. $x^2 + y$.
29. $x(a - 3b)$.
30. $2x + 1$.
31. $x^2(3x + 2)$.

XVIII. b. PAGE 133.

1. x^2-3x+2.
2. $x^2-13x+5$.
3. x^2-8.
4. x^2-5.
5. x^2+2x+1.
6. $x+3$.
7. $a^2-2ax+x^2$.
8. $x+1$.
9. x^2-3x+7.
10. $2x^2-7$.
11. $3x^2+1$.
12. $2x^2-3$.
13. $3x^2+2a^2$.
14. x^2-ax+a^2.
15. $x^2+2ax-a2$.
16. $3a^2-ax-2x^2$.
17. $xy(2x^2+xy-3y^2)$.
18. $2x^2a^2(2x-3a)$.
19. $2x^2(2x+7)$.
20. $6(3x-5a)$.
21. $3x^2-2xy+y^2$.
22. x^4+x^3-1.
23. $1+x^3-x^4$.
24. $1+a$.
25. $x(3+4x)$.
26. x^2-2x+1.
27. $2x^2-7$.

XIX. a. PAGE 137.

1. $\dfrac{3}{2b}$.
2. $\dfrac{b}{c}$.
3. $\dfrac{1}{ax-1}$.
4. $\dfrac{3b^2c}{20(a-b)}$.
5. $\dfrac{2x-3y}{2x}$.
6. $4(x-y)$.
7. $\dfrac{1}{2a+3x}$.
8. $\dfrac{x}{x^2-2y^2}$.
9. $\dfrac{x-3y}{x^2+3xy+9y^2}$.
10. $\dfrac{x}{x+1}$.
11. $\dfrac{3x}{x+2}$.
12. $\dfrac{5a}{3b}$.
13. $\dfrac{xy}{x-2}$.
14. $\dfrac{3(a+b)}{a-b}$.
15. $\dfrac{x^2-17}{x^2-5}$.
16. $\dfrac{x+2y}{x^2+xy+y^2}$.
17. $\dfrac{2x+3}{3x+5}$.
18. $\dfrac{a(x-4)}{x+5}$.
19. $\dfrac{x+7}{x+13}$.
20. $\dfrac{3+a}{2}$.

XIX. b. PAGE 139.

The expression in [ ] is in each case the H.C.F. of the numerator and the denominator.

1. $\dfrac{a-2b}{a+2b}$ $[a^2+ab+b^2]$.
2. $\dfrac{x-3}{x+2}$ $[(x-1)^2]$.
3. $\dfrac{a+5}{a+4}$ $[(a-1)(a-2)]$.
4. $\dfrac{x^2+4xy-9y^2}{2x^2+3xy+7y^2}$ $[2x-3y]$.
5. $\dfrac{2a+5b}{3a+5b}$ $[(2a+3b)(a-b)]$.
6. $\dfrac{1-x+2x^2}{1-x+3x^2}$ $[1+x+x^2]$.
7. $\dfrac{x-1}{3x^2+3x+10}$ $[x-1]$.
8. $\dfrac{3a^2+b^2}{4a-b}$ $[a-b]$.
9. $\dfrac{4x^2-ax+a^2}{x^3+a^3}$ $[x+a]$.
10. $\dfrac{2(2x^2-3x-1)}{3x^3+x^2+x-2}$ $[x-1]$.
11. $(2x-3a)^2$ $[(2x+3a)^2]$.
12. $\dfrac{3x^2-x-2}{3x^2+x-2}$ $[2x+1]$.

ANSWERS. 463

13. $\dfrac{5x+2}{7x-4}$ $[x^2-3]$. 14. $\dfrac{2x^2+3x+5}{2x^2+3x-5}$ $[2x^2-3x+5]$.

15. $\dfrac{3(x-3a)(x-4a)}{2(x+3a)(x+4a)}$ $[x-2a]$. 16. $\dfrac{a(x+8a)}{x(x+7a)}$ $[x^2-13ax+5a^2]$.

XIX. c. Page 142.

1. $\dfrac{7}{12}$. 2. $\dfrac{ab}{2a-1}$. 3. 2. 4. $\dfrac{a-11}{a-2}$.

5. $\dfrac{4x+3a}{x+2}$. 6. $\dfrac{5a-b}{x(3a-2)}$. 7. $\dfrac{x+2}{x-1}$. 8. $\dfrac{x+1}{x+5}$.

9. $\dfrac{x}{x-2}$. 10. $\dfrac{2x-1}{2x-3}$. 11. $\dfrac{x+1}{x+5}$. 12. $\dfrac{x-1}{4x+7}$.

13. b^2+3b+9. 14. $\dfrac{1}{x+7}$. 15. $8pq-z^2$. 16. x.

17. $\dfrac{x+1}{x-1}$. 18. $\dfrac{x-5}{x-1}$. 19. $\dfrac{x}{y}$. 20. x.

21. $\dfrac{2x-1}{2x-5}$. 22. 1. 23. $\dfrac{1}{b}$. 24. $\dfrac{a+b-c}{b-c-a}$.

25. $\dfrac{1}{x-8}$. 26. $\dfrac{a-x}{a+x}$. 27. $\dfrac{m}{n}$. 28. $x(2+x)$.

29. $\dfrac{x+4}{x(x-4)}$. 30. 1. 31. $a+x$. 32. $\dfrac{a^2}{16a^2+4ab+b^2}$.

XX. a. Page 144.

1. $x(x+1)$. 2. $x^2(x-3)$. 3. $12x^2(x+2)$. 4. $21x^3(x+1)$.
5. $x(x+1)(x-1)$. 6. $ab(a+b)$. 7. $xy(2x+1)(2x-1)$.
8. $6x(3x-1)$. 9. $x(x+1)(x+2)$. 10. $(x+1)(x-1)(x-2)$.
11. $(x+2)^2(x+3)$. 12. $(x-1)(x-2)(x-4)$.
13. $(x-3)(x-1)(x+2)$. 14. $(x+5)(x-4)(x-6)$.
15. $(x+7)(x-6)(x-5)$. 16. $(x+1)(x+2)(2x+1)$.
17. $(x+2)(x+3)(3x+2)$. 18. $(x+2)(x+3)(5x+1)$.
19. $(x+2)(x+8)(2x-1)$. 20. $(x+2)(x-2)(3x-7)$.
21. $12x(x+2)(2x+1)(4x-7)$. 22. $6x^2(x+7)(3x+5)(3x-2)$.
23. $20x^2y(3x+1)(5x+1)(4x-1)$. 24. $(x+y)(2x-7y)(4x-5y)$.
25. $(x-y)(3x-2y)(4x-5y)$. 26. $3a^2x(3x-a)(2x+3a)(x+5a)$.
27. $2axy^3(x+3)(4x-1)(3x-2)$. 28. $x^2(3-5x)^2(2+x)^2$.
29. $42a^4b^2(a-b)^3(a+b)(a^2+ab+b^2)$.
30. $m^3n(m^6-n^6)(m-n)^2$. 31. $8c^2(2c-3d)^2(8c^3-27d^3)$.

XX. b. Page 146.

1. H.C.F. $x-2$. L.C.M. $(x+1)^2(x+2)(x-2)(x-3)$.
2. $(ax+b)(ax-b)(bx+a)$.   3. $xy(x-a)(y-b)(y-2b)$.
4. H.C.F. $x(x+3)$. L.C.M. $x(x-1)(x+3)(2x-1)$.
5. $(1+x)^3(1-x)^2$.   6. $(x-2)(x-4)(x-6)$.
7. H.C.F. $2x+1$. L.C.M. $(2x+1)(x+1)(x-1)(3x+2)(3x-2)$.
8. $ab^2c^2(c+a)^2(c-a)^2$.
9. L.C.M. $y^2(x-y)^2(x^2+xy+y^2)$. H.C.F. $x-y$.
10. H.C.F. $2x-3$. L.C.M. $(2x-3)(3x-2)(x+4)(3x+4)$.
11. $(x+a)^2(x^2+ax+a^2)(x^2-ax+a^2)$.
12. H.C.F. $3x-y$. L.C.M. $(3x-y)(x+y)^2(x-y)^2$.
13. $x-1$.   14. $(a+b)(a-b)(a-2b)(a^2+ab+b^2)$.
15. H.C.F. a^2+xy. L.C.M. $(a^2+xy)(2x+3y)(2x-3y)$.
16. H.C.F. $(x-3)(x-4)$. L.C.M. $(x-2)(x-3)(x-4)(x-5)$.
17. $x-8a$.   18. $105x^2y^2(x+y)^2(x-y)^2$.

XXI. a. Page 150.

1. $\dfrac{4(x+1)}{5}$. 2. $\dfrac{13(x-2)}{12}$. 3. $\dfrac{25x-61}{56}$. 4. $\dfrac{17x}{36}$.
5. $\dfrac{19x-201}{225}$. 6. $\dfrac{12x^2+28x-27}{8x^2}$. 7. 0. 8. $\dfrac{3(a+3b)}{8a}$.
9. $\dfrac{6b^2c+6bc^2+3ac^2+3a^2c-4a^2b+4ab^2}{12abc}$. 10. $\dfrac{a^2+3x^2}{2ax}$.
11. $\dfrac{5x+31}{102x}$. 12. $\dfrac{a^4b^2-b^4c^2+a^2c^4}{a^2b^2c^2}$. 13. $\dfrac{11x^3-18x^2-27x-16}{30x^3}$.
14. $\dfrac{x^3+y^3}{x^2y^3}$. 15. $\dfrac{3y+2z}{yz}$. 16. $\dfrac{a^3+b^3+c^3-3abc}{abc}$.

XXI. b. Page 151.

1. $\dfrac{2x+5}{(x+2)(x+3)}$. 2. $\dfrac{x+5}{(x+3)(x+4)}$. 3. $\dfrac{1}{(x-4)(x-5)}$.
4. $\dfrac{2(x+6)}{(x-6)(x+2)}$. 5. $\dfrac{(a-b)x}{(x+a)(x+b)}$. 6. $\dfrac{(a+b)x-2ab}{(x-a)(x-b)}$.
7. $\dfrac{2}{(x+2)(x+4)}$. 8. $\dfrac{4ax}{a^2-x^2}$. 9. $\dfrac{8x}{x^2-4}$.
10. $\dfrac{6}{(x-2)(x-5)}$. 11. $\dfrac{ax}{x^2-a^2}$. 12. $\dfrac{5x+9}{x^2-9}$.
13. $\dfrac{x+2y}{4x^2-9y^2}$. 14. $\dfrac{3ax}{x^2-4a^2}$. 15. $\dfrac{4ab}{4a^2-b^2}$.

ANSWERS. 465

16. $\dfrac{2xy}{x^2-y^2}$.   17. $\dfrac{2x^3}{1-x^4}$.   18. $\dfrac{x^2+y^2}{xy(x^2-y^2)}$.

19. $\dfrac{5x^2}{25x^2-y^2}$.   20. $\dfrac{x^4+y^4}{xy(x^4-y^4)}$.   21. $\dfrac{4a^2}{x(x+2a)}$.

22. $\dfrac{2x^3}{x^2-y^2}$.   23. $-\dfrac{2ax}{a^3-8x^3}$.   24. $2b$.

25. $\dfrac{4(x-1)}{(x-2)^2(x+2)}$.   26. $\dfrac{x^2+a^2}{ax(x-a)(x+a)^2}$.

XXI. c. PAGE 153.

1. $\dfrac{2}{x+y}$.   2. $\dfrac{x}{4x^2-y^2}$.   3. $\dfrac{1-6x^2}{1-4x^2}$.   4. $\dfrac{4a^2+b^2}{4a^2-9b^2}$.

5. $\dfrac{1+a}{9-a^2}$.   6. $\dfrac{4x-5}{6(x^2-1)}$.   7. 0.   8. $\dfrac{12a^2-4a+7}{3(4a^2-9)}$.

9. $\dfrac{2(13x+7)}{3(x^2-4)}$.   10. $\dfrac{x^2+y^2}{x^4+x^2y^2+y^4}$.   11. $\dfrac{2}{(x-4)(x-6)}$.

12. $\dfrac{2}{(x-2)(x-3)(x-4)}$.   13. $\dfrac{2}{(x-1)(2x+1)(2x+3)}$.

14. $\dfrac{1}{(x-1)(2x+1)(3x-2)}$.   15. $\dfrac{17a}{(1-2a)(4+a)(3+5a)}$.

16. $\dfrac{23x}{(1+2x)(2+x)(5-9x)}$.   17. $\dfrac{x+2}{(x+1)(x+3)}$.

18. $\dfrac{1}{x+1}$.   19. $\dfrac{1}{a+b}$.   20. $\dfrac{3x+2}{(x-2)(x-1)(x+1)}$.

21. $\dfrac{1}{x+2}$.   22. $\dfrac{8x^2+4x-3}{(x-1)(x+1)(2x+1)}$.   23. $\dfrac{1}{2x+1}$.

24. $\dfrac{x^2+11}{(x-1)(x+2)(x+3)}$.   25. $\dfrac{2x+13}{(x+3)(x+4)(x-4)}$.

26. $\dfrac{32a^2}{(1-2a)^2(1+2a)}$.   27. $\dfrac{96x^2}{(3-2x)^2(3+2x)}$.

28. $\dfrac{4x^3}{81-x^4}$.   29. $\dfrac{72a}{16a^4-81}$.   30. $\dfrac{1}{1-x^4}$.

31. $\dfrac{a(a^2+2ax+3x^2)}{4(a^4-x^4)}$.   32. $\dfrac{16x}{16-x^4}$.   33. $\dfrac{x(37+172x^2)}{6(1-16x^4)}$.

34. $\dfrac{2a(a^2+32x^2)}{3(a^4-256x^4)}$.   35. $\dfrac{7a^2+45}{6(a^4-81)}$.   36. $\dfrac{x^5}{1-x^8}$.

37. $\dfrac{36a^4}{a^8-6561}$.   38. $\dfrac{2}{x^2(x^2-4)}$.   39. $\dfrac{1}{(3x-y)(x-3y)}$.

40. $\dfrac{2}{(x-1)(x+1)^2}$.   41. 1.   42. 0.   43. $\dfrac{4x}{x^2-1}$.

XXI. d. Page 156.

1. $\dfrac{x-11}{20(x^2-1)}$.
2. $\dfrac{1}{1-a^2}$.
3. $\dfrac{x+3a}{x+a}$.
4. $\dfrac{2x-a}{x+a}$.
5. 0.
6. $\dfrac{7x}{1-x^2}$.
7. $\dfrac{1}{x-3}$.
8. $\dfrac{12(2x+1)}{4x^2-9}$.
9. $\dfrac{61-21b}{12(1-b^2)}$.
10. $\dfrac{2}{3(1-a^2)}$.
11. $\dfrac{y^5}{x^6-y^6}$.
12. $\dfrac{x}{y}$.
13. $\dfrac{2x}{x+y}$.
14. $\dfrac{2x^3}{x^2-4}$.
15. $\dfrac{a}{4a^2-25b^2}$.
16. $\dfrac{b(a+b)}{x^2-b^2}$.
17. $\dfrac{2bx}{4x^2-1}$.
18. $\dfrac{x+c}{(x-a)(x-b)}$.
19. $\dfrac{x-c}{(x-a)(x-b)}$.
20. $\dfrac{2a}{(x-a)(x-b)}$.
21. 0.
22. $\dfrac{4a^3}{x^4-a^4}$.
23. $\dfrac{48a^3}{(x^2-a^2)(x^2-9a^2)}$.
24. $\dfrac{x^4}{a^8-x^8}$.
25. 0.
26. $\dfrac{a^6}{a^8-b^8}$.
27. $\dfrac{a-x}{a+x}$.
28. $\dfrac{a^3}{(a-b)(a^3+b^3)}$.
29. $\dfrac{2+x+3x^2}{2(1-x^4)}$.
30. $\dfrac{2(x^2+1)}{x(x^2-1)}$.
31. 0.
32. $\dfrac{4ab}{a^2-b^2}$.

XXI. e. Page 159.

1. 0.
2. $\dfrac{bc+ca+ab-a^2-b^2-c^2}{(a-b)(b-c)(c-a)}$.
3. $\dfrac{x^2+y^2+z^2-yz-zx-xy}{(x-y)(y-z)(z-x)}$.
4. 0.
5. $\dfrac{2(bc+ca+ab-a^2-b^2-c^2)}{(a-b)(b-c)(c-a)}$.
6. 0.
7. 0.
8. 0.
9. $\dfrac{2(qr+rp+pq-p^2-q^2-r^2)}{(p-q)(q-r)(r-p)}$.
10. 0.
11. 0.
12. $\dfrac{p(y-z)+q(z-x)+r(x-y)}{(y-z)(z-x)(x-y)}$.

XXII. a. Page 163.

1. $\dfrac{m^2-nl}{na-mb}$.
2. $\dfrac{x+y}{y-x}$.
3. $\dfrac{ad+b}{dx-y}$.
4. $\dfrac{x+c}{b-x}$.
5. $\dfrac{3}{4b}$.
6. $\dfrac{a}{c}$.
7. $\dfrac{x^2-y^2}{x^2+y^2}$.
8. $\dfrac{c}{ac+b}$.
9. $\dfrac{ad}{bd+c}$.
10. $\dfrac{nx}{nx-m}$.
11. $\dfrac{pn(ad+bc)}{bd(pm+kn)}$.
12. $x-1$.
13. $\dfrac{x(x+3)}{x+4}$.
14. $-\dfrac{x+1}{x^2(x+3)}$.
15. $-\dfrac{x^2(2x+3)}{x+2}$.
16. $\dfrac{1}{x}$.

ANSWERS. 467

17. $\dfrac{a^2-b^2}{2}$. 18. 2. 19. $\dfrac{y^4}{x^2+y^2}$. 20. $\dfrac{1}{2x^2-1}$.

21. $\dfrac{2(a+b)}{a-b}$. 22. $a+x$. 23. $\dfrac{4}{x^2}$. 24. $\dfrac{1+x}{1+x^2}$.

25. $\dfrac{a^2-a+1}{2a-1}$. 26. $\dfrac{6+x+2y}{8x(y+6)}$. 27. $\dfrac{a(yz+n)}{xyz+nx+mz}$. 28. $1-x$.

29. $\dfrac{x^2-3x+1}{x^2-4x+1}$. 30. $\dfrac{2}{x^3}$. 31. $\dfrac{a-c}{1+ac}$. 32. $\dfrac{b}{a}$.

33. $\dfrac{1}{a+x}$. 34. 4. 35. $8x^2-1$. 36. $2x^2$.

XXII. b. Page 167.

1. $\dfrac{x}{3}+\dfrac{y}{9}-\dfrac{y^2}{9x}$. 2. $\dfrac{a^2}{4}-\dfrac{ax}{3}+\dfrac{x^2}{2}$. 3. $\dfrac{a^2}{2b}-\dfrac{3a}{2}+\dfrac{3b}{2}+\dfrac{b^2}{2a}$.

4. $\dfrac{1}{bc}+\dfrac{1}{ca}+\dfrac{1}{ab}$. 5. $\dfrac{1}{a}+\dfrac{1}{b}+\dfrac{1}{c}$. 6. $\dfrac{a^2}{6}-\dfrac{b^2}{2}+\dfrac{1}{3}$.

7. $x-x^2+x^3-x^4$; Rem. x^5.   8. $1+\dfrac{b}{a}+\dfrac{b^2}{a^2}+\dfrac{b^3}{a^3}$; Rem. $\dfrac{b^4}{a^3}$.

9. $1+2x+2x^2+2x^3$; Rem. $2x^4$.   10. $1+x-x^3-x^4$; Rem. x^6.

11. $x-3+\dfrac{9}{x}-\dfrac{27}{x^2}$; Rem. $\dfrac{81}{x^2}$.   12. $1+2x+3x^2+4x^3$; Rem. $5x^4-4x^5$.

17. $\dfrac{x-3}{x-4}$. 18. $3(a-2x)^2$. 19. $\dfrac{b^2-3b-2}{b-6}$.

20. $\dfrac{a^2-4b^2}{a+3b}$. 21. $\dfrac{(2x-3)(2x+7)}{6}$.

XXII. c. Page 168.

1. $\dfrac{4(c-x)}{3(a+x)}$. 2. $\dfrac{x(x+a)}{2}$. 3. $\dfrac{1}{a^2+ab-2b^2}$.

4. $\dfrac{8xy(x^2+y^2)}{(x^2-y^2)^2}$. 5. $\dfrac{4x(2-x)}{(x-1)(x^3+1)}$. 6. $\dfrac{x^4}{1-x^8}$.

7. $\dfrac{1}{x(x+1)^2(1+x+x^2)}$.   8. $\dfrac{1-x+x^2}{1+x+x^2}$.

9. $\dfrac{2x+3}{3(x+6)}$. 10. $\dfrac{bx+a}{ax+b}$. 11. $\dfrac{ax^3(x^2+a^2)}{x^3+a^3}$.

12. $\dfrac{a^6}{(a-x)(a+x)^2}$. 13. $\dfrac{(x+1)^2}{3x^3+6x^2-x-8}$. 14. $\dfrac{a+y}{a^2-y}$.

15. $\dfrac{2(a^2+a+1)}{a(a+1)(a+2)}$. 16. $\dfrac{1}{2(3-2x)}$. 17. $\dfrac{4}{1-x^4}$.

468 ALGEBRA.

18. 1. **19.** $\dfrac{(2x-1)(x+1)}{(x+2)(x-1)}$ **20.** $\dfrac{x-2}{4x^2-5x-5}$.

21. $\dfrac{x}{(x-2a)^2}$. **22.** $\dfrac{a(a^2+x^2)}{(x-a)(a+x)^2}$. **23.** 1. **24.** 1.

25. $9x-\dfrac{1}{x}$. **26.** $\dfrac{a}{2x^2}$. **27.** $\dfrac{1}{2x(2x-1)}$. **28.** $\dfrac{b^4}{b^2+a^2}$.

29. $\dfrac{ab}{a+b}$. **30.** x. **31.** x. **32.** bx. **33.** $\dfrac{a^2-b^2}{2}$.

34. 1. **35.** $\left(x-\dfrac{1}{x}\right)^2$. **36.** 1. **37.** 1.

38. $\dfrac{x(x+1)}{x^2+4x+1}$. **39.** $\dfrac{12}{(a^4-4)(a^4-1)}$. **40.** $\dfrac{3n^2}{(3m+2n)(9m^2-n^2)}$.

41. $\dfrac{1+x+x^2}{(1+x)(1+x^2)(1-x)^2}$. **42.** $\dfrac{2x}{(x-2)(x+1)^2}$. **43.** $\dfrac{1}{x+y}$.

44. 1. **45.** 1. **46.** 1. **47.** 0.

48. $\dfrac{a^2+b^2+c^2-bc-ca-ab}{(b-c)(c-a)(a-b)}$. **49.** 1. **50.** 1. **51.** 0.

52. $2y+a+b$. **53.** $\dfrac{(2a^2+x^2)(a-x)}{a^2x}$. **54.** $\dfrac{28(x+4)}{9(x+3)}$.

55. $\dfrac{7(x-4)}{4(x-1)}$. **56.** $x+3$. **57.** $1+a-a^3$. **58.** $-\dfrac{c}{c}$.

Miscellaneous Examples IV. PAGE 172.

1. $-\dfrac{1}{22}$. **2.** 6. **3.** $abc(b-c)$; -6. **4.** 7. **5.** $\dfrac{5}{3}$.

6. (1) 232; (2) -29. **7.** (1) -19; (2) 0. **8.** 1. **9.** $-\dfrac{3}{10}$.

10. (1) -12; (2) 1. **11.** 1. **12.** $8\tfrac{1}{2}$. **13.** $98x-2y$; $19\tfrac{1}{3}$.
14. (1) 1; (2) 21. **15.** $(x+9)(x+12)$. **16.** $(a-7)(a+13)$.
17. $(x-8y)(x-12y)$. **18.** $(ab-17)(ab+3)$. **19.** $c(c+13)(c-12)$.
20. $n(m-3n)(m-3n)$. **21.** $(p^2+7q^2)(p^2-8q^2)$.
22. $(d^2+5c^2)(d+3c)(d-3c)$. **23.** $xy(x+6y)(x-7y)$.
24. $(m+13)(m+15)$. **25.** $(14-a)(15+a)$. **26.** $(19-pq)(3+pq)$.
27. $(x^2+16)(x^2+11)$. **28.** $(a^2+14)(a^2-7)$. **29.** $(c+27)(c+27)$.
30. $(9-xy)(8+xy)$. **31.** $(a^2+2x^2)(a^2+7x^2)$.
32. $(p-12q)(p+9q)$. **33.** $2(a^3+12)(a^3-11)$.
34. $x^2(x-9)(x+7)$. **35.** $(bc+12)(bc-7)$. **36.** $(z+17)(z+17)$.
37. $(a-3c)(a-19c)$. **38.** $yz(y-7)(y+13)$.
39. $(2+3x^3)(1-x)(1+x+x^2)$. **40.** $(2ab-5)(ab+3)$.

ANSWERS.   469

41. $(3p-4)(3p-4)$.   42. $(5+mn)(7+mn)$.   43. $(17+c)(7-c)$.
44. $x^3(2-x)(3-x)$.   45. $(2m+3)(3m-1)$.   46. $(2a-5b)(2a+b)$.
47. $(6p-q)(p-2q)$.   48. $(5x+4z)(4x-5z)$.   49. $(2x^2+3)(4x^2-5)$.
50. $6(2y-1)(y-2)$.   51. $(3ab+4)(4ab-3)$.
52. $(2a^2b-5)(a^2b-2)$.   53. $(7x+8y)(3x-2y)$.
54. $(9m-5n)(2m+3n)$.   55. $(c+a-b)(c-a+b)$.
56. $(a+b-c)(a-b+c)$.   57. $(5x+3y)(25x^2-15xy+9y^2)$.
58. $(ab+7)(a^2b^2-7ab+49)$.   59. $(8b-a^2)(64b^2+8ba^2+a^4)$.
60. $(a+2x-2y)(a-2x+2y)$.   61. $(m+n+1)(m+n-1)$.
62. $2c^2(3c+d)(c-d)$.   63. $(a^2b^2-1+x-y)(a^2b^2-1-x+y)$.
64. $(1+2m)(1-2m)(1-2m+4m^2)(1+2m+4m^2)$.
65. $p^3(1+10q)(1-10q+100q^2)$.   66. $(81+a^2)(9+a)(9-a)$.
67. $(x^2-1+y-z)(x^2-1-y+z)$.   68. $(a+4b-4c)(a-4b+4c)$.
69. $(c-d)(1+2c-2d)(1-2c+2d)$.   70. $(p-4q)(p+4q+1)$.
71. $2[1+4a+4b][1-4(a+b)+16(a+b)^2]$.
72. $(x+3y)(1+x^2-3xy+9y^2)$.   73. $(x+y)(x^2+y^2)$.
74. $(cx-d)(ax+b)$.   75. $(7+a)(2-a)$.
76. $(14x^2+y^2)(7x^2-y^2)$.   77. $(17+a)(3-a)$.
78. $(1+m+p)(1-m-p)$.   79. $(bx-a)(ax-b)$.
80. $(3b-c+4)(3b-c-4)$.   81. $(c+1)(c^2-c+1)(x+1)(x-1)$.
82. $(3x-b)(x+2a)$.   83. $(m-n)(m+n+x)(m+n-x)$.
84. $(a+b)(c+a-b)(c-a+b)$.
85. $(x+2)(x^2-2x+4)(x^2+1)(x+1)(x-1)$.   86. $(x+1)(x+7)(2x-3)$.
87. $(2x+5y)(x-3y)(2x-5y)$.   88. $325a^3b^3(x^2-a^2)^2(x+2a)$.
89. $2x^2-9x+9$.   90. $2x^3(x^2-4)(x^2-16)$.
91. H.C.F. $=a+b+c$, L.C.M. $=(a+b+c)(a-b)(b-c)(c-a)$.
92. $a+b-c$.   93. $(a-b)^2(a+b)$.   95. $(a^4-b^4)(a+b-2c)$.
97. H.C.F. $=(x-7)(x-3)$,
    L.C.M. $=(x-1)(x-2)(x-3)(x-4)(x-5)(x-7)$.
98. $\dfrac{1}{(1-x)^2}$.   99. $\dfrac{x-9}{(x^2-9)(x-3)}$.   100. $\dfrac{6x+1}{(2x+1)^2(2x-1)}$.
101. $\dfrac{2}{x}$.   102. $\dfrac{4}{(1-x^2)^2}$.   103. 1.   104. 0.   105. $\dfrac{x}{9}$.
106. $\dfrac{2x-y}{x^2-y^2}$.   107. $y-x$.   108. ab.   109. $2(ac+bd)(ad+bc)$.
110. 1.   111. 1.   112. $\dfrac{(x^2+2)(x^4+1)}{x}$.   113. $\dfrac{1}{f-g}$.
114. $\dfrac{1}{x+1}$.   115. $x(1+x-x^2)$.   116. $\dfrac{3abc}{a+b}$.   117. $a+b$.   118. 1.

XXIII. a. PAGE 180.

1. 6. 2. $1\frac{3}{10}$. 3. $\frac{1}{5}$. 4. 1. 5. 20.
6. 2. 7. $\frac{7}{17}$. 8. 0. 9. 2. 10. $-6\frac{5}{6}$.
11. 5. 12. 6. 13. $-\frac{8}{11}$. 14. $-\frac{7}{18}$. 15. 1.
16. -10. 17. -4. 18. $3\frac{3}{5}$. 19. 3. 20. 4.
21. 6. 22. 13. 23. -7. 24. 2. 25. $2\frac{1}{2}$.
26. 4. 27. $1\frac{1}{2}$. 28. 14. 29. $\frac{1}{4}$. 30. $2\frac{1}{4}$.
31. $\frac{1}{6}$. 32. 3. 33. 20.

XXIII. b. PAGE 182.

1. $\dfrac{2b-3a}{a-5b}$. 2. $a+b$. 3. $\dfrac{b^2-a^2}{2b}$. 4. $\dfrac{a^2-ab+b^2}{a-b}$.
5. 3. 6. $m-n$. 7. $-\dfrac{ab}{a+b+c}$. 8. $\dfrac{a^2-2ab+bc}{c-b}$.
9. a. 10. $\dfrac{7bc}{9b+4c}$. 11. $\dfrac{2ab}{a+b}$. 12. $17a$.
13. $\dfrac{1}{c}$. 14. $3a+2b$. 15. $\dfrac{a+b}{2}$. 16. $\dfrac{a^2-2b^2}{3a-4b}$.
17. $\dfrac{a}{17}$. 18. $\dfrac{a^2}{b}$. 19. $\dfrac{bc^2}{a^2}$. 20. a.
21. $a+b$. 22. $\dfrac{2a}{21}$. 23. $\dfrac{a+2b}{2}$. 24. $\dfrac{a}{3}$. 25. $\dfrac{b(2a-b)}{a}$.

XXIII. c. PAGE 185.

1. $x=\dfrac{al-bm}{a^2-b^2},\ y=\dfrac{am-bl}{a^2-b^2}$. 2. $x=\dfrac{nq-mr}{lq-mp},\ y=\dfrac{lr-np}{lq-mp}$.
3. $x=\dfrac{bc}{a^2+b^2},\ y=\dfrac{ac}{a^2+b^2}$. 4. $x=\dfrac{a^2+ab+b^2}{a+b},\ y=-\dfrac{ab}{a+b}$.
5. $x=\dfrac{a'^2-a}{a'-a^2},\ y=\dfrac{1-aa'}{a'-a^2}$. 6. $x=\dfrac{q^2-pr}{qr-p^2},\ y=\dfrac{pq-r^2}{qr-p^2}$.
7. $x=\dfrac{a+a'}{a'b+ab'},\ y=\dfrac{b'-b}{a'b+ab'}$. 8. $x=2a,\ y=2b$.
9. $x=\dfrac{2}{3}a,\ y=\dfrac{1}{2}b$. 10. $x=\dfrac{pa}{q},\ y=\dfrac{rb}{p}$.

ANSWERS.

11. $x = \dfrac{mm'(m+m')}{m^2+m'^2}$, $y = \dfrac{mm'(m-m')}{m^2+m'^2}$.

12. $x = \dfrac{qn}{ql-pm}$, $y = \dfrac{pn}{mp-lq}$.
13. $x = \dfrac{c(a+b)}{2a}$, $y = \dfrac{c(a-b)}{2a}$.

14. $x = a+b$, $y = a-b$.
15. $x = 3a$, $y = -2b$.

16. $x = \dfrac{2aa'b}{ab'+a'b}$, $y = \dfrac{2abb'}{ab'+a'b}$.
17. $x = a$, $y = 0$.

18. $x = m+l$, $y = m+l$.
19. $x = \dfrac{a}{b}$, $y = \dfrac{b}{c}$.

20. $x = a+b$, $y = a-b$.
21. $x = a^3 - b^3$, $y = a^3 + b^3$.

XXIV. PAGE 188.

1. 40.   2. 60.   3. 55.   4. £2. 12s.
5. Silk 9s. Calico 9d. per yard.   6. 54.   7. 42.
8. 48, 23.   9. $21\tfrac{9}{11}'$ past one.   10. $17\tfrac{5}{11}'$ past three.
11. $32\tfrac{8}{11}'$ past six.   12. $5\tfrac{10}{11}'$ past two.   13. 378, 216.
14. 15 persons; 5 shillings.   15. 8 yards at 4s. 6d.; 16 yards at 4s.
16. 17, 15.   17. 3 miles per hour.
18. 54.   19. $2\tfrac{1}{2}$ miles per hour.
20. $21\tfrac{9}{11}'$ and $54\tfrac{6}{11}'$ past seven. At $5\tfrac{5}{11}'$ past.   21. $\dfrac{8}{12}$.
22. 10 P.M.; halfway.   23. $1\tfrac{1}{3}$ hours.   24. £200.
25. 30 miles.   26. £36000.   27. £200.
28. 4 and 3 gallons.   29. $\tfrac{3}{5}$ and $\tfrac{2}{5}$ of a pint.   30. $\dfrac{pa}{p+q}$ miles.
31. 111 and 126 miles.   32. Coffee to chicory as 7 to 2.
33. $c-b$ and $a-c$ lbs.   34. $\dfrac{c}{2a}, \dfrac{2c}{b}$ yards.   36. 60 miles.

XXV. a. PAGE 194.

1. ± 5.   2. ± 4.   3. 3, -25.   4. 1, -25.
5. 3, 7.   6. ± 8.   7. 3, -6.   8. 2, -7.
9. 9, -4.   10. 9, -8.   11. 31, -11.   12. 20, -11.
13. 4, -17.   14. 13, -12.   15. 11, -17.   16. 8, 15.
17. 7, 6.   18. 23, -1.   19. 6, $-\dfrac{16}{3}$.   20. $\dfrac{1}{3}, -\dfrac{3}{5}$.
21. $\dfrac{3}{2}, -\dfrac{1}{3}$.   22. $\dfrac{1}{5}, -4$.   23. ± 9.

XXV. b. PAGE 197.

1. $\frac{11}{5}, -5.$ 2. $11, \frac{11}{3}.$ 3. $3, \frac{7}{6}.$ 4. $\frac{15}{8}, -2.$

5. $\frac{7}{3}, 5.$ 6. $2, -\frac{11}{6}.$ 7. $\frac{3}{4}, -5.$ 8. $\frac{7}{2}, -3.$

9. $\frac{13}{3}, -\frac{11}{3}.$ 10. $\frac{7}{4}, \frac{2}{3}.$ 11. $\frac{3}{4}, -\frac{4}{5}.$ 12. $\frac{5}{8}, -3.$

13. $-\frac{5}{7}, -\frac{1}{3}.$ 14. $\frac{9}{10}, -\frac{3}{5}.$ 15. $\frac{13}{6}, -\frac{2}{3}.$ 16. $3, -\frac{7}{5}.$

17. $\frac{a}{3}, -\frac{a}{5}.$ 18. $\frac{3a}{7}, -\frac{a}{3}.$ 19. $\frac{7k}{3}, -\frac{k}{2}.$ 20. $-\frac{5k}{4}, -\frac{2k}{3}.$

21. $\frac{4c}{3}, -\frac{5c}{4}.$ 22. $3, -\frac{4}{3}.$ 23. $5, -\frac{5}{2}.$ 24. $4, \frac{7}{5}.$

25. $3, -1.$ 26. $2, \frac{1}{3}.$ 27. $4, \frac{11}{2}.$ 28. $7, 2.$

29. $11, 2.$ 30. $4, \frac{4}{3}.$ 31. $13, \frac{2}{3}.$ 32. $6, \frac{40}{13}.$

33. $2, \frac{39}{8}.$ 34. $3, -\frac{1}{2}.$ 35. $12, -2.$ 36. $5, \frac{23}{7}.$

37. $3a, \frac{3a}{2}.$ 38. $2c, \frac{11c}{14}.$ 39. $a, \frac{ab}{a-2b}.$

XXV. c. PAGE 201.

1. $\frac{5}{3}, -3.$ 2. $\frac{3}{2}, -5.$ 3. $1, \frac{7}{2}.$ 4. $\frac{3 \pm \sqrt{29}}{2}.$

5. $-4, -\frac{1}{5}.$ 6. $\frac{7 \pm \sqrt{5}}{2}.$ 7. $1, -\frac{7}{8}.$ 8. $\frac{17 \pm \sqrt{89}}{10}.$

9. $7, -\frac{5}{2}.$ 10. $\frac{1 \pm \sqrt{13}}{6}.$ 11. $\frac{1}{3}, -2.$ 12. $3, -\frac{11}{2}.$

13. $\frac{7}{6}, -1.$ 14. $\frac{7}{4}, \frac{3}{2}.$ 15. $\frac{7}{11}, -3.$ 16. $-\frac{6}{5}, -4.$

17. $\frac{9}{10}, -\frac{5}{6}.$ 18. $\frac{8}{3}, -\frac{3}{4}.$ 19. $\frac{2}{7}, -14.$ 20. $\frac{5}{12}, -\frac{3}{8}.$

21. $\frac{3}{5}, -\frac{2}{5}.$ 22. $\frac{5}{2}, -\frac{7}{2}.$ 23. $\frac{9a}{4}, -\frac{4a}{3}.$ 24. $\frac{9a}{4}, \frac{4a}{3}.$

25. $\frac{5b}{3}, -\frac{7b}{3}.$ 26. $\frac{7b}{6}, -\frac{5b}{6}.$ 27. $2a, 2b.$ 28. $2a, -8.$

ANSWERS. 473

29. $0, \dfrac{2a+b}{3}$. 30. $0, \dfrac{b-2}{a}$. 31. $\pm 2, \pm 1$. 32. $\pm 2, \pm 3$.

33. $1, -2$. 34. $3, -2$. 35. $\pm 4, \pm \dfrac{1}{4}$. 36. $\pm a, \pm b$.

37. $2, -3$. 38. $\pm 3, \pm 4$. 39. $3, -2, 4, -3$. 40. $4a, -2a, a$.

XXV. d. PAGE 201$_{\text{E}}$.

1. $1, -1, -1$. 2. $1, -1, 2$. 3. $1, 2, -2$. 4. $1, -3, -5$.
5. $2, -1, -1$. 6. $0, 1, 1, -2$. 7. $3, 2, -5$. 8. $5, 2, -7$.
9. $7, -3, -4$. 10. $-2a, -2a, 4a$. 11. $0, 6a, 6a, -12a$.
12. $1\cdot05, -3\cdot05$. 13. $3\cdot89, -\cdot89$. 14. $\cdot66, -1\cdot66$.
15. $18\cdot55, 17\cdot45$. 16. $5\cdot99, 1\cdot01$. 17. $3\cdot18, 2\cdot32$.
18. $\cdot55, -\cdot22$. 19. $1\cdot4, \cdot6$.
20. $\dfrac{a}{2}(\sqrt{5}-1), -\dfrac{a}{2}(\sqrt{5}+1)$. $7\cdot416, -19\cdot416$.
21. $\dfrac{1}{2}(a \pm \sqrt{a^2 - 4c^2})$. $13\cdot292, 2\cdot708$.

XXVI. a. PAGE 203.

1. $x=17, 11$; $y=11, 17$. 2. $x=37, 14$; $y=14, 37$.
3. $x=53, 21$; $y=21, 53$. 4. $x=14, -9$; $y=9, -14$.
5. $x=27, -19$; $y=19, -27$. 6. $x=43, -25$; $y=25, -43$.
7. $x=71, 13$; $y=13, 71$. 8. $x=33, -41$; $y=41, -33$.
9. $x=52, -74$; $y=74, -52$. 10. $x=43, -51$; $y=-51, 43$.
11. $x=29, -47$; $y=47, -29$. 12. $x=22, -87$; $y=-87, 22$.
13. $x=\pm 8, \pm 5$; $y=\pm 5, \pm 8$. 14. $x=\pm 13, \pm 1$; $y=\pm 1, \pm 13$.
15. $x=\pm 4, \pm 7$; $y=\pm 7, \pm 4$. 16. $x=13, 3$; $y=3, 13$.
17. $x=10, 5$; $y=5, 10$. 18. $x=9, -5$; $y=5, -9$.
19. $x=12, -6$; $y=6, -12$. 20. $x=11, -8$; $y=8, -11$.
21. $x=9, 4$; $y=4, 9$. 22. $x=5, 4$; $y=4, 5$.
23. $x=7, -4$; $y=4, -7$. 24. $x=10, 4$; $y=4, 10$.
25. $x=12, -2$; $y=2, -12$. 26. $x=1$; $y=1$.
27. $x=4, 3$; $y=3, 4$. 28. $x=\dfrac{1}{a}$; $y=\dfrac{1}{b}$.
29. $x=\pm 1$; $y=\pm 1$.

XXVI. b. Page 205.

1. $x=7, 4;$ $y=4, 7.$
2. $x=8, 5;$ $y=5, 8.$
3. $x=14, 9;$ $y=9, 14.$
4. $x=7, -5;$ $y=5, -7.$
5. $x=11, -7;$ $y=7, -11.$
6. $x=13, 0;$ $y=0, -13.$
7. $x=\pm 6, \pm 4;$ $y=\pm 4, \pm 6.$
8. $x=\pm 7, \pm 3;$ $y=\pm 3, \pm 7.$
9. $x=\pm 9, \pm 5;$ $y=\pm 5, \pm 9.$
10. $x=\pm 9, \pm 3;$ $y=\pm 3, \pm 9.$
11. $x=\frac{6}{5}, \frac{8}{3};$ $y=\frac{8}{3}, \frac{6}{5}.$
12. $x=\pm 6, \pm 5;$ $y=\pm 5, \pm 6.$
13. $x=4, 2;$ $y=2, 4.$
14. $x=7, -3;$ $y=3, -7.$
15. $x=5, 3;$ $y=3, 5.$
16. $x=4, -2;$ $y=2, -4.$
17. $x=8, -2;$ $y=2, -8.$
18. $x=5, 1;$ $y=1, 5.$
19. $x=5, 1;$ $y=1, 5;$
20. $x=\frac{1}{6}, -\frac{1}{5};$ $y=\frac{1}{5}, -\frac{1}{6}.$

XXVI. c. Page 208.

1. $x=4, -\frac{3}{5};$ $y=3, -20.$
2. $x=\pm 3;$ $y=\pm 2.$
3. $x=12, 8;$ $y=2, -2.$
4. $x=2, \frac{10}{3};$ $y=5, 3.$
5. $x=4, 7;$ $y=1, 10.$
6. $x=4, -3;$ $y=1, -\frac{4}{3}.$
7. $x=1, -\frac{71}{17};$ $y=4, \frac{112}{17}.$
8. $x=\pm 2, \pm \frac{4}{\sqrt{5}};$ $y=\pm 1, \pm \frac{3}{\sqrt{5}}.$
9. $x=2, \frac{5}{8};$ $y=-7, -\frac{1}{8}.$
10. $x=\pm 4, \pm 6;$ $y=\pm 2, \pm 4.$
11. $x=\pm 3, \pm 4;$ $y=\pm 2, \pm 5.$
12. $x=\pm \frac{3}{2}, \pm \frac{1}{2};$ $y=\pm \frac{1}{2}, \pm \frac{3}{2}.$
13. $x=\pm 2, \pm 1;$ $y=\pm 1, \pm 2.$
14. $x=\pm 2, \pm 5;$ $y=\pm 3, \pm 6.$
15. $x=\pm 7, \pm \sqrt{3};$ $y=\pm 2, \mp 3\sqrt{3}.$
16. $x=\pm 3, \pm 36;$ $y=\pm 5, \mp \frac{23}{2}.$
17. $x=5, 3;$ $y=3, 5.$
18. $x=7, -6;$ $y=6, -7.$
19. $x=6, -2;$ $y=2, -6.$
20. $x=7, 1, 4\pm\sqrt{28};$ $y=1, 7, 4\mp\sqrt{28}.$
21. $x=4, 3, 6, 2;$ $y=\frac{3}{2}, 2, 1, 3.$
22. $x=2, \frac{2}{3}, 4, \frac{1}{3};$ $y=2, 6, 1, 12.$

ANSWERS.   475

XXVII. Page 212.

1. 13.   2. 45, 9.   3. 7, 8.   4. 3.   5. 15, 12.
6. 9.   7. 7 hours.   8. 7, 5.   9. 90 yards, 160 yards.
10. 55 feet, 30 feet.   11. 36′, 60′.   12. 6.
13. 5 shillings.   14. 12.   15. Ninepence.   16. 3 feet.
17. 4 inches.   18. 121 square feet.   19. Fourpence.
20. 40, 12; 30, 16 yards.   21. 56.   22. 50.   23. 25.
24. $6\tfrac{2}{3}$ miles.   25. 75.   26. 20, 30 miles an hour.
27. 40 and 45 miles an hour.   28. 10 gallons.
29. A, 16; B, 14.   30. Distance, 12 miles; rate, 8 miles an hour.
31. $\dfrac{a}{2}(-1 \pm \sqrt{5})$.   32. 3·7 cm., 2·3 cm.
33. $AP = 20\cdot9$ cm., $BP = 12\cdot9$ cm.   35. 8·4 cm.
36. 2·6 cm., 1·6 cm.   37. 9 cm., 4 cm.
39. (i) 3, 4; (ii) 5, 6; (iii) 5·2, 0·8; (iv) 5·7, 2·3.

XXVIII. a. Page 216.

1. $(x^2 + 4x + 16)(x^2 - 4x + 16)$.   2. $(9a^2 + 3ab + b^2)(9a^2 - 3ab + b^2)$.
3. $(x^2 + 3xy + y^2)(x^2 - 3xy + y^2)$.   4. $(m^2 + 4mn - n^2)(m^2 - 4mn - n^2)$.
5. $(x^2 + 2xy - y^2)(x^2 - 2xy - y^2)$.
6. $(2x^2 + 9xy - 3y^2)(2x^2 - 9xy - 3y^2)$.
7. $(2m^2 + 6mn + 3n^2)(2m^2 - 6mn + 3n^2)$.
8. $(3x^2 + xy + 2y^2)(3x^2 - xy + 2y^2)$.
9. $(x^2 + 3xy - 5y^2)(x^2 - 3xy - 5y^2)$.
10. $(4a^2 - 6ab + b^2)(4a^2 + 6ab + b^2)$.
11. $\left(\dfrac{3}{ab} - 1\right)\left(\dfrac{9}{a^2b^2} + \dfrac{3}{ab} + 1\right)$.   12. $\left(6a - \dfrac{b}{2}\right)\left(36a^2 + 3ab + \dfrac{b^2}{4}\right)$.
13. $\left(\dfrac{x}{5} + y\right)\left(\dfrac{x^2}{25} - \dfrac{xy}{5} + y^2\right)$.   14. $\left(\dfrac{mn}{9} - 1\right)\left(\dfrac{m^2n^2}{81} + \dfrac{mn}{9} + 1\right)$.

15. $\left(\dfrac{ab}{5}+10\right)\left(\dfrac{a^2b^2}{25}-2ab+100\right)$. 16. $\left(\dfrac{x}{8}-\dfrac{4}{x}\right)\left(\dfrac{x^2}{64}+\dfrac{1}{2}+\dfrac{16}{x^2}\right)$.

17. $(y-3x)(x+y)(x-y)$.

18. $(m-5n)(2n+3m)(2n-3m)$.

19. $(ax+b)(bx+a)$.

20. $(x^2z^2+y^2)(xy+z)(xy-z)$.

21. $(a^2+bx)(a+x)$.

22. $(mn-p)(pm-n)$.

23. $(3ab-2x)(2ax-3b)$.

24. $(2x+3y)(a^2+xy)$.

25. $(2x-3y)(a^2+xy)$.

26. $\{ax+(a+1)\}\{(a-1)x+a\}$.

27. $(x-a)(3x-a-2b)$.

28. $\{ax+2(b-c)y\}\{2ax-(3b-4c)y\}$.

29. $\{(a-1)x+a\}\{(a-2)x+(a-1)\}$.

30. $\{(a+1)x-(b-1)y\}(ax+by)$.

31. $(b+c-1)(b^2+c^2+1-bc+c+b)$.

32. $(a+2c+1)(a^2+4c^2+1-2ac-a-2c)$.

33. $(a+b+2c)(a^2+b^2+4c^2-ab-2bc-2ca)$.

34. $(a-3b+c)(a^2+9b^2+c^2+3ab+3bc-ca)$.

35. $(a-b-c)(a^2+b^2+c^2+ab-bc+ca)$.

36. $(2a+3b+c)(4a^2+9b^2+c^2-6ab-3bc-2ca)$.

37. $(x^4-9x^2+81)(x^2+3x+9)(x^2-3x+9)$.

38. $(a^4-4a^2b^2-b^4)(a^2+b^2)(a+b)(a-b)$.

39. $(a+b+c-d)(a+b-c+d)(c+d+a-b)(c+d-a+b)$.

40. $\left(x^4+\dfrac{1}{16}\right)\left(x^2+\dfrac{1}{4}\right)\left(x+\dfrac{1}{2}\right)\left(x-\dfrac{1}{2}\right)$.

41. $(x^8+y^8)(x^4+y^4)(x^2+y^2)(x+y)(x-y)$.

42. $(x^6+x^3y^3+y^6)(x^6-x^3y^3+y^6)(x^2+xy+y^2)(x^2-xy+y^2)(x+y)(x-y)$

43. $\left(\dfrac{1}{x}+1\right)\left(\dfrac{1}{x}-1\right)(a-2x)(a^2+2ax+4x^2)$.

44. $(x^2+y^2)(x^4-x^2y^2+y^4)(x-2y)(x^2-2xy+4y^2)$.

45. $(x^2+4)(x^4-4x^2+16)(x+1)(x^2-x+1)$.

46. $\left(\dfrac{2}{a}+\dfrac{3}{b}\right)\left(\dfrac{2}{a}-\dfrac{3}{b}\right)(a+b)(a^2-ab+b^2)$.

ANSWERS. 477

47. $\left(\dfrac{1}{3x}+\dfrac{y}{2}\right)\left(\dfrac{1}{3x}-\dfrac{y}{2}\right)\left(\dfrac{xy}{2}-1\right)\left(\dfrac{x^2y^2}{4}+\dfrac{xy}{2}+1\right)$.

48. $(x^2+5)(x^2-5)\left(x+\dfrac{1}{2}\right)\left(x-\dfrac{1}{2}\right)$.

49. $(x+1)(x^2-x+1)(x^2+4)(x+2)(x-2)$.

50. $(x-1)(x^2+x+1)(4x^2+9)(2x+3)(2x-3)$.

XXVIII. b. Page 220.

1. $4x^2-49y^2+42yz-9z^2$.   2. $9x^4+26x^2y^2+49y^4$.
3. $25x^4-115x^2y^2+81y^4$.   4. $49x^4-64x^2y^2+48xy^3-9y^4$.
5. x^6-y^6.   6. $(x+y)^4+4(x+y)^2+16$.   7. $16x^2(1-4x^2)$.
8. $48a^2(a^4-1)$.   9. x^6-64.   10. x^6-729a^6.
11. $\dfrac{a^4}{x^2}-3a^2-x^2-\dfrac{x^4}{a^2}$.   12. $64x^4(9x^2-1)$.   13. $x^8+a^4x^4+a^8$.
14. $1+x^8+x^{16}$.   15. $a^{12}-3a^8x^4+3a^4x^8-x^{12}$.
16. $1-2x^8+x^{16}$.   17. $x^6-14x^4+49x^2-36$.
18. $x^6-14x^4+49x^2-36$.   19. x^6-64.   20. $a^4-18a^2b^2+81b^4$.
21. $a^3+b^3+c^3-3abc$.   22. $7x+y+z$.
23. $(x^4-4a^2x^2+16a^4)(x^2-2ax+4a^2)$.   24. $5x+7y-6z$.
25. $x+5$.   26. $2x(x+1)$.   27. $5(x-13)$.
28. $(x+3)(x^2+2x+4)$.   29. $(7x-3)(x-1)$.   30. $a-b$.
31. $x^3-3x^2y-3xy^2+y^3$.   32. $x^4-4x^2yz+7y^2z^2$.
33. $1+9x^2+4y^2+6xy+2y-3x$.   34. $(x+1)(x-3)$.
35. $(2a-5)(2a-7)$.   36. $(x-a)(x-b)$.
37. $a^2+9x^2+4y^2-6xy+2ay+3ax$.
38. $9+4x^2+16y^2-8xy+12y+6x$.
45. $\dfrac{m(m^2+3n^2)}{4}$.   47. $(3a^2+b^2)(a^2+3b^2)$.
49. $\dfrac{1}{16}(9p^2-5q^2)(9q^2-5p^2)$.   50. $16ab^3$.

XXIX. a. Page 223.

1. $x+c$.   2. x^2-ax+b.   3. $x^2+2bx-ax-2ab$.
4. $x^2-(p+q)x+2q(p-q)$.   5. $x^2-(m+n)x+m(m-n)$.
6. $ax+a+1$.   7. x^2+bx+a^2.   8. $2lx-(3m-4n)y$.
9. $(a+2)x+(a+1)y$.   10. $x+b$.
11. $(x+1)^6+3(x+1)^4+3(x+1)^2+1$.
12. $(m+1)b^2x^2+(n+1)(m+1)abx+(n+1)a^2$.

478 ALGEBRA.

13. $(m-1)x+m$. 14. $mx-n$. 15. $ap-bq$.
16. $ax+b$. 17. $2ax-3$. 18. $x+2ab$.
19. $(x^2-1)(x^2-px+q)(x^2-qx+p)$.
20. $\{px-(p-1)\}\{(p+1)x+p\}\{(p+2)x+p+1\}$.
21. $\{(a-3)x+a+1\}\{(a-2)x-a\}\{ax-(a+4)\}$.

XXIX. b. PAGE 227.

1. $x-7$. 2. $2-\dfrac{1}{m}$. 3. $a+2x$.
4. $1+x-x^2$. 5. $1+2x-x^2$. 6. $1+x$.
7. $x-2a$. 8. $a-3x$. 9. $x-y$.
10. $x^2+(p-1)x-1$. 11. $1+\dfrac{x}{2}-\dfrac{x^2}{8}+\dfrac{x^3}{16}$. 12. $1-x-\dfrac{x^2}{2}-\dfrac{x^3}{2}$.
13. $2+\dfrac{x}{2}-\dfrac{x^2}{16}+\dfrac{x^3}{64}$. 14. $1-\dfrac{x}{2}-\dfrac{5x^2}{8}-\dfrac{5x^3}{16}$.
15. $a-\dfrac{x}{2a}-\dfrac{x^2}{8a^3}-\dfrac{x^3}{16a^5}$. 16. $x+\dfrac{a^2}{2x}-\dfrac{a^4}{8x^3}+\dfrac{a^6}{16x^5}$.
17. $a^2-\dfrac{3x^2}{2a^3}-\dfrac{9x^4}{8a^6}-\dfrac{27x^6}{16a^{10}}$. 18. $3a+2x-\dfrac{2x^2}{3a}+\dfrac{4x^3}{9a^2}$.
19. $x-\dfrac{a^3}{3x^2}-\dfrac{a^6}{9x^5}$. 20. $2+\dfrac{x}{12}-\dfrac{x^2}{288}$.
21. $\dfrac{1}{a}+3a^2x-9a^5x^2$. 22. $1-2x+3x^2$.
23. $3x^2-x-1$. 24. $4-x-\dfrac{x^2}{16}$.

XXIX. c. PAGE 230.

19. 0. 20. 0. 26. 0.

XXIX. d. PAGE 232.

1. 0. 2. 1. 3. 1. 4. $a+b+c$.
5. 1. 6. $\dfrac{1}{abc}$. 7. $\dfrac{1}{abc}$. 8. 1.
9. d. 10. $\dfrac{1}{(x-a)(x-b)(x-c)}$. 11. $\dfrac{x^2}{(x+a)(x+b)(x+c)}$.
12. $(a+b+c)^2$. 13. $-\dfrac{a+b+c}{3}$. 14. $\dfrac{1}{3}$. 15. $a+b+c$.
16. $bc+ca+ab$. 17. abc. 18. $(b+c)(c+a)(a+b)$.

ANSWERS. 479

XXIX. e. Page 238.

1. 5. 2. 10. 3. $\dfrac{b}{a}$. 4. $\dfrac{p}{16q}$.

5. $\dfrac{5c}{b}$. 6. $\dfrac{d-a^4}{2a^3-c}$. 7. $a=c$, $b=\dfrac{a^2}{4}+2$. 8. 6.

9. $\pm 3a$. 10. $\pm\sqrt{\dfrac{2n}{m}}$. 11. $b^3=27c^2$.

12. $c=a(b-a^2)^2$, $d=(b-a^2)^3$, whence $c^3=a^3d^2$. 13. 32.

15. $(x-1)(x-2)(x-3)$. 16. $(x+2)(x-3)(x-4)$.
17. $(x+2)(x+3)(x+4)$. 18. $(x-3)(x-5)(x+7)$.
19. $(x-2)(x-5)(x+7)$. 20. $(x+1)(x+2)(x-11)$.
21. $(x+1)(3x+2)(2x-1)$. 22. $(x+2)(3x-1)(2x-3)$.
23. $x^6-x^5y+x^4y^2-x^3y^3+x^2y^4-xy^5+y^6$.
24. $x^7-x^6y+x^5y^2-x^4y^3+x^3y^4-x^2y^5+xy^6-y^7$.
25. $x^5+x^4y+x^3y^2+x^2y^3+xy^4+y^5$.
26. $x^8+x^7y+x^6y^2+x^5y^3+x^4y^4+x^3y^5+x^2y^6+xy^7+y^8$.
27. $x^2+(a-2)x+a$. 28. $(a+1)x^2+ax+a-3$. 29. 6 or $\dfrac{2}{3}$.
30. 13. 34. 3005. 35. $-37a^3$.

XXX. a. Page 244.

1. $\dfrac{2}{x^{\frac{1}{4}}}$. 2. $\dfrac{3}{a^{\frac{2}{3}}}$. 3. $\dfrac{4a^3}{x^2}$. 4. $3a^2$. 5. $\dfrac{a^2}{4}$.

6. $\dfrac{x^{\frac{1}{2}}}{5}$. 7. $\dfrac{3c^4x^2}{5a^3y^3}$. 8. $\dfrac{x^ab^a}{y^b}$. 9. $\dfrac{6}{x^{\frac{1}{2}}}$. 10. $\dfrac{a^{\frac{1}{3}}}{2}$.

11. y^2. 12. $\dfrac{1}{3a^2x^2}$. 13. $\dfrac{1}{x^{\frac{7}{2}}}$. 14. $\dfrac{x^{\frac{3}{5}}}{4}$. 15. $2y^{\frac{3}{2}}$.

16. $x^{\frac{5}{4}}$. 17. $\dfrac{a}{x^{\frac{1}{2}}}$. 18. $\dfrac{1}{a^{\frac{2}{3}}}$. 19. $\dfrac{1}{a^2}$. 20. $\sqrt[5]{x^3}$.

21. $\dfrac{1}{\sqrt{a}}$. 22. $\dfrac{5}{\sqrt{x}}$. 23. $\dfrac{2}{\sqrt[x]{a}}$. 24. $\dfrac{1}{2\sqrt[3]{a}}$. 25. $2\sqrt[4]{b^3}$.

26. $\dfrac{1}{2\sqrt[3]{c}}$. 27. $\sqrt[x]{x}$. 28. $\dfrac{2}{\sqrt[6]{a^5}}$. 29. $\dfrac{\sqrt{a}}{2\sqrt[3]{x^2}}$. 30. $\dfrac{21}{\sqrt{a^3}}$.

31. $\dfrac{2}{\sqrt{a}}$. 32. $\dfrac{1}{3\sqrt{a^3}}$. 33. $\dfrac{4}{\sqrt[3]{x^2}}$. 34. $\dfrac{1}{\sqrt[3]{x^{a+2}}}$. 35. $\sqrt[6]{a^{13}}$.

480   ALGEBRA.

36. $\sqrt[5]{a^x}$.   37. $\sqrt[2a]{x^5}$.   38. $\dfrac{1}{\sqrt[2a]{x}}$.   39. $\dfrac{1}{\sqrt[x]{a}}$.   40. $\sqrt[6]{a^n}$.

41. 8.   42. $\dfrac{1}{32}$.   43. 25.   44. $\dfrac{1}{4}$.   45. $\dfrac{1}{216}$.

46. 625.   47. 9.   48. $\dfrac{3}{2}$.   49. $\dfrac{27}{8}$.   50. $\dfrac{2187}{128}$.

XXX. b. Page 247.

1. $a^6 b^9$.   2. $\dfrac{x^{\frac{4}{3}}}{y}$.   3. $\dfrac{1}{y^{2a+3b}}$.   4. $\dfrac{1}{2x^{\frac{1}{2}}y^{\frac{1}{2}}}$.   5. $\dfrac{4}{9a^2 x^2}$.

6. $16ac^4$.   7. $\dfrac{x^{\frac{1}{3}}}{y^{\frac{1}{4}}}$.   8. $x^{\frac{1}{6}}$.   9. $\dfrac{3ax}{2}$.   10. x^{n-1}.

11. $\dfrac{1}{x^{n+1}}$.   12. $\dfrac{1}{x^{\frac{1}{a}}}$.   13. $\dfrac{1}{a^{\frac{2}{3}}b^{\frac{1}{2}}}$.   14. $a^{\frac{1}{2}}$.   15. x^{b+1}.

16. $\dfrac{1}{x^{\frac{1}{2}}}$.   17. $\dfrac{1}{x^2}$.   18. ab^2.   19. $a+b$.   20. $\dfrac{1}{(x^2-y^2)^{3n}}$.

21. $\dfrac{1}{a^5}$.   22. $b^{\frac{2}{3}}$.   23. $x^{\frac{1}{9}}$.   24. $\dfrac{a+b}{(a-b)^{\frac{1}{2}}}$.   25. $c^{\frac{7}{2}}$.

26. $\dfrac{x^2}{a^3}$.   27. $\dfrac{1}{a^{\frac{8}{3}}}$.   28. $ab(b^6-a^6)^{\frac{1}{3}}$.   29. $a^{n(n-1)} \div a$.

30. $x^{n(n-1)} + x^{n-1}$.   31. $a^{4n(p-q)}$.   32. x^5.   33. $\dfrac{x^7}{y^7}$.

34. $x^{\frac{1}{7}} y^{\frac{35}{6}}$.   35. $2n^2$.   36. $\dfrac{1}{4}$.   37. 4.   38. 1.

XXX. c. Page 250.

1. $12x^{\frac{2}{3}} - 20x^{\frac{1}{3}} + 41 - 15x^{-\frac{1}{3}} + 24x^{-\frac{2}{3}}$.

2. $9a^{\frac{4}{5}} - 9a^{\frac{2}{5}} - 25 + 23a^{-\frac{2}{5}} + 6a^{-\frac{4}{5}}$.

3. $2c^{2x} - 9c^x - 34 + 31c^{-x} - 6c^{-2x}$.   4. $8x^{3a} + 14x^a - 3x^{-a} - 9x^{-3a}$.

5. $7x^{\frac{2}{3}} - 2x^{\frac{1}{3}} + 1$.   6. $3a^{\frac{1}{3}} - 3a^{-\frac{1}{3}} + 2a^{-1}$.

7. $8a^{-2} + 7a^{-1} + 6$.   8. $5b^{\frac{1}{2}} + 4b^{\frac{1}{6}} + 3b^{-\frac{1}{6}} + 2b^{-\frac{1}{2}}$.

9. $7a^{2x} + 3a^x - 4$.   10. $c^{2n} - 1 + c^{-2n}$.

11. $3x^{\frac{1}{2}} - 2 + x^{-\frac{1}{2}}$.   12. $5a^{\frac{2}{3}} - 3a^{\frac{1}{3}} + 4$.

ANSWERS. 481

13. $2x^{\frac{n}{2}} - 4 + 3x^{-\frac{n}{2}}$.
14. $a^{2x} - 3a^x - 2$.
15. $a^2 + 2a - 16a^{-2} - 32a^{-3}$.
16. $1 - x^{\frac{1}{6}} - 2x^{\frac{1}{3}} + 2x^{\frac{2}{3}}$.
17. $4a^{\frac{8}{3}} - 8a^{\frac{4}{3}} - 5 + 10a^{-\frac{4}{3}} + 3a^{-\frac{8}{3}}$.
18. $x^{\frac{1}{2}} - 2x^{\frac{1}{3}} + 4x^{-\frac{1}{6}} - 8x^{-\frac{1}{2}}$.
19. $1 - 2a - 2a^{\frac{3}{2}}$.
20. $2x^{\frac{1}{4}} - 3x^{-\frac{1}{12}} - x^{-\frac{5}{12}}$.
21. $3x^{-2} - 3x^{-1}y^{\frac{1}{2}} + y$.
22. $2x^{\frac{3}{4}} - 3y^{\frac{1}{4}} + 4x^{-\frac{3}{4}}y^{\frac{1}{2}}$.
23. $9x^{\frac{2}{3}}y^{-1} + 2x^{\frac{1}{3}}y^{-\frac{1}{2}} - 9$.
24. $\frac{1}{4}x^{-1} + 1 - 3y^{\frac{1}{3}}$.

XXX. d. Page 252.

1. $x - 4x^{\frac{1}{2}} - 21$.
2. $16x^2 - 8 - 15x^{-2}$.
3. $49x^2 - 81y^{-2}$.
4. $x^m y^{-n} - x^{-m} y^n$.
5. $a^{2x} - 4 + 4a^{-2x}$.
6. $a^{2x} + 2a^{x + \frac{1}{x}} + a^{\frac{2}{x}}$.
7. $x^a - x^{-\frac{a}{2}} + \frac{1}{4}x^{-2a}$.
8. $20x^{2a}y^{2b} + 13 - 15x^{-2a}y^{-2b}$.
9. $\frac{1}{9}a^{\frac{2}{3}} - \frac{2}{3} + a^{-\frac{2}{3}}$.
10. $9x^{2a} + 15y^{2b} - 15y^{-2b} - 25x^{-2a}$.
11. $a^{2x} - a^x - \frac{7}{4} + a^{-x} + a^{-2x}$.
12. $x^{\frac{2}{a}} + x^{-\frac{2}{a}} + x^2 - 2 + 2x^{1+\frac{1}{a}} - 2x^{1-\frac{1}{a}}$.
13. $2a + 2(a^2 - b^2)^{\frac{1}{2}}$.
14. $a + b + (a-b)^{-1} - 2(a+b)^{\frac{1}{2}}(a-b)^{-\frac{1}{2}}$.
15. $x^{\frac{1}{2}} - 3a^{\frac{1}{2}}$.
16. $x + 3x^{\frac{1}{2}} + 9$.
17. $a^x + 4$.
18. $x^{2a} - 2x^a + 4$.
19. $c^x + c^{-\frac{x}{2}}$.
20. $1 + 2a^{-1} + 4a^{-2}$.
21. $a^{2x} - x^3$.
22. $x^{-3} - x^{-2} + x^{-1} - 1$.
23. $x^{\frac{4}{3}} + x + x^{\frac{2}{3}} + x^{\frac{1}{3}} + 1$.
24. $x^{4n} - 2x^{3n} + 4x^{2n} - 8x^n + 16$.
25. $x^2 + 2x^{\frac{3}{2}} + x - 16$.
26. $4x^{\frac{2}{3}} + 16x^{\frac{1}{3}} + 16 - 9x^{-\frac{2}{3}}$.
27. $4 - x^{\frac{2}{3}} + 4x + x^2$.
28. $a^{2x} - 49 - 42a^{-x} - 9a^{-2x}$.
29. $a^{\frac{1}{3}}(a^{\frac{1}{3}} - 2b^{\frac{1}{3}})$.
30. 1.
31. $\dfrac{x^{\frac{2}{3}} - 2}{x^{\frac{2}{3}} + 2}$.
32. $\dfrac{a^{\frac{1}{2}}}{b}$.

XXXI. a. Page 255.

1. $\sqrt[12]{x^4}$.
2. $\dfrac{1}{\sqrt[12]{a^6}}$.
3. $\sqrt[12]{\dfrac{x}{a}}$.
4. $\sqrt[12]{a^9}$.
5. $\sqrt[12]{a^{21}}$.
6. $\sqrt[12]{a^4}$.
7. $\sqrt[n]{x^{\frac{2n}{3}}}$.
8. $\sqrt[n]{x^{an}}$.

9. $\sqrt[n]{a^{\frac{n}{2}}}$.   10. $\dfrac{1}{\sqrt[n]{a^{\frac{1}{2}}}}$.   11. $\sqrt[n]{x^{\frac{n^2}{3}}y^{\frac{1}{3}}}$.   12. $\sqrt[n]{a^n}$.

13. $\dfrac{1}{\sqrt[n]{x^{\frac{n}{2}}y^{2n}}}$.   14. $\sqrt[n]{a^{\frac{n}{2}}x^{n^2}}$.   15. $\sqrt[18]{a^9}$, $\sqrt[18]{a^{10}}$.   16. $\sqrt[10]{a^6}$, $\sqrt[10]{a^5}$.

17. $\sqrt[24]{x^9}$, $\sqrt[24]{x^{18}}$, $\sqrt[24]{x^6}$.   18. $\sqrt[12]{x^3}$, $\sqrt[12]{x^{10}}$.   19. $\sqrt[21]{a^3b^4}$, $\sqrt[21]{a^3b^3}$.

20. $\sqrt[26]{a^{13}x^{26}}$, $\sqrt[26]{a^6x^4}$.   21. $\sqrt[6]{125}$, $\sqrt[6]{121}$, $\sqrt[6]{13}$.

22. $\sqrt[8]{64}$, $\sqrt[8]{81}$, $\sqrt[8]{6}$.   23. $\sqrt[3]{2}$, $\sqrt[3]{2}$, $\sqrt[3]{2}$.

XXXI. b. Page 256.

1. $12\sqrt{2}$.   2. $7\sqrt{3}$.   3. $4\sqrt[3]{4}$.   4. $6\sqrt[3]{2}$.
5. $15\sqrt{6}$.   6. $24\sqrt{5}$.   7. $35\sqrt{5}$.   8. $7\sqrt[3]{3}$.
9. $5\sqrt[4]{5}$.   10. $-9\sqrt[3]{3}$.   11. $6a\sqrt{a}$.   12. $3ab^2\sqrt{3ab}$.
13. $-3xy\sqrt[3]{4x}$.   14. $x^3y^2\sqrt[n]{y^5}$.   15. $xy^2\sqrt[p]{x^a}$.   16. $(a+b)\sqrt{a}$.
17. $2(x-y)\sqrt[3]{xy}$.   18. $\sqrt{242}$.   19. $\sqrt{980}$.
20. $\sqrt[3]{864}$.   21. $\sqrt[3]{750}$.   22. $\sqrt{\dfrac{14}{11}}$.   23. $\sqrt{5b}$.
24. $\sqrt{9a^2y}$.   25. $\sqrt{\dfrac{3a}{x}}$.   26. $\sqrt[3]{8ax}$.   27. $\sqrt[4]{2a}$.
28. $\sqrt[n]{a^2b^2}$.   29. $\sqrt[p]{ab}$.   30. $\sqrt{\dfrac{x}{y}}$.   31. $\sqrt{x^2-y^2}$.
32. $\sqrt{\dfrac{a+x}{a-x}}$.   33. $14\sqrt{5}$.   34. $\sqrt{7}$.   35. $-12\sqrt{11}$.
36. $-15\sqrt{3}$.   37. $7\sqrt[3]{7}$.   38. $11\sqrt[3]{3}$.   39. 0.
40. $17\sqrt[3]{2}$.   41. $20\sqrt{3}-13\sqrt{2}$.   42. $3\sqrt{6}$.
43. $6\sqrt{7}-15\sqrt{6}$.   44. $\dfrac{181\sqrt{3}}{9}$.

XXXI. c. Page 259.

1. $14\sqrt{6}$.   2. $12\sqrt{3}$.   3. $10\sqrt{3a}$.   4. $30\sqrt{3}$.
5. $288\sqrt{2}$.   6. $\sqrt[3]{x^2-4}$.   7. $3\sqrt{3}$.   8. $\dfrac{5\sqrt{2}}{4}$.
9. $-\sqrt{13}$.   10. $14\sqrt[3]{9}$.   11. $240\sqrt[3]{4}$.   12. $\sqrt{6}$.
13. $ab^3\sqrt{ab}$.   14. $\dfrac{33}{10}$.   15. $\dfrac{1}{10}\sqrt{2}$.   16. $\dfrac{2\sqrt{2}}{a}$.

ANSWERS. 483

17. $\dfrac{a-b}{x}$. 18. $9\cdot 8995$. 19. $11\cdot 1804$. 20. $3\cdot 7797$.
21. $19\cdot 5959$. 22. $26\cdot 8328$. 23. $58\cdot 7878$. 24. $\cdot 8165$.
25. $\cdot 2887$. 26. $\cdot 0447$. 27. $\cdot 2566$. 28. $1\cdot 5749$.
29. $\cdot 4032$.

XXXI. d. Page 260.

1. $6x - 10\sqrt{x}$. 2. $2x - 2\sqrt{ax}$. 3. $a\sqrt{b} + b\sqrt{a}$.
4. $x + y - \sqrt{x+y}$. 5. $30 + 12\sqrt{6}$. 6. $6\sqrt{21} - 46$.
7. $6 + \sqrt{10}$. 8. $6a - 6x + 5\sqrt{ax}$. 9. $x - 1 + \sqrt{x^2 - x}$.
10. $x + a - \sqrt{x^2 - a^2}$. 11. $5a + x - 4\sqrt{a^2 + ax}$.
12. $1 + 8a - 4\sqrt{a + 4a^2}$. 13. $2a - 2\sqrt{a^2 - x^2}$.
14. $a + x + 2 - 3\sqrt{a + x}$. 15. $2\sqrt{6}$. 16. $16 + 6\sqrt{10}$.
17. $4x - 2\sqrt{4x^2 - a^2}$. 18. $2x^2 + 2\sqrt{x^4 - 4y^4}$.
19. $2m + 2\sqrt{m^2 - n^2}$. 20. $13a^2 + 5b^2 - 12\sqrt{a^4 - b^4}$.
21. $63 - 18x\sqrt{14 - 4x^2}$. 22. $8x^2 - 2\sqrt{16x^4 - 1}$.

XXXI. e. Page 262.

1. 113. 2. -166. 3. 172. 4. -6.
5. $a - 4b$. 6. $9c^2 - 4x$. 7. x. 8. $2p - q$.
9. $2x$. 10. $25(x^2 - 3y^2) - 49a^2$. 11. $\dfrac{11 - 3\sqrt{7}}{2}$.
12. $\dfrac{3\sqrt{7} - 2\sqrt{3}}{3}$. 13. $\dfrac{19 - 6\sqrt{2}}{17}$. 14. $2 + \sqrt{6}$.
15. $\dfrac{\sqrt{xy}}{y}$. 16. $\dfrac{\sqrt{5}}{5}$. 17. $\dfrac{\sqrt{ax}}{a - x}$. 18. $4 + \sqrt{15}$.
19. $5 + \sqrt{6}$. 20. $8 - \sqrt{42}$. 21. $\dfrac{\sqrt{7} - \sqrt{2}}{5}$.
22. $3\sqrt{2} - 2\sqrt{3}$. 23. $x - \sqrt{x^2 - y^2}$. 24. $\sqrt{x^2 + a^2} - a$.
25. $\dfrac{1 - \sqrt{1 - x^4}}{x^2}$. 26. $\dfrac{7a + b + 8\sqrt{a^2 - b^2}}{3a + 5b}$.
27. $\dfrac{18 + x^2 - 6\sqrt{9 + x^2}}{x^2}$. 28. $\sqrt{3}$.
29. $2 - \sqrt{3} = \cdot 26795$. 30. $11 + 5\sqrt{5} = 22\cdot 18035$.
31. $\sqrt{5} - \sqrt{3} = \cdot 50402$. 32. $\sqrt{5} + 2 = 4\cdot 23607$.
33. $\dfrac{\sqrt{5}}{2} = 1\cdot 11803$. 34. $\dfrac{3\sqrt{3} - 5}{2} = \cdot 09807$.

XXXI. f. Page 266.

1. $\sqrt{5}-\sqrt{2}$.
2. $\sqrt{10}+\sqrt{3}$.
3. $\sqrt{7}-1$.
4. $\sqrt{3}+\sqrt{2}$.
5. $3\sqrt{7}+2\sqrt{3}$.
6. $\sqrt{10}-2\sqrt{2}$.
7. $4\sqrt{2}-3$.
8. $2\sqrt{5}+3\sqrt{7}$.
9. $2\sqrt{11}-\sqrt{3}$.
10. $\dfrac{1}{2}\sqrt{5}+1$.
11. $2-\dfrac{1}{3}\sqrt{3}$.
12. $5\sqrt{\dfrac{1}{2}}+\sqrt{\dfrac{7}{2}}$.
13. $\sqrt[4]{3}(\sqrt{2}+1)$.
14. $\sqrt[4]{2}(\sqrt{3}-1)$.
15. $\sqrt[4]{5}(\sqrt{2}+1)$.
16. $\sqrt{2}+1$.
17. $\sqrt{5}+1$.
18. $\dfrac{1}{2}(\sqrt{5}+1)$.
19. $\dfrac{1}{\sqrt[4]{2}}(\sqrt{3}+1)$.
20. $\sqrt{3}-\sqrt{2}$.
21. $\sqrt[4]{2}(\sqrt{5}+\sqrt{3})$.
22. $\sqrt{2}-1$.
23. $\sqrt{3}+1$.
24. $\sqrt{5}-1$.
25. $4+\sqrt{3}$.
26. $\sqrt{5}+\sqrt{3}$.
27. $\sqrt{7}-\sqrt{2}$.
28. $2\sqrt{2}+\sqrt{3}$.
29. $2\sqrt{2}-\sqrt{7}$.
30. $\sqrt{11}+3\sqrt{2}$.

XXXI. g. Page 268.

1. 14.
2. 33.
3. 20.
4. 44.
5. 13.
6. $\dfrac{6}{5}$.
7. $\dfrac{17}{6}$.
8. 9.
9. 7.
10. $\dfrac{56}{5}$.
11. 144.
12. 2.
13. $\dfrac{121}{25}$.
14. $\dfrac{25}{16}$.
15. 5.
16. 12.
17. 1.
18. 9.
19. 8.
20. 12.
21. $\dfrac{1}{51}$.
22. 1.
23. 2.
24. $(b-a)^2$.
25. $\dfrac{(a-b)^2}{2a-b}$.
26. $0, a-b$.
27. 10.
28. $\dfrac{5}{2}$.
29. 2.
30. ± 1.

XXXI. h. Page 269.

1. 49.
2. 4.
3. 49.
4. $\dfrac{121}{9}$.
5. 16.
6. 64.
7. $\dfrac{64}{9}$.
8. $\dfrac{1}{3}$.
9. $\dfrac{4}{3}$.
10. 9.
11. 4.
12. 1.
13. 4.
14. 50.
15. 11.
16. 3.
17. 6.
18. 25.
19. $\dfrac{1}{4}$.
20. $\dfrac{8}{5}$.
21. 6.
22. 361.

ANSWERS. 485

XXXII. a. Page 274.

1. 6 : 1. 2. 1 : 2. 3. 1 : 5. 4. 9 : 32.
5. $2x : 3y$. 6. $3b : 4a$. 7. 6 : 1. 8. $\dfrac{1}{5}$.
9. 4 : 1. 10. 17 : 7. 11. 3 : 4. 12. 5 : 4.
16. 21, 28. 17. 11. 18. 27.

XXXII. b. Page 280.

1. bc. 2. $\dfrac{6b^3}{a}$. 3. $5y^2$. 4. b.
5. $4x$. 6. $12xy^2$. 7. x^2. 8. ab.
9. $4x^2$. 10. $6a^2x$. 11. $9ab^2$. 19. 8 or $\dfrac{2}{3}$.
20. $x=17, y=11$. 21. 2 or 0. 22. 5 or 0.

XXXII. c. Page 285.

1. 54. 2. 27. 3. 35. 4. 21. 5. 10.
6. $\dfrac{1}{5}$. 7. 16. 8. 18. 12. $3a=5b$. 13. $5x=7y$.
14. $25x^3=27y^2$. 15. $a^3=b^2$. 16. ± 6. 17. 20.
19. $y=6x-3x^2+x^3$. 20. $346\frac{1}{2}$ square feet.
22. 20 miles per hour. 23. 9 : 4. 24. $1\frac{5}{9}$ feet. 25. 4 feet.
26. £1960.

XXXIII. a. Page 289.

1. 161, 245. 2. 59, -37. 3. 34, $89\frac{1}{2}$. 4. 16, 9.
5. $574\frac{1}{2}$, $93\frac{1}{2}$. 6. 98, 243·6. 7. 43. 8. -49.
9. $-40\frac{1}{2}$. 10. 7·2. 11. 9·7. 12. $25x$.
13. $a+57d$. 14. $80a-79b$. 15. 964, 9780. 16. 3·2, 25·2.
17. $-387, -18900$. 18. $-9\frac{3}{4}, -99\frac{3}{4}$. 19. $-41\frac{1}{2}, -361$.
20. 544, 4864. 21. 779. 22. -483. 23. $980\frac{1}{2}$.
24. $-5569\frac{1}{2}$. 25. 493. 26. 140. 27. p^3.
28. $a^2(4-a)$. 29. $\dfrac{a^2(3-a)}{2}$. 30. $pq(p-4)$. 31. 30, 3.
32. 25, -3. 33. 16, -1. 34. 24, $2\frac{1}{2}$. 35. 14, -1.
36. 20, 4. 37. 7, $2a$. 38. 20, $-2x$.

XXXIII. b. Page 293.

1. 4, 11, 18, ...   2. 13, 10, 7, ...   3. 3, 1, −1, ...
4. 1, $-\frac{1}{2}$, −2, ...   5. 4, $5\frac{1}{2}$, 7, ...   6. −11, 4, 19, ...
7. 43.   8. −95.   9. $-6\frac{1}{2}$.
10. 68, 65, ... 26.   11. $91\frac{2}{3}$, $90\frac{1}{3}$, ... $70\frac{1}{3}$.
12. $-6\frac{13}{15}$, $-6\frac{8}{15}$, ... $-2\frac{8}{15}$.   13. 6·4, 5·6, ... −5·6.
14. $8\frac{1}{3}$, $8\frac{1}{8}$, ... $2\frac{1}{2}$.   15. 14 or 15.   16. 8 or 25.
17. 9 or 86.   18. 13 or 20.   19. 7 or 8.
20. 11 or 24.   21. 12, 13, 14.   22. 1, 4, 7.
23. 7, 11, 15, 19, 23.   24. 2, 5, 8, 11, 14.   25. 131.

XXXIV. a. Page 297.

1. 48, 384.   2. $\frac{1}{2}$, $\frac{1}{128}$.   3. 1, $\frac{1}{16}$.
4. $-\frac{1}{27}$, $-\frac{1}{2187}$.   5. 128, 1.   6. 1, 625.
7. 512.   8. −4374.   9. $\frac{243}{16}$.
10. -3^{2n}.   11. x^{2p-1}.   12. $\frac{1}{x^{28}}$.
13. 162, 54, 18.   14. $\frac{1}{2}$, 2, 8, 32.   15. −28, 14, ... $\frac{7}{8}$.
16. $\frac{16}{27}$, $\frac{8}{9}$, ... 3.   17. 384, 765.   18. −1458, −1092.
19. $\frac{1}{8}$, $127\frac{7}{8}$.   20. $\frac{1}{90}$, $12\frac{13}{90}$.   21. $30\frac{3}{8}$, $45\frac{5}{9}$.
22. $\frac{1}{1458}$, $6\frac{1093}{1458}$.   23. $2\frac{20}{81}$.   24. $1\frac{601}{1458}$.
25. $\frac{1281}{2560}$.   26. $\frac{1365}{2048}$.   27. $5\frac{58}{81}$.
28. $\frac{4369}{8192}$.   29. $\frac{1}{2}(3^p - 1)$.   30. $\frac{2}{3}(1 - 2^{2p})$.
31. $\frac{40(3+\sqrt{3})}{3}$.   32. $\frac{\sqrt{a}(a^a - 1)}{a-1}$.   33. $\frac{585\sqrt{2} - 292}{2}$.
34. $364(\sqrt{6} + \sqrt{2})$.

ANSWERS.

XXXIV. b. PAGE 300.

1. 27. 2. 24. 3. 1. 4. $\dfrac{1}{3}$. 5. 1. 6. $\dfrac{64}{65}$.

7. $\dfrac{27}{29}$. 8. $\dfrac{8}{15}$. 9. $\dfrac{1}{3}$. 10. $\dfrac{1}{6}$. 11. $\dfrac{8}{33}$. 12. $\dfrac{25}{66}$.

13. $\dfrac{1}{27}$. 14. $\dfrac{5}{8}, \dfrac{5}{4}, \dfrac{5}{2}, \ldots$. 15. $\dfrac{2187}{256}, \dfrac{729}{128}, \dfrac{243}{64}, \ldots$.

16. $\dfrac{1}{25}, -\dfrac{1}{5}, 1, \ldots$. 17. $\dfrac{9}{64}, -\dfrac{9}{32}, \dfrac{9}{16}, \ldots$. 18. 75, 60, 48.

20. $\dfrac{3}{4}, \dfrac{1}{4}, \dfrac{1}{12}, \ldots$. 21. $\dfrac{y^2(y^{2n}-1)}{y^2-1} + bn(n+1)$. 22. $\dfrac{140+99\sqrt{2}}{8}$.

23. $\dfrac{9(3\sqrt{6}+2\sqrt{2})}{46}$. 24. $2n^2(2n+1) - \dfrac{3}{8}\left(1 - \dfrac{1}{3^{2n}}\right)$.

XXXV. PAGE 304.

1. $\dfrac{2}{3}$. 2. $\dfrac{5}{14}$. 3. -4. 4. $\dfrac{3}{n}$. 5. $-\dfrac{1}{3}, -1, 1, \ldots$.

6. $4, 2, 1\tfrac{1}{3}, \ldots$. 7. $-\dfrac{1}{27}, -\dfrac{1}{26}, -\dfrac{1}{25}, \ldots$. 8. $2\tfrac{2}{3}$. 9. $1\tfrac{6}{7}$.

10. $\dfrac{1}{7}$. 11. $\dfrac{2}{a+b}$. 12. $\dfrac{1}{x}$. 13. $\dfrac{x^2-y^2}{x}$. 14. $5\tfrac{1}{7}, 7\tfrac{1}{5}$.

15. 3, 4, 6. 16. $1\tfrac{1}{5}, 1\tfrac{1}{2}, 2, 3$. 19. $36\tfrac{987}{1024}$. 20. $17\tfrac{1}{4}$.

21. $\dfrac{p}{2}\{(p+3)a - (p-3)x\}$. 22. $1\tfrac{46}{1215}$. 23. -18. 31. $\dfrac{a^{2n}-b^{2n}}{b^{2n-2}(a^2-b^2)^2}$.

37. $\dfrac{2}{3}$. 38. $\dfrac{n(n+1)}{2} \cdot x + \dfrac{1}{x}\left(1 - \dfrac{1}{2^n}\right)$. 39. $n(3n+2)$.

Miscellaneous Examples V. PAGE 307.

1. $\dfrac{c^{\frac{1}{12}}}{a^{\frac{1}{3}}b^{\frac{1}{4}}}$; 1. 7. $4\sqrt{2}$. 8. 5. 10. 52, 78, 91 yards.

12. (1) $x^{2b} + x^{-2b}$. (2) $(a+b)^2$. 13. $-\dfrac{32}{7a}$. 14. (1) $\dfrac{b^4}{a^2}$. (2) $\dfrac{7}{8}$.

15. $\dfrac{a+b}{x^2-y^2}$. 21. 2. 23. $-5, -2, 1, 4, 7, 10, 13$.

24. $1, 4, 7, \ldots$. 25. 1. 26. (1) 275; (2) -1705.

28. $-2, 0, 2, 4, 6$, 29. $18\left[1 - \left(\dfrac{5}{6}\right)^n\right]$. 30. 1. 31. $a+b$.

32. $\dfrac{3n^2}{2}$. $\dfrac{63}{2}, \dfrac{69}{2}, \dfrac{75}{2}, \dfrac{81}{2}, \dfrac{87}{2}$. 34. (1) $9(19a + 64x)$. (2) $\dfrac{2315}{81}$.

488   ALGEBRA.

37. 10. 38. (1) $s=\dfrac{1-3^{2n}}{4}$; $l=-3^{2n-1}$. (2) $s=-2n$; $l=1-4n$.

39. 1 and 9. 40. $1+\dfrac{1}{3}+\dfrac{1}{9}+\ldots\ldots$. 41. 8 and 2.

XXXVI. Page 317.

1. Rational. 2. Rational. 3. Equal, but opposite in sign.
4. Imaginary. 5. Imaginary. 6. Equal, but opposite in sign.
7. $x^2-2x-15=0$. 8. $x^2+20x+99=0$.
9. $x^2-2ax+a^2-b^2=0$. 10. $12x^2-28x+15=0$.
11. $15x^2+2ax-8a^2=0$. 12. $8x^2-7x=0$. 13. $-\tfrac{1}{2}$.
15. Sum $\tfrac{4}{3}$, difference $\dfrac{2\sqrt{7}}{3}$, sum of squares $\dfrac{22}{9}$. 17. $x^2-6x+4=0$.
18. $x^2+4x+1=0$. 19. $30x^2+(6a-5b)x-ab=0$. 20. $4x^2-16x+9=0$.
21. $(a^2-b^2)x^2-2(a^2+b^2)x+a^2-b^2=0$. 22. $4abx^2-2(a^2+b^2)x+ab=0$.
23. $\dfrac{q^2-2pr}{p^2}$. 24. $\dfrac{q^2-4pr}{p^2}$. 25. $-\dfrac{qr}{p^2}$. 26. $\dfrac{q^4-4prq^2+2p^2r^2}{p^4}$.
27. $\dfrac{qr^2(3pr-q^2)}{p^5}$. 28. $\dfrac{q(3pr-q^2)}{p^2r}$. 29. $P=p(p^2-3q)$, $Q=q^3$.
30. $b^2x^2-(a^3-3ab)x+b=0$. 31. $2b^2=9ac$.
32. $8x^2-20a^3x-a^6=0$. 35. $2px^2-(p^2+4q)x+2pq=0$.

XXXVII. a. Page 323.

1. 120, 5040, 56, 300. 2. (1) 2520. (2) 5040. 3. 8.
4. 126. 5. 6. 6. 36. 7. 7 or 8. 8. 2100.
9. 455, 816; $(r=15)$. 10. 242880. 11. 1596000. 12. 504000.

XXXVII. b. Page 328.

1. (1) 9979200. (2) 151200. (3) 166320. 2. 420, 360. 3. 18.
4. 1023. 5. m^n. 6. 168168. 7. 34650.
8. 120, 144. 9. 1296. 10. 180. 11. 11520.
13. 78. 14. $(n-2)(n-3)\underline{|n-2}$.

XXXVIII. a. Page 333.

1. $x^4+8x^3+24x^2+32x+16$.
2. $x^5+15x^4+90x^3+270x^2+405x+243$.
3. $a^7+7a^6x+21a^5x^2+35a^4x^3+35a^3x^4+21a^2x^5+7ax^6+x^7$.

ANSWERS. 489

4. $a^5 - 5a^4x + 10a^3x^2 - 10a^2x^3 + 5ax^4 - x^5$.
5. $1 - 10y + 40y^2 - 80y^3 + 80y^4 - 32y^5$.
6. $16x^4 + 16x^3y + 6x^2y^2 + xy^3 + \dfrac{y^4}{16}$.
7. $64 - 96x + 60x^2 - 20x^3 + \dfrac{15x^4}{4} - \dfrac{3x^5}{8} + \dfrac{x^6}{64}$.
8. $a^7 - \dfrac{21a^6}{b} + \dfrac{189a^5}{b^2} - \dfrac{945a^4}{b^3} + \dfrac{2835a^3}{b^4} - \dfrac{5103a^2}{b^5} + \dfrac{5103a}{b^6} - \dfrac{2187}{b^7}$.
9. $a^9x^9 + 9a^7x^8y + 36a^5x^7y^2 + 84a^3x^6y^3 + 126ax^5y^4$
   $+ \dfrac{126x^4y^5}{a} + \dfrac{84x^3y^6}{a^3} + \dfrac{36x^2y^7}{a^5} + \dfrac{9xy^8}{a^7} + \dfrac{y^9}{a^9}$.
10. $220x^3$.   11. $-448y^5$.   12. $21875a^3b^4$.
13. $5440x^3$.   14. $\dfrac{210}{x^6}$.   15. $\dfrac{5103x^4a^5}{16}$.
16. -20.   17. $\dfrac{2300b^{22}}{x^{16}}$.   18. $-24310x^{25}$.
19. $2x^4 + 36x^2 + 18$.   20. $32 - 40x^2 + 10x^4$.   21. 11520.
22. $-\dfrac{1001}{256}a^9$.   23. 7920.   24. $\dfrac{1025024}{81}$.

XXXVIII. b. Page 338.

1. The 8th.   2. The 9th.   3. The 2nd and 3rd.
4. The 7th.   5. The 11th.   6. The 6th and 7th.
7. $r = 7$; excluding the value $r = 4$, which makes the terms the same.
8. $n = 40$.   9. $3r = 3n + 2$.   10. $\dfrac{\lfloor 2m}{\lfloor m \lfloor m}$.   11. $\dfrac{\lfloor 2n}{\lfloor n \lfloor n} x^n$.
12. 65536.   13. 262144.
14. $\dfrac{n(n-1)\ldots\ldots(n-r+2)}{\lfloor r-1} a^{n-r+1}(2x)^{r-1}$,

    $\dfrac{n(n-1)\ldots\ldots(n-r+2)}{\lfloor r-1} a^{r-1}(2x)^{n-r+1}$.
15. $a^6 + 6a^5 + 15a^4 + 20a^3 + 15a^2 + 6a + 1$;
    $x^6 - 12x^5 + 54x^4 - 112x^3 + 108x^2 - 48x + 8$.
16. $1 + \dfrac{1}{3}x - \dfrac{1}{9}x^2 + \dfrac{5}{81}x^3 - \ldots$.   17. $1 + \dfrac{3}{4}x - \dfrac{3}{32}x^2 + \dfrac{5}{128}x^3 - \ldots$.
18. $1 + \dfrac{2}{5}x - \dfrac{3}{25}x^2 + \dfrac{8}{125}x^3 - \ldots$.   19. $1 - 6x + 27x^2 - 108x^3 + \ldots$.
20. $1 + 3x^2 + 6x^4 + 10x^6 + \ldots$ 21. $1 - 12x + 90x^2 - 540x^3 + \ldots$.

490   ALGEBRA.

22. $\dfrac{1}{8} - \dfrac{3}{16}x + \dfrac{3}{16}x^2 - \dfrac{5}{32}x^3 + \ldots$ 23. $1 - x + \dfrac{3}{2}x^2 - \dfrac{5}{2}x^3 + \ldots$

24. $\dfrac{1}{a^{\frac{3}{2}}}\left(1 + \dfrac{3x}{a} + \dfrac{15x^2}{a^2} + \dfrac{35x^3}{2a^3} + \ldots\right).$ 25. $\dfrac{315}{128}x^4,\quad -\dfrac{230945}{65536}x^9.$

26. $\dfrac{99}{2}x^2,\ \dfrac{77}{256}x^{10}.$ 27. $-4x^3,\ (-1)^r(r+1)x^r.$

28. $-\dfrac{21}{1024}x^6,\ -\dfrac{1.3.5\ldots(2r-3)}{2^r\lfloor r}x^r.$

29. $\dfrac{b^r}{a^{r+1}}x^r,\ -\dfrac{(n-1)(2n-1)\ldots\{(r-1)n-1\}}{\lfloor r}x^r.$

30. 4·95967.   31. 4·98998.   32. 1·98734.   33. ·100504.

XXXIX. a. Page 347.

1. $\dfrac{15}{2},\ -15,\ -1,\ \dfrac{5}{2}.$ 2. $\dfrac{9}{2},\ -4,\ -\dfrac{3}{4},\ \dfrac{3}{2}.$

3. $5,\ \dfrac{1}{9},\ 125000,\ 1,\ -\dfrac{1}{4},\ 2·89,\ ·01.$ 4. $\dfrac{1}{2}\log a.$

5. $-5\log y.$ 6. $3,\ 2,\ 0,\ -1,\ -1,\ -4,\ 1.$

7. 1·5705780, $\bar{5}$·5705780, 8·5705780.

8. 7·623, ·000007623, 76230000.

9. 2·8627278.   10. 3·9242793.   11. $\bar{1}$·4082400.

12. ·7658178.   13. ·8644286.   14. $\bar{1}$·4841414.

16. $\log 7 + 4\log 3 = 2·7535832.$ 17. $6\log 2 + \dfrac{43}{6}\log 3 - \dfrac{11}{6} = 3·3922160.$

18. $\dfrac{1}{3}\log 2 + \dfrac{1}{2}\log 3 + \dfrac{1}{6}\log 7 = ·4797536.$

19. $\dfrac{1}{5}(7\log 2 - 3\log 3 - \log 7) = \bar{1}·9661496.$

20. Sixty-nine.   22. 176.   23. ·398742.

24. ·500977.   25. $2 - \log 2 - \log 3 - \log 7 = ·3767507.$

26. $\dfrac{1 + 2\log 3 - \log 2}{\log 3} = 3·46.$ 27. $\dfrac{5 - 7\log 2}{3 - 2\log 2} = 1·206.$

28. $\dfrac{3}{1 - \log 2} = 4·29.$ 29. $\dfrac{2\log 2 - 4\log 3}{4\log 2 - \log 3} = -1·8$ very nearly.

XXXIX. b. Page 348c.

1. 481·9.   2. 46·22.   3. ·1396.   4. ·008684.

5. 342·9.   6. ·03892.   7. 8·119.   8. 44·22.

9. ·6797.   10. 7·446.   11. ·03055.   12. 3·361.

ANSWERS. 491

13.	·7783.	14.	1·923.	15.	1·443.	16.	19·97.
17.	·2008.	18.	·00008855.	19.	415.	20.	·7142.
21.	2·887.	22.	1·997.	23.	1·935.	24.	1·973.
25.	1·291.	26.	9·29 ; 2560.	27.	9·076	28.	178·1.
29.	16.	30.	4·616.	31.	9·528.	32.	£73.
33.	£514.	34.	20 yrs.	35.	19.	36.	·89.

XL. Page 353.

1. 30215.
2. 25566556.
3. 244332343.
4. 36641tt.
5. 3245.
6. 14320241.
7. 123807 ; 1101122.
8. $3e7580$.
9. 32099.
10. 30523.
11. 10000011.
12. $2^9+2^8+2^7+2^6+2^5+2^3+1$; $6e^4+9e^3+e^2+4e+t$.
13. 1736 ; 1t5 ; 328108.
14. 667.
15. 203·71.
16. 31·573.
17. ·100133.
18. ·50213 ; ·404052.
19. $\dfrac{7}{9}$.
20. Five.
21. Nine.
22. Nine.
23. Seven.
24. 444 ; 1425 ; 3333.

XLI. Page 359.

2. $\dfrac{1}{2}\left(x-\dfrac{x^2}{2}+\dfrac{x^3}{3}-\dfrac{x^4}{4}+\ldots\right)$.
8. $b=a+\dfrac{a^2}{\underline{2}}+\dfrac{a^3}{\underline{3}}+\dfrac{a^4}{\underline{4}}+\ldots$
9. 1·6487.
10. $\dfrac{(-1)^{r-1}2^r-1}{r}x^r$.
13. $\dfrac{2^r+1}{r}x^r$.

XLII. a. Page 362.

1. $2, -3, \dfrac{-1\pm\sqrt{-27}}{2}$.
2. $2, -\dfrac{1}{2}, \dfrac{1}{3}(1\pm\sqrt{10})$.
3. $\dfrac{-1\pm\sqrt{-3}}{2}, \dfrac{5\pm\sqrt{21}}{2}$.
4. $-2, -\dfrac{1}{2}$.
5. $16, -\dfrac{4}{3}$.
6. $3, 0$.
7. $2, \dfrac{15}{4}$.
8. $9, \dfrac{1}{9}$.
9. $3, -2, 1, -6$.
10. $1, \dfrac{1}{81}$.
11. $20, 11$.
12. $2, -1$.
13. $-8, -1, 0$.
14. 0.
15. $2, \dfrac{3}{2}, \dfrac{7\pm\sqrt{33}}{4}$.
16. $1, 1\pm 2\sqrt{15}$.
17. $7, -1, 3\pm 2\sqrt{2}$.
18. $3, -\dfrac{7}{2}, \dfrac{-3\pm\sqrt{1357}}{12}$.

19. $\dfrac{3\pm\sqrt{5}}{2}, \dfrac{9\pm\sqrt{-83}}{6}$. 20. $2, -\dfrac{1}{2}, \dfrac{3\pm\sqrt{505}}{4}$. 21. $\dfrac{2}{13}, \dfrac{8}{7}$.

22. $a^3, \dfrac{1}{a}$. 23. $1, \dfrac{c-a}{a-b}$. 24. $1, \dfrac{c(a-b)}{a(b-c)}$.

25. a, b. 26. $c, \dfrac{a^2+b^2-ac-bc}{a+b-2c}$. 27. $p+q$.

28. $8, -2$. 29. $\dfrac{5}{2}, -\dfrac{13}{2}$. 30. a, b.

XLII. b. PAGE 365.

1. $x=5, \quad y=2$.   2. $x=4, 3; y=3, 4$.
3. $x=7, -2; y=2, -7$.   4. $x=8, 2; y=2, 8$.
5. $x=12, 3; y=3, 12$.   6. $x=3, \quad y=6$.
7. $x=\dfrac{1}{7}, y=\dfrac{1}{3}$.   8. $x=4, 1, \dfrac{1}{2}(-5\pm\sqrt{41}); y=1, 4, \dfrac{1}{2}(-5\mp\sqrt{41})$.
9. $x=\pm 10, 0; y=\pm 1, \pm\dfrac{\sqrt{34}}{2}$.   10. $x=7, -\dfrac{35}{4}; y=3, -\dfrac{15}{4}$.
11. $x=5, 2, 1\pm\sqrt{6}; y=-2, -5, -1\pm\sqrt{6}$.
12. $x=\dfrac{3}{4}, 1, 0; y=\dfrac{3}{4}, 0, 1$.
13. $x=-1, 5\pm\sqrt{6}; y=-1, 1\pm\sqrt{\dfrac{2}{3}}$. [It may be shewn that
$(x+1)^3=27(y+1)^3$.]
14. (1) $3, 2; 2, 3$.   (2) $x=3, -1; y=1, -3$.
15. $x=16, 1; y=1, 16$.   16. $x=17, y=\pm 8$.
17. $x=\pm 3, y=\pm 2, z=\pm 4$.   18. $x=5, y=6, z=1, u=4$.
19. $x=\pm 6, y=\pm\dfrac{8}{3}, z=\pm\dfrac{3}{2}$.   20. $x=\pm 5, y=\pm 7, z=\pm 11$.
21. $x=-a\pm 2, y=b\pm 1, z=-c\pm 3$.   22. $x=\pm 5, y=\pm 2, z=\pm 4$.
23. $x=\pm 7, y=\pm 3, z=\pm 2$.   24. $x=\pm 3, y=\mp 4, z=\pm 2$.
25. $x=2, -4; y=3, -5; z=0, -2$.   26. $x=7, y=2, z=4$.
27. $x=\pm\dfrac{1}{2}, 0; y=\pm\dfrac{1}{3}, 0; z=\pm 1, 0$.   28. $x=8, 2; y=2, 8; z=4$.
29. $x=2, -6; y=5; z=6, -2$.   30. $x=-5; y=3, 1; z=1, 3$.
31. $x=6; y=9, 4; z=4, 9$.   32. $x=\dfrac{30}{11}, y=\dfrac{26}{11}$.

ANSWERS. 493

33. $x=3,\ y=9,\ z=6.$

34. $x=\pm\dfrac{a(b^2+c^2)}{2bc},\ y=\pm\dfrac{b(c^2+a^2)}{2ca},\ z=\pm\dfrac{c(a^2+b^2)}{2ab}.$

XLIII. Page 370.

1. £128855000.
2. 87 years nearly.
3. £1287. 10s.
4. 20·63 per cent. nearly.
5. £2001. 17s. 6d.
6. 4 per cent.
7. £3360.
8. £11708. 8s.
9. £1604.
10. 20 years nearly.

XLIV. a. Page 374.

7. 36. 8. 32. 9. 25. 11. 1·2 sq. cm.

12. $y=3x.$ Any point whose ordinate is equal to three times its abscissa.

14. The lines are $x=5,\ y=8.$ The point (5, 8).

15. A circle of radius 13 whose centre is at the origin.

XLIV. b. Page 377.

21. 32 units of area.
22. 1 sq. in.
23. 72 units of area.
24. 0·64 sq. cm.

XLIV. c. Page 380.

1. $x=1,\ y=5.$
2. $x=2,\ y=10.$
3. $x=3,\ y=12.$
4. $x=3,\ y=-2.$
5. $x=4,\ y=2.$
6. $x=6,\ y=8.$
7. $x=-2,\ y=4.$
8. $x=0,\ y=-3.$
9. $x=-3,\ y=0.$
10. At the point (0, 21).
11. $3x+4y=7.$

XLIV. d. Page 385.

1. $y=x.$
2. (0, 0), (−4, 2).
5. (2, 1).

6. (i) 1·46, −5·46; (ii) 3·24, −1·24; (iii) 3·32, 0·68.

7. −5; 7. 8. $-\dfrac{1}{4}$; 3·79, −0·79; 4·62, −1·62.

9. $x=8,$ or 6; $y=6,$ or 8.

10. The straight line $3x+2y=25$ *touches* the circle $x^2+y^2=25$ at the point (3, 4).

XLIV. e. Page 394.

3. Each axis is an asymptote to the curve, which approaches the axis of y much less rapidly than it does the axis of x.

7. $x=2,\ \dfrac{10}{3};\ y=5,\ 3.$ 8. $x=3,\ -3;\ y=2,\ -2.$

9. $x=2;\ y=-1.$ 31. −1, 1, 2. 32. −2, 4·41, 1·59.

XLIV. f. PAGE 400.

1. 0·52, 2·9, 11·6 ; 2·75, 2·3, 3·1.   3. 2·080, 2·140.   4. 2·4.
5. −2, 4 ; −9.   9. −5 and 1.   10. −2, 1, 4.
12. 26·9, 38, 3·58.   13. 0·477, 0·225, 0·350, 1·538.
14. 3 in. from the point of suspension.
15. 22 lbs., $16\frac{1}{2}$ lbs., $14\frac{3}{4}$ lbs., $13\frac{1}{4}$ lbs., 11 lbs., $9\frac{1}{2}$ lbs., 8 lbs., $7\frac{1}{4}$ lbs. The curve is a rectangular hyperbola whose equation is $xy = 22 \times 12$.
16. 1·5.   17. $x = -\frac{1}{4}$.   18. (i) 1. (ii) 1, 2, −3.
19. (i) $x=8$, $y=1\frac{1}{2}$; or $x=1\frac{1}{2}$, $y=8$.   (ii) $x=6\cdot43$, $y=3\cdot07$.
    (iii) $x=6$, $y=2$; or $x=-6$, $y=-2$.
20. 2, 2, −1.
22. 1, 1·732, −1·732. Negative for values of $x < -1\cdot732$; positive between −1·732 and 1 ; negative between 1 and 1·732 ; positive for values of $x > 1\cdot732$.

XLIV. g. PAGE 410.

2. (i) 53·7 grains ; (ii) 0·2.   3. 39·3 ; 91·6 ; $y = 0\cdot393x$.
4. 3·85 in. ; 17·6 in.   5. 54·5° F. 86·9° F. $F = 32 + \frac{9}{5}C$.
6. $y = 100 + \frac{x}{10}$; £350 ; 4250.   7. 45·96 ; 39·40.
8. £2. 12s. ; £3. 8s.   9. 8·1 in. ; 24·375 oz.
10. (i) £320 ; (ii) £580.   11. $y = \frac{10}{11}x - 70$. 112 ; 168 ; 78.
12. 5 ft. per sec. ; $5\frac{3}{4}$ secs. ; $v = 5 + 4t$.   13. 2·49 sq. ft.
14. (i) 52 ft. ; (ii) 160 ft.   15. max. height = 64 ft. ; 4 secs.
16. $P = 0\cdot6 G - 14\cdot4$; 24.   17. 26s. ; 36s. 6d.
18. 93·5° E.   20. $y = 0\cdot21x + 1\cdot37$.
21. $y = 0\cdot4x + 1\cdot6$; 9·2 ; 3.
22. $a = 45\cdot7$, $b = 118$. Error = 8·43 in defect.
23. 8·6 ; $P = 0\cdot14W + 0\cdot2$; 225 lbs.
24. $a = \frac{1}{4}$; $b = 3$. 2 ; 12.   25. $a = 3$, $b = 2$. 7 ; 4·25.
26. $n = 3$, $c = 27 \times 10^5$.   27. $n = 1\cdot5$, $c = 79500$.

ANSWERS. 495

Miscellaneous Examples VI. PAGE 416.

1. 0.
2. $-6a-2b-4d$.
3. $\frac{1}{2}x^2+\frac{1}{2}xy-\frac{2}{9}y^2$.
4. 3.
5. $4-12x+13x^2-6x^3+x^4$.
6. $4\frac{1}{3}$.
7. $a-2$.
8. $\frac{a^2+b^2}{a^2-b^2}$.
9. $x=15, y=16$.
10. 1, 3.
11. $\frac{4}{9}$.
12. $\frac{13}{12}x^2-\frac{11}{20}y^2+\frac{5}{3}z^2$.
13. $x^3+24x^2y+192xy^2+512y^3$.
14. x^2-y^2.
15. 11.
16. H.C.F. $(x+2)(x-1)$. L.C.M. $(x-1)(x+2)^2(x^2+2)$.
17. $2a^2-3a+3$.
18. $x=\frac{8b+7a}{9}, y=\frac{8a+7b}{9}$.
19. $\frac{x}{a-x}$.
20. $\frac{5}{3}$.
22. $x-\frac{3}{4}y$.
23. $-35x+18y+17z$.
24. (1) $(10x-1)(x+8)$.
    (2) $(3x-y)(3x+y)(9x^2+3xy+y^2)(9x^2-3xy+y^2)$.
25. 13.
26. 2.
27. x^2-1.
28. $\frac{x-3y}{x+3y}$.
29. $x=\frac{2}{a}, y=3b$.
30. $21\frac{9}{11}'$ and $54\frac{6}{11}'$ past 7.
31. 1.
32. $\frac{1}{8}x^6-8y^6$.
33. $2a^4+12a^2+2$.
34. 884.
35. $\frac{1}{x^2-y^2}$.
36. $x=14, y=17$.
37. $2x^3-3x+7$.
38. $x=\frac{2}{3}$.
39. $x(x^3+y^3)(3x-y)$.
40. 21 crowns, 40 half-guineas.
41. $8ab$.
42. $2ab^3+3b^4$.
43. 36.
44. $4x-5$.
45. $\frac{a^2+b^2}{ab(a-b)^2}$.
46. $2x-\frac{y}{6}$.
47. $x=\frac{1}{2}, y=\frac{1}{3}$; or $x=0, y=0$.
48. $\frac{2(x-7)(2x-7)}{(x-2)(x-3)(x-4)(x-5)}$.
49. $(2c+3)(4x+5)(3x-5)(x+2)(x-2)$.
50. 3s. 9d.
51. $-5605x+5589$.
52. $4a^2-9b^2+24bc-16c^2$.
54. $6(x+1)(x-3)(x-4)$.
55. $2(a^2+b^2)(x^2+y^2)$.
56. $x=3, y=2, z=1$.
57. 0.
58. $x=-5$.
59. $\frac{x^4+2y^4}{y^2(x-y)^2(x^2+xy+y^2)}$.
60. 24 days.
61. $\frac{3}{2}x^3-5x^2+\frac{x}{4}+9$.

ALGEBRA.

62. 94. **63.** $\dfrac{1}{x^2-1}$. **64.** $\dfrac{ac}{b}x^2-\dfrac{b}{c}x$.

65. $x=2\tfrac{1}{5}$. **66.** (1) $(x^2+1)(x+5)$. (2) $(x-19y)(x+17y)$.

67. $x=24, y=9, z=5$. **68.** $\dfrac{2x}{x+5y}$. **69.** $(2a-3b+2c)^2$.

70. 3. **71.** $x^2+y^2+z^2$. **72.** $-2ab$.

73. 6. **74.** (1) $3x(x+9)(x-7)$. (2) $(a+b+1)(a+b)$.

75. $x=\dfrac{qr}{p^2+q^2}$, $y=\dfrac{pr}{p^2+q^2}$. **76.** $\dfrac{2x^2}{8x^3-y^3}$.

77. $x=8$. **78.** $\dfrac{x^3+x^2-2}{2x^2+2x+1}$. **79.** $\dfrac{2}{(1-x^2)^2}$.

80. 640. **81.** $1-4x-\dfrac{46}{15}x^2+\dfrac{10}{3}x^3$.

82. $\dfrac{1}{2}$. **83.** $\dfrac{3}{4}x^5-4x^4+\dfrac{77}{8}x^3-\dfrac{43}{4}x^2-\dfrac{33}{4}x+27$.

84. $2a^2b^2(a-2b)^2(2a+b)^2$. **85.** $x=5$.

86. $\dfrac{5x^2-4x-8}{3x^2+4x+24}$. **87.** $2a^2+3a+\dfrac{3}{a}$. **88.** $x=2ab,\ y=3ab$.

89. $3(2x-y)(5x+4y)$. **90.** 25 shillings, 30 half-crowns. **91.** 0.

92. $x^3-x^2+\dfrac{5}{3}x-\dfrac{23}{9}$. Rem. $-\dfrac{163}{9}$. **93.** $60(p^6-q^6)$.

94. (1) $(a-2b^5)(a^2+2ab^5+4b^{10})$. (2) $(x^2+x-1)(x^2-x+1)$.

95. $x=a-2b$. **96.** (1) $\dfrac{7bc}{13a^3}$. (2) $\dfrac{y(y^2-y+2)}{2(y+5)}$.

97. $x=1,\ y=-1,\ z=0$. **98.** $\dfrac{(y+1)(y-5)}{(y-1)(y-6)}$.

99. $\dfrac{2a-3b+c}{2a-3c}$. **100.** Twelve minutes past four.

101. (1) $5\tfrac{1}{6}$. (2) -1. **102.** (1) $\dfrac{2(a+x)}{a^2+ax+x^2}$. (2) $1+x-x^3$.

103. $a^3-3a+\dfrac{3}{a}-\dfrac{1}{a^3}$; $a-\dfrac{1}{a}$. **104.** $1-5x+15x^2-45x^3$.

105. £20. **106.** $(2a-3b)(a+b)$.

107. (1) 2 or $\dfrac{7}{5}$. (2) $\dfrac{5}{2}$ or $\dfrac{4}{3}$. **109.** $7x^2-\dfrac{x}{5}+3$.

110. $\dfrac{3}{2a^2}$. **111.** $\dfrac{3x^2+7x-12}{(x^2-9)(x^2-16)}$; $\dfrac{1}{(x-3)(x-4)}$.

112. H.C.F. $x-5b$. L.C.M. $6(x+3a)(x-3a)(x-5b)$.

ANSWERS.

113. (1) $\dfrac{ab}{2c}$ or $2d$. (2) 9 or -3.   114. 177.   115. 18 miles.

116. $(3x+2y)^2 + (3x+2y)(2x+3y) + (2x+3y)^2 = 19x^2 + 37xy + 19y^2$.

118. (1) $(x+y)(x+y)(x+y)$. (2) $mn(m-n)$.

119. (1) $\left.\begin{array}{l} x=3,1 \\ y=-1,-3 \end{array}\right\}$. (2) $\left.\begin{array}{l} x=7,-5 \\ y=2,-2 \end{array}\right\}$.   120. a^2+b^2.

121. (1) 1. (2) $\dfrac{x-1}{x^2}$.   122. $a-b$.   123. 435.

124. (1) $x = a \pm b$. (2) $x=3, y=2$.

125. (1) $\dfrac{x(x+y+z)}{z(x-y+z)}$. (2) $\dfrac{3x^2+1}{4x(x^2+1)}$.   126. $x^2+(a+2)x+3$.

127. (1) $(x-3y)(x+8y+1)$. (2) $x\left(x+\dfrac{2}{x}\right)\left(x-\dfrac{2}{x}\right)$.

128. $p-\dfrac{3q}{2p}-\dfrac{9q^2}{8p^3}$.   129. (1) $x=4\tfrac{1}{2}$. (2) $x=\dfrac{b}{a^2-ab+b^2}, y=\dfrac{a}{a^2-ab+b^2}$.

130. $x+a$.   131. (1) x^{a+b+c}. (2) $x^{1\frac{5}{12}} y^{1\frac{3}{12}}$.

132. 3 shillings.   134. $x-4+\dfrac{2}{x}$.

135. H.C.F. x^2+a^2. L.C.M. $(x^2+a^2)(x^2-4a^2)$.

136. (1) $-2y$. (2) 3.   137. $(x-2a)(x^2+2ax+4a^2)(2a+3b)(2a-3b)$.

138. (1) 3. (2) $x=105, y=210, z=420$.   139. The difference is 3.

140. $x^3 - 4y^3 - 9z^3 - 12y^{\frac{3}{2}}z^{\frac{3}{2}}$.   141. 3 hrs. 36 min.

142. (1) $\dfrac{3}{2}$. (2) $\dfrac{8}{9}$.   143. $\sqrt{\dfrac{x}{y}} + \sqrt{\dfrac{y}{x}} - 1$.

144. (1) $\dfrac{1}{x^3+1}$. (2) $\dfrac{a^2+y^2}{a-y}$.   145. (1) $4(\sqrt{2}+\sqrt{3})$. (2) $\sqrt{21}+\sqrt{14}$.

146. (1) $x=\dfrac{a^2+b^2}{a+b}$. (2) $x=2\tfrac{1}{2}$ or $-1\tfrac{3}{4}$, $y=-1\tfrac{1}{6}$ or $1\tfrac{2}{3}$.

148. $a^{\frac{1}{2}} - 2a^{\frac{1}{4}}x^{\frac{1}{4}} + x^{\frac{1}{2}}$.   149. a^2+1.   150. Six Shillings.

152. 0.   153. $\dfrac{x^2}{(7x+4)(4x-3)}$.

155. (1) $x=7$ or $-\dfrac{77}{2}$. (2) $x=\pm 5$ or $\pm 2\sqrt{3}$, $y=\pm 3$ or $\pm\dfrac{\sqrt{3}}{3}$.

156. (1) 0. (2) $x^{\frac{7}{12}}y^{\frac{5}{6}}$.   157. $(p+1)x-(p-1)$.   158. $\dfrac{3a}{5b(x^2y^2-1)}$.

159. (1) $1-2^{2n}$. (2) $3^n - 2^n$.   160. 5 hrs. $57'$, $47\tfrac{1}{2}''$.

161. 12.   162. (1) $\left(\dfrac{ab}{2a+b}\right)^2$. (2) $\dfrac{3}{5}$.

163. 1.   164. 1, $5+\sqrt{7}$.
165. $(3y+2x)(3y-2x)(x+2)(x^2-2x+4)(x-2)(x^2+2x+4)$.
166. (1) 1.   (2) $\dfrac{32}{3}$.   167. (1) 1.   (2) $\sqrt{11}-\sqrt{3}$.
168. H.C.F. $5x^2-1$. L.C.M. $=(5x^2-1)^2(4x^2+1)(5x^2+x+1)$.
169. (1) $\dfrac{(a-b)^2}{2b}$.   (2) $-2\frac{5}{8}, -8\frac{5}{8}$.   170. 9s. & 12s. a dozen.
172. 0.   173. (1) $\dfrac{x^3+y^3+z^3}{3xyz}$.   (2) 1.
174. 11.   175. (1) n.   (2) $2-\sqrt{3}$.
176. (1) $x=\dfrac{69}{20}a$; (2) $x=\pm\dfrac{1}{2}, \pm\dfrac{3}{2}$; $y=\pm\dfrac{3}{2}, \pm\dfrac{1}{2}$.
177. $\dfrac{ax+b^2x^2}{a^m+x^n}$.   178. (1) $c^{1\frac{3}{12}}$.   (2) 27.   179. 4 miles an hour.
181. $\dfrac{ac \quad b^2}{a+c-2b}$.   182. (1) $x^0+y^0+xy=1$.   (2) $\dfrac{1}{2x-1}$.
183. $3-2x^2$.   184. (1) $2x$.   (2) 10.
185. (1) 5, 1.   (2) $x=6, 2, 4$; $y=2, 6, 4$.   186. 0.
187. (1) $20\frac{3}{4}$ or $16\frac{3}{4}$.   (2) $x=3$ or $\dfrac{1}{2}$, $y=-1$ or $\dfrac{2}{3}$.
189. $(2a+1)x-a$.   190. $2x-\dfrac{1}{2x}$.   191. $(b+c)(c+a)(a+b)$.
192. (1) $\dfrac{3x-4}{(x+1)(x^3-1)}$.   (2) $\dfrac{2a}{\sqrt{x+a}}$.
194. Began at $16\frac{4}{11}'$ past 3, and ended $27\frac{3}{11}'$ past 5; walked 2 hours, $10\frac{10}{11}$ minutes.
196. $\left(\dfrac{p}{q}\right)^{p+q}$.   197. (1) 47.   (2) b.   199. 20.   200. $9x^2+y^2$.
202. 17 years.   203. $q=4$. The other root is 1.   204. $1\frac{5}{6}$ hours.
206. $a=\dfrac{5}{3}, r=\dfrac{3}{5}$; or $a=\dfrac{20}{3}, r=-\dfrac{3}{5}$.   207. $(-1)^m\left(\dfrac{x+y}{x-y}\right)^{m-n}$.
208. 30240.   209. £1. 15s. 6d.; £377. 10s.   211. $tte0121$.
212. $\dfrac{x}{2}=\dfrac{y}{3}=\dfrac{z}{4}$ [zero values are excluded].   214. 9979200; 7560.
215. $32a^5-240a^4x+720a^3x^2-1080a^2x^3+810ax^4-243x^5$. The 3rd and 4th terms.   216. 1.   217. -1.
218. (1) 2, 3, $\dfrac{5\pm\sqrt{37}}{2}$; (2) $\pm 1, \pm\sqrt{-3\pm 2\sqrt{2}}$.
220. 13104000.   222. 2400; 4032000.

ANSWERS.

223. Number of terms $=6$; common difference $=2$.

224. (1) $32 - 60a + 45a^2 - \dfrac{135}{8}a^3 + \dfrac{405}{128}a^4 - \dfrac{243}{1024}a^5$;

(2) $1 - x + \dfrac{1}{6}x^2 + \dfrac{1}{54}x^3 + \dfrac{1}{216}x^4$.

225. Senary. **226.** $4\tfrac{2}{3}$; $1\cdot 412$. **227.** $5\cdot 039684$. **228.** 315.

229. (1) $\sqrt{5x}$; (2) $x - \dfrac{1}{x}$. **230.** $x^2 - 10x + 19 = 0$.

231. $2\cdot 71405$; *tet*. **232.** $1\cdot 75$; $1\cdot 75$; -2; $\bar{5}\cdot 2375439$.

233. B overtakes A at the end of the 8th day; then A overtakes B at the end of the 15th day.

234. (1) $x=21$, $y=6$; (2) $x=4$, $\dfrac{56}{9}$; $y=3$, $-\dfrac{11}{3}$; $z=9$, $\dfrac{121}{9}$.

235. $a = 2d$. **236.** The 4th and 5th terms. **237.** 120; 60.

238. $\bar{3}\cdot 698970$; $\cdot 799340$; $\bar{1}\cdot 785248$; $x = \dfrac{22}{17}$. **239.** $8x = \sqrt{yz} + 2\sqrt[3]{yz}$.

240. $\log 2$. **241.** 14. **244.** (1) 3; (2) $\bar{1}\cdot 36564$; (3) 22.

245. $\dfrac{5 \cdot 2 \cdot 1 \cdot 4 \cdot 7 \ldots (3r-8)}{3^r \lfloor r}(2a)^{\frac{5}{3}-r}x^{\frac{10}{3}+r}$. **246.** 2880.

247. $x^2 - 2(a+b)x + 2ab = 0$. **248.** $2(n-1)$ hours.

249. (1) a, $\dfrac{a(a-b)}{b-c}$, $\left[\text{one root is evidently } a, \text{ and the product of the roots is } \dfrac{a^2(a-b)}{b-c};\right]$ (2) q, $p-q$.

250. The series is the expansion of $\left(1 - \dfrac{1}{2}\right)^{-\frac{3}{2}}$.

Mathematical Works

BY

Messrs. HALL and KNIGHT,

PUBLISHED BY

Macmillan and Co., Limited.

EIGHTH EDITION, Revised and Enlarged. Now Ready.

ELEMENTARY ALGEBRA FOR SCHOOLS (containing a chapter on Graphs). By H. S. HALL, M.A., formerly Scholar of Christ's College, Cambridge; and S. R. KNIGHT, B.A., M.B., Ch.B., formerly Scholar of Trinity College, Cambridge. Globe 8vo. (bound in maroon-coloured cloth), 3s. 6d. With Answers (bound in green-coloured cloth), 4s. 6d.

The distinctive features of the Eighth Edition are:—

(1) A full treatment of Graphs, occupying more than 40 pages.

(2) A new set of easy examples on Substitution in Chapter I.

(3) The greater part of Chapter VIII., on Simple Equations, has been re-written so as to bring the use of the fundamental axioms into greater prominence, and to urge the importance of verifying solutions.

(4) Chapter IX., on Symbolical Expression, has been enlarged. In particular, the section on Formulæ has been illustrated by a new set of Examples.

(5) A section on Square Root by inspection has been inserted in Chapter XVI.

(6) In Chapter XVII., on Factors, a section on factorisation of trinomials, by completing the square, has been introduced. Also a large number of *easy* miscellaneous examples now take the place of the Exercise XVII. 1 of earlier editions.

MACMILLAN AND CO., LIMITED, LONDON.

Works by H. S. HALL, M.A., and S. R. KNIGHT, B.A.

(7) Considerable additions to the chapters on Quadratic Equations. In particular a set of examples involving applications to Geometry will be found at the end of Chapter XXVII.

(8) The chapter on Logarithms has been re-written so as to introduce and explain the use of Four-Figure Tables. The Tables of Logarithms and Antilogarithms have been taken, with slight modifications, from those published by the Board of Education, South Kensington.

(9) An easy first course has been mapped out enabling teachers to postpone, if they wish, the harder cases of 'Long' Multiplication and Division, and the rules dependent on these processes.

OPINIONS OF THE PRESS.

SCHOOLMASTER—" . . . **Has so many points of excellence as compared with its predecessors, that no apology is needed for its issue.** The plan always adopted by every good teacher, of frequently recapitulating and making additions at every recapitulation, is well carried out."

NATURE—" . . . **We confidently recommend it to mathematical teachers, who, we feel sure, will find it the best book of its kind for teaching purposes.**

ACADEMY—"We will not say that this is the best Elementary Algebra for school use that we have come across, but we can say that **we do not remember to have seen a better.** . . . **It is the outcome of a long experience of school teaching, and so is a thoroughly practical book.**"

EDUCATIONAL TIMES—" . . . A very good book. The explanations are concise and clear, and the examples both numerous and well chosen."

EDUCATIONAL NEWS—"A book of exceptional value."

OPINIONS OF TEACHERS.

"**I think it decidedly the best of all books on Elementary Algebra yet published.** The great merit seems to me to be that, while it is quite simple and elementary, there are no misleading and inaccurate statements which must afterwards be unlearned. I shall certainly make use of it in my classes, and hope it may come into general use throughout the country."—A. J. WALLIS, M.A., *Fellow and Lecturer of Corpus Christi College, Cambridge.*

"We have examined your Algebra very carefully; **and we agree that it is as perfect as a book can be.** I will introduce it at St. Paul's as soon as I can."—C. PENDLEBURY, M.A., *Senior Mathematical Master, St. Paul's School.*

"After employing it with my evening class this term, **I feel it to be quite the best Elementary Algebra that has yet appeared.**"—R. A. HERMAN, M.A., *Fellow of Trinity College, Cambridge; Late Professor of Mathematics at University College, Liverpool.*

MACMILLAN AND CO., LIMITED, LONDON.

Works by H. S. HALL, M.A., and S. R. KNIGHT, B.A.

KEY TO ELEMENTARY ALGEBRA FOR SCHOOLS. Crown 8vo. 8s. 6d.

ANSWERS TO THE EXAMPLES IN ELEMENTARY ALGEBRA. Fcap. 8vo. Sewed. 1s.

By H. S. HALL.

ALGEBRAICAL EXAMPLES. Supplementary to Hall and Knight's Algebra for Beginners, and Elementary Algebra. (Chaps. I-XXVII.) With or Without Answers. Globe 8vo. 2s.

By H. S. HALL.

A SHORT INTRODUCTION TO GRAPHICAL ALGEBRA. Revised Edition. Globe 8vo. 1s.

KEY TO THE SHORT INTRODUCTION TO GRAPHICAL ALGEBRA. Crown 8vo. 3s. 6d.

EASY GRAPHS. Globe 8vo. 1s.

OPINIONS OF THE PRESS.

SCHOOL WORLD—"May be recommended without reservation."

TEACHERS' AID—"The second edition revised and enlarged, of this small but admirable sketch of Graphical Algebra lies before us, and we are constrained to accord its appearance with unstinted praise."

EDUCATIONAL TIMES—"An excellent little pamphlet. . . . From a careful study of Mr. Hall's descriptions and explanations a student without any previous knowledge of the subject will be able to obtain an intelligent grasp of its elements."

SCHOOLMASTER—"This little book is invaluable. . . . Aspirants to the engineer's office, the laboratory or workshop will find it all but indispensable."

MACMILLAN AND CO., LIMITED, LONDON.

Works by H. S. HALL, M.A., and S. R. KNIGHT, B.A.

HIGHER ALGEBRA. A Sequel to Elementary Algebra for Schools. By H. S. HALL, M.A., and S. R. KNIGHT, B.A. Fifth Edition, revised and enlarged. Crown 8vo. 7s. 6d.

The Fifth Edition contains a collection of three hundred Miscellaneous Examples, which will be found useful for advanced students. These Examples have been selected mainly from recent Scholarship or Senate House Papers.

SCHOOL GUARDIAN—"**We have no hesitation in saying that, in our opinion, it is one of the best books that have been published on the subject.** . . . The authors have certainly added to their already high reputation as writers of mathematical text-books by the work now under notice, which is remarkable for clearness, accuracy, and thoroughness."

"It is a splendid sequel to your *Elementary Algebra*, and I am very pleased to see you have introduced the essential parts of the Theory of Equations in Chap. XXXV., which contains all that is required of the subject for ordinary practical purposes."—A. G. GREENHILL, M.A., *Professor of Mathematics, to the Senior Class of Artillery Officers, R.A. Institution, Woolwich.*

ATHENÆUM—"The *Elementary Algebra* by the same authors, which has already reached a third edition, is a work of such exceptional merit that those acquainted with it will form high expectations of the sequel to it now issued. Nor will they be disappointed. Of the authors' *Higher Algebra* as of their *Elementary Algebra*, **we unhesitatingly assert that it is by far the best work of its kind with which we are acquainted. It supplies a want much felt by teachers.**"

ACADEMY—"Is as admirably adapted for College students as its predecessor was for schools. It is a well-arranged and well-reasoned-out treatise, and contains much that we have not met with before in similar works. For instance, we note as specially good the articles on Convergency and Divergency of Series, on the treatment of Series generally, and the treatment of Continued Fractions. . . . **The book is almost indispensable, and will be found to improve upon acquaintance.**"

SATURDAY REVIEW—"They have presented such difficult parts of the subject as Convergency and Divergency of Series, Series generally, and Probability with great clearness and fulness of detail. . . . **No student preparing for the University should omit to get this work in addition to any other he may have, for he need not fear to find here a mere repetition of the old story.** We have found much matter of interest and many valuable hints. . . . We would specially note the examples, of which there are enough, and more than enough, to try any student's powers."

KEY. Crown 8vo. 10s. 6d.

MACMILLAN AND CO., LIMITED, LONDON.

Works by H. S. HALL, M.A., and S. R. KNIGHT, B.A.

ALGEBRA FOR BEGINNERS. By H. S. HALL, M.A., and S. R. KNIGHT, B.A., M.B., Ch.B. Globe 8vo. 2s. With Answers. 2s. 6d.

EDUCATIONAL TIMES—"*Algebra for Beginners* is dealt with on the same lines as the earlier and somewhat more advanced book. The learner is introduced as soon as possible to the practical and more interesting side of the subject, such as equations and problems, while work which largely consists in the manipulation and simplification of elaborate expressions is postponed till later on. The examples for practice are copious, and have been newly composed for this particular book; and, as heretofore, the explanations are clear, concise, and simply expressed. **Indeed, without hesitation we pronounce this book the best of its size which we have seen.**"

UNIVERSITY CORRESPONDENT—"Those masters who have already adopted Messrs. Hall and Knight's *Elementary Algebra* in their schools, will welcome this new work for the use of their junior classes. . . . The numerous exercises for the student are excellent in quality and entirely new. **We can unhesitatingly recommend the book to the notice of both teachers and students.**"

SCHOOLMASTER—"To teachers who have had experience of either the Elementary or the Higher Algebra it will only be necessary to say that this book is marked by the same qualities which have brought these works into such deserved repute. **To those who are still in ignorance of these books, we can say that for clear, simple, and concise explanation, convenient order of subject-matter, and copious and well-graduated exercises, these books have, to say the least, no superiors.** Quite early the student is introduced to easy problem work, which can only be looked upon as an advantage. The very numerous exercises are entirely new, so that the book might easily serve as a companion and supplement to the elementary work."

GUARDIAN—"It possesses the systematic arrangement and the lucidity which have gained so much praise for the works previously written by the authors in collaboration."

ALGEBRAICAL EXERCISES AND EXAMINATION PAPERS. With or without Answers. By H. S. HALL, M.A., and S. R. KNIGHT, B.A. Third Edition, revised and enlarged. Globe 8vo. 2s. 6d.

This book has been compiled as a suitable companion to the *Elementary Algebra* by the same authors. It consists of one hundred and twenty progressive Miscellaneous Exercises, followed by a comprehensive collection of papers set at recent examinations.

SATURDAY REVIEW—"To the exercises, one hundred and twenty in number, are added a large selection of examination papers set at the principal examinations which require a knowledge of algebra. These papers are intended chiefly as an aid to teachers, who no doubt will find them useful as a criterion of the amount of proficiency to which they must work up their pupils before they can send them in to the several examinations with any certainty of success."

SCHOOLMASTER—"We can strongly recommend the volume to teachers seeking a well-arranged series of tests in algebra."

MACMILLAN AND CO., LIMITED, LONDON.

Works by H. S. HALL, M.A., and S. R. KNIGHT, B.A.

ELEMENTARY TRIGONOMETRY. By H. S. HALL, M.A., and S. R. KNIGHT, B.A. Third Edition, containing 300 Additional Miscellaneous Examples. Globe 8vo. 4s. 6d.

EDUCATIONAL REVIEW—"The authors have that instinctive knowledge of the needs, both of the pupil and of the teacher, which only belongs to the practical teacher. . . . **On the whole it is the best elementary treatise on Trigonometry we have seen.**"

GUARDIAN—"They are lucid and concise in exposition, their methods are simple, and the examples are judiciously selected."

LYCEUM—"**It is not too much to say of this book, that it is the very best class-book that can be placed in the hands of beginners.**"

SPEAKER—"They here present Elementary Trigonometry so far as it can well be treated without infinite series and imaginary quantities. The authors lay a solid foundation by insisting on the thorough comprehension of trigonometrical ratios before passing on to other subjects. **Logarithms and heights and distances have been treated with special care.** . . . **The full table of contents is a useful feature of the book.**"

CAMBRIDGE REVIEW—"Messrs. Hall and Knight's Algebra has won them a reputation which we think their Trigonometry will sustain."

NATURE—"This book can safely be recommended to beginners, and it may, besides imparting to them a sound elementary knowledge of the subject, ingraft an intelligent interest for more advanced study."

KEY. Crown 8vo. 8s. 6d.

ARITHMETICAL EXERCISES AND EXAMINATION PAPERS. With an Appendix containing Questions in LOGARITHMS AND MENSURATION. With or without Answers. By H. S. HALL, M.A., and S. R. KNIGHT, B.A. Third Edition, revised and enlarged. Globe 8vo. 2s. 6d.

"In the Second Edition, the Appendix has been increased by a new series of examples, which are intended to be worked by the aid of Logarithmic Tables. In view of the increasing importance of logarithmic calculation in many examinations, this last section will be found especially useful."—*From the Preface.*

CAMBRIDGE REVIEW—"All the mathematical work these gentlemen have given to the public is of genuine worth, and these exercises are no exception to the rule. The addition of the logarithm and mensuration questions adds greatly to the value."

MACMILLAN AND CO., LIMITED, LONDON.

By H. S. HALL, M.A., and F. H. STEVENS, M.A.

A SCHOOL GEOMETRY, based on the recommendations of the Mathematical Association, and on the recent report of the Cambridge Syndicate on Geometry. Cr. 8vo.

    Parts I. and II.—*Part I.* Lines and Angles. Rectilineal Figures. *Part II.* Areas of Rectilineal Figures. Containing the substance of Euclid Book I. 1s. 6d. **Key**, 3s. 6d.

    Part I.—Separately. 1s.

    Part II.—Separately. 6d.

    Part III.—Circles. Containing the substance of Euclid Book III. 1-34, and part of Book IV. 1s.

    Parts I., II., III. in one volume. 2s. 6d.

    Part IV.—Squares and Rectangles. Geometrical equivalents of Certain Algebraical Formulae. Containing the substance of Euclid Book II., and Book III. 35-37. 6d.

    Parts III. and IV. in one volume. 1s. 6d.

    Parts I.-IV. in one volume. 3s. **Key**, 6s.

    Part V.—Containing the substance of Euclid Book VI. 1s. 6d.

    Parts IV. and V. in one volume. 2s.

    Parts I.-V. in one volume. 4s.

    Part VI.—Containing the substance of Euclid Book XI. 1-21, together with Theorems relating to the Surfaces and Volumes of the simpler Solid Figures. 1s. 6d.

    Parts IV., V., VI. in one volume. 2s. 6d.

    Parts I.-VI. in one volume. 4s. 6d. **Key**, 8s. 6d.

LESSONS IN EXPERIMENTAL AND PRACTICAL GEOMETRY. Crown 8vo, 1s. 6d.

A SCHOOL GEOMETRY, PARTS I. AND II. AND LESSONS IN EXPERIMENTAL AND PRACTICAL GEOMETRY. In one volume. Crown 8vo. 2s. 6d.

A TEXT-BOOK OF EUCLID'S ELEMENTS, including Alternative Proofs, together with Additional Theorems and Exercises, classified and arranged. By H. S. HALL, M.A., and F. H. STEVENS, M.A., Masters of the Military Side, Clifton College. Books I.-VI., XI., and XII., Props. 1 and 3. Globe 8vo. 4s. 6d.

    Also in parts separately as follows:—

Book I.	- 1s.	Books III. and IV.	- 2s.
Books I. and II.	1s. 6d.	Books III.—VI.	- 3s.
Books II. and III.	2s.	Books IV.—VI.	- 2s. 6d.
Books I.—III.	- 2s. 6d.	Books V., VI., XI. and XII. 1 and 3.	- 2s. 6d.
Books I.—IV.	- 3s.		
Sewed, 2s. 6d.		Book XI.	- 1s.

MACMILLAN AND CO., LIMITED, LONDON.

By H. S. HALL, M.A., and F. H. STEVENS, M.A.

A KEY TO THE EXERCISES AND EXAMPLES CONTAINED IN A TEXT-BOOK OF EUCLID'S ELEMENTS. Books I.-VI. and XI. By H. S. HALL, M.A., and F. H. STEVENS, M.A., Masters of the Military Side, Clifton College. Crown 8vo. 8s. 6d. Books I.-IV., 6s. 6d. Books VI. and XI., 3s. 6d.

AN ELEMENTARY COURSE OF MATHEMATICS, comprising Arithmetic, Algebra, and Euclid. Globe 8vo. 2s. 6d.

AN ELEMENTARY COURSE OF MATHEMATICS, comprising Arithmetic, Algebra, and Geometry. Globe 8vo. 2s. 6d.

By H. S. HALL and R. J. WOOD.

ALGEBRA FOR ELEMENTARY SCHOOLS. Globe 8vo. Parts I., II., and III., 6d. each. Cloth, 8d. each. Answers, 4d. each.

By F. H. STEVENS, M.A.

ELEMENTARY MENSURATION. Globe 8vo. 3s. 6d.

NATURE—"The large number of original examples will be found of great assistance by teachers, and the questions, selected from papers set by the principal examining bodies, will prove of service as tests of the student's capabilities in working out mensuration problems."

MENSURATION FOR BEGINNERS, with the Rudiments of Geometrical Drawing. Globe 8vo. 1s. 6d.

EDUCATIONAL TIMES—"A considerable amount of ground is covered, and **the whole is written with rare judgment and clearness.**"

GUARDIAN—"Mr. Stevens seems to us to have chosen just the elements that are of use in every-day life, and in the notes and the examples he has worked out, to have given sufficient illustration of right methods.'

MACMILLAN AND CO., LIMITED, LONDON.